RETINAL DEGENERATIONS

Mechanisms and Experimental Therapy

ADVANCES IN EXPERIMENTAL MEDICINE AND BIOLOGY

Recent Volumes in this Series

Volume 524
DIPEPTIDYL AMINOPEPTIDASES IN HEALTH AND DISEASE
Edited by Martin Hildebrandt, Burghard F. Klapp, Torsten Hoffmann, and Hans-Ulrich Demuth

Volume 525
ADVANCES IN PROSTAGLANDIN, LEUKOTRIENE, AND OTHER BIOACTIVE LIPID RESEARCH: Basic Science and Clinical Applications
Edited by Zeliha Yazıcı, Giancarlo Folco, Jeffrey M. Drazen, Santosh Nigam, and Takao Shimizu

Volume 526
TAURINE 5: Beginning the 21st Century
Edited by John B. Lombardini, Stephen W. Schaffer, and Junichi Azuma

Volume 527
DEVELOPMENTS IN TRYPTOPHAN AND SEROTONIN METABOLISM
Edited by Graziella Allegri, Carlo V. L. Costa, Eugenio Ragazzi, Hans Steinhart, and Luigi Varesio

Volume 528
ADAMANTIADES-BEHÇET'S DISEASE
Edited by Christos C. Zouboulis

Volume 529
THE GENUS *YERSINIA*: Entering the Functional Genomic Era
Edited by Mikael Skurnik, José Antonio Bengoechea, and Kaisa Granfors

Volume 530
OXYGEN TRANSPORT TO TISSUE XXIV
Edited by Jeffrey F. Dunn and Harold M. Swartz

Volume 531
TROPICAL DISEASES: From Molecule to Bedside
Edited by Sangkot Marzuki, Jan Verhoef, and Harm Snippe

Volume 532
NEW TRENDS IN CANCER FOR THE 21ST CENTURY
Edited by Antonio Llombart-Bosch and Vicente Felipo

Volume 533
RETINAL DEGENERATIONS: Mechanisms and Experimental Theory
Edited by Matthew M. LaVail, Joe G. Hollyfield, and Robert E. Anderson

---

A Continuation Order Plan is available for this series. A continuation order will bring delivery of each new volume immediately upon publication. Volumes are billed only upon actual shipment. For further information please contact the publisher.

RETINAL DEGENERATIONS

Mechanisms and Experimental Therapy

Edited by

Matthew M. LaVail

Beckman Vision Center
University of California, San Francisco
San Francisco, California

Joe G. Hollyfield

Cole Eye Institute
The Cleveland Clinic Foundation
Cleveland, Ohio

and

Robert E. Anderson

Dean A. McGee Eye Institute
University of Oklahoma Health Sciences Center
Oklahoma City, Oklahoma

Kluwer Academic / Plenum Publishers
New York, Boston, Dordrecht, London, Moscow

Library of Congress Cataloging-in-Publication Data

International Symposium on Retinal Degenerations (10th: 2002: Bürgenstock, Switzerland)
Retinal degenerations: mechanisms and experimental therapy/edited by Matthew M. LaVail, Joe G. Hollyfield, and Robert E. Anderson.
p. ; cm. — (Advances in experimental medicine and biology; v. 533)
"X International Symposium on Retinal Degeneration (also known as RD2002) was held on September 30–October 5, 2002 ... Bürgenstock region of Switzerland"—Pref.
Includes bibliographical references and index.
ISBN 0-306-47781-5
1. Retinal degeneration—Congresses. I. LaVail, Matthew M. II. Hollyfield, Joe G. III. Anderson, Robert E. (Robert Eugene) IV. Title. V. Series.
[DNLM: 1. Retinal Degeneration—genetics—Congresses. 2. Disease Models, Animal—Congresses. 3. Gene Therapy—Congresses. 4. Photoreceptors—Congresses. 5. Retinal Degeneration—physiopathology—Congresses. 6. Retinal Degeneration—therapy—Congresses. WW 270 I6252r 2003]
RE661.D3157 2002
617.7'35—dc22

2003058902

Proceedings of the X International Symposium on Retinal Degeneration, held September 30–October 5, 2002, in Bürgenstock, Switzerland

ISBN 0-306-47781-5

233 Spring Street, New York, New York 10013

http://www.wkap.nl/

10 9 8 7 6 5 4 3 2 1

A C.I.P. record for this book is available from the Library of Congress

Printed in the United States of America

Christina Fasser

This book is dedicated to Christina Fasser for her untiring efforts for more than 20 years in the advocacy for patients with retinal degenerations, for fostering interactions among research scientists, clinicians and patients, and for leading an international effort to seek the causes, treatments and cures for retinal degenerative diseases. Diagnosed with retinitis pigmentosa at the age of 13 and blinded by the disease at the age of 42, Christina has worked as an occupational therapist and served in many capacities in her country, Switzerland, to assist and empower those with visual handicaps. She has been the long-time President of Retina International, formerly the International Retinitis Pigmentosa Foundation. Most significantly, Christina has served as a truly remarkable role model for those with visual impairments and as a friend and inspiration to vision scientists worldwide.

PREFACE

Since 1984, we have organized International Symposia on Retinal Degenerations that are held in conjunction with the biennial International Congress of Eye Research (ICER). The timing and location of our Retinal Degeneration Symposia have allowed scientists and clinicians from around the world to convene and present their exciting new findings. The Symposia have been arranged to allow ample time for discussions and one-on-one interactions in a relaxed atmosphere, where international friendships and collaborations could be established.

The X International Symposium on Retinal Degeneration (also known as RD2002) was held on September 30-October 5, 2002 in the beautiful Bürgenstock region of Switzerland. The meeting brought together approximately 150 scientists, retinal specialists in ophthalmology and trainees in the field from all parts of the world. In the course of the meeting, 40 platform and 70 poster presentations were given, and a majority of these are presented in this proceedings volume. As in the past, the most recent findings of most research areas in the field of retinal degenerations were presented. In this particular meeting, the areas of cellular and molecular mechanisms of retinal degeneration, as well as Usher syndrome, were heavily represented. The RD2002 meeting was highlighted by two Special Lectures, the first by Prof. Alan Bird of London, who discussed the potential developing biological therapies for retinal dystrophies and what clinicians should do to be prepared for the advent of treatments. The second Special Lecture was by Dr. Jean Bennett of Philadelphia, who considered many aspects of gene therapy for retinal degenerations, including an update on the successful restoration of vision in dogs lacking RPE65, a model of Leber's congenital amaurosis in young human patients.

We want to give special recognition to our local hosts and co-organizers of the Symposium, Professor Charlotte Remé, Andreas Wenzel and Christian Grimm from Zurich. They and their administrative assistant, Ulli Busse, worked tirelessly for months preparing an extraordinarily smooth meeting and an exhilarating excursion and social program. We also thank the staff of the Bürgenstock Resort and Hotels for their help with the meeting. In addition, we want to recognize the extensive efforts of Professor Yusuf Durlu of Istanbul, Turkey, who had prepared a meeting site and program in the Antalya region of Turkey for the RD2002 meeting. While the venue of the meeting had to be moved in the aftermath of the September 11th tragedy, Professor Durlu continued to serve as a co-organizer of the RD2002 Symposium.

The Symposium received financial support from a number of organizations. We are particularly pleased to thank the Foundation Fighting Blindness, Owings Mills, Maryland, for its continuing support of this and the previous biennial Symposia, without which we could not have held these important meetings. In addition, for the first time the National Eye Institute of the National Institutes of Health contributed substantially to the meeting. This additional funding allowed the meeting to cost attendees less than several previous Symposia, despite being in an expensive resort locale. Moreover, it allowed us to provide 23 Travel Awards to young investigators and trainees working in the field of retinal degenerations. The response to the Travel Awards program was overwhelming; 52 applicants competed for the 23 Awards, which were selected by the meeting organizers and members of the National Eye Institute and the Foundation Fighting Blindness. We are especially grateful to Dr. Andrew Mariani of the National Eye Institute for his help, guidance, and encouragement with this meeting. The Foundation Fighting Blindness also contributed to the Travel Awards program.

We want to acknowledge the diligent and outstanding efforts of Ms. Holly Whiteside, who carried out most of the administrative aspects of the RD2002 Symposium, designed and maintained the meeting website, and organized the production of this volume. Holly is the Administrative Manager of Dr. Anderson's laboratory at OUHSC, and she has become the permanent Coordinator for the Retinal Degeneration Symposia. Her dedicated efforts with the Symposia since 2000 have provided continuity heretofore not available, and we are deeply indebted to her.

We want to thank Dr. Michael T. Matthes, the Co-Director of Dr. LaVail's laboratory for his tireless effort in the final compilation of the manuscript of this book.

We also thank Kluwer Academic/Plenum Publishers for publishing this volume.

Matthew M. LaVail
Joe G. Hollyfield
Robert E. Anderson

CONTENTS

DIAGNOSTIC, CLINICAL, CYTOPATHOLOGICAL AND PHYSIOLOGIC ASPECTS OF RETINAL DEGENERATION

AGE-RELATED MACULAR DEGENERATION

ANIMAL MODELS OF RETINAL DEGENERATION

MECHANISMS OF RETINAL DEGENERATION

GENE THERAPY

IDENTIFICATION OF THE RP1 AND RP10 (IMPDH1) GENES CAUSING AUTOSOMAL DOMINANT RP

Stephen P. Daiger[1], Lori S. Sullivan[1], Sara J. Bowne[1], Avril Kennan[2], Peter Humphries[2], David G. Birch[3], John R. Heckenlively[3], and the RP1 Consortium[4]

1. INTRODUCTION

In a project begun more than 15 years ago we used linkage mapping and positional candidate gene cloning to identify two genes causing autosomal dominant retinitis pigmentosa (adRP). The genes are RP1, which maps to chromosome 8q12.1, and RP10 (IMPDH1), which maps to 7q32.1. Although different families and different regions of the genome are involved, the studies were done in parallel because the methods used, and the underlying genetic concepts, are identical. Now that the genes have been identified, though, the functional properties of the two proteins are strikingly different, so divergent approaches are required to better understand the pathophysiology of mutations in each gene. This report is a summary of our current understanding of these two genes, contrasting the similar approaches to identifying the genes and mutations with the dissimilar strategies for functional analysis. In microcosm, this serves as a reminder that retinal disease genes associated with similar clinical phenotypes may have very different biological roles.

Research to identify and characterize disease-causing genes now follows a traditional methodology which has developed over the past 15 years. Although there are many nuances, the standard steps are ...

- ascertainment and characterization of families,
- linkage mapping,
- positional candidate gene cloning,
- mutation screening,
- gene sequence analysis, and
- functional analysis.

[1]Human Genetics Center and Dept. of Ophthalmology, The Univ. of Texas, Houston; TX, USA; [2]Ocular Genetics Unit, Trinity College, Dublin, Ireland; [3]Retina Foundation of the Southwest, Dallas, TX, USA; [4]Jules Stein Eye Inst., The Univ. of California, Los Angeles, CA, USA; [4]S.P. Daiger, L.S. Sullivan, S.J. Bowne, J.R. Heckenlively, D.S. Farber, E.A. Pierce, and J. Zou. Corresponding author: S.P. Daiger, E-mail: stephen.p.daiger@uth.tmc.edu

The first steps are part of the process originally called "reverse genetics", with emphasis on identification of the disease locus independent of the biological or clinical details of the disease. Subsequent steps are focused on a biological understanding of the gene and gene product. This summary describes the current status of these steps in understanding the RP1 and RP10 loci.

2. ASCERTAINMENT OF FAMILIES AND LINKAGE MAPPING

2.1 The RP1 locus

In 1982 Field *et al.* described a large, six-generation adRP family located predominantly in the Kentucky - West Virginia area (Figure 1). This was named the "RP1" locus based on preliminary linkage mapping to chromosome 1 (later shown to be incorrect). Subsequent linkage testing assigned the disease locus in this family to human chromosome 8q11-q12 (Blanton *et al.*, 1991). Linkage testing in an Australian adRP family led to identification of a second RP1 family. Combining data from the two families reduced the maximum non-recombinant region to 4 cM, or roughly 4 Mb (Xu *et al.*, 1996). An additional British RP1 family was also identified by linkage mapping (Inglehearn *et al.*, 1999). Data from this family did not help refine the critical region for the RP1 locus.

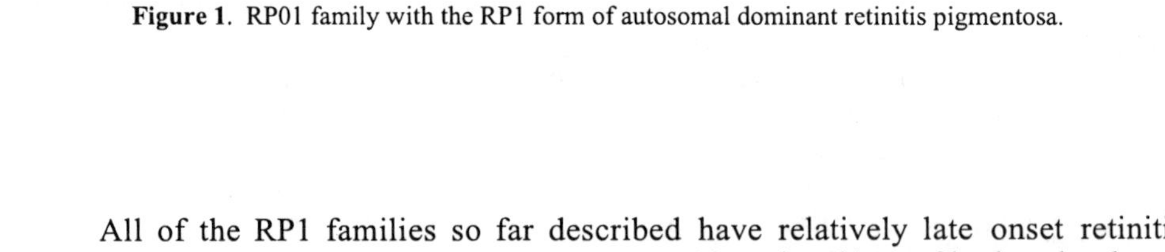

Figure 1. RP01 family with the RP1 form of autosomal dominant retinitis pigmentosa.

All of the RP1 families so far described have relatively late onset retinitis pigmentosa, usually described as type 2 or "R" type with regional loss of both rod and cone photoreceptor activities (Field *et al.*, 1982; Inglehearn *et al.*, 1999; Xu *et al.*, 1996). There is wide variation in severity and age of onset among affected members of large RP1 families, with a few reported instances of unaffected "carriers" over age 50. One notable clinical feature of the Kentucky family (RP01) is the presence of two affected individuals who are homozygous for an RP1 mutation as indicated by haplotype analysis and, more recently, by mutation testing (Sullivan *et al.*, 1999). These individuals have severe, congenital or early-onset retinal degeneration, but are otherwise apparently unaffected by these mutations.

2.2 The RP10 locus

The RP10 locus was first assigned to chromosome 7q in 1993 by Jordan *et al.* based on linkage mapping in a Spanish adRP family. Identification of a second RP10 family by linkage mapping localized the RP10 gene to 7q31-q35 (McGuire *et al.*, 1995), later refined to a 5 cM (4.9 Mb) region at 7q32.1 (McGuire *et al.*, 1996). The second RP10 family is of American origin, with at least 6 known, affected generations (Figure 2). Two additional RP10 families were identified by linkage mapping, one of Spanish origin (Millán *et al.*, 1995) and the second of Scottish origin (Mohamed *et al.*, 1996).

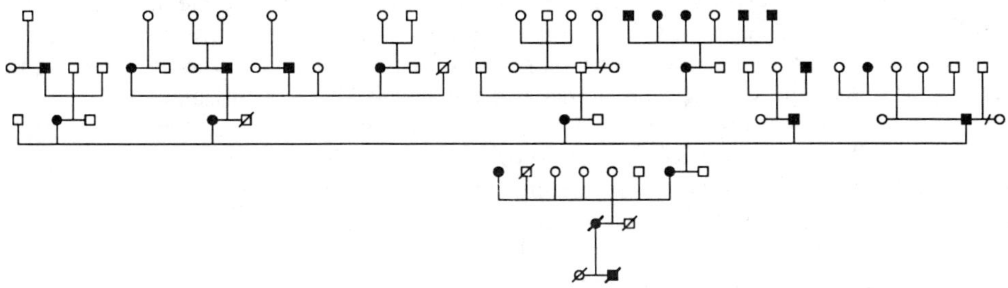

Figure 2. UTAD045 family with the RP10 form of autosomal dominant retinitis pigmentosa.

The clinical findings in RP10 families are similar to RP1 families, except that RP10 has earlier onset and is more consistent with type 1or "diffuse" retinitis pigmentosa (Bowne *et al.*, 2002). Like RP1, patients with the RP10 form of adRP have equal reduction in both rod and cone ERGs, though with earlier onset. There is also considerable variation in clinical findings among affected members of families with the RP10 form of adRP but "skipped generations" (unaffected carriers) have not been reported.

3. POSITIONAL CANDIDATE CLONING

3.1 The RP1 gene

Fine structure linkage mapping and physical mapping of flanking markers reduced the RP1 genetic region to 4.0 Mb (Sullivan *et al.*, 1999). Candidate genes were screened by sequencing in heterozygous and homozygous members of the RP01 family, with emphasis on genes preferentially expressed in photoreceptors. Among these, a large gene mapping to 8q12.1 was found to contain a nonsense substitution in codon 677 producing an Arg677ter mutation tracking in phase with the disease haplotype in all affected family members (Sullivan *et al.*, 1999). The transcribed gene produces a 6,471 bp mRNA composed of 4 exons, 3 of which code for a novel protein 2,156 amino acids in length (Figure 3). Published ESTs of the RP1 gene are exclusively from retinal cDNA libraries suggesting expression is limited to the retina. Because sequence analysis did not immediately suggest a function for the RP1 gene product, the gene symbol and protein name have remained "RP1".

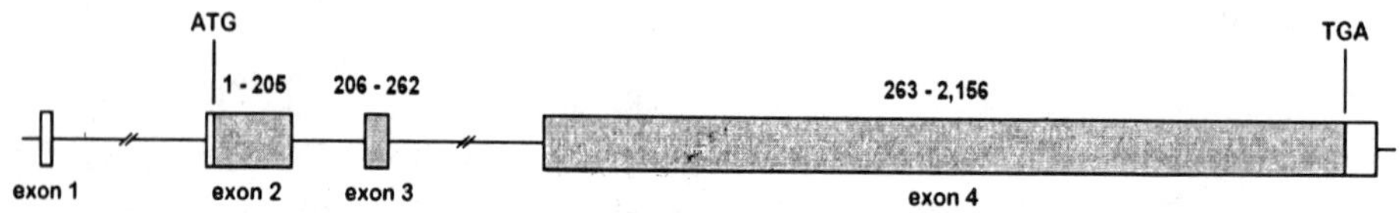

Figure 3. Structure of the RP1 gene.

Two other research groups independently identified the RP1 gene at the same time as Sullivan *et al.* Pierce *et al.* (1999) used differential display to identify mouse retinal genes that show significant changes in expression following retinal hypoxia. The human homolog of one of these, oxygen response protein 1 (ORP1), was subsequently shown to map to the RP1 critical region on chromosome 8. Sequencing of members of the RP01 family showed this to be the RP1 gene. Additionally, Guillonneau *et al.* (1999) identified a retinal-expressed gene in the mouse chromosomal region (proximal 1) syntenic to human 8q. A disease-causing mutation was identified in the human homolog in an adRP family previously linked to 8q and, subsequently, the mouse gene proved to be identical to ORP1.

Since these initial reports, the research groups who independently identified the RP1 gene have joined to form the RP1 Consortium (see footnote 4, page 1). The Consortium members meet regularly to share research results and plan future RP1-related projects.

3.2 The RP10 gene

Two laboratories also reported independent identification of the RP10 gene. Bowne *et al.* (2002) identified additional RP10 families by linkage mapping and used this information to reduce the physical region containing the gene to three non-contiguous regions totaling 3.45. Dr. C. L. Cepko provided information on retinal genes whose expression is reduced in *crx-/crx-* knockout mice in comparison to normal mice, that is, genes whose retinal expression might be limited to photoreceptors (Blackshaw *et al.*, 2001). The human homologs of three of these genes map within the RP10 physical region. Sequencing of one of these genes, IMPDH1, identified a missense mutation, Asp226Asn, segregating with disease in the UTAD045 family (Bowne *et al.*, 2002). Missense mutations were also found in two other RP10 families identified earlier by linkage mapping and in 3 of 60 other, unrelated adRP families. The RP10 gene thus identified codes for the enzyme inosine monophosphate dehydrogenase 1, which catalyzes the rate limiting step in guanine biosynthesis. Because the RP10 gene has been known by the symbol "IMPDH1" for many years, this has become the new symbol for the adRP locus on 7q32.1.

Kennan *et al.* (2002) also identified IMPDH1 as the RP10 gene using a similar approach. Retinal cDNAs from *rho-/rho-* knockout mice were compared with retinal cDNAs from wild-type retinas using microarray analysis. As was true for the *crx-/crx-* mice, the IMPDH1 transcript was significantly reduced in the *rho-/rho-* knockout retinas. Sequencing in the original, Spanish RP10 family (Jordan *et al.*, 1993) revealed a missense mutation, Arg224Pro, segregating with disease. Thus two different approaches to identifying photoreceptor transcripts led to identification of the same RP10 gene, confirming the power of expression analysis to prioritize potential candidate genes.

The IMPDH1 gene spans 18 kb and is comprised of 17 exons, of which 14 are coding exons (Figure 4) (Gu *et al.*, 1997). The gene uses alternate 5' sites for initiation of transcription but the variant mRNAs produce identical proteins.

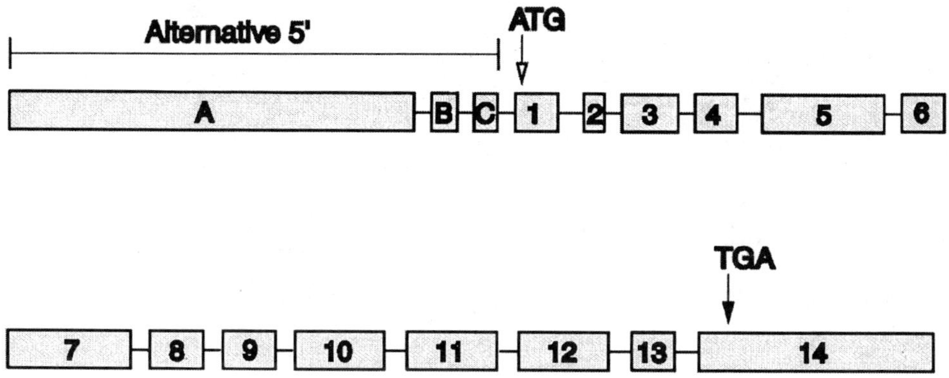

Figure 4. Structure of the IMPDH1 gene.

4. MUTATION SCREENING

4.1 RP1

At least 20 distinct disease-causing mutations have been identified in RP1 in families with adRP (Table 1). To date, mutations in RP1 appear to cause from 6 to 8% of adRP cases among Americans of European origin and Europeans (Berson *et al.*, 2002; Bowne *et al.*, 1999; Dietrich *et al.*, 2002; Payne *et al.*, 2000; Pierce *et al.*, 1999; Sullivan *et al.*, 1999). All of the mutations with convincing evidence of pathogenicity cause premature termination of translation, either by introduction of a premature stop codon or a frame shift leading to early termination. Interestingly, these mutations all fall within a relatively narrow region of exon 4, and apparently escape nonsense mediated mRNA decay because they are located within the terminal exon. Although many missense changes have been reported, both in isolation and as polymorphic variants, none have been shown to cause retinal disease, at least not in heterozygotes. Finally, mutation screening of RP1 in patients with other forms of retinopathy has failed to identify non-adRP mutations

4.2 RP10

Mutation screening of IMPDH1 in adRP families has been less extensive than RP1 screening but, nonetheless, at least 6 distinct pathogenic mutations have been reported, including unpublished data from S.J.B. (Table 2) (Bowne *et al.*, 2002; Kennan *et al.*, 2002). Mutations in IMPDH1 account for from 3 to 5% of adRP cases among Americans

Table 1. Disease-causing mutations in RP1

Mutation	No. of families	Reference
Met500 ins 2	1	Payne *et al.*, 2000
Pro658 ins 1	1	Berson *et al.*, 2002
Arg677ter	24	Berson *et al.*, 2002; Bowne *et al.*, 1999; Payne *et al.*, 2000; Sullivan *et al.*, 1999
Arg677 del 1	1	Bowne *et al.*, 1999
Gln679ter	1	Sullivan *et al.*, 1999
Glu700ter	2	Bowne *et al.*, 1999; Payne *et al.*, 2000
Gly723ter	1	unpublished
Gly724 del 14	3	Bowne *et al.*, 1999; Payne *et al.*, 2000
Ile725 ins 1	1	Bowne *et al.*, 1999
Glu729 del 1	1	unpublished
Thr736 ins 1	1	Bowne *et al.*, 1999
Cys744ter	2	Bowne *et al.*, 1999; Payne *et al.*, 2000
Ser747 del 1	1	Berson *et al.*, 2002
Leu762 del 5	9	Berson *et al.*, 2002; Bowne *et al.*, 1999; Payne *et al.*, 2000; Pierce *et al.*, 1999
Asn763 del 4	3	Payne *et al.*, 2000; Pierce *et al.*, 1999
Ser768 del 1	1	Bowne *et al.*, 1999
Lys778ter	1	Dietrich *et al.*, 2002
Thr865 del 2	1	Payne *et al.*, 2000
Arg872 ins 1	1	Payne *et al.*, 2000
Tyr1053 del 1	1	Berson *et al.*, 2002

of European origin. In contrast to RP1, all disease-causing variants of IMPDH1 are missense mutations, and IMPDH1 displays no polymorphic amino acid variation. Pathogenic mutations in IMPDH1have not been reported in other forms of retinopathy.

Table 2. Disease-causing mutations in IMPDH1

Mutation	No. of families	Protein domain	Reference
Thr116Met	1	CBS	Bowne *et al.*, 2002
Arg224Pro	1	CBS	Kennan *et al.*, 2002
Arg226Asn	5	CBS	Bowne *et al.*, 2002
Val268Ile	1	unknown	unpublished
Gly324Asp	2	active site	unpublished
His372Pro	1	active site	unpublished

5. GENE SEQUENCE AND FUNCTIONAL ANALYSIS

5.1 RP1

Analysis of the human RP1 coding sequence demonstrates very little sequence similarity to other human and non-human genes (Sullivan *et al.*, 1999). The first 10% of the amino-terminus of the protein, residues 1 through 228, shares 25% identity (39% similarity) to the human doublecortingene (DCX) and 15 to 40% identity to other members of the doublecortin family. Mutations in DCX, an X-linked gene in humans, cause lissencephaly, a congenital malformation of the brain resulting from impaired neuronal migration. The DCX protein interacts with microtubules so the cognate region of the RP1 gene may do likewise. Unfortunately, this sheds little light on the functional properties of the RP1protein because the remaining 90% contains no extended protein motifs, at least none conserved across mammalian species.

Northern analysis and *in situ* hybridization suggest that RP1 is expressed in the retina only and, specifically, in photoreceptors (Pierce *et al.*, 1999; Sullivan *et al.*, 1999). Immunolocalization indicates that the RP1 protein, 240 kDa in size, accumulates within the connecting cilium of rods and cones (Pierce *et al.*, 1999). Thus the doublecortin-like domain of RP1 may interact directly with the cilium. Finally, two different lines of RP1 knockout mice have been developed (Gao *et al.*, 2002; Liu *et al.*, 2002). The heterozygous knockouts show subtle retinal abnormalities but homozygous animals have overt retinal degeneration at a young age, consistent with the human phenotypes.

There is one exception to the absence of genes similar to RP1 in the human genome. A second gene, with the symbol RP1L1 for "RP1-like protein 1", has the same gene structure and overall length of RP1 and is 31% identical (52% similar) over the first 15%, that is, residues 28 through 355 (Figure 5) (unpublished). The remaining 85% of the two proteins are not statistically similar. Preliminary evidence suggests that RP1L1 is expressed exclusively in photoreceptors. Thus RP1 and RP1L1 appear to be true paralogs.

It remains to be determined what roles the two proteins play in photoreceptors and how closely their biological functions are interconnected.

```
RP1L1: 26    SVTKVTPAKKITFLKRGDPRFAGVRLAVHQRAFKTFSALMDELSQRVPLSFGVRSVTTPR 85
             S+T    AK+I+F K GDP+F GVR+ V+ R+FK+F AL+D LS++VPL FGVR+++TPR
RP1:   28    SLTHPVVAKRISFYKSGDPQFGGVRVVVNPRSFKSFDALLDNLSRKVPLPFGVRNISTPR 87

RP1L1: 86    GLHSLSALEQLEDGGCYLCSDKKPPKTPSGPGRPQERNPTAQQLRDVEGQREAPGTSSSR 145
             G HS++ LE+LEDG  YLCS   +  +       +   P              P   ++
RP1:   88    GRHSITRLEELEDGESYLCSHGRKVQPVDLDKARRRPRPWLSSRAISAHSPPHPVAVAAP 147

RP1L1: 146   KSLKTPRRILLIKNMDPRLQQTVVLSHRNTRNLAAFLGKASDLLRFPVKQLYTTSGKKVD 205
                + PR +++ +N DP+ ++ V+LS R T++  AFL   +++++ PV +LY T G++V
RP1:   148   GMPRPPRSLVVFRNGDPKTRRAVLLSRRVTQSFEAFLQHLTEVMQRPVVKLYATDGRRVP 207

RP1L1: 206   SLQALLHSPSVLVCAGHEAFRTPAMKNARRSEAETLSGLTSR--NKNGSWGPKTKPSVIH 263
             SLQA++ S   +V AG E F+        +      L G++ R   K  +     K S
RP1:   208   SLQAVILSSGAVVAAGREPFKPGNYDIQKYLLPARLPGISQRVYPKGNAKSESRKISTHM 267

RP1L1: 264   SRSPPGSTPRLPERPGPSNP-----PVGPAPGRHPQDTPAQSGPLV-AGDDMKKKVRMNE 317
             S S       +      +N         P      +   +Q+ P+  + DD++K +  N+
RP1:   268   SSSSRSQIYSVSSEKTHNNDCYLDYSFVPEKYLALEKNDSQNLPIYPSEDDIEKSIIFNQ 327

RP1L1: 318   DGSLSVEMKVRFHLVGEDTLLWSRRMGR 345
             DG+++VEMKVRF +  E+T+ W+  + +
RP1:   328   DGTMTVEMKVRFRIKEEETIKWTTTVSK 355
```

Figure 5. Human RP1 vs. RP1LI sequence comparison (first 15%).

5.2 RP10 (IMPDH1)

In sharp contrast to the RP1 gene and protein, the IMPDH1 gene is expressed in all tissues, is highly conserved in all living organisms, and has functional properties that are well-understood (Bowne *et al.*, 2002; Gu *et al.*, 1997; Kerr and Hedstrom, 1997). Inosine monophosphate dehydrogenase (EC 1.1.1.205) is an essential enzyme, required for *de novo* synthesis of guanine nucleotides, found in all species, including bacteria. In humans there are two paralogs, IMPDH1 and IMPDH2, which are 84% identical at an amino acid level and share identical kinetic properties (Figure 6). Across species, IMPDH genes are 30 to 85% identical. Thus IMPDH1 is a member of a family of highly-conserved, widely-expressed enzymes which, taken together, are essential for life. The mystery, of course, is why dominant-acting missense mutations in IMPDH1 cause retinal degeneration, and retinal degeneration only (as far as is known). One possible explanation is suggested by the *crx-/crx-* and *rho-/rho-* knockout retinas: IMPDH1 is significantly downregulated but not IMPDH2, suggesting that IMPDH1 is expressed in photoreceptors, but not IMPDH2.

6. CONCLUSIONS

Thus similar approaches were used to identify the RP1 and RP10 (IMPDH1) genes; mutations in both of these genes cause a substantial fraction of cases of autosomal dominant retinitis pigmentosa; and the clinical phenotypes are similar. At a functional level, though, the two gene products are strikingly different. The RP1 protein has very little sequence similarity to other genes, is limited to rod and cone photoreceptors, and plays an unknown role in the biology of the interconnecting cilium. In contrast, IMPDH1 is ubiquitously expressed, is a member of a highly-conserved family of enzymes, and has well-characterized biochemical properties. Efforts to identify retinal disease genes are undertaken

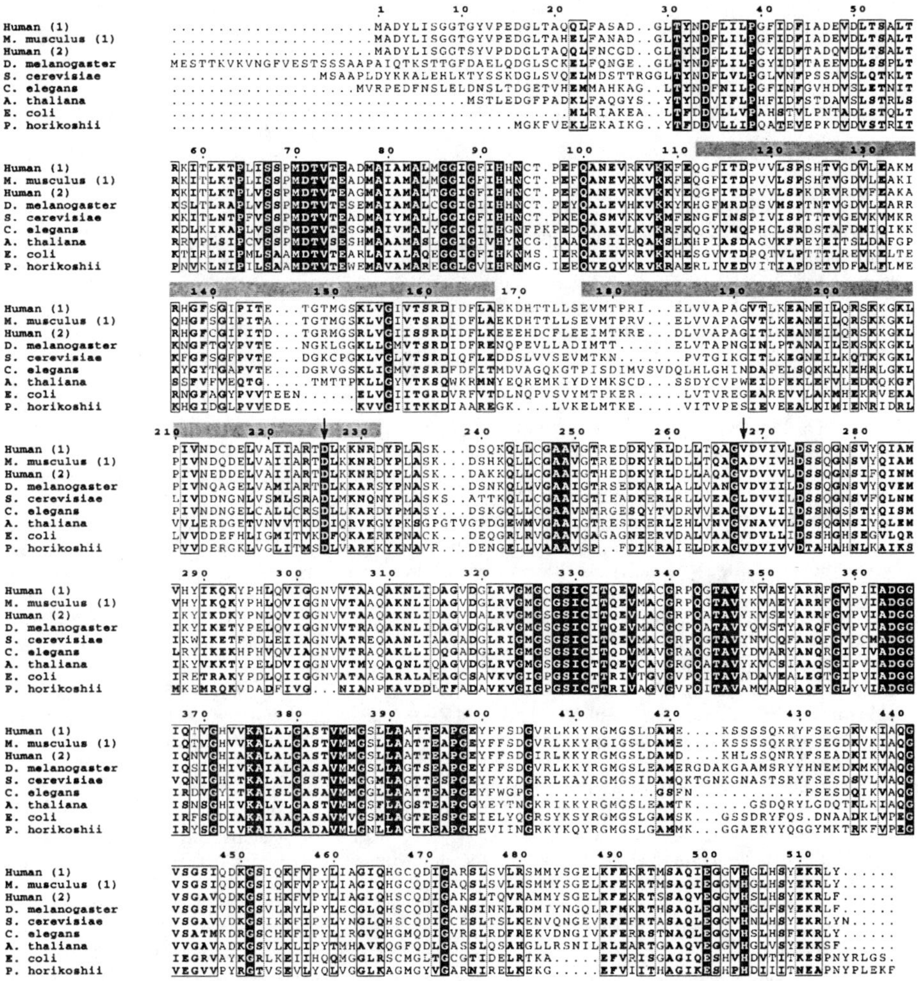

Figure 6. IMPDH sequence comparisons across species.

to help patients affected with these disorders, but characterization of these genes often reveals new biological processes.

7. ACKNOWLEDGMENTS

Supported by grants from the Foundation Fighting Blindness and the George Gund Foundation, the William Stamps Farish Fund, the M.D. Anderson Foundation, the John S. Dunn Foundation, and Alfred W. Lasher III; and by grants EY05235 (D.G.B), EY07142 (S.P.D.), and EY14170 (S.P.D.) from the National Eye Institute-National Institutes of Health.

8. REFERENCES

Berson, E. L., Grimsby, J. L., Adams, S. M., McGee, T. L., Sweklo, E., Pierce, E. A., Sandberg, M. A., and Dryja, T. P., 2002, Clinical features and mutations in patients with dominant retinitis pigmentosa-1 (RP1), *Invest. Ophthalmol. Vis. Sci.* **42**,2217-2224.

Blackshaw, S., Fraioli, R. E., Furukawa, T., and Cepko, C. L., 2001, Comprehensive analysis of photoreceptor gene expression and the identification of candidate retinal disease genes, *Cell* **107**,579-589.

Blanton, S. H., Heckenlively, J. R., Cottingham, A. W., Friedman, J., Sadler, L. A., Wagner, W., Friedman, L. H., and Daiger, S. D., 1991, Linkage mapping of autosomal dominant retinitis pigmentosa (RP1) to the pericentric region of human chromosome 8, *Genomics* **11**,857-869.

Bowne, S. J., Daiger, S. P., Hims, M. M., Sohocki, M. M., Malone, K. A., McKie, A. B., Heckenlively, J. R., Birch, D. G., Inglehearn, C. F., Bhattacharya, S. S., Bird, A., and Sullivan, L. S., 1999, Mutations in the RP1 gene causing autosomal dominant retinitis pigmentosa, *Hum. Mol. Genet.* **11**:2121-2128.

Bowne, S. J., Sullivan, L. S., Blanton, S. H., Cepko, C. L., Blackshaw, S., Birch, D. G., Hughbanks-Wheaton, D., Heckenlively, J. R., and Daiger, S. P., 2002, Mutations in the inosine monophosphatedehydrogenase 1 gene (IMPDH1) cause the RP10 form of autosomal dominant retinitis pigmentosa, *Hum. Mol. Genet.* **11**,559-568.

Dietrich, K., Jacobi, F. K., Tippmann, S., Schmid, R., Zrenner, E., Wissinger, B., and Apfelstedt-Sylla, E., 2002, A novel mutation of the RP1 gene (Lys778ter) associated with autosomal dominant retinitis pigmentosa, *Brit. J. Ophthalmol.* **86**,328-332.

Field, L. L., Heckenlively, J. R., Sparks, R. S., Garcia, C. A., Farson, C., Zedalis, D., Sparkes, M. C., Crist, M., Tideman, S., and Spence, M. A., 1982, Linkage analysis of five pedigrees affected with typical autosomal dominant retinitis pigmentosa, *J.Med. Genet.* **19**,266-270.

Gao, J., Cheon, K., Nusinowitz, S., Liu, Q., Bei, D., Atkins, K., Azimi, A., Daiger, S. P., Farber, D. B., Heckenlively, J. R., Pierce, E. A., Sullivan, L. S., and Zuo, J., 2002, Progressive photoreceptor degeneration, outer segment dysplasia and rhodopsin mis-localization in mice with targeted disruption of the retinitis pigmentosa-1 (*Rp1*) gene, *Proc. Nat. Acad. Sci. USA.* **99**,5698-5703.

Gu, J. J, Spychala, J., and Mitchell, B. S., 1997, Regulation of the human inosine monophosphate dehydrogenase type I gene. Utilization of alternative promoters, *J. Biol. Chem.* **272**,4458-466.

Guillonneau, X., Piriev, N. I., Danciger, M., Kozak, C. A., Cideciyan, A. V., Jacobson, S. G., and Farber, D. B., 1999, A nonsense mutation in a novel gene is associated with retinitis pigmentosa in a family linked to the RP1 locus, *Hum. Mol. Genet.* **8**,1541-1546.

Inglehearn, C. F., McHale, J. C., Keen, T. J., Skirton, H., and Lunt, P. W., 1999, A New family linked to the RP1 dominant retinitis pigmentosa locus on chromosome 8q, *J. Med. Genet.* **36**,646-648 .

Jordan, S. A., Farrar, G. J., Kenna, P., Humphries, M. M., Sheils, D. M., Kumar-Singh, R., Sharp, E. M., Ayuso, C., Benitez, J., and Humphries, P., 1993, Localization of an autosomal dominant retinitis pigmentosa gene to chromosome 7q, *Nat. Genet.* **4**,54-58.

Kennan, A., Aherne, A., Palfi, A., Humphries, M., McKee,, A. Stitt, A., Simpson, D. A. C., Demtroder, K., Orntoft, T., Ayuso, C., Kenna, P. F., Farrar, G. J., and Humphries, P., 2002, Identification of an IMPDH1 mutation in autosomal dominant retinitis pigmentosa (RP10) revealed following comparative microarray analysis of transcripts derived from retinas of wild-type and Rho-/- mice, *Hum. Mol. Genet.* **11**,547-557.

Kerr, K. M. and Hedstrom, L., 1997, The roles of conserved carboxylate residues in IMP dehydrogenase and identification of a transition state analog, *Biochem.* **36**,13365-13373.

Liu, Q., Zhou, J., Daiger, S. P., Farber, D. B., Heckenlively, J. R., Smith, J. E., Sullivan, L. S., Zuo, J., Milam, A. H., and Pierce, E. A., 2002, Identification and subcellular localization of the RP1 protein in human and mouse photoreceptors, *Invest. Ophthalmol. Vis. Sci.* **43**,22-32.

McGuire, R. E., Gannon, A. M., Sadler-Sullivan, L. A., Rodriguez, J. A., and Daiger, S. P., 1995, Evidence for a major gene (RP10) for autosomal dominant retinitis pigmentosa on chromosome 7q: linkage mapping in a second, unrelated family, *Hum. Genet.* **95**,71-74.

McGuire, R. E., Jordan, S. A., Braden, V. V., Bouffard, G. G., Humphries, P., Green, E. D., and Daiger, S. P., 1996, Mapping the RP10 locus for autosomal dominant retinitis pigmentosa on 7q: refined genetic positioning and localization within a well-defined YAC contig, *Genome Res.* **6**,255-266.

Millán, J. M., Martínez, F., Vilela, C., Beneyto, M., Prieto, F., and Nájera, C., 1995, An autosomal dominant retinitis pigmentosa family with close linkage to D7S480 on 7q, *Hum. Genet.* **96**,216-218.

Mohamed, Z., Bell, C., Hammer, H. M., Converse, C. A., Esakowitz, L., and Haites, N. E., 1996, Linkage of a medium sized Scottish autosomal dominant retinitis pigmentosa family to chromosome 7q, *J. Med. Genet.* **33**,714-715.

Payne, A., Vithana, E., Khaliq, S., Hameed, A., Deller, J., Abu-Safieh, L., Kermani, S., Leroy, B. P., Mehdi, S. Q., Moore, A. T., Bird, A. C., and Bhattacharya, S. S., 2000, RP1 protein truncating mutations predominate at the RP1 adRP locus, *Invest. Ophthalmol. Vis. Sci.* **41**,4069-4073.

Pierce, E. A., Quinn, T., Meehan, T., McGee, T. L., Berson, E. L., and Dryja, T. P., 1999, Mutations in a gene encoding a new oxygen-regulated photoreceptor protein cause dominant retinitis pigmentosa, *Nat. Genet.* **22**,248-254.

Sullivan, L. S., Heckenlively, J. R., Bowne, S. J., Zuo, J., Hide, W. A., Gal, A., Denton, M., Inglehearn, C. F., Blanton, S. H., and Daiger, S. P., 1999, Mutations in a novel retina-specific gene cause autosomal dominant retinitis pigmentosa, *Nat. Genet.* **22**,255-259.

Xu, S. - Y., Denton, M., Sullivan, L. S., Daiger, S. P., and Gal, A., 1996, Genetic mapping of RP1 on 8q11-q21 in an Australian family with autosomal dominant retinitis pigmentosa reduces the critical region to 4 cM between D8S601 and D8S285, *Hum. Genet.* **98**,741-743.

ON THE ROLE OF IMPDH1 IN RETINAL DEGENERATION

Avril Kennan[1], Aileen Aherne[1], Sara J. Bowne[2], Stephen P. Daiger[2], G. Jane Farrar[1], Paul F. Kenna[1] and Pete Humphries[1]

1. INTRODUCTION

To date twelve loci have been associated with the autosomal dominant form of Retinitis Pigmentosa (adRP). The genes at eleven of these have now been identified (http://www.sph.uth.tmc.edu/Retnet/). One such locus (RP10) is on chromosome 7q and was original identified as a result of a large linkage study undertaken at this laboratory (Jordan et al., 1993). Affected individuals of this family show the classic clinical symptoms of RP, including bone spicule pattern pigmentary deposits, optic disc pallor and retinal vascular attenuation (for a detailed clinical description of this family see Jordan et al., 1993). An unrelated American adRP family, UTAD045, exhibiting a later onset and a slower progression of symptoms, was also subsequently revealed to have a disease gene on the same region of 7q (McGuire et al., 1995).

Recently, we have shown that mutations within a gene of the *de novo* guanine nucleotide biosynthesis pathway, inosine monophosphate dehydrogenase type 1 (IMPDH1), cause the RP10 form of adRP (Kennan et al., 2002). IMPDH1 is the first protein of a nucleotide biosynthesis pathway to be associated with retinal disease and it shows a widespread pattern of expression. The enzyme catalyses the NAD-dependant conversion of inosine monophosphate (IMP) to xanthosine monophosphate (XMP), which is subsequently converted to GMP. GMP gives rise to one of the building blocks of DNA (dGTP) and guanine nucleotides also play an essential role in intracellular signalling pathways.

It is intriguing that a mutation in a gene with such an apparently critical function and widespread pattern of expression causes a disease with a clinical manifestation only in the retina. Detailed analysis of the expression of IMPDH1 in the retina has not previously been undertaken. In attempting to understand the tissue specific nature of the disease it is

[1]The Ocular Genetics Unit, Department of Genetics, Trinity College, Dublin 2, Ireland; [2]Human Genetics Center, School of Public Health, The Univ. of Texas HSC, Houston, TX 77030, USA. Corresponding author: A. Kennan, E-mail: akennan@tcd.ie

also important to consider that, in mammals, there are two alternative routes that may be used in the production of guanine. The first of these involves the isoenzyme of IMPDH1, IMPDH2. IMPDH2 is 84% homologous to the type 1 form of the gene and the two proteins are indistinguishable in terms of their catalytic activity and substrate affinities (Carr et al., 1993). Most tissues are also capable of synthesizing GMP from the degradation products of nucleic acids and catabolism by means of a salvage pathway. This salvage pathway is less energetically expensive than the *de novo* pathway and in fact exerts feedback control on the *de novo* pathway. The key enzyme of the purine salvage pathway is hypoxanthine phosphoribosyl transferase (HPRT) which phosphoribosylates the purine bases hypoxanthine and guanine for reutilisation by the cell. The enzymes involved in purine metabolism and the control of these enzymes is not the same for all tissues and can in fact be cell type specific (Stone and Simmonds, 1991). The relative contributions of the *de novo* and salvage pathways to the synthesis of guanine nucleotides in the retina have not been previously determined. Analysis of the expression of IMPDH1, IMPDH2 and HPRT in the mouse retina, using *in situ* hybridisation and relative quantitative RT-PCR may shed some light on the pathology of RP10 and the results of such analysis are presented here.

2. MATERIALS AND METHODS

2.1. *In situ* Hybridisations

The *in situ* hybridisation technique employed in this study involved the use of digoxygenin (DIG)-labelled riboprobes on frozen cryosections. Sense and antisense DIG-labelled probes were generated from PCR templates which had incorporated T7 and T3 promoters. Sections were fixed in para-formaldehyde post cutting, treated with active DEPC and hybridisation with both sense and antisense probes allowed to proceed overnight at 58°C. Following stringent washes with SSC, incubation with an AP coupled, anti-DIG antibody was carried out. Binding of the probes was detected using NBT/BCIP solution. Sections were mounted and analysed using a Zeiss Axioplan 2 microscope.

2.2. Quantitative RT-PCR

Relative quantitative RT-PCR was used to quantify the levels of transcript in the retina for each of the three genes analysed in this study. First strand cDNA synthesis was carried out using 2µg of total RNA from the pooled wildtype and rho-/- RNA samples with 4 units Omniscript RT (Qiagen) and 1µM of random hexamers. Quantitative PCR was subsequently performed on a Roche LightCycler using the Quantitech PCR kit (Qiagen). Serial dilutions of cDNA were used to generate standard curves of crossing cycle number versus the logarithmus of concentration for each gene of interest. A linear regression line calculated from the standard curves allowed relative transcript levels in both RNA samples to be determined. Values were normalised to the relative amounts of GAPDH present in the same cDNA preparations.

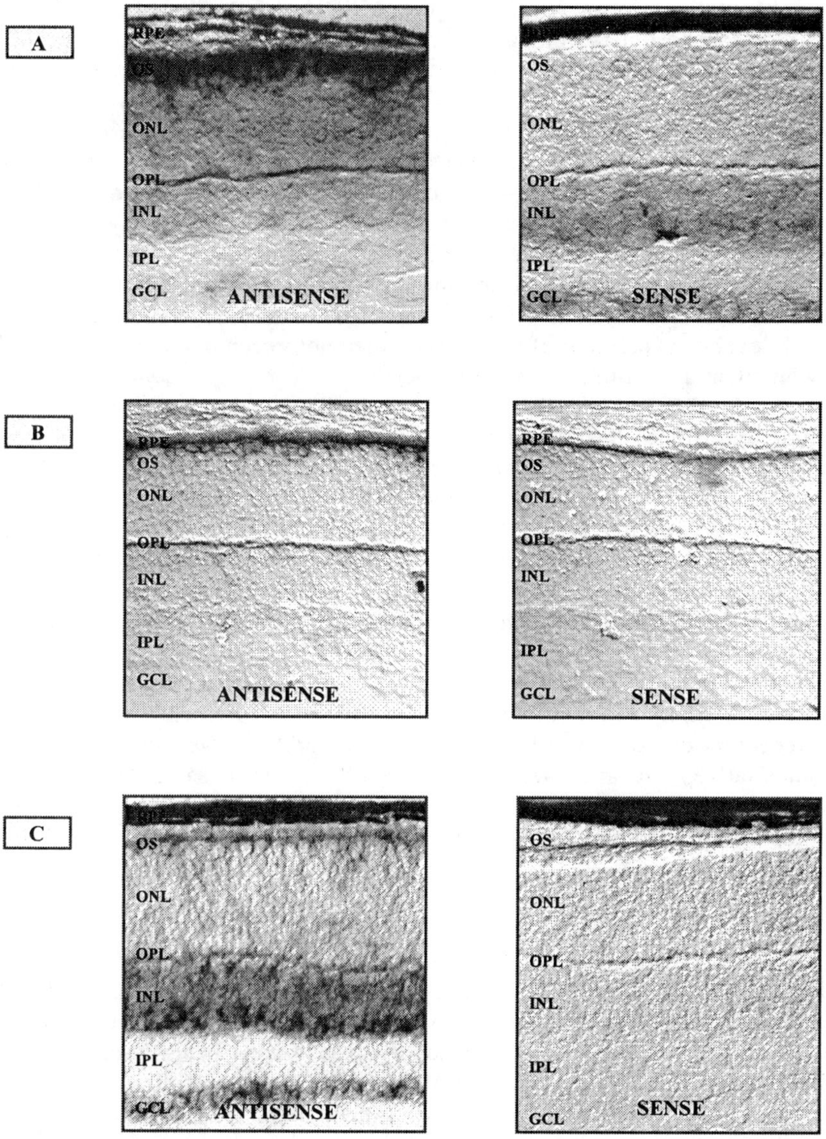

Figure 1. *In situ* hybridisations on mouse retinal cryosections using DIG-labelled riboprobes. Results for antisense and sense probes for (A) IMPDH1, (B) IMPDH2 and (C) HPRT are shown. The IMPDH2 *in situs* are performed on CD1 mouse retinas which lack pigment in the RPE while the IMPDH1 and HPRT *in situs* are performed on wild type pigmented retinas. (RPE) retinal pigment epithelium, (OS) outer segments, (ONL) outer nuclear layer, (OPL) outer plexiform layer, (INL) inner nuclear layer, (IPL) inner plexiform layer, (GCL) ganglion cell layer. 200X Magnification.

3. RESULTS

RP-causing mutations were identified in the IMPDH1 gene following a microarray study which was carried out to compare levels of transcript for over 6,000 genes and ESTs in wild type and rhodopsin knockout (rho-/-) mouse retinas. When compared to wild type the IMPDH1 gene showed a down-regulation of 5.8-fold in 4 month rho-/- mouse samples, which have essentially no photoreceptors remaining (Kennan et al., 2002). Relative quantitative PCR carried out to compare levels of IMPDH1 transcript in wild type and rho-/- retinas, also at 4 months of age, confirmed this result. Such a decrease in transcript levels of a gene is an indication that it may be highly expressed in photoreceptors and the pattern observed with IMPDH1 is similar to that of other genes known to be specifically expressed in photoreceptors, within the retina. *In situ* hybridisation carried out using IMPDH1 riboprobes on mouse retinal cryosections revealed a high level of expression of IMPDH1 in the photoreceptors, with only very low levels, if any, in other layers of the retina (Fig. 1A).

Analysis of IMPDH2 ESTs in the UniGene database (http://www4.ncbi.nlm.nih.gov/UniGene) has indicated that this isoenzyme is expressed in retinal pigment epithelium (RPE) libraries. *In situ* hybridisations were carried out with IMPDH2 probes on retinal sections from CD1 mice, which do not posess pigment in the RPE, so as not to mask a positive result. This confirmed that there are transcripts in the RPE cells and no transcription in the neuronal cells of the retina was detected (Fig. 1B). Quantitative PCR carried out on cDNA from retinas at four months of age shows that, despite the lack of photoreceptors, levels of transcript remained constant in the rho-/- retina in comparison to wild type.

Analysis of the guanine nucleotide salvage pathway gene encoding HPRT was also performed using *in situ* hybridisation and revealed that it is expressed only at low levels in the photoreceptors but at higher levels in the inner nuclear layer and ganglion cells (Fig. 1C). Quantitative PCR on cDNA from four month retinas revealed that the levels of HPRT transcript in rho-/- mice are the same as in wild type, providing supporting evidence for the lack of high level HPRT expression in photoreceptors.

4. DISCUSSION

The identification of IMPDH1 as the causitive gene in the RP10 form of adRP has, for the first time, implicated a nucleotide biosynthesis pathway in a degeneration of the retina. The IMPDH1 gene encodes a protein subunit of 514 amino acid residues, with the active IMPDH1 enzyme consisting of a homotetramer of these subunits. Each monomer of the protein possesses two domains; the larger forms a barrel which contains the active site loop and the smaller is comprised of two tandem CBS (cystathionine-beta-synthase) dimer domains (Nimmesgern et al., 1999). The established disease causing mutations identified to date lie in the CBS 2 domain of the protein (Kennan et al., 2002 and Bowne et al., 2002). CBS domains are motifs found in a number of different proteins which code for modules of unknown function. It has been suggested that these domains may attach to a wide range of other protein domains and in this way play a regulatory role (Sintchak et al., 2000). It is not clear at this stage how mutations in IMPDH1 bring about the degeneration of retinal photoreceptors. In addressing this however, there are two related questions that must be answered. Firstly, what effect are the mutations having on the

IMPDH1 protein and secondly, why does this effect only reveal itself in the form of retinal disease? It is the latter question which we have attempted to address here.

The two IMPDH isoforms have previously been shown to exhibit considerable differences in their regulation and expression. The *IMPDH2* gene exhibits inducible expression and is highly up-regulated in proliferating cells, whereas the *IMPDH1* gene is more widely expressed but shows variability of expression in different tissues (Senda et al., 1994). Three different promoters have been identified in the human type 1 form of the protein which give rise to three distinct transcripts. This may be an indication that IMPDH1 expression is regulated in a complex cell type-specific manner (Gu et al., 1997). Investigations into the expression patterns of IMPDH1, IMPDH2 and HPRT within the retina have not been previously undertaken. The results from this study suggest that IMPDH1 is preferentially expressed in photoreceptor cells within the retina, as opposed to secondary retinal neurons. Conversely, no IMPDH2 expression was detected in photoreceptors and only low levels of HPRT were observed. The results therefore suggest that IMPDH1 is the enzyme primarily responsible for the production of guanine nucleotides within the photoreceptors. A previous study has shown that in mice lacking the IMPDH2 gene there is no observed compensatory mechanism for regulation of expression of IMPDH1 and HPRT (Gu *et al.*, 2000). It is possible therefore, that the other guanine biosynthesis enzymes may not be able to substitute for a loss of IMPDH1 activity in RP10.

As IMPDH1 is the first gene of its kind to be linked to a retinal degeneration it is difficult to speculate how precisely mutations within the gene may bring about the pathology. One possibility is that, while not in the active site of the enzyme, the mutation may cause a reduction in the levels of guanine nucleotides available to photoreceptors which it appears, may be relying almost solely on IMPDH1 for maintaining their guanine nucleotide reserves. Photoreceptors are among the most physiologically active cells of the body and require high levels of guanine. GDP is required for the activation of transducin during visual transduction and the circadian shedding of outer segment disc membranes requires a very high level of transcription. As the *in situ* hybridisations shown here have been performed on cyrosections it is difficult to say definitively which region of the photoreceptor cells are showing the higest levels of IMPDH1 transcription. It does appear however that the outer segments of the photoreceptors, where visual transduction takes place, are showing very high levels of staining. Large quantities of IMPDH1 transcript may be translocated to this region of the photoreceptors in order to satisfy the demand for guanine nucleotides during phototransduction.

A reduction in levels of guanine as a result of IMPDH1 haploinsufficiency in patients may, in theory, affect the viability of photoreceptors and result in apoptosis of these cells. Interestingly, perturbations in intracellular pools of dNTPs have been shown to result in DNA fragmentation and apoptosis in various systems (Oliver et al., 1998). Other tissues of the body may not be similarly affected due to a lower requirement for guanine nucleotides and/or a supply of guanine from IMPDH2 activity or the salvage pathway. Alternatively, the mutations may have a dominant negative effect on the IMPDH1 enzyme which is more detrimental to photoreceptors than other cells types due to the high level of IMPDH expression from the type 1 form of the gene. We are currently investigating the effect of the RP mutations on the activity and properties of the IMPDH1 enzyme in an attempt to futher expand our knowledge of the role of IMPDH1 in both normal and diseased photoreceptors.

5. ACKNOWLEDGEMENTS

The Ocular Genetics Unit of TCD is supported by grants from the Wellcome Trust, the Foundation Fighting Blindness (USA), the British RP Society, Fighting Blindness Ireland, the Health Research Board of Ireland and the European Union 5th Framework Programme. The Unit is a member of the HEA-Ireland-sponsored Biopharmaceutical Science Network. We thank the National Eye Institute (USA) and the Foundation Fighting Blindness (USA) for the travel award provided to AK to attend this meeting.

6. REFERENCES

Bowne, S.J., Sullivan, L.S., Blanton, S.H., Cepko, C.L., Blackshaw, S., Birch, D.G., Hughbanks-Wheaton, D., Heckenlively, J.R., and Daiger, S.P., 2002, Mutations in the inosine monophosphate dehydrogenase 1 gene (IMPDH1) cause the RP10 form of autosomal dominant retinitis pigmentosa. *Hum Mol Genet.* **11**:559.

Carr, S.F., Papp, E., Wu, J.C., and Natsumeda, Y., 1993, Characterization of human type I and type II IMP dehydrogenases. *J Biol Chem.* **268**: 27286.

Gu, J.J., Spychala, J., and Mitchell, B.S., 1997, Regulation of the human inosine monophosphate dehydrogenase type I gene. Utilization of alternative promoters. *J Biol Chem.* **272**: 4458.

Gu, J.J., Stegmann, S., Gathy, K., Murray, R., Laliberte, J., Ayscue, L., and Mitchell, B.S., 2000, Inhibition of T lymphocyte activation in mice heterozygous for loss of the IMPDH2 gene. *J Clin Invest.* **106**:599.

Kennan, A., Aherne, A., Palfi, A., Humphries, M., McKee, A., Stitt, A., Simpson, D.A., Demtroder, K., Orntoft, T., Ayuso, C., Kenna, P.F., Farrar, G.J., and Humphries, P., 2002, Identification of an IMPDH1 mutation in autosomal dominant retinitis pigmentosa (RP10) revealed following comparative microarray analysis of transcripts derived from retinas of wild-type and Rho(-/-) mice. *Hum Mol Genet.* **11**:547.

Jordan, S.A., Farrar, G.J., Kenna, P., Humphries, M.M., Sheils, D.M., Kumar-Singh, R., Sharp, E.M., Ayuso, C., Benitez, J., and Humphries, P., 1993, Localization of an autosomal dominant retinitis pigmentosa gene to chromosome 7q. *Nat Genet.* **4**:54.

McGuire, R.E., Gannon, A.M., Sullivan, L.S., Rodriguez, J.A., and Daiger, S.P., 1995, Evidence for a major gene (RP10) for autosomal dominant retinitis pigmentosa on chromosome 7q: linkage mapping in a second, unrelated family. *Hum Genet.* **95**:71.

Oliver, F.J., Muñoz-Pinedo, C., Collins, M.K.L., and López Rivas, A., 1998, Regulation of apoptosis by survival factors: the role of metabolism. IV Metabolismo de la Metionina. Mecanismos moleculares e implicaciones clinicas. Boehringer Ingelheim. http:www.boehringer-ingelheim.es/workshop-methionina/anglesa/cap11

Nimmesgern, E., Black, J., Futer, O., Fulghum, J.R., Chambers, S.P., Brummel, C.L., Raybuck, S.A., and Sintchak, M.D., 1999, Biochemical analysis of the modular enzyme inosine 5'-monophosphate dehydrogenase. *Protein Expr Purif.,* **17**:282.

Senda, M., and Natsumeda, Y., 1994, Tissue-differential expression of two distinct genes for human IMP dehydrogenase (E.C.1.1.1.205). *Life Sci.* **54**: 1917.

Stone, T. W., and Simmonds H. A., 1991, Purines: Basic and Clinical Aspects. Kluwer, London.

AN INTEGRATED GENETIC APPROACH TO IDENTIFY CANDIDATE GENES FOR HUMAN CHROMOSOME 6q-LINKED RETINAL DISORDERS

Pamela S. Lagali[1], Radha Ayyagari[2], and Paul W. Wong[1]

1. INTRODUCTION

In recent years the field of retinal gene discovery has experienced an explosion. An estimated 10,000-30,000 genes are thought to be expressed in the human retina (Shimizu-Matsumoto et al., 1997), with nearly 5,000 retinal transcripts catalogued from expressed sequence tag (EST) datasets (Bortoluzzi et al., 2000) and over 26,000 different transcripts represented in retina serial analysis of gene expression (SAGE) libraries (Sharon et al, 2002). A wide variety of methods have been developed and utilized to define novel retinal genes, particularly in an effort to identify genes targeted in human retinopathies. To date there are 144 diseases of the retina described that have been mapped to human chromosomal locations, and an additional 41 human disorders involving some form of retinal dystrophy for which the genetic map locations have yet to be determined (RetNet Retinal Information Network, http://www.sph.uth.tmc.edu/retnet/; Online Mendelian Inheritance in Man (OMIM), http://www.ncbi.nlm.nih.gov/omim/). Despite recent efforts to discover novel genes by analyzing increasingly greater volumes of information arising from human genomic sequence data organization and expression studies, the disease genes for less than one-half of these conditions have at present been cloned. While some of these disorders may be attributed to allelic heterogeneity, a significant number of genes that account for retinal degenerative phenotypes remain to be defined.

It is our aim to identify and characterize novel retinal genes that may be considered as candidate genes for human retinal disease. Typically, previous studies have focussed on isolating a specific gene that underlies a particular pathological condition whose phenotype is shared amongst related individuals, but for which there is little or no knowledge of the biochemical defect that causes the disease. This procedure is commonly known as the positional cloning method and has resulted in the identification of a number

[1]Department of Biological Sciences, University of Alberta, Edmonton, Alberta, Canada T6G 2E7; [2]Department of Ophthalmology and Visual Sciences, Kellogg Eye Centre, University of Michigan, Ann Arbor, MI 48105, USA. Corresponding author: P. S. Lagali, E-mail: plagali@ualberta.ca

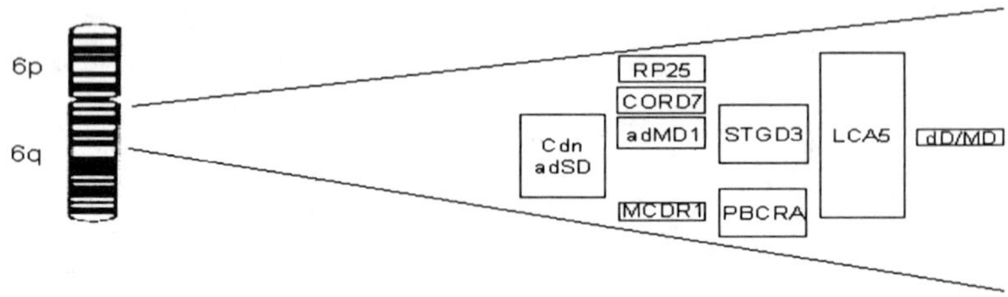

Figure 1. Human chromosome 6q is a genomic hotspot for retinal disease genes. Disease loci abbreviations: Cdn adSD, Canadian autosomal dominant Stargardt-like disease (Lagali et al., 2000); RP25, retinitis pigmentosa 25 (Khaliq et al., 1999); CORD7, cone-rod dystrophy 7 (Kelsell et al., 1998); adMD1, autosomal dominant macular dystrophy 1 (Griesinger et al., 2000); MCDR1, North Carolina macular dystrophy (Sauer et al., 1997); STGD3, autosomal dominant Stargardt-like disease (Stone et al., 1994); PBCRA, progressive bifocal chorioretinal atrophy (Kelsell et al., 1995); LCA5, Leber's congenital amaurosis 5 (Dharmaraj et al., 2000); dD/MD, dominant drusen with macular degeneration (Kniazeva et al., 2000).

of retinal disease genes including *CHM* for choroideremia (Cremers et al., 1990), *NDP* for Norrie's disease (Berger et al., 1992; Chen et al., 1992), *RPGR* and *RP2* for X-linked retinitis pigmentosa (RP) (Meindl et al., 1996; Roepman et al., 1996; Schwahn et al., 1998), *ABCA4* for recessive Stargardt's macular dystrophy (Allikmets et al., 1997), and *VMD2* for Best's disease (Petrukhin et al., 1998), as well as several others. With the completion of the Human Genome Project, recent studies have alternatively taken a more global approach to identify putative candidate retinal disease genes by analyzing a large number of expressed sequences that ultimately comprise the entire retinal transcriptome through the use of techniques such as the generation and characterization of retina-enriched cDNA libraries (Bernstein et al., 1996; Shimizu-Matsumoto et al., 1997; Sinha et al., 2000), the assessment of differential expression of retinal and sub-retinal transcripts (Schoen et al., 1995; Bernstein et al., 1995; Bernstein et al., 1997; Yoshida et al., 2002), and *in silico* analysis of mRNA sequences, EST clusters (Sohocki et al., 1999; Malone et al., 1999; Bortoluzzi et al., 2000; Stohr et al., 2000; Katsanis et al., 2002), and SAGE tags (Blackshaw et al., 2001; Sharon et al., 2002). The approach we have adopted is a combination of these methods, enabling us to take advantage of well established procedures for family studies and linkage mapping of disease loci towards the positional cloning of disease genes as well as exploiting genetic sequence and expression analysis resources that are now available in the post-genome era.

2. GENETIC MAPPING OF A MACULAR DYSTROPHY LOCUS TO A HUMAN RETINAL DISEASE HOTSPOT

A five-generation Canadian pedigree in which a form of autosomal dominant macular dystrophy is segregating was genetically evaluated using polymorphic markers and linkage of the disease locus with the long arm of human chromosome 6 was established (Lagali et al., 2000). Closer inspection of this genomic region revealed that multiple retinal disease loci had been mapped to chromosome 6q, including those for cases of RP, Leber's congenital amaurosis, cone-rod dystrophy, and multiple macular degenerative phenotypes (Fig. 1). The existence of these numerous and in some cases genetically distinct disease loci in addition to the inability to detect mutations in the retinal genes already known to lie within the corresponding genetic intervals (Gehrig et al., 1998; Kniazeva et al., 1999) suggested that multiple genes relevant for retinal function are within this genomic region. The course of study we have pursued was thus undertaken to both identify the disease gene in the Canadian family under study as well as to uncover additional novel genes that may serve as candidates for the other retinopathies that map to within a more broad genomic interval on chromosome 6q.

·Linkage analysis of the Canadian pedigree was performed with a total of 43 microsatellite markers located on human chromosome 6, spanning approximately 38 centimorgans (cM) from 6p12.3 (marker D6S271) to 6q16.3 (marker D6S475), to define the boundaries of the genetic interval containing the disease locus (Lagali et al., 2000). A maximum LOD score of 5.50 at a recombination fraction of 0.00 was obtained for marker D6S300. For the purposes of narrowing our search for the disease gene, we defined the disease haplotype for the family and subsequently determined that the genetic lesion lay within a 9.4-cM minimal critical interval on chromosome 6q. This interval excluded the genetic locus for North Carolina macular dystrophy however still overlapped the seven other loci for retinal disorders then known to exist between 6q13-q16.

3. CONSTRUCTION OF A CHROMOSOME 6q13-q16 RETINAL TRANSCRIPT MAP

To search for genes localized to the chromosome 6q interval of interest, we utilized the GeneMap'99 human genome database, available through the National Centre for Biotechnology Information (http://www.ncbi.nlm.nih.gov/genemap), as a resource to identify all known non-overlapping ESTs that mapped to the 6q13-q16 region and obtained 137 clones for expression studies. Dot blot analysis was performed to categorize the EST clones according to their expression profile within the human retina and retinal pigment epithelium (RPE), defining possible candidates for retinal disease (Fig. 2).

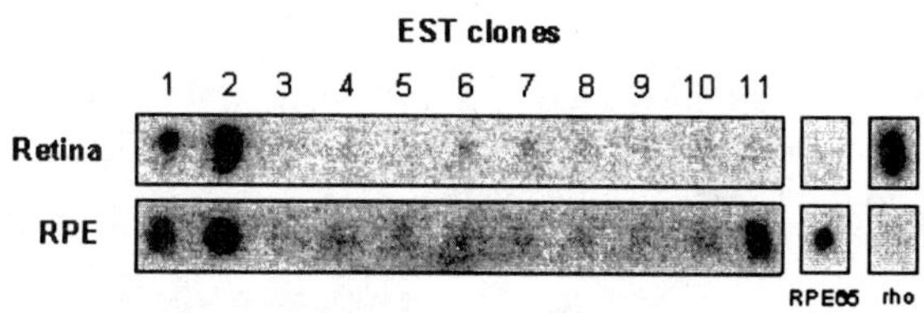

Figure 2. A sample of the duplicate sets of dot blots of chromosome 6q EST clones probed with human retina or RPE cDNA. Examples of retina-enriched (clone 1), RPE-enriched (clone 11), and equivalently expressed (clone 2) ESTs are given. Control blots using tissue-specific markers are shown on the right. *rho*, rhodopsin.

Ninety-five (69%) of the clones were found to be expressed in either the retina or RPE, with 55/95 (40% of the total number of ESTs screened) exhibiting preferential retinal expression and 8/95 (6% of the clones) demonstrating RPE-enriched expression relative to corresponding transcript levels in retinal tissue. ESTs of interest were subsequently further characterized by confirming and extending their expression status through Northern blot analysis, defining the corresponding gene structure using genomic sequence analysis tools, and assessing biochemical features of the predicted proteins through both *in vitro* and bioinformatic analyses. Exon sequences of defined genes were screened for mutations in affected members of the Canadian macular dystrophy pedigree as well as in other individuals and families with retinal disorders mapping to chromosome 6q13-q16.

4. ISOLATION AND ANALYSIS OF NOVEL CANDIDATE RETINAL DISEASE GENES

4.1 Chromosome 6 open reading frame 37 (C6orf37)

Two independent, non-overlapping ESTs that exhibited significant expression in the human fovea and peripheral retina and comparatively lower expression in the RPE based on the dot blot analysis were suspected to reside within the same gene as each was observed to yield identical results when used to probe Northern blots. This was confirmed upon construction of the 5.6-kilobase full-length mRNA transcript from EST cluster and chromosome 6 genomic sequence data (Lagali et al., 2002). The corresponding gene, given the name chromosome 6 open reading frame 37 (*C6orf37*), was found to consist of three exons and two introns spanning roughly 7 kilobases of genomic DNA. The gene is highly conserved as homologous sequences were detected in DNA isolated from a variety of eukaryotic species including mammals, birds, and yeast via Southern blot analysis, and protein homologues of unknown function were identified in rat, mouse, fruit fly, and nematode worm through sequence database searches (Lagali et al., 2002). The approximately 50-kilodalton protein encoded by the gene appears to contain a number of conserved phosphorylation sites and is predicted to be a soluble cytoplasmic factor, suggestive of a putative role in signal transduction within retinal cells. At the present time, no further functional information can be extrapolated based on protein sequence analysis methods. The *C6orf37* mRNA transcript was detected in a wide variety of human tissues, however a preference for retinal expression was observed upon analysis of ocular tissues (Fig. 3).

Taken together, these findings, in addition to its chromosomal location, classify the *C6orf37* gene as a candidate gene for retinal disease. To date there have yet to be mutations identified in the *C6orf37* sequence that are associated with a disease phenotype.

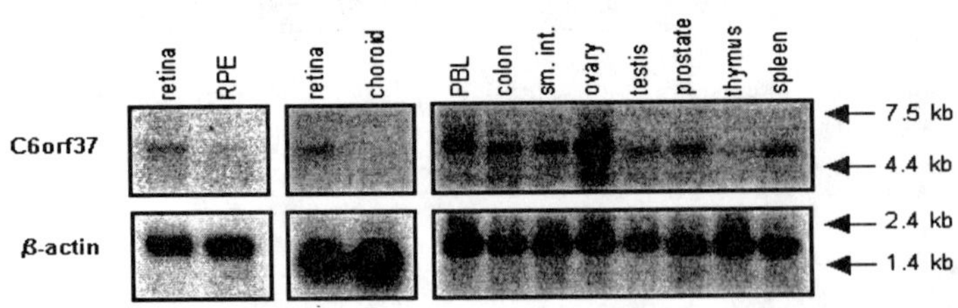

Figure 3. Northern blots of human RNA samples probed with C6orf37 and control (β-actin) cDNA probes. Retina, RPE, and choroid lanes contain 5 micrograms of total RNA while all other lanes contain 2 micrograms of poly(A)+ RNA. Positions of RNA size markers are shown on the right. *PBL*, peripheral blood leukocytes; *sm. int.*, small intestine; *kb*, kilobases. Modified from Lagali et al., 2002 with permission from Elsevier Science.

4.2 Elongation of very long chain fatty acids 4 (ELOVL4)

During the course of performing our gene isolation procedure, we entered into collaboration with other groups studying chromosome 6q-linked autosomal dominant macular dystrophy phenotypes in families that were ultimately found to be related to the Canadian pedigree. The pooling of our genetic data enabled additional refinement of the chromosome 6q minimal critical interval, resulting in the isolation of the disease gene amongst all five pedigrees under study (Zhang et al., 2001). Sequencing of the exons of a novel gene predicted from the genomic sequence found within the refined critical interval revealed the presence of a 5-base pair deletion within all the affected individuals but absent in controls (Zhang et al., 2001; Edwards et al., 2001). Since then, another mutation has been identified in this gene within individuals from an unrelated pedigree affected with additional clinical forms of macular dystrophy (Bernstein et al., 2001). Such information on the specific mutations associated with disease has immediate diagnostic value to other family members that may be affected as well as providing targets for mutation screening in other unrelated individuals that share clinical features of the disease.

Concurrently, we had also identified this gene as a candidate for retinal disease based on our chromosome 6q13-q16 gene isolation strategy. The dot blot results for EST AA993222 revealed expression in both the human fovea and peripheral retina, and a lower level of expression was observed for RPE tissue, which was confirmed by Northern blot analysis. Follow-up studies that we and others have performed have yielded additional information on this new retinal disease gene. Sequence analysis of the encoded protein reveals similarity to enzymes that have been isolated from other species and that have been shown to function in catalyzing the elongation of fatty acids, thus describing the name chosen for the gene, elongation of very long chain fatty acids 4 *(ELOVL4)* (Zhang et al., 2001).

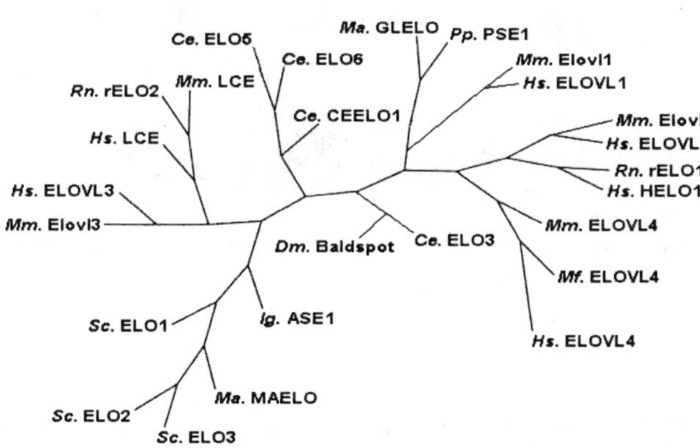

Figure 4. Unrooted phylogenetic tree generated using the PHYLIP (version 3.5) software package depicting the evolutionary relationship between known proteins homologous to human ELOVL4. Species abbreviations: *Hs*, Homo sapiens (human); *Mf*, Macaca fascicularis (monkey); *Mm*, Mus musculus (mouse); *Rn*, Rattus norvegicus (rat); *Dm*, Drosophila melanogaster (fruit fly); *Pp*, Physcomitrella patens (moss); *Ma*, Mortierella alpina (fungus); *Sc*, Saccharomyces cerevisiae (yeast); *Ce*, Caenorhabditis elegans (nematode); *Ig*, Isochrysis galbana (micro-algae).

Both Southern zooblot hybridization and sequence database analyses indicate that the *ELOVL4* gene and its protein product are members of a highly conserved gene family, exhibiting at least 25% amino acid sequence identity with 25 currently known proteins found in 10 different species (Fig. 4), and significant homology is detected amongst over 50 additional proteins predicted to date from a total of 18 different organisms (P.S. Lagali, unpublished results). The most highly conserved amino acids correspond to a HXXHH redox centre motif, predicted transmembrane segments, and a carboxy-terminal endoplasmic reticulum (ER) targeting signal, all believed to be critical functional domains imparting the defining features of fatty acid elongases (Toke and Martin, 1996; Oh et al., 1997; Cinti et al., 1992).

Sub-retinal expression studies of the *ELOVL4* gene at both the mRNA and protein levels shows that the gene products are predominantly localized to the inner segments of rod and cone photoreceptor cells (Zhang et al., 2001; Lagali et al., 2003), as would be predicted for a cellular biosynthetic enzyme found in the ER membrane. This expression pattern is conserved across a wide range of mammalian species including human, monkey, cow, mouse, rat, and cat. Based on both EST tissue source data and Northern analysis, ELOVL4 mRNA expression is not retina-specific, however only the neural retina appears to express all of the ELOVL4 transcripts detectable by hybridization when compared with other human ocular and non-ocular tissues (Fig. 5).

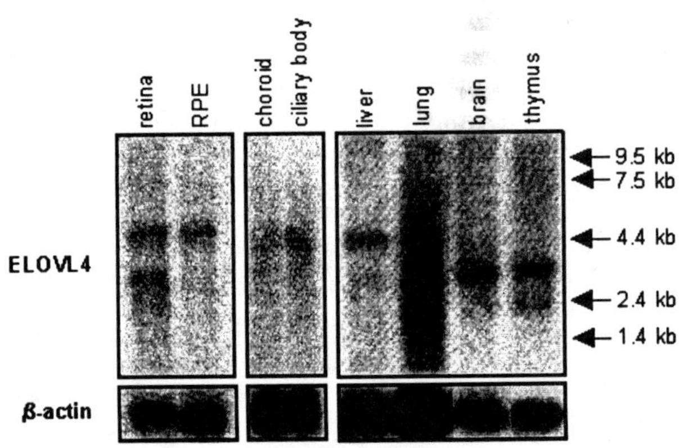

Figure 5. Northern blots of human RNA samples probed with ELOVL4 and control (β-actin) cDNA probes. Ocular tissue lanes contain 5 micrograms of total RNA while non-ocular tissue lanes contain 2 micrograms of poly(A)+ RNA. Positions of RNA size markers are shown on the right. *kb*, kilobases.

The association of a putative function with the *ELOVL4* gene allows us to proceed with functional studies to confirm and characterize this proposed biochemical role of the protein and also provides clues to the pathogenic mechanism underlying the retinal degenerative phenotype observed in affected pedigrees. Furthermore, it suggests possible treatment approaches that can be tested, such as supplementation of patients with fatty acids that may be deficient owing to the effects of *ELOVL4* gene mutations.

5. CONCLUSION

The combination of genetic approaches involving family and linkage studies where available and gene identification methods utilizing the wealth of information provided by the Human Genome Project and global expression studies can greatly facilitate the process by which retinal disease genes can be identified. Indeed, some method of prioritizing the characterization of the roughly 30,000 retinal genes estimated to exist is required to progress towards an understanding of the roles of these genes in the proper functioning of the human retina. Drawing on resources and tools that have independently been used to search for disease genes by a positional cloning strategy and to define putative candidate genes by a genome-wide retinal gene screening strategy, we have begun to characterize a human genomic hotspot for retinal degeneration on chromosome 6q. All of the ESTs exhibiting retinal expression and the corresponding genes defined in our analysis essentially comprise a comprehensive tissue-selective transcript map for the genetic interval investigated. By following an analogous approach, similar maps can be designed for other genomic regions suspected to harbour multiple loci for diseases affecting a particular tissue, including other hotspots for retinal conditions, such as intervals on chromosomes 17q and Xp (RetNet, http://www.sph.uth.tmc.edu/retnet/), and also those for non-retinal disorders (e.g. Becker et al., 1998). This can serve as a platform from which the constituent genes can be used for both mutation screening and for further biological characterization.

Our efforts have thus far contributed to both the isolation of the *ELOVL4* macular disease gene as well as the identification of *C6orf37*, a novel candidate retinal disease gene, thereby validating the utility of our methodological approach and paving the way for further gene discovery. The characterization of these genes beyond their classification as putative or *bona fide* disease genes is in addition valuable to our study and understanding of retinal biology and can suggest many avenues for future experimental work. Conservation of the *C6orf37* and *ELOVL4* genes across diverse species indicates their importance in the retina through evolution and also provides model organisms for further biological investigation. Continued study of the function of these genes may in turn uncover additional disease gene candidates that genetically and/or biochemically interact with the molecular processes in which the chromosome 6q genes identified play a role. Such investigation into the mechanisms of normal retinal biology and the etiology of retinal pathogenesis will ultimately provide us with the information needed to devise effective treatments for the progressive and irreversible blindness resulting from retinal degenerative diseases.

6. ACKNOWLEDGEMENTS

We thank the families and their clinicians for their contributions to these studies. This research was financially supported by the Canadian Institutes of Health Research (PSL), the E.A. Baker Foundation of the Canadian National Institute for the Blind (PSL, PWW), the Alberta Heritage Foundation for Medical Research (PSL, PWW), the Foundation Fighting Blindness-Canada (PWW), the Foundation Fighting Blindness USA (RA), the National Institutes of Health (research grant EY13198 and core grant EY07003) (RA), and a Research to Prevent Blindness Harrington Scholar Award (RA).

We thank the National Eye Institute for the travel award provided to PSL to attend this meeting.

7. REFERENCES

Allikmets, R., Singh, N., Sun, H., Shroyer, N.F., Hutchinson, A., Chidambaram, A., Gerrard, B., Baird, L., Stauffer, D., Peiffer, A., Rattner, A., Smallwood, P., Li, Y., Anderson, K.L., Lewis, R.A., Nathans, J., Leppert, M., Dean, M., and Lupski, J.R., 1997, A photoreceptor cell-specific ATP-binding transporter gene (*ABCR*) is mutated in recessive Stargardt macular dystrophy. *Nat. Genet.* **15**:236-46.

Becker, K.G., Simon, R.M., Bailey-Wilson, J.E., Freidlin, B., Biddison, W.E., McFarland, H.F., and Trent, J.M., 1998, Clustering of non-major histocompatibility complex susceptibility candidate loci in human autoimmune diseases. *Proc. Natl. Acad. Sci. U.S.A.* **95**:9979-84.

Berger, W., Meindl, A., van de Pol, T.J., Cremers, F.P., Ropers, H.H., Doerner, C., Monaco, A., Bergen, A.A., Lebo, R., Warburg, M., Zergollern, L., Lorenz, B., Gal, A., Bleeker-Wagemakers, E.M., and Meitinger, T., 1992, Isolation of a candidate gene for Norrie disease by positional cloning. *Nat. Genet.* **1**:199-203.

Bernstein, S.L., Borst, D.E., and Wong, P.W., 1995, Isolation of differentially expressed human fovea genes: candidates for macular disease. *Mol. Vis.* **1**:4.

Bernstein, S.L., Borst, D.E., Neuder, M.E., and Wong, P., 1996, Characterization of a human fovea cDNA library and regional differential gene expression in the human retina. *Genomics* **32**:301-8.

Bernstein, S.L., Borst, D.E., and Wong, P., 1997, Region specific mitochondrial gene expression in the human retina. *Brain Res.* **744**:143-6.

Bernstein, P.S., Tammur, J., Singh, N., Hutchinson, A., Dixon, M., Pappas, C.M., Zabriskie, N.A., Zhang, K., Petrukhin, K., Leppert, M., and Allikmets, R., 2001, Diverse macular dystrophy phenotype caused by a novel complex mutation in the *ELOVL4* gene. *Invest. Ophthalmol. Vis. Sci.* **42**:3331-6.

Blackshaw, S., Fraioli, R.E., Furukawa, T., and Cepko, C.L., 2001 Comprehensive analysis of photoreceptor gene expression and the identification of candidate retinal disease genes. *Cell* **107**:579-89.

Bortoluzzi, S., d'Alessi, F., and Danieli, G.A., 2000, A novel resource for the study of genes expressed in the adult human retina. *Invest. Ophthalmol. Vis. Sci.* **41**:3305-8.

Chen, Z.Y., Hendriks, R.W., Jobling, M.A., Powell, J.F., Breakefield, X.O., Sims, K.B., and Craig, I.W., 1992, Isolation and characterization of a candidate gene for Norrie disease. *Nat. Genet.* **1**:204-8.

Cinti, D.L., Cook, L., Nagi, M.N., and Suneja, S.K., 1992, The fatty acid chain elongation system of mammalian endoplasmic reticulum. *Prog. Lipid Res.* **31**:1-51.

Cremers, F.P., van de Pol, D.J., van Kerkhoff, L.P., Wieringa, B., and Ropers, H.H., 1990, Cloning of a gene that is rearranged in patients with choroideraemia. *Nature* **347**:674-7.

Dharmaraj, S., Li, Y., Robitaille, J.M., Silva, E., Zhu, D., Mitchell, T.N., Maltby, L.P., Baffoe-Bonnie, A.B., and Maumenee, I.H., 2000, A novel locus for Leber congenital amaurosis maps to chromosome 6q. *Am. J. Hum. Genet.* **66**:319-26.

Edwards, A.O., Donoso, L.A., and Ritter, R., 3rd., 2001, A novel gene for autosomal dominant Stargardt-like macular dystrophy with homology to the SUR4 protein family. *Invest. Ophthalmol. Vis. Sci.* **42**:2652-63.

Gehrig, A., Felbor, U., Kelsell, R.E., Hunt, D.M., Maumenee, I.H., and Weber, B.H., 1998, Assessment of the interphotoreceptor matrix proteoglycan-1 (IMPG1) gene localised to 6q13-q15 in autosomal dominant Stargardt-like disease (ADSTGD), progressive bifocal chorioretinal atrophy (PBCRA), and North Carolina macular dystrophy (MCDR1). *J. Med. Genet.* **35**:641-5.

Griesinger, I.B., Sieving, P.A., and Ayyagari, R., 2000, Autosomal dominant macular atrophy at 6q14 excludes CORD7 and MCDR1/PBCRA loci. *Invest. Ophthalmol. Vis. Sci.* **41**:248-55.

Katsanis, N., Worley, K.C., Gonzalez, G., Ansley, S.J., and Lupski, J.R., 2002, A computational/functional genomics approach for the enrichment of the retinal transcriptome and the identification of positional candidate retinopathy genes. *Proc. Natl. Acad. Sci. U.S.A.* **99**:14326-31.

Kelsell, R.E., Godley, B.F., Evans, K., Tiffin, P.A., Gregory, C.Y., Plant, C., Moore, A.T., Bird, A.C., and Hunt, D.M., 1995, Localization of the gene for progressive bifocal chorioretinal atrophy (PBCRA) to chromosome 6q. *Hum. Mol. Genet.* **4**:1653-6.

Kelsell, R.E., Gregory-Evans, K., Gregory-Evans, C.Y., Holder, G.E., Jay, M.R., Weber, B.H., Moore, A.T., Bird, A.C., and Hunt, D.M., 1998, Localization of a gene (CORD7) for a dominant cone-rod dystrophy to chromosome 6q. *Am. J. Hum. Genet.* **63**:274-9.

Kniazeva, M.F., Chiang, M.F., Cutting, G.R., Zack, D.J., Han, M., and Zhang, K., 1999, Clinical and genetic studies of an autosomal dominant cone-rod dystrophy with features of Stargardt disease. *Ophthalmic Genet.* **20**:71-81.

Kniazeva, M., Traboulsi, E.I., Yu, Z., Stefko, S.T., Gorin, M.B., Shugart, Y.Y., O'Connell, J.R., Blaschak, C.J., Cutting, G., Han, M., and Zhang, K., 2000, A new locus for dominant drusen and macular degeneration maps to chromosome 6q14. *Am. J. Ophthalmol.* **130**:197-202.

Lagali, P.S., MacDonald, I.M., Griesinger, I.B., Chambers, M.L., Ayyagari, R., and Wong, P.W., 2000, Autosomal dominant Stargardt-like macular dystrophy segregating in a large Canadian family. *Can. J. Ophthalmol.* **35**:315-24.

Lagali, P.S., Kakuk, L.E., Griesinger, I.B., Wong, P.W., and Ayyagari, R., 2002, Identification and characterization of *C6orf37*, a novel candidate human retinal disease gene on chromosome 6q14. *Biochem. Biophys. Res. Commun.* **293**:356-65.

Lagali, P.S., Liu, J., Ambasudhan, R., Kakuk, L.E., Bernstein, S.L., Seigel, G.M., Wong, P.W., and Ayyagari, R., 2003, Evolutionarily conserved *ELOVL4* gene expression in the vertebrate retina. *Invest. Ophthalmol. Vis. Sci.* (in press).

Malone, K., Sohocki, M.M., Sullivan, L.S., and Daiger, S.P., 1999, Identifying and mapping novel retinal-expressed ESTs from humans. *Mol. Vis.* **5**:5.

Meindl, A., Dry, K., Herrmann, K., Manson, F., Ciccodicola, A., Edgar, A., Carvalho, M.R., Achatz, H., Hellebrand, H., Lennon, A., Migliaccio, C., Porter, K., Zrenner, E., Bird, A., Jay, M., Lorenz, B., Wittwer, B., D'Urso, M., Meitinger, T., and Wright, A., 1996, A gene (*RPGR*) with homology to the RCC1 guanine nucleotide exchange factor is mutated in X-linked retinitis pigmentosa (RP3). *Nat. Genet.* **13**:35-42.

Oh, C.S., Toke, D.A., Mandala, S., and Martin, C.E., 1997, *ELO2* and *ELO3*, homologues of the *Saccharomyces cerevisiae ELO1* gene, function in fatty acid elongation and are required for sphingolipid formation. *J. Biol. Chem.* **272**:17376-84.

Petrukhin, K., Koisti, M.J., Bakall, B., Li, W., Xie, G., Marknell, T., Sandgren, O., Forsman, K., Holmgren, G., Andreasson, S., Vujic, M., Bergen, A.A., McGarty-Dugan, V., Figueroa, D., Austin, C.P., Metzker, M.L., Caskey, C.T., and Wadelius, C., 1998, Identification of the gene responsible for Best macular dystrophy. *Nat. Genet.* **19**:241-7.

Roepman, R., van Duijnhoven, G., Rosenberg, T., Pinckers, A.J., Bleeker-Wagemakers, L.M., Bergen, A.A., Post, J., Beck, A., Reinhardt, R., Ropers, H.H., Cremers, F.P., and Berger, W., 1996, Positional cloning of the gene for X-linked retinitis pigmentosa 3: homology with the guanine-nucleotide-exchange factor RCC1. *Hum. Mol. Genet.* **5**:1035-41.

Khaliq, S., Hameed, A., Ismail, M., Mehdi, S.Q., Bessant, D.A., Payne, A.M., and Bhattacharya, S.S., 1999, Refinement of the locus for autosomal recessive Retinitis pigmentosa (RP25) linked to chromosome 6q in a family of Pakistani origin. *Am. J. Hum. Genet.* **65**:571-4.

Sauer, C.G., Schworm, H.D., Ulbig, M., Blankenagel, A., Rohrschneider, K., Pauleikhoff, D., Grimm, T., and Weber, B.H., 1997, An ancestral core haplotype defines the critical region harbouring the North Carolina macular dystrophy gene (MCDR1). *J. Med. Genet.* **34**:961-6.

Schoen, T.J., Mazuruk, K., Chader, G.J., and Rodriguez, I.R., 1995, Isolation of candidate genes for macular degeneration using an improved solid-phase subtractive cloning technique. *Biochem. Biophys. Res. Commun.* **213**:181-8.

Schwahn, U., Lenzner, S., Dong, J., Feil, S., Hinzmann, B., van Duijnhoven, G., Kirschner, R., Hemberger, M., Bergen, A.A., Rosenberg, T., Pinckers, A.J., Fundele, R., Rosenthal, A., Cremers, F.P., Ropers, H.H., and Berger, W., 1998, Positional cloning of the gene for X-linked retinitis pigmentosa 2. *Nat. Genet.* **19**:327-32.

Sharon, D., Blackshaw, S., Cepko, C.L., and Dryja, T.P., 2002, Profile of the genes expressed in the human peripheral retina, macula, and retinal pigment epithelium determined through serial analysis of gene expression (SAGE). *Proc. Natl. Acad. Sci. U.S.A.* **99**:315-20.

Shimizu-Matsumoto, A., Adachi, W., Mizuno, K., Inazawa, J., Nishida, K., Kinoshita, S., Matsubara, K., and Okubo, K., 1997, An expression profile of genes in human retina and isolation of a complementary DNA for a novel rod photoreceptor protein. *Invest. Ophthalmol. Vis. Sci.* **38**:2576-85.

Sinha, S., Sharma, A., Agarwal, N., Swaroop, A., and Yang-Feng, T.L., 2000, Expression profile and chromosomal location of cDNA clones, identified from an enriched adult retina library. *Invest. Ophthalmol. Vis. Sci.* **41**:24-8.

Sohocki, M.M., Malone, K.A., Sullivan, L.S., and Daiger, S.P., 1999, Localization of retina/pineal-expressed sequences: identification of novel candidate genes for inherited retinal disorders. *Genomics* **58**:29-33.

Stohr, H., Mah, N., Schulz, H.L., Gehrig, A., Frohlich, S., and Weber, B.H., 2000, EST mining of the UniGene dataset to identify retina-specific genes. *Cytogenet. Cell. Genet.* **91**:267-77.

Stone, E.M., Nichols, B.E., Kimura, A.E., Weingeist, T.A., Drack, A., and Sheffield, V.C., 1994, Clinical features of a Stargardt-like dominant progressive macular dystrophy with genetic linkage to chromosome 6q. *Arch. Ophthalmol.* **112**:765-72.

Toke, D.A. and Martin, C.E., 1996, Isolation and characterization of a gene affecting fatty acid elongation in *Saccharomyces cerevisiae*. *J. Biol. Chem.* **271**:18413-22.

Yoshida, S., Yashar, B.M., Hiriyanna, S., and Swaroop, A., 2002, Microarray analysis of gene expression in the aging human retina. *Invest. Ophthalmol. Vis. Sci.* **43**:2554-60.

Zhang, K., Kniazeva, M., Han, M., Li, W., Yu, Z., Yang, Z., Li, Y., Metzker, M.L., Allikmets, R., Zack, D.J., Kakuk, L.E., Lagali, P.S., Wong, P.W., MacDonald, I.M., Sieving, P.A., Figueroa, D.J., Austin, C.P., Gould, R.J., Ayyagari, R., and Petrukhin, K., 2001, A 5-bp deletion in *ELOVL4* is associated with two related forms of autosomal dominant macular dystrophy. *Nat. Genet.* **27**:89-93.

MOUSE GENETIC APPROACHES TO ACCESS PATHWAYS IMPORTANT IN RETINAL FUNCTION

Patsy M. Nishina and Jürgen K. Naggert[1]

1. INTRODUCTION

Mouse models have been shown to be invaluable for studying ocular diseases and for furthering our understanding of these diseases. With their anatomical, physiological, and genetic similarities to humans, mouse models have proven to be important for identification of genes underlying inherited forms for retinal disease and for providing much needed tissue to study the pathological consequences of disease throughout the lifetime of the animal. In addition, to obtain these disease models we need not wait for spontaneous mutations to occur, new models can be created through the use of induced mutagenesis, transgenesis and homologous recombination. More recently, mouse models have been used to identify pathways in which particular molecules function through the use of genetic modifier screens and the comparison of gene expression or protein levels between mutant and wild-type mice. Such approaches, that are in some cases unique to mouse models, are described briefly in further detail below.

2. IDENTIFYING INTERACTING PARTNERS BY GENETIC MODIFIER SCREENS TO DETERMINE THE PATHWAY IN WHICH PARTICULAR MOLECULES FUNCTION

Variability in disease phenotypes is a common occurrence in both human and animal models (Haider et al. 2002). While these variable phenotypes may be due to allelic differences or environmental influences, genetic modifiers can affect the phenotypic outcome of a given genotype by interacting in the same or parallel biological pathway as a disease gene. In the case of mouse strains, which are inbred and genetically identical and in which experiments can be carried out in a controlled environment, the existence of genetic modifiers has been demonstrated. Thus genetic crosses between inbred strains can be used to identify such modifier genes. In mice, the first report of variable expressivity in photoreceptor degeneration as a result of genetic background was for retinal

[1]The Jackson Laboratory, Bar Harbor, ME 04609, USA. Corresponding author: P. M. Nishina, E-mail: pmn@jax.org

Retinal Degenerations: Mechanisms and Experimental Therapy
Edited by LaVail *et al.*, Kluwer Academic/Plenum Publishers, 2003

degeneration 3 (*rd3*) mutant mice, in which the onset and progression of degeneration differed among strains RBF/Dn, Meta-In(1)Rk, Rb(11.13)4Bnr, and In-30 (Heckenlively et al.,1993). In addition to affecting the severity or onset of a mutant phenotype, modifiers that suppress or reduce the mutant phenotype even to the extent of completely restoring the wildtype condition have been described in *rd7/rd7* mice in which retinal spotting was completely abolished in some F2 mice homozygous for the *rd7* mutation in an intercross between (B6.Cg-*rd7/rd7* and CAST/Ei)F1 mice (Akhmedov et al., 2000). Subsequent studies revealed that the suppression of the retinal spotting correlates with suppression of both the retinal whorls and folds, and of the photoreceptor degeneration observed histologically (Haider et al., personal communication). Modifiers may also alter the pleiotropy of a given disease, resulting in different combinations of traits. The best example of this was reported in a family in which members all carrying the same *peripherin/RDS* mutation developed either RP, pattern dystrophy, or fundus flavimaculatus (Weleber et al., 1993).

Although modifier effects on retinal phenotypes can be due to the a single gene at a locus independent of the mutant gene (Danciger et al., 2000; Ikeda et al., 2002; Ikeda et al., 1999), they may also be due to a combination of modifier genes that act together to create a cumulative effect on the expression of a phenotype, as is the case with quantitative traits (Ikeda et al., 2002). In mice we are able to quantitate measurable phenotypes such as the thickness of the photoreceptor layer, the number of cells in the photoreceptor layer, or the amplitude of ERG waveforms, and ask whether the particular phenotype is affected when placed in combination with different genetic backgrounds. For example, C57BL/6 (B6)-*tub/tub* mice are known to loose almost all of their photoreceptors by 24 weeks of age. However, when intercrossed with the AKR or CAST strains, some B6/AKR F2 and B6/CAST F2 *tub/tub* mice still have 75-100% of their photoreceptors remaining. This would suggest that something in the AKR and CAST backgrounds is able to delay photoreceptor degeneration. We have found 3 chromosomal regions on Chrs. 2, 8 and 11 and one chromosomal region on Chr. 9, which are able to delay retinal degeneration in *tub/tub* mice in the AKR and CAST intercrosses, respectively (Ikeda et al., 2002; Ikeda et al. personal communication). When we have identified the modifying genes underlying the protection for photoreceptor degeneration, we will determine if they are protective for only *tub/tub* mice or whether their protection can be generalized to different forms of photoreceptor degeneration (see discussion below). If the protection is specific to *tub*, it may help us to understand the function of tubby in the retina. That is, if the gene is a known gene, then we may be able to, by inference, obtain some clues as to what the function of the tubby gene is. If the protection is more generalized, it may help us to develop therapeutics that could be used to treat photoreceptor degeneration regardless of the cause. In either case, when the gene is cloned, important information will be obtained about the pathway leading from the *tub* mutation to photoreceptor degeneration.

3. THE USE OF MOUSE MODELS TO IDENTIFY COMMON PATHWAYS INVOLVED IN PHOTORECEPTOR DEGENERATION

One pathway, which appears to be associated with retinal, and potentially, macular degeneration is programmed cell death (PCD) or apoptosis (Wong 1994; Portera-Cailliau et al., 1994). In a few human studies in which histological sections of eyes could be obtained, apoptosis also appears to be an essential component of photoreceptor and retinal pigmented epithelial cell death (Hinton et al., 1998; Xu et al., 1996; Milam et al., 1998). Little is known about what triggers PCD, what the actual pathway is in the retina or even if all mutations identified utilize the same apoptotic pathway.

Studies designed, in part, to address these issues are ongoing in our laboratory. We have chosen three genetically determined mutants that exhibit different clinical phenotypes and/or rates of retinal degeneration. *rd3* develops panretinal degeneration by 8 weeks, *rd5* (*tub*) by 12 weeks, and *rd7* by of 6 months of age. *rd3* and *tub* exhibit similar loss of pigment epithelium and patches of pigment deposits (Heckenlively et al., 1993; Heckenlively et al., 1995), whereas, *rd7* mice present with small dots across the retina (Akhmedov et al., 2000). While the *rd3* gene has yet to be identified, *tub* is caused by a mutation in a novel gene, which is a member of small gene family, *Tulps* (tubby like proteins) with unknown function (Ikeda et al., 2002). Mutations within *Tulp1* have been also shown to lead to retinal degeneration in both humans and mouse (Hagstrom et al., 1998; Hagstrom et al. 1999; Ikeda et al., 2000). *rd7* is caused by mutations within a gene (Akhmedov et al., 2000), *Nr2e3*, which appears to be important in specification of both total cone number and cone-type (Haider et al., 2001; Milam et al., 2002). Mutations within *Nr2e3* cause enhanced S-cone syndrome (ESCS) in human patients.

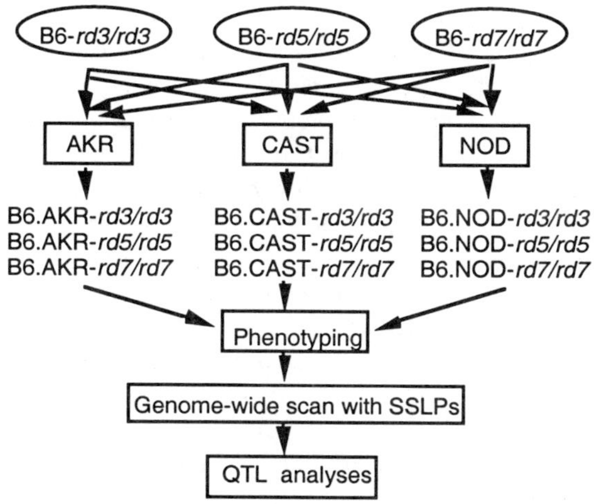

Figure 1. Schematic of the experimental design for identification of common pathways involved in photoreceptor degeneration.

Each of the mutants above are being intercrossed with three genetically diverse strains, CAST/EiJ, a wild derived strain, and two standard inbred strains, AKR/J and NOD.NON^{H2K}, a stock originating from Furth and a Swiss derived stock originating in Japan, respectively. Photoreceptor function in F2 progeny from each cross will be assessed for morphological alterations by histological examination. Genome-wide scans with simple sequence length polymorphic (SSLP) markers are being carried out and QTL that delay or suppress retinal degeneration identified (Figure 1). Our results to date show that each strain is able to delay or suppress retinal degeneration in one or more of the mutants. It is interesting to note that multiple loci have been identified within each intercross, indicating the complexity of the modification of the retinal degenerative phenotype. We are also identifying QTL traits that appear to overlap between mutations within the same or different crosses. If we identify *commonly shared* genes that are able to suppress retinal degeneration among these different mutations and crosses, then we can hypothesize that there is a common PCD pathway involved. The crosses described above will allow us to dissect and separate the genetic factors that lead to delayed or suppressed photoreceptor degeneration and may provide access into the pathways through which they function.

4. IDENTIFYING MOLECULES AFFECTED BY PARTICULAR MUTATIONS IN ORDER TO DEFINE PATHWAYS THAT LINK THE MUTATION TO A PHENOTYPE

The use of mouse models to identify molecules affected by particular mutations has been demonstrated in both microarray and SAGE expression analysis approaches by Livesey et al. (2000) and Blackshaw et al. (2001). Comparison of gene expression patterns in retinas of wildtype and $Crx^{-/-}$ mutant mice lead these authors to identify a number of genes that were expressed differentially. Therefore, by inference, these genes must interact with the mutant gene, be misregulated by the loss of its function, or be altered due to secondary effects of the disease phenotype. Careful examination of the promoter regions of some of these differentially expressed genes led these authors (Livesey et al 2000) to identify a putative, novel CRX binding motif.

A similar approach of comparing protein levels in mutant Cpe^{fat} and wild-type mice to obtain factors normally processed by Carboxypeptidase E (CPE) is currently being applied in our laboratory. Carboxypeptidase E (CPE) is an exopeptidase necessary for full maturation of many neuroendocrine/endocrine prohormones. In recent years, interest in novel neuropeptides and peptide hormones has increased because their receptors are attractive drug targets. In the course of the genome project, a large number of orphan receptors, particularly the G-coupled seven transmembrane spanning receptors, have been discovered, suggesting the existence of currently unknown peptide ligands. The search for endogenous agonists for these orphan receptors has become a new paradigm for the identification of novel bioactive peptides (Civelli et al., 1998; Hinuma et al., 1999). Neuroactive peptides in the retina are thought to be regulated, not at the level of transcription or translation but by further processing from prepropeptides to mature peptides by post-translational modification (Morgan and Chubb, 1991). In fact, Nickells et al. (Nickells et al., 1993) suggested that CPE is the enzyme that regulates the turnover of neuroactive peptides by processing them from large pools of prepropeptides to meet the physiological needs of the retina.

CPE is highly expressed in rod photoreceptors and to a lesser extent in the inner nuclear and ganglion cell layers of retinas from Long-Evans rats (Nickells et al., 1993). Interestingly, there is no concrete evidence that indicates that photoreceptor cells synthesize neuropeptides. However, as only those tissues that synthesize neuropeptides express CPE, this would suggest that photoreceptors must be synthesizing and using those peptides for normal function. Interestingly, we have observed abnormal ERG responses in Cpe^{fat} mice in the presence of normal retinal histology, indicating that indeed normal photoreceptor signal transduction requires neuropeptides. It was proposed that these neuropeptides may either be of low abundance that has made their detection difficult or alternatively that they are novel and yet, undescribed (Schlamp and Nickells, 1996). It has also been proposed that CPE is responsible for final processing of trophic factors that may be important in photoreceptor survival (Schlamp and Nickells, 1996).

Cpe itself appears to be regulated quite dramatically in the retina (Schlamp and Nickells, 1996). In the dark adapted rat, *Cpe* expression is primarily observed in the inner segments of the photoreceptor cells while exposure to light for as short as 5 minutes leads to a rapid increase in *Cpe* mRNA levels in the ganglion cell layer. By 4 hours of light exposure, *Cpe* mRNA levels are barely detectable in the photoreceptors. Fifteen minutes of dark adaptation reverses the localization of *Cpe* from ganglion cells to the photoreceptors. Interestingly, *Cpe* expression is constant in the inner nuclear layer during dark/light cycle changes. CPE protein appears to follow much the same pattern as mRNA expression, except that it exhibits a slower rate of accumulation at dark/light transitions, and it is expressed not only in the inner segments of the photoreceptor cells but also in the outer segments and outer plexiform layer.

In the Cpe^{fat} mouse model, a point mutation in the *Cpe* gene abolishes CPE activity and leads to an abnormal accumulation of peptide hormone precursors (Naggert et al., 1995. Comparison of protein profiles of wildtype and Cpe^{fat} retina will allows us the unique opportunity to identify both known as well as potentially novel retinal neuropeptides that are normally processed by CPE.

5. SUMMARY

The laboratory mouse is a premier research tool for understanding human biology and disease. With the availability of the first complete draft sequences of the mouse genome and ongoing large scale screens for new mouse mutations to serve as novel models, the usefulness of the mouse will only increase. The vision research field, in particular, which has identified many human diseases for which no models are yet available, will benefit from the newly developed tools in mouse molecular genetics.

6. REFERENCES

Akhmedov, N. B., Piriev, N. I., Chang, B., Rapoport, A. L., Hawes, N. L., et al., 2000, A deletion in a photoreceptor-specific nuclear receptor mRNA causes retinal degeneration in the *rd7* mouse, *Proc Natl Acad Sci U S A.* **97**:5551-6.

Blackshaw, S., Fraioli, R. E., Furukawa, T., and Cepko, C. L., 2001, Comprehensive analysis of photoreceptor gene expression and the identification of candidate retinal disease genes, *Cell.* **107**:579-89

Chang, B., Hawes, N. L., Hurd, R. E., Davisson, M. T., Nusinowitz, S., and Heckenlively, J. R., 2002, Retinal degeneration mutants in the mouse, *Vision Research.* **42**:517-525

Civelli, O, Nothacker, HP, and Reinscheid, R., 1998, Reverse physiology: discovery of the novel neuropeptide, orphan in FQ/nociceptin, *Crit Rev Neurobiol.* **12**: 163-76.

Danciger, M., Matthes, M. T., Yasamura, D., Akhmedov, N. B., Rickabaugh, T., et al. 2000, A QTL on distal chromosome 3 that influences the severity of light- induced damage to mouse photoreceptors, *Mamm Genome.* **11**:422-7.

Hagstrom, S. A., Duyao, M., North, M. A., Li, T., 1999, Retinal degeneration in tulp1-/- mice: vesicular accumulation in the interphotoreceptor matrix, *Invest Ophthalmol Vis Sci.* **40**: 2795-802.

Hagstrom, S. A., North, M. A., Nishina, P. L., Berson, E. L., Dryja, T. P., 1998, Recessive mutations in the gene encoding the tubby-like protein TULP1 in patients with retinitis pigmentosa, *Nat Genet.* **18**: 174-6.

Haider, N. B., Jacobson, S. G., Cideciyan, A. V., Swiderski, R., Streb, L. M., et al., 2000, Mutation of a nuclear receptor gene, NR2E3, causes enhanced S cone syndrome, a disorder of retinal cell fate. *Nat Genet.* **24**:127-31.

Haider, N. B., Naggert, J. K., Nishina, P. M. 2001. Excess cone cell proliferation due to lack of a functional NR2E3 causes retinal dysplasia and degeneration in *rd7/rd7* mice, *Hum Mol Genet* **10**:1619-26.

Haider, N. B., Ikeda, A., Naggert, J. K., Nishina, P. M., 2002, Genetic modifiers of vision and hearing, *Hum Mol Genet.* **11**: 1195-206.

Heckenlively, J.R., Chang, B., Peng, C., Hawes, N.L., and Roderick, T.H., 1993, Variable expressivity of *rd-3* retinal degeneration dependent on background strain, in: *Retinal Degeneration,* J.G. Hollyfield et al. (eds), Plenum Press, New York, pp. 273-280.

Heckenlively, J. R., Chang, B., Erway, L. C., Peng, C., Hawes, N. L., Hageman, G. S., Roderick, T. H., 1995, Mouse model for Usher syndrome: linkage mapping suggests homology to Usher type I reported at human chromosome 11p15, *Proc Natl Acad Sci U S A.* **92**: 11100-4.

Hinton, D.R., He, S., and Lopez, P.F., 1998, Apoptosis in surgically excised choroidal neovascular membranes in age-related macular degeneration, *Arch Opthalmol.* **116**: 203-209.

Hinuma, S, Onda, H, and Fujino, M., 1999, The quest for novel bioactive peptides utilizing orphan seven-transmembrane-domain receptors, *J Mol Med* **77**: 495-504.

Ikeda, A., Zheng, Q. Y., Rosenstiel, P., Maddatu, T., Zuberi, A. R., et al., 1999, Genetic modification of hearing in tubby mice: evidence for the existence of a major gene (*moth1*) which protects tubby mice from hearing loss, *Hum Mol Genet.* **8**:1761-7.

Ikeda, S., Shiva, N., Ikeda, A., Smith, R. S., Nusinowitz, S., Yan, G., Lin, T. R., Chu, S., Heckenlively, J. R., North, M. A., Naggert, J. K., Nishina, P. M., Duyao, M. P., 2000, Retinal degeneration but not obesity is observed in null mutants of the tubby-like protein 1 gene, *Hum Mol Genet.* **9**: 155-63.

Ikeda, A., Naggert, J. K., Nishina, P. M. 2002. Genetic modification of retinal degeneration in tubby mice. *Exp Eye Res* **74**:455-61.

Ikeda, A., Nishina, P. M., Naggert, J. K., 2002a, The tubby-like proteins, a family with roles in neuronal development and function, *J Cell Sci.* **115**: 9-14.

Livesey, F. J., Furukawa, T., Steffen, M. A., Church, G. M., Cepko, C. L., 2000, Microarray analysis of the transcriptional network controlled by the photoreceptor homeobox gene Crx, *Curr Biol.* **10**:301-10

Milam, A., Li, Z., and Fariss, R. 1998, Histopathology of the human retina in retinitis pigmentosa, *Prog Retin Eye Res* **17**:175-205.

Milam, A. H., Rose, L., Cideciyan, A. V., Barakat, M. R., Tang, W. X., et al., 2002, The nuclear receptor NR2E3 plays a role in human retinal photoreceptor differentiation and degeneration, *Proc Natl Acad Sci USA.* **99**:473-8.

Morgan, I.G. and Chubb, I.W., 1991, How peptidergic neurons cope with variation in physiological stimulation, *Neurochem Res.* **16**: 705-14.

Naggert, J.K., Fricker, L.D., Varlamov, O., Nishina, P.M., Rouille, Y., Steiner, D.F., Carroll, R.J., Paigen, B.J., and Leiter, E.H., 1995, Hyperproinsulinaemia in obese *fat/fat* mice associated with a carboxypeptidase E mutation which reduces enzyme activity. *Nat Genet.* **10**: 135-42.

Nickells, RW, Schlamp, CL, Newton, AC, and Williams, DS., 1993, Gene expression of the neuropeptide-processing enzyme carboxypeptidase E in rat photoreceptor cells, *J Neurochem.* **61**: 1891-900.

Portera-Cailliau, C., Sung, C.H., Nathans, J., and Adler, R., 1994, Apoptotic photoreceptor cell death in mouse models of retinitis pigmentosa, *Proc Natl Acad Sci USA.* **91**: 974-978.

Schlamp, C.L. and Nickells, R.W., 1996, Light and dark cause a shift in the spatial expression of a neuropeptide-processing enzyme in the rat retina, *J Neurosci.* **16**: 2164-71.

Weleber, R. G., Carr, R. E., Murphey, W. H., Sheffield, V. C., and Stone, E. M., 1993, Phenotypic variation including retinitis pigmentosa, pattern dystrophy, and fundus flavimaculatus in a single family with a deletion of codon 153 or 154 of the peripherin/RDS gene, *Arch Ophthalmol.* **111**:1531-42.

Wong, P., 1994, Apoptosis, retinitis pigmentosa, and degeneration, *Biochem Cell Biol.* **72**: 489-498.

Xu, G.Z., Li, W.W., and Tso,M.O., 1996, Apoptosis in human retinal degenerations, *Trans Am Ophthal Soc.*: 411-430.

Retinal Degenerative Disorders in Southern Africa: A Molecular Genetic Approach

R.S. Ramesar, L. Roberts, G. Rebello, R. Goliath, A. Vorster, A. September, L. Ehrenreich, D. Gama, J. Greenberg[1]

1. INTRODUCTION

For genetics researchers, the DNA from our populations is our most precious commodity. We have previously indicated that the population stratifications in South Africa are likely to be a major asset in our RDD investigations. The most recent published population statistics are from the South African (SA) Census 1996 and this revealed that there were 40.58 million people in SA (http:www.gov.za/yearbook/2001/landpeople.html), classified as follows:

76.7% Indigenous African: The Nguni-speaking people i.e., Zulu, Xhosa and Swazi make up two thirds of the SA population; other subpopulations in the indigenous group include Sotho-Tswana; Tsonga; Venda; Khoi and San

10.9% White: Afrikaners and English. Afrikaners are descendents of the original settlers from the Netherlands, Germany and France, and logically the English are descendents of settlers from the UK.

8.9% Mixed Ancestry: The SA mixed ancestry population group is an admixture of Caucasian, Khoi-San, Malay (Javanese, Sumatran) and indigenous African. The majority of this group live in the Western Cape Province.

2.6% Indian/Asian: The largest component of this group represents the settlers whose ancestors were from the Indian subcontinent. However, 'Asian' is used to denote that a substantial influx of other subpopulations e.g. Chinese, Malay, and other groups, has contributed to the present population census.

1% Immigrants: From Africa, Asia and Europe, who maintain strong cultural identities also exist.

South Africans thus provide a novel gene pool with which we can study inherited disorders, for while it is apparent that the population is **highly heterogeneous**, it is also made up of discrete subpopulations, many of which may be regarded as genetically endogamous. The **Afrikaners**, for example are a relatively homogeneous population.

[1]MRC Human Genetics Research Unit, Faculty of Health Sciences, University of Cape Town, South Africa. Corresponding author: R. S. Ramesar, E-mail: rr@cormack.uct.ac.za

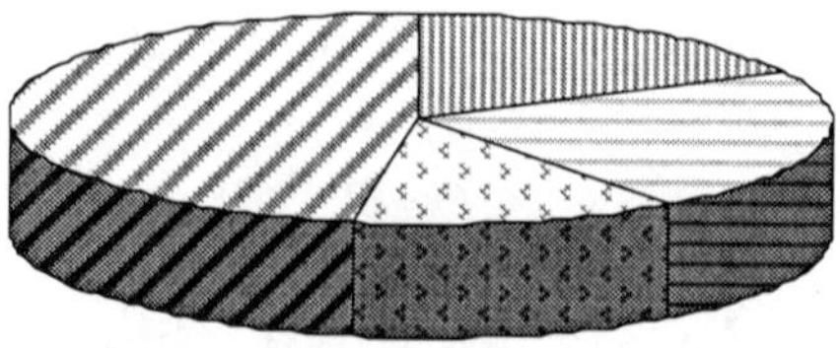

Figure 1. The frequencies of inherited RDD subtypes internationally.

They have been discovered to carry founder mutations in the protoporphyrinogen oxidase gene causing variegate porphyria (VP) in this country (Groenewald et al., 1998). VP is one of the most common genetic disorders in South Africa, whilst it is fairly rare in most other populations. Similarly, Hereditary Haemachromatosis (HH) has a high prevalence in the Afrikaners (1 in 100 compared to 1 in 300 SA individuals of Northern European descent) (de Villiers et al., 1999). Black SA patients with oesophageal cancer have fewer p53 gene mutations than elsewhere, suggesting a different disease aetiology (Hendricks et al., 2002).

Our research into genetic blindness in SA involves the determination of the molecular defects in inherited retinal degenerative disorders (RDD), including retinitis pigmentosa (RP), macular dystrophy, Usher Syndrome, and Stargardt disease. The establishment of the genetic profile underlying retinal degenerative diseases in SA is the first step in developing therapies to treat them.

Retinitis Pigmentosa affects 1 in 4000 people worldwide (Yeung et al., 2001), and multiple studies have shown that frequencies generally fall in the ranges as shown in Figure 1.

SA families with RDDs have been documented and investigated in our department since 1972 (Oswald, A. H., Goldblatt et al., 1985; Greenberg et al., 1993). Historically, the racial distribution of DNA samples collected by this laboratory has shown an ascertainment bias, which we have been attempting to address. This will be discussed and the data we have accumulated throughout this study will be reviewed. Our findings will be compared to those reported internationally.

2. PATIENTS, METHODS AND MATERIALS

A breakdown of the different genetic categories of RDD is shown in Figure 2. The majority of the RDD patients and families have been referred to our department by Retina South Africa (the lay support group), however, referrals by ophthalmologists also occur.

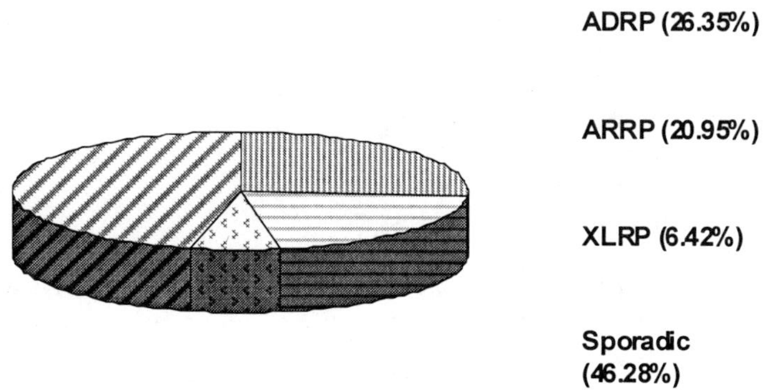

Figure 2. The frequencies of RDD subtypes in SA.

In collaboration with the Department of National Health and Population Development, genetic nurses throughout South Africa obtain blood specimens and pedigrees. A dedicated research nurse, based in our department and employed by Retina South Africa, maintains contact with families and assists in the extension and clarification of pedigrees and the recruitment of further family members. She also refers selected individuals to ophthalmologists for clinical assessment and confirmation of diagnosis. Confirmation of diagnosis is usually supported by electroretinograms, fundus examinations, and tests to determine visual fields and acuity. Informed consent is obtained before any samples or investigations are performed.

DNA from these families is the raw material for a host of molecular investigative techniques.

i. **Linkage analysis** is performed on suitable families, using candidate loci. This approach is done both manually and/or using the ABI 377 Automated Sequencer *(Applied Biosystems)*.

ii. **Mutation screening** of appropriate primary candidate genes is performed on affected individuals selected from the available families. The mode of inheritance normally associated with mutations in the candidate gene is used to determine the subset of patients examined. For example, the Rhodopsin gene is screened in simplex cases, as well as autosomal dominant RP (ADRP) and autosomal recessive RP (ARRP) families; the Peripherin/RDS gene is screened for ADRP families and simplex cases; the RPGR and RP3 genes for X-linked RP (XLRP) families and simplex cases, where the affected person is a male; and the ABCA4 gene for families with Stargardt disease.

Several techniques are used for this mutation screening, including single stranded polymorphism (SSCP) analysis, restriction fragment length polymorphism (RFLP) analysis, automated sequencing on the ABI 377 and denaturing high-performance liquid chromatography (dHPLC) using the WAVE machine *(Transgenomic)*.

In addition, DNA samples have been sent to collaborators in laboratories around the world for screening of genes implicated in Usher syndrome, Leber congenital amaurosis, Alstrom syndrome, Choroideraemia, Sorsby Fundus dystrophy, optic atrophy, Leber optic neuropathy, Laurence-Moon syndrome, and Bardet-Biedl syndrome.

iii. **Positional candidate genes** have been extensively investigated in the regions specified in our fine mapping studies for the two novel "South African" ADRP loci on chromosomes 17p and 17q

iv. **Bioinformatics** is a tool which is employed in support of all of the aforementioned techniques to enhance and improve our research.

3. RESULTS

By 2002, this project had recruited 1956 individuals in 582 families with RDD. Within the research cohort (Table 1), there are 296 families with retinitis pigmentosa, and the remainder are affected with macular degeneration, Stargardt disease, Usher syndrome, Leber Congenital Amaurosis and other RDDs.

Two novel loci for ADRP were mapped by this group, on chromosome 17p (RP13) and 17q (RP17), the disorder/loci were originally identified in two large immigrant families in SA; one of British stock and the other of German origin, respectively. One of these novel genes has been identified as PRP8 (McKie et al, 2001).

To date, mutations have been detected in a total of 11 genes in SA RDD families. Furthermore, within these genes, mutations have been found in 21 families. These include the first reported Rhodopsin mutation in a Black African family (Greenberg et al., 1999). The genes involved, and the number of families detected with each are summarised in Table 2.

Table 1. Breakdown of families and disorders in the SA research cohort

Disorder	Number of Individuals	Number of Families
Sporadic	280	137
ADRP	647	78
RPL	28	9
RPM	177	101
RPO	45	20
ARRP	238	62
RPS	316	110
RPU	135	46
XLRP	90	19
TOTAL	1956	582

RPL = Leber congenital Amaurosis, RPU = Usher syndrome, RPM = Macular Degeneration, RPS = Stargardt disease, RPO = Other RDD.

Due to historical bias, these tables may not necessarily reflect absolute population prevalence. We have attempted to address these issues as can be seen by the distribution of ethnic groups in new samples per year.

Table 2. Disease genes with mutations in SA families

Gene	Number of Families
ALMS1	1
CDH23	1
CHM	1
Myo7A	2
PRP8	1
RDS	1
RHO	6
RP2	2
RPGR	1
TIMP3	1
USH2A	4

4. DISCUSSION

As stated earlier the power of research in human genetics relies on the DNA resources available for analysis. We are fortunate to have a large biological materials archive which correlates with the distribution of RDD subtypes published internationally (compare Figures 1 and 2). The non-representative nature of our cohort in terms of ethnic distribution is a limitation we are trying to address. We are now actively canvassing for patients in the under-represented groups (Table 3). The performance of many thousands of mutation detection experiments in 611 individuals has yielded 21 mutations in 11 different genes in SA patients. It is our ongoing goal to extend this number to cover as many as possible of the growing numbers of families in our RDD cohort.

Table 3. Distribution of ethnic origin **for samples received each year**

	1985 -89	**1990**	**1991**	**1992**	**1993**	**1994**	**1995**	**1996**	**1997**	**1998**	**1999**	**2000**	**2001**	**2002**
B	0	3	7	23	13	21	1	3	12	29	14	13	12	31
O	13	88	11	131	150	193	74	82	137	147	64	74	93	73
I	0	0	8	6	3	11		2	1	1	3	2	1	8
M	1	2	26	33	2	13	8	2	11	14	3	13	9	5
U	1	0	3	1	12	18	17	24	37	9	7	24	3	5

B= Black, C= Caucasian, M=Mixed Ancestry, I=Indian/Asian, U=Unknown

5. REFERENCES

http:www.gov.za/yearbook/2001/landpeople.html

de Villiers, J.N.P., Hillerman, R., de Jong, G., Langenhoven, E., Rossouw, H., Marx, M., and Kotze, M., 1999, High prevalence of the Cys282Tyr HF mutation facilities an improved diagnostic service for hereditary haemochromatosis in South Africa, *S. Afr. Med. J.* **89**:279-282.

Greenberg, J., Bartmann, L., Ramesar, R., and Beighton, P., 1993, Retinitis pigmentosa in South Africa, *Clin Genet.* **44**:232-235.

Greenberg, J., Franz, T., Goliath, R., and Ramesar, R., 1999, A photoreceptor gene mutation in an indigenous black African family with retinitis pigmentosa identified using a rapid screening approach for common rhodopsin mutations, S. *Afr. Med. J.,* **89**:877-878.

Groenewald, J.Z., Liebenberg, J., Groenewald, I.M., and Warnich, L., 1998, Linkage disequilibrium analysis in a recently founded population: Evaluation of the variegate porphyria in Southern African Afrikaners, *Am. J. Hum. Genet.* **62**:1254-1258.

Hendricks, D., and Parker, M. I., 2002, Esophageal cancer in Africa, *IUBMB Life*, **53**:263-268.

McKie, A. B., McHale, J. C., Keen, T. J., Tarttelin, E. E., Goliath, R. R., Greenberg, J., Ramesar, R., Cremers, F. P. M., Hoyng, C. B., van Lith-Verhoeven, J.J.C., Mackey, D.A., Bhattacharaya, S.S., Bird, A.C., Markham, A.F., and Inglehearn, C.F., 2001, Mutations in the pre-mRNA splicing factor gene PRPC8 in autosomal dominant retinits pigmentosa, *Hum. Mol. Gen.* **10**:1555-1562.

Oswald, A. H., Goldblatt, J., Sampson, G., Clokie, R., and Beighton, P., 1985, Retinitis pigmentosa in South Africa, *S. Afr. Med. J.* **68**:863-866.

Yeung, K. Y., Baum, L., Chan, W.M.C., Lam, D.S.C., Kwok, A.K.H., and Pang, C.P., 2001, Molecular diagnostics for retinitis pigmentosa, *Clinica Chimica Acta.* **313**:209-215.

COMPARING ROD AND CONE FUNCTION WITH FUNDUS AUTOFLUORESCENCE IMAGES IN RETINITIS PIGMENTOSA

Anthony G. Robson[1], Catherine Egan[1], Graham E. Holder[1], Alan C. Bird[1], and Fred W. Fitzke[2]

1. INTRODUCTION

In vivo recording of human retinal pigment epithelium (RPE) autofluorescence (AF) using a scanning laser ophthalmoscope has proved a useful technique in the investigation of inherited and acquired retinal disease (Von Ruckman et al., 1995; 1999; Delori et al., 1995a; Holz et al., 1999; 2001; Lois et al., 1999; 2001; Kurz-Levin et al., 2002). The emission spectrum of AF resembles that of lipofuscin (Eldred and Katz, 1988; Delori et al., 1995b) and the technique uniquely allows visualisation of lipofuscin at the level of the RPE. Abnormal high density AF results from accumulation of lipofuscin and may be due to disrupted metabolism at the level of the RPE (Kennedy et al., 1995). Lipofuscin may itself have deleterious effects: the dominant fluorophore, A2E, may cause short-wavelength induced death of the RPE and detergent like disruption of cell membranes (Sparrow et al., 2002). The absence of AF may indicate photoreceptor cell death leading to RPE atrophy. However, the presence of abnormal AF suggests continuing metabolic demand possibly representing compromised RPE function before cell death.

We observed that some patients with retinitis pigmentosa (RP) and normal visual acuity have abnormal fundus AF in the form of a para-foveal ring of high density. Preliminary studies on these patients revealed that the size of the high density ring was related to the degree of macular dysfunction, as indicated by the Pattern ERG P50 component (Robson et al., 2002). The object of this study was to compare photopic and scotopic sensitivity with both the distribution and intensity of fundus AF abnormalities.

[1]Moorfields Eye Hospital, City Road, London, UK; [2]Institute of Ophthalmology, Bath Street, London, UK.
Corresponding author: A. G. Robson, E-mail: Anthony.Robson@moorfields.nhs.uk

2. METHODS

Patients were selected according to the following criteria; Snellen visual acuity of 6/9 or better, a clinical diagnosis of RP, abnormal AF of the posterior pole in the form of a high density para-foveal ring. Full-field ERGs were performed according to extended testing protocols incorporating the ISCEV minimum standard (Marmor and Zrenner, 1998) in order to assess generalised retinal function (see Fishman et al., 2001 for a recent review). All patients were dilated before full-field ERG testing using tropicamide (1%) and/or phenylephrine hydrochloride (2.5%). Pattern ERGs evoked by high contrast checkerboard reversal were recorded according to ISCEV recommendations before mydriasis (Bach et al. 2000) using standard parameters. The pattern ERG P50 component was used as an index of macular function (Holder, 2001).

All patients underwent AF imaging according to previously described techniques (von Ruckman et al., 1995). Thirty two individual frames were averaged over 6 seconds. The topographic distribution of AF across the retina was measured using a grey-scale index of intensity (0-255 units) of single pixels, arranged diagonally across the whole image and at retinal locations tested psychophysically.

Calibration of the grey scale was performed using a fluorescence standard (Avalon Technologies SFS-225) excited and recorded through the high-pass barrier filter (transmission 1×10^{-5} at 495nm) by a Zeiss confocal scanning laser ophthalmoscope and simultaneously measured using a UDT radiometer by reflecting a fraction from a glass slip. The calibration curve could be described by a modified logistic function.

Fine matrix maps were obtained by testing sensitivity at 1 ° intervals within 10 °x10° areas (Westcott et al., 1997; Guymer et al., 1997) under photopic conditions and under scotopic conditions following 40 minutes of dark adaptation. Areas tested psychophysically were chosen to correspond with regions of abnormal autofluorescence.

3. RESULTS

Representative ISCEV-standard full-field ERGs from a normal subject and from 3 patients with one or more clinical signs of RP are shown in Figure 1. ERGs reveal generalised retinal dysfunction consistent with a diagnosis of rod-cone dystrophy. Pattern ERG P50 components varied between subnormal and normal (>2uV).

The abnormal para-foveal annulus of high density varied in diameter between patients (Figure 1, right column). The grey scale profile across an AF image of a patient with a large high density ring is shown in Figure 2. Two major peaks in the distribution profile correspond with the spatial distribution of the high density ring. Blood vessels, manifest as low density shadows in AF images, result in localised troughs along the grey-scale intensity profiles (arrows in Figure 2).

Figure 3 shows fine matrix maps obtained under photopic and scotopic conditions. In the normal subject (Figures 3A and 3B); there is a relatively flat profile across the whole of the area tested. Photopic and scotopic threshold values in the patient, tested over a

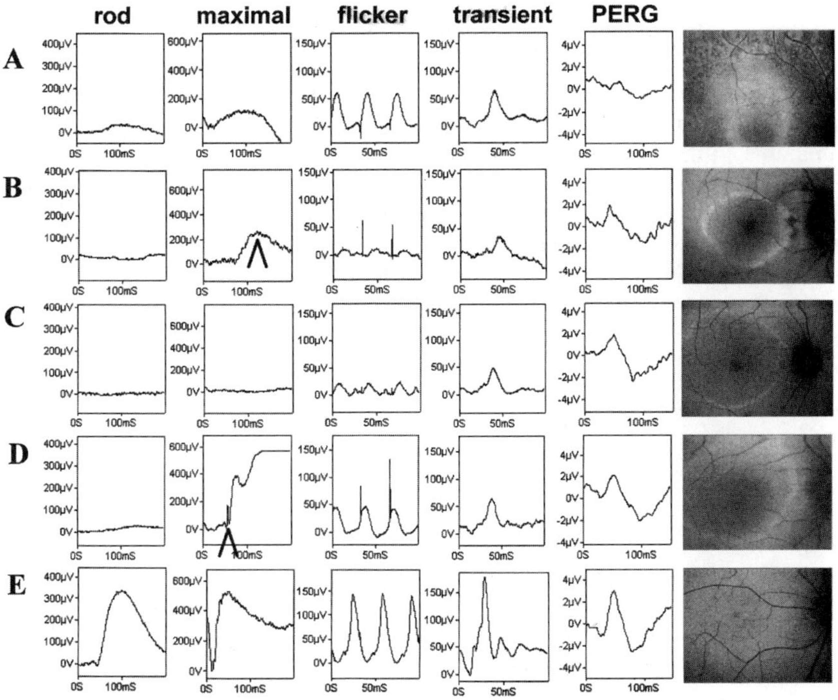

Figure 1. ISCEV-standard full-field ERGs, Pattern ERGs and fundus autofluorescence images from four patients with RP (A to D) and one normal subject (E). Arrows in B and D mark eye movement artefacts.

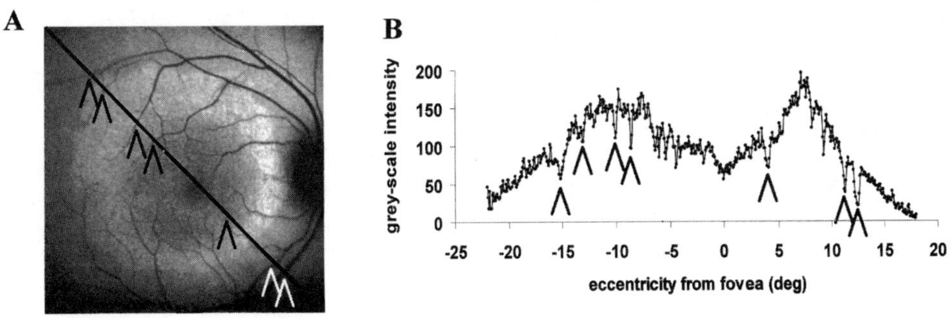

Figure 2. Fundus autofluorescence image (A) and corresponding grey-scale profile (B) from a patient with RP. Arrows mark intrersection with the retinal blood vessels.

similar inferior-nasal area, show relative preservation of photopic and scotopic sensitivity in areas within the high density ring but there is marked elevation of thresholds over eccentric areas (Figures 3C and 3D).

Figure 4 illustrates the relationship between the grey scale intensity of autofluorescence and rod and cone sensitivity over 10 retinal locations. In the normal subject, autofluorescence intensity and sensitivity show slight fluctuations across the area tested (see discussion). In the patient, values of autofluorescence intensity and sensitivity are comparable to those seen in the normal subject within the central 7-degrees. Outside the high density ring, illustrated by the peak in the autofluorescence distribution profile shown in Figure 4D, there is a reduction in photopic and scotopic sensitivity, concordant with the reduction in autofluorescence intensity. Essentially the same relationship was seen in a second patient with a smaller high density ring (data not shown).

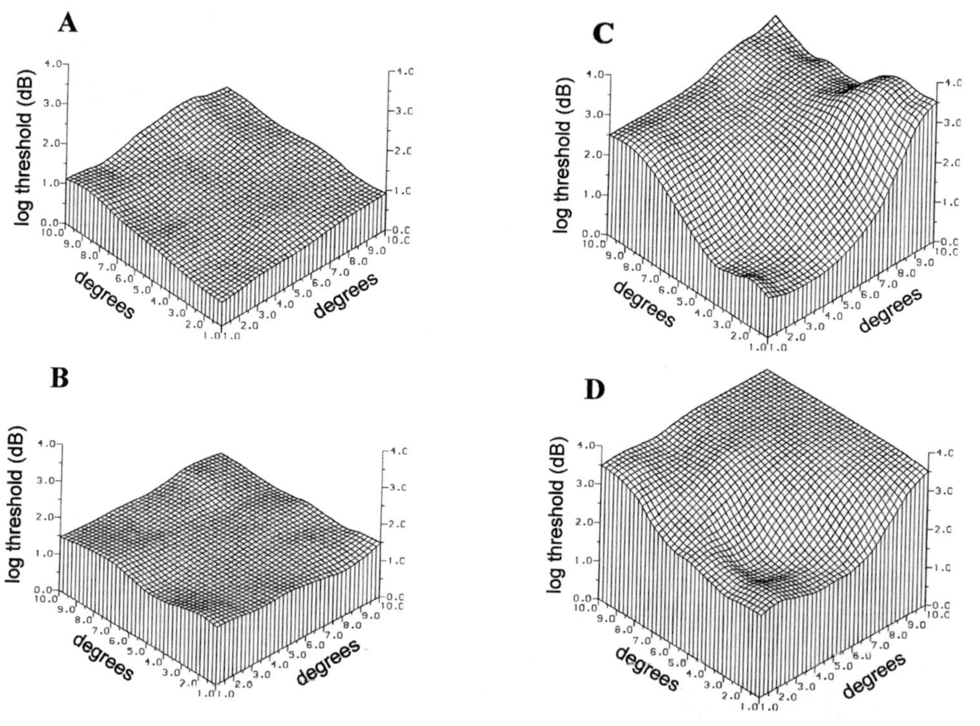

Figure 3. Fine matrix maps from a normal subject (A and B) and from a patient with RP and a ring of high density autofluorescence (C and D). Photopic thresholds are shown in A and C, scotopic thresholds following 40-minutes dark adaptation in B and D. The inferior-nasal areas tested in the patient are shown in Figure 4, corresponding areas were tested at the same eccentricity in the normal subject.

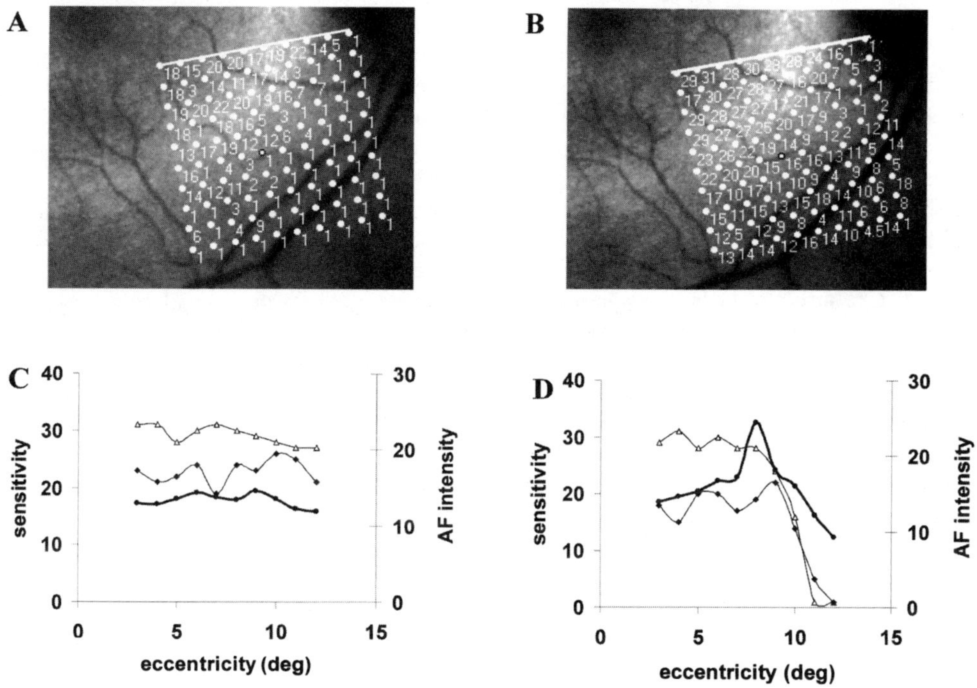

Figure 4. Sensitivity values in a patient with RP, tested under photopic (A) and scotopic (B) conditions. Comparison of photopic (open triangles) and scotopic (filled diamonds) sensitivity values with calibrated grey-scale values of autofluorescence intensity (bold line) in a normal subject (C) and in the same patient (D). The solid white line (in A and B) intersects the 10 retinal locations compared in D. Corresponding inferior-nasal areas were tested in the normal subject.

4. DISCUSSION

This preliminary study shows correspondence between the distribution of high density abnormal fundus AF and the spatial limitation of preserved central macular function in patients with retinitis pigmentosa and preserved visual acuity. The topographic distribution of abnormal AF correlates with the loss in rod and cone sensitivity seen over identical retinal locations (Figures 3 and 4). The intensity of abnormal AF also has functional significance, as suggested in previous studies of macular lesions (Lois et al., 2000). The grey-scale intensity of abnormal AF decreases with the reduction in rod and cone sensitivity, as measured psychophysically over identical retinal areas.

The approximately circular rings of high density AF appear to demarcate central areas of preserved cone and rod sensitivity in some patients with RP. This corroborates preliminary electrophysiological findings in which predominantly cone-driven PERGs (Holder, 2001) were found to correlate in amplitude with the size of the high density ring

(Robson et al., 2002). An advantage of the psychophysical method employed in this study is that localized rod sensitivity within the macula may also be tested.

Qualitative assessment of fundus AF distribution has proved a useful tool in the evaluation of macular abnormalities (Von Ruckman et al., 1995; 1999; Delori et al., 1995a; Holz et al., 1999; 2001; Lois et al., 1999; 2001, Kurz-Levin et al., 2002).

Quantification of AF intensity has received less attention and requires a more stringent approach e.g. calibration of grey-scale intensity (see methods). Close inspection of Figures 1 to 4 reveals slight illumination gradients and "hot spots" in some AF images. The retinal vasculature, manifest as low density shadows in AF images, may also distort intensity profiles, as shown in Figure 2. A recent analysis of normal AF images suggested that under routine recording conditions, radial asymmetry across the central 16 degrees may be up to 8% (Moreland et al., 2002). Small eye movements may also result in misaligned frames which if not excluded, may degrade averaged images. Automated exclusion of vessel artifacts from AF profiles and post-acquisition correction of illumination gradients may improve future accuracy. The low density area of AF evident at the fovea of most subjects varies in its lateral extent (van Kuijk et al., 2002) and is a normal feature, predominantly due to absorption of the incident 488nm laser radiation by luteal pigment (Delori et al., 1995, Wustemeyer et al., 2002, Van Kuijk et al., 2002) accentuated by reduced AF efficiency at the fovea (Losch et al., 2001). Single pixel analysis of AF intensity has specific advantages in terms of spatial resolution of localized lesions (Lois et al., 2000) but since rod and cone sensitivity is measured with a larger stimulus (approximately 30.7'), averaging of pixel intensity over a corresponding larger area may reduce the variation seen in Figures 4C and 4D.

Abnormal AF intensity is related to rod and cone sensitivity and may prove useful in the assessment of photoreceptor function in vivo. The current preliminary study suggests that accurate quantification may be of value in assessing disruption of metabolism at a stage preceding apoptosis. The current study identifies some of the complications associated with quantification of fundus AF with a view to future refinement.

5. CONCLUSIONS

Some patients with retinitis pigmentosa manifest a ring of high density on autofluorescence imaging. The distribution of the high density para-foveal ring corresponds with differences in rod and cone sensitivity over the macula. Abnormal autofluorescence intensity may be related to dysfunction over para-foveal areas, providing confounding optical and anatomical factors are considered.

6. REFERENCES

Bach, M., Hawlina, M., Holder, G.E., Marmor, M.F., Meigen, T., Vaegan and Miyake, Y., 2000, Standard for pattern electroretinography, *Doc Ophthalmol.* **101**:11-18.

Delori, F.C., Staurenghi, G., Arend, O., Dorey, C.K., Goger, D.G. and Weiter, J.J., 1995a, In vivo measurement of lipofucscin in Stargardts' disease-Fundus flavimaculatus. *Invest Ophthalmol Vis Sci.* **36**:2327-2331.

Delori, F.C., Dorey, C.K., Staurenghi, G., Arend, O., Goger, D.G. and Weiter, J.J., 1995b, In vivo fluorescence of the ocular fundus exhibits retinal pigment epithelium lipofuscin characteristics. *Invest Ophthalmol Vis Sci.* **36**:718-729.

Eldred, G.E. and Katz, M.L., 1988, Fluorophores of the human retinal pigment epithelium: separation and spectral characterization. *Exp Eye Res.* **47**:71-86.

Fishman, G.A., Birch, D.G., Holder, G.E. and Brigell, M.G. 2001. *Electrophysiologic testing in Disorders of the Retina, Optic nerve and Visual Pathway.* 2nd Ed. Ophthalmology Monographs No.2. The Foundation of the American American Academy of Ophthalmology.

Guymer, R.H., Gross-Jendroska, M., Owens, S.L., Bird, A.C. and Fitzke, F.W., 1997, Laser treatment in subjects with high-risk clinical features of age-related macular degeneration. Posterior pole appearance and retinal function. *Arch Ophthalmol.* **115**:595-603

Holder, G.E., 2001,Pattern ERG and an integrated approach to visual pathway diagnosis. *Prog Retin Eye Res.* **20**:531-561.

Holz, F.G., Bellman, C., Margaritidis, M., Schutt, F., Otto, T.P. and Volcker, H.E., 1999, Patterns of increased fundus autofluorescence in the junctional zone of geographic atrophy of the retinal pigment epithelium associated with age related macular degeneration. *Graefes Arch Clin Exp Ophthalmol.* **237**:145-152.

Holz, F.G., Bellman, C., Staudt, S., Schutt, F. and Volcker, 2001, Fundus auofluorescence and development of age related macular degeneration. *Invest Ophthalmol Vis Sci.* **42**:1051-1056.

Lois, N., Halfyard, A.S., Bird, A.C. and Fitzke, F.W., 2000, Quantitative evaluation of fundus autofluorescence imaged "in vivo" in eyes with retinal disease. *Br J Ophthalmol.* **84**:741-745.

Kennedy, C.J., Rakoczy, P.E. and Constable, I.J., 1995, Lipofuscin of the retinal pigment epithelium: a review. *Eye.* **9**:763-771.

Kurz-Levin, M.M., Halfyard, A.S., Bunce, C., Bird, A.C. and Holder, 2002, Clinical variations in assessment of bull's-eye maculopathy. *Arch Ophthalmol.* **120**:567-75.

Lois, N., Holder, G.E., Bunce, C., Fitzke, F.W. and Bird, A.C., 2001, Phenotypic subtypes of Stargardt macular dystrophy-fundus flavimaculatus. *Arch Ophthalmol.* **119**:359-69.

Lois, N., Holder, G.E., Fitzke, F.W., Plant, C. and Bird AC., 1999, Intrafamilial variation of phenotype in Stargardt macular dystrophy-Fundus flavimaculatus. *Invest Ophthalmol Vis Sci.* **40**:2668-75.

Losch, A.J., Jorzik, J. and Holz, F.G., 2001, Mapping of macular pigment distribution and RPE lipofuscin in human donor eyes with a confocal scanning laser ophthalmoscope (cSLO). *Invest Ophthalmol Vis Sci* **42**, (Suppl.), 221.

Marmor, M.F. and Zrenner, E. 1998, Standard for clinical electroretinography. *Doc Ophthalmol.* **97**:143-156.

Moreland, J.D., Robson, A.G., Pauleikhoff, D., Morrissey, T., Holder, G.E., Fitzke, F.W., van Kuijk, F.J.G.M. and Bird, A.C., 2002, Assessments of macular pigment density and distribution using motion photometry and fundus autofluorescence. *Perception,* **31**;165.

Sparrow, J.R., Zhou, J., Ben-Shabat, S., Vollmer, H., Itagaki, Y. and Nakanishi, K., 2002,. Involvement of oxidative mechanisms in blue-light-induced damage to A2E-laden RPE. *Invest Ophthalmol Vis Sci.* **43**;1222-7

Robson, A.G., El-Amir, A., Bailey, C., Egan, C., Bird, A.C., Fitzke, F.W. and Holder, G.E. 2002, Pattern ERG correlates and fine matrix maps of fundus autofluorescence abnormalities in patients with retinitis pigmentosa and good visual acuity. *Invest Ophthalmol Vis Sci. B781. ARVO Abstract no. 1771*

Van Kuijk, F.J.G.M. , Robson, A.G., Moreland, J.D., Pauleikhoff, D., Morrissey, T., Holder, G.E., Fitzke, F.W., Bird, A.C., 2002, Macular pigment density and distribution: comparison of a mimimum motion method with autofluorescence. *Invest Ophthalmol Vis Sci. B545. ARVO Abstract no. 2548*

von Rückmann, A., Fitzke, F.W. and Bird, A.C., 1995, Distribution of fundus autofluorescence with a scanning laser ophthalmoscope. *Br J Ophthalmol.* **79**:407-412

von Rückmann, A., Fitzke, F.W., Bird, A.C., 1999, Distribution of pigment epithelium autofluorescence in retinal disease state recorded in vivo and its change over time. *Graefes Arch Clin Exp Ophthalmol.* **237**;1-9.

Westcott, M.C., McNaught, A.I., Crabb, D.P., Fitzke, F.W., Hitchings, R.A., 1997, High spatial resolution automated perimetry in glaucoma. *Br J Ophthalmol.* **81**:452-9.

A MODIFIED PROTOCOL FOR THE ASSESSMENT OF VISUAL FUNCTION IN PATIENTS WITH RETINITIS PIGMENTOSA

N. Lodha[1], C. A. Westall[1,3], M. Brent[1,2,3], M. Abdolell[4], and E. Héon[1,3]

1. ABSTRACT

1.1. Background

The assessment of visual function for retinitis pigmentosa routinely includes: electroretinography, visual acuity and visual field-testing. Patients with retinitis pigmentosa sometimes complain of changes in visual function, which are not paralleled by routine eye tests.

1.2. Aims and Objectives

To determine which visual function test or group of tests can predict reliably perceived visual function in patients with retinitis pigmentosa

1.3. Methods

Subjects with progressive retinitis pigmentosa are recruited from the Ocular genetics program of The Hospital for Sick Children and Mount Sinai Hospital, Toronto. Subjects will be tested four times over the over the period of one year. On each visit they undergo following tests- 1) Central visual acuity (VA) using the crowded logMAR acuity chart, 2) Contrast Sensitivity (CS) using Pelli-Robson contrast sensitivity chart, 3) Visual field test (VF) using Humphrey (10-2), 4) Color vision using Mollon-Reffin 'minimalist' test and 5) Subjective visual function questionnaire testing near and global perceived visual function respectively.

Departments of Ophthalmology, [1]The Hospital for Sick Children, 555 University Avenue, Toronto, Ontario, Canada, M5G 1X8; [2]Mount Sinai Hospital; [3]The University of Toronto, Toronto, Ontario, Canada; [4] Research Institute, Population Health Sciences, The Hospital for Sick Children, Toronto, Ontario, Canada. Corresponding author: E. Héon, E-mail: eheon@attglobal.net

1.4. Results

Phase I (baseline and visit I measure) results are reported. Total of sixty-eight patients with mean age of 41 years, age range of twelve to sixty seven were tested. Of these thirty-one were males and thirty-seven were females. Repeat testing correlation was high ($r>0.8$, $p<0.05$) for all parameters between baseline visit and visit I. The near perceived visual function correlated best with the combination of visual acuity and contrast sensitivity. The global perceived visual function correlated best with combination of visual field and visual acuity. Objective measure of central visual function (HVF 10-2) correlated best with contrast sensitivity.

1.5. Discussion

The addition of contrast sensitivity and Humphrey visual field to routine visual assessment should improve the quality of the longitudinal data of visual function

recorded on these patients. Patients will be re- tested at six months and one-year interval. To date of the sixty-eight subjects twenty-seven have returned for their six-month visit (phase II).

2. INTRODUCTION

Retinitis pigmentosa refers to a group of progressive hereditary disorders that diffusely and primarily affect photoreceptors and retinal pigment epithelial function. Clinically these disorders can be characterized by the following: night blindness, peripheral visual field restrictions and/or scotomata, abnormalities in the electroretinogram (ERG) and ophthalmoscopic evidence of retinal degeneration. (Marmor et al., 1983). Functional abnormalities of rod and cone systems typically are most apparent in the peripheral visual field, but foveal visual acuity is often reduced as well (Farber et al., 1985; Alexander et al., 1992).

The assessment of visual function for Retinitis Pigmentosa routinely includes: electroretinography, visual acuity assessment, color vision assessment and visual field testing (Goldmann visual field). Many patients with retinitis pigmentosa complain of changes in visual function, which are not paralleled by changes in routine eye tests such as visual acuity and visual fields. This may be explained by the fact that patients with retinitis pigmentosa show disproportionate loss of acuity at lower contrasts levels (Spellman et al., 1989). This is most often noted in patients with a restricted field of vision.

Obtaining an objective measurement, which corresponds to subjective changes, is important, as future treatments are being designed to modulate the outcome of the disease.

The purpose of our study is to determine which visual function test or group of tests can predict changes in the visual functions reliably in patients with retinitis pigmentosa. Results shown here are at the baseline measure.

3. METHODS

A modified protocol for assessment of visual function in patients with retinitis pigmentosa is tested currently with the patient populations attending the ocular genetics program at The Hospital for Sick Children and the eye clinic Mount Sinai Hospital, Toronto.

3.1. Subjects

Sixty-eight patients aged ten years to sixty-five years, with well-established progressive retinitis pimentosa (Non-record able ERG, abnormal Fundus and Goldmann visual fields) with visual acuity of 20/100 or better were included in phase I of the study. Patients were excluded from the study if they had anterior segment anomaly affecting vision, stationary retinal dystrophies (e.g. CSNB, achromatopsia) and any neurological problem affecting vision or capacity to perform tests. Sixty age-matched controls with 20/20 or better-corrected vision and normal eye examination were included in the study. Neurological illness or other systemic problem affecting vision or their ability to perform tests resulted in exclusion.

3.2. Testing Procedure

All patients will be seen four times over a period of one year. Phase I includes baseline visit and first-visit. These are one to four weeks apart and provide knowledge of learning effects and inter visit variability of the tests. Phase II and III will include six month and one year visits respectively.

For the following tests of visual function one eye are tested at a time (first eye to be tested was selected randomly) the other eye was occluded with an eye patch.

3.2.1. Visual Fields (VF)

The central ten degrees of the visual field was tested using a Humphrey perimeter (Zeiss Humphrey systems) using 10-2 SITA standard software program with white and size III stimulus. Full refractive correction was used. The Mean deviation for each central visual field was calculated, which was the sum of loss of sensitivity at all tested points. Sensitivity loss was measured in db (Hirohido et al., 1999). Mean standard deviation was used to define the better eye. Results from eye with less mean standard deviation was used for analysis for all the tests.

3.2.2. Central Visual Acuity (VA)

Visual acuity was tested using a crowded log MAR acuity chart at three meters. The logMAR crowded test (Vistech Consultants, Dayton, Ohio) comprises of books with one acuity level per page graduates in 0.1 logMAR slops. The acuity ranges from 0.8 logMAR to –0.3 logMAR at the testing distance of three meters. For each acuity level test there are four letters and the letters are surrounded by a boundary creating equal crowding. The patient's visual acuity was calculated by scoring each letter read correctly (0.025 log units added for each individual letter read correctly) (McGraw et al., 2000).

3.2.3. Color Vision (CV)

The Mollon-Reffin 'Minimalist' test (Shute and Westall, 2000) is used to test color vision. Subjects were tested under standardized lighting in a darkened room. As a preliminary demonstration trial, the subject was presented with 5 grey chips and an orange test chip. The distracter chips (grey chips) were used to ensure that the task couldn't be solved on the basis of perceived lightness. The subject followed the same procedure for the three color confusion lines: Proton, Deutan and Triton. There were six chips in each color series denoted from 1 (least saturation) to 6 (most saturated). In each case chip number 3 was presented first. If the response was correct then chip number 2 was presented and finally chip number 1. In the event of error, saturation of the chip was increased. For each color set, the number of the least saturated chip identified was noted as the patient's score. The scores for each color series were added; final score of 4 or less was considered be a normal color vision and score of 18 as sever color vision defect.

3.2.4. Contrast sensitivity (CS)

Pelli-Robson contrast sensitivity charts (Clement Clarke-Haag-Street International, Essex, United Kingdom) were used to measure letter contrast sensitivity. Testing was performed at one meter and at three meters with luminance of 85-cd/m^2. Eyes were tested monocularly, followed by binocular testing. Each eye was tested three times before recording the final result. Contrast sensitivity increases by 0.15 log units each set of three letters The patient's contrast sensitivity is calculated by applying a score for each letter read correctly (0.05 log units for each individual letter) (Elliott et al., 1990).

3.2.5. Subjective/Perceived visual function assessment (PVF)

All the subjects who are part of the study are asked to complete the self-report questionnaire consisting of 39 questions. These are designed to target daily activities that are potentially problematic for people with Retinitis Pigmentosa. Twenty-nine of these questions are related to central visual function and ten are related to global visual function. The format of the questionnaire is based on the studies of Szlyk et al (2001), and Turano et al (1999), the subjects are first asked whether or not she/he can perform a specific activity. If yes, then he/she is asked to rate himself/herself on a 5-point Likert scale (Likert, 1932), which describes the difficulty of the test (5 = none, 4 = mild, 3 moderate, 2 = extreme, and 1 = not able to perform task). This questionnaire is repeated at each visit. If the subject responded that he/she is unable to perform an activity for reasons other than visual ones, that specific item will be excluded from the analysis. The outcome score is assessed as the sum total of the level of difficulty scores, for central and global visual functions. Scores of 140-135 for near perceived visual function and 55-50 for global perceived function were considered normal and scores of 50 or less for near perceived visual function and sores of 20 or less for global perceived function were considered sever loss of function.

3.3. Statistical analysis:

For the phase I of this study the primary questions were to determine if the measure of visual functions correlate with perceived central visual function and objective visual function (Humphrey visual field) respectively. Analysis of covariance (ANCOVA) determined if the measured visual functions (visual acuity, color vision and or contrast sensitivity) correlate with perceived central visual function. Similarly, an ANCOVA was conducted to determine if these visual functions correlated with objective visual function loss as measured with the Humphrey perimeter. A secondary ANCOVA was performed to determine if measured visual functions correlate with perceived global visual functions.

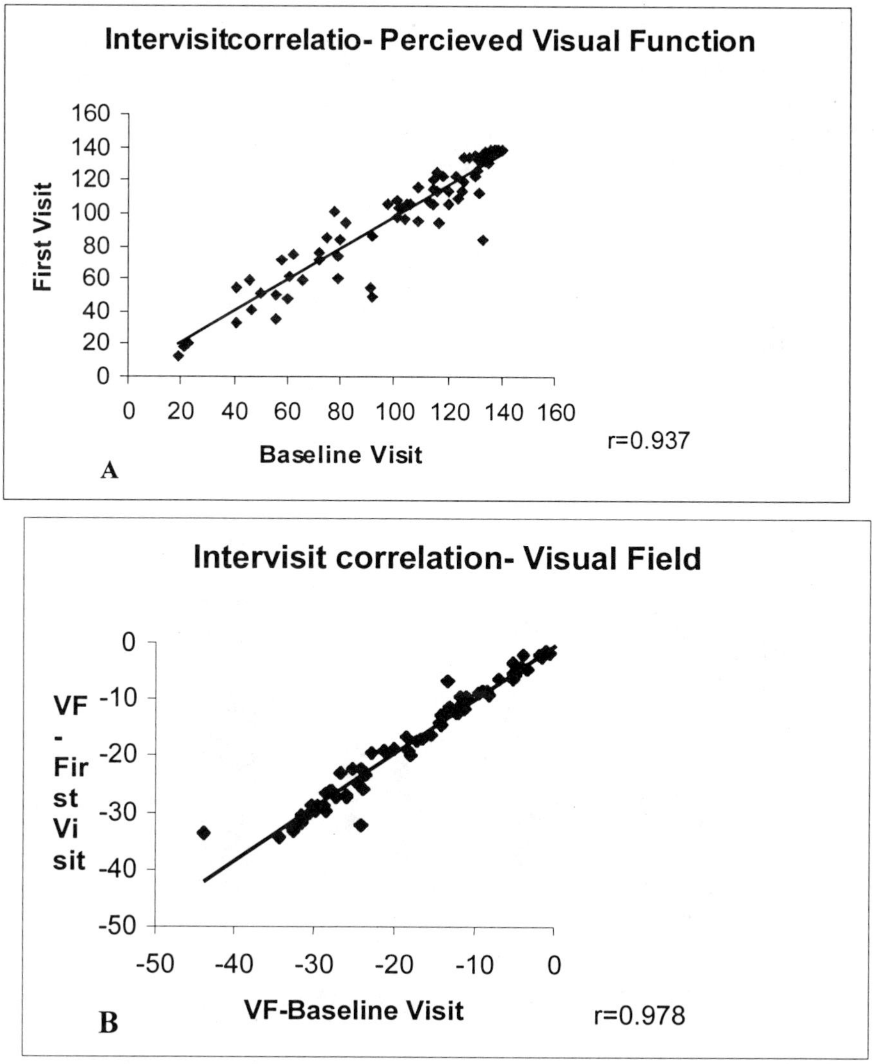

Figure 1. Inter visit repeatability between baseline visit and first visit. a) Inter visit repeatability of perceived visual function (near and global combined). Perceived visual function score was sum of the grading of thirty-nine different activities on questionnaire by Likert scale (5 no difficulty and 1 extreme difficulty). X axis shows baseline visit scores and y axis first visit scores. b) Inter visit repeatability of the Humphrey 10-2 visual fields. Mean standard deviation in db of baseline visit (x axis) was compared with that of visit one (y axis).

Further more Mallow's C (p) analysis was done to find model with best fit. Mallow's C (p) is a statistic that measures how well a regression model fits the data. C (p) adds penalty values when the number of test parameters keeps increasing. Its use is slightly more complicated than the R^2 as it incorporates a penalty for the number of covariates. Lower values of C (p) are associated with better fits of the model.

4. RESULTS

In phase one of the study a total of sixty-eight subjects were tested. Of these, thirty-one were male and thirty-seven were female, ranging in age from twelve to sixty-seven years (mean age - forty years).

Repeat testing correlation was high ($r>0.8$, $p<0.05$) for all parameters between baseline visit and visit one (Phase I visits one –four weak apart) (Figure 1).

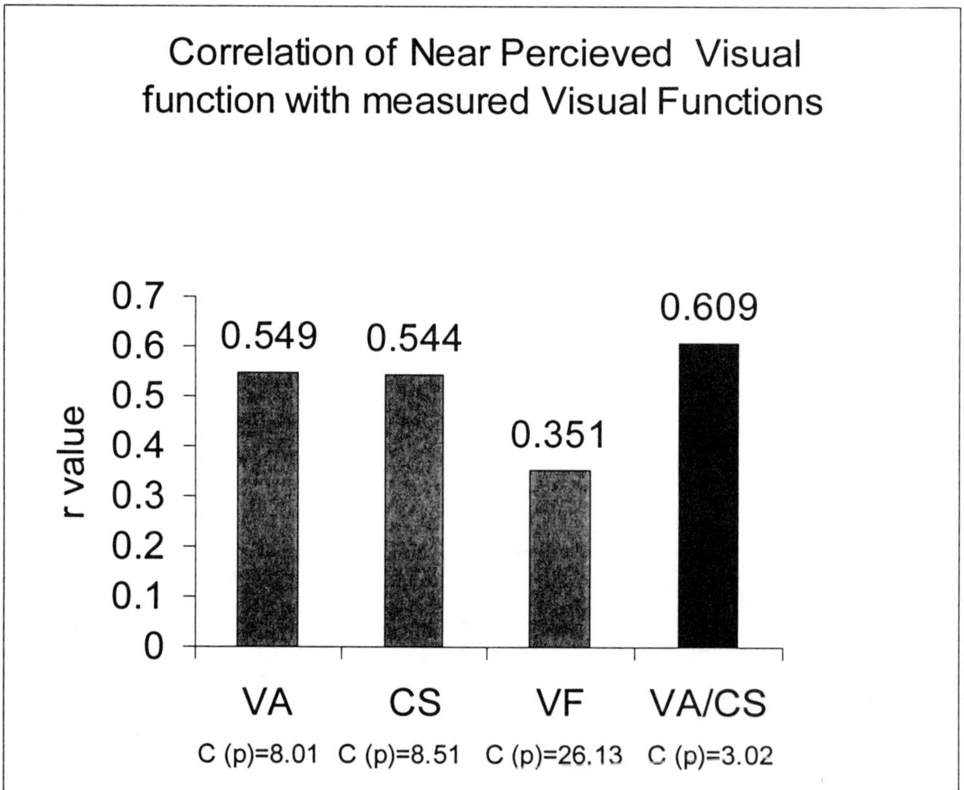

Figure 2. Correlation of near perceived visual function (questionnaire) with measured visual functions and their combination. X axis shows the visual functions and y-axis the r-value. C (p) =Mallow's C (p) statistic. Low scores represents a good fit.

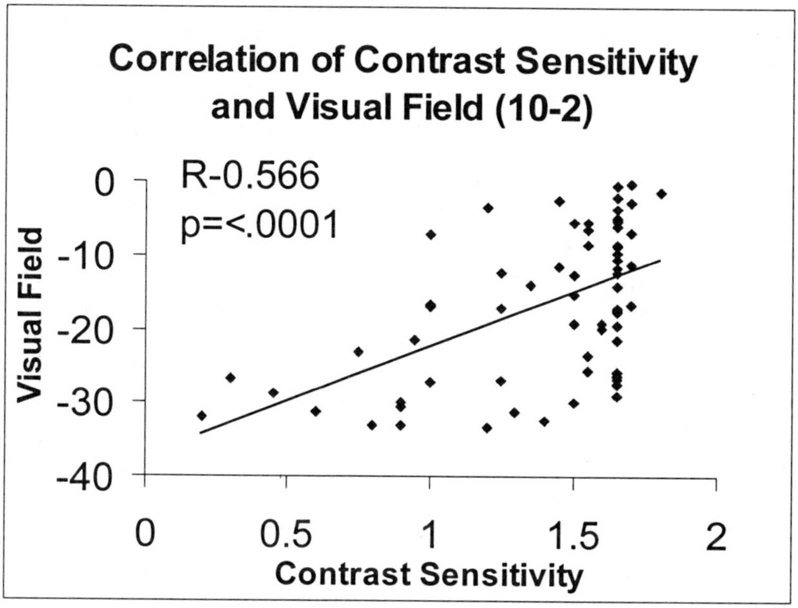

Figure 3. Correlation between contrast sensitivity and visual fields. X axis shows the contrast sensitivity (log units) and y axis shows visual field mean standard deviation (db). Results are from first visit.

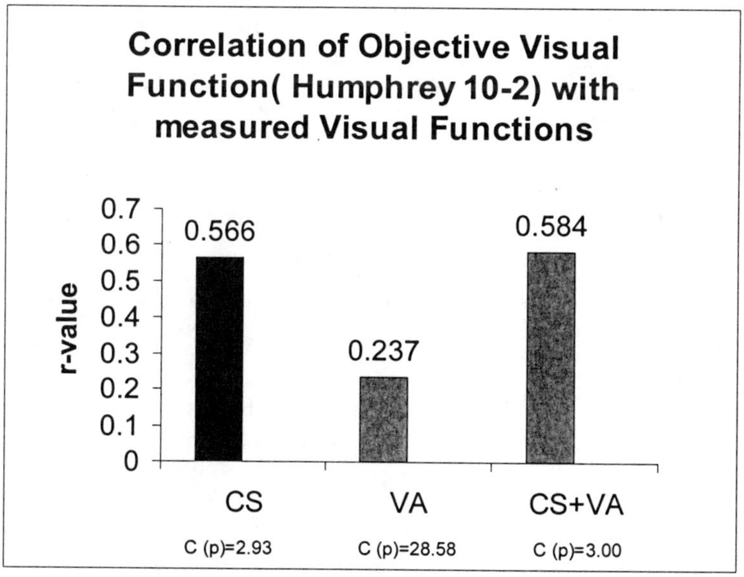

Figure 4. Correlation of objective visual function as defined by Humphrey 10-2 visual field with other visual function and their combination (visual acuity and contrast sensitivity). X axis shows visual functions and y axis shows the r-value.

Near perceived visual function correlated best with the combination of visual acuity and contrast sensitivity (r = 0.609, C (p) =3.02). The correlation of the combined tests was stronger than the value of each test taken individually, or any other combination of tests. Near perceived visual function correlate poorly with Visual Fields (HVF 10-2) individually (r=0.351, C (p) =26.13) or in combination with other visual function (Figure 2).

Visual fields (HVF 10-2) result was correlated best with the combination of visual acuity and contrast sensitivity (r = 0.584, C (p) =3.00) however, multiple regression analysis revealed a high penalization factor which diminished the importance of this correlation. The most important correlation of visual field was with contrast sensitivity (r = 0.566, C (p) =2.93 and p<0.0001) (Figure 3). Visual acuity correlated poorly with visual fields (r = 0.237, C (p) = 28.58) (Figure 4).

Global visual function correlated best with a combination of visual field and visual acuity (r = 0.545, C (p) =2.06). The correlation contrast sensitivity was weak (r = 0.421, C (p) =10.84) (Figure 5).

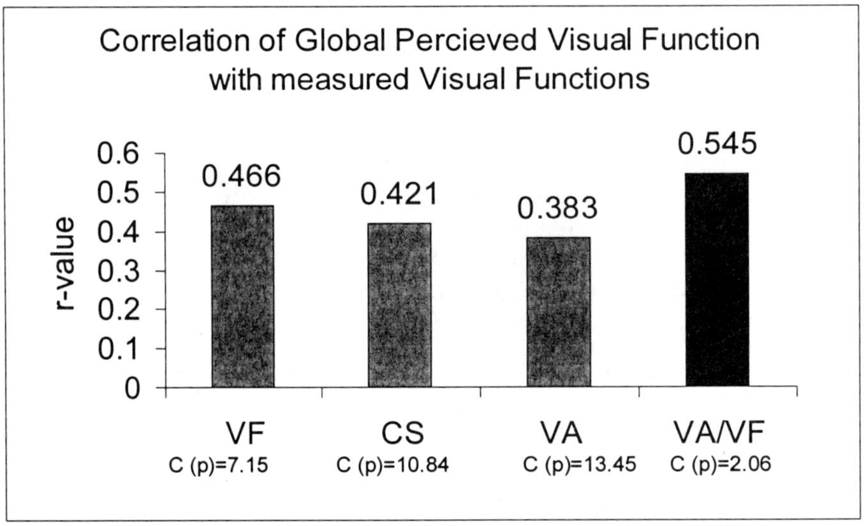

Figure 5. Correlation of global perceived visual function with measured visual function and their combination (visual field, visual acuity and contrast sensitivity). X axis shows the measured visual functions and their combination and y axis shows r-value.

Addition the color vision in the statical analysis model decreased the significance of correlation model. Hence was not used for statical analysis.

5. DISCUSSION

Patients with retinitis pigmentosa perform poorly in recognizing letters lower contrasts levels. This disproportionate loss of acuity at lower levels of contrast may be responsible for many subjective/perceived visual changes and complains of patients. Visual acuity measured routinely in the eye clinic does not measure vision at lower contrasts. Measurement of letter contrast sensitivity provides better correlation with the patient's report of visual performance.

The phase I results shows that correlation of measured visual function to patient's subjective visual perception improves significantly by adding contrast sensitivity assessment. Contrast sensitivity correlated very significantly with visual fields. The variability in the contrast sensitivity in relation with visual field observed may reflect the variability in field constriction included in the study. This includes the presence or absence of central or paracentral scotomata.

High repeatability of the tests between visits of phase proves the reliability of the tests used in the- study.

The addition of Contrast Sensitivity and Humphrey Visual Field testing (10-2) to the routine visual assessment should improve the quality of the longitudinal data of visual function obtained from these patients. This is especially true for patients with visual fields measuring less then ten degrees, where a Goldman visual field does not give as much information about field changes.

All the patients will be seen again at six month (phase II) and one year (phase III). Twenty-seven have returned for six-month visit (phase II).

6. ACKNOWLEDGEMENTS:

We acknowledge Dr Jose Antonio Vega for helping with visual field testing and providing helpful suggestions and Linda Petty and Deborah Wilson for formulating the perceived visual function questionnaire.

7. REFERENCES

Alexander K.R., et al., 1992, Grating, vernier, and letter acuity in retinitis pigmentosa. *Invest Ophthalmol Vis Sci.* **33**:3400-3406.

Elliott D.B., K. Sanderson, and A. Conkey, 1990, The reliability of the Pelli-Robson contrast sensitivity chart. *Ophthalmic Physiol Opt.* **10**:21-24.

Farber M.D., G.A Fishman, and R.A. Weiss, 1985, Autosomal dominantly inherited retinitis pigmentosa. Visual acuity loss by subtype. *Arch Ophthalmol.* **103**:524-528.

Hirohido H., Hiroyuki Iijima., Takashi Gohdo., Masahito Imai.,and Shigeo Tsukahara., 1999, Progression of defect in the central 10-degree visual field of patients with retinitis pigmentosa and choroideremia. *Am J Ophthalmology* **127**:436-442.

Likert R.A., 1932, Technique for the measurment of attitude. *Arch Psycology* **140**:1-55.

Marmor M.F., Aguirre G., Aden G., Berson E.L. et al., 1983, Retinitis Pigmentosa A symposium on terminology and methods of examinaton. *Ophthamology* **90**:126-131.

McGraw P.V., et al., 2000, Improving the reliability of visual acuity measures in young children. *Ophthalmic Physiol Opt.* **20**:173-184.

Shute R.H. and C.A. Westall, 2000, Use of the Mollon-Reffin minimalist color vision test with young children. *J Aapos.* **4**:366-372.

Spellman D.C., et al., 1989, Letter contrast sensitivity in retinitis pigmentosa patients assessed by Regan charts. *Retina* **9**:287-291.

Szlyk J.P., Seiple W., Fishman G.A., Alexander K.R., Grover S., Mahler C.L., 2001, Perceived and actual performance of daily tasks: Relationship to visual function tests in individuals with retinitis pigmentosa. *Ophthalmology* **108**:65-75.

Turano K.A., et al., 1999, Perceived visual ability for independent mobility in persons with retinitis pigmentosa. *Invest Ophthalmol Vis Sci.* **40**:865-877.

PRENATAL HUMAN OCULAR DEGENERATION OCCURS IN LEBER'S CONGENITAL AMAUROSIS (LCA1 and 2)

Fernanda B. O. Porto[1,3], Isabelle Perrault[2], David Hicks[1], Jean-Michel Rozet[2], Noëlle Hanoteau[1], Sylvain Hanein[2], Josseline Kaplan[2], and José A. Sahel[1]

1. INTRODUCTION

Leber's Congenital Amaurosis (LCA, MIM204000) accounts for at least 5% of all inherited retinal disease and is the most severe form of inherited retinal dystrophy, responsible for congenital blindness with the earliest age of onset. It is diagnosed at birth or in the first few months of life because of severely impaired vision or blindness, pendular nystagmus, eye poking and an abnormal or extinguished electroretinogram (ERG)(Kaplan et al., 1990). At birth the fundi of most patients with LCA appear normal, although fundus changes characteristic of retinitis pigmentosa may develop later during the course of the disease (Perrault et al., 1996). Both clinical and genetic heterogeneity in LCA have been demonstrated, currently identified mutations occuring in genes coding for photoreceptor-specific retinal guanylate cyclase 1 (RetGC-1) (LCA1), RPE-specific RPE65 (LCA2), photoreceptor-specific cone rod homeobox (Crx) transcription factor (LCA3), photoreceptor-pineal gene-specific aryl interacting-like protein 1 (AILP1) (LCA4), retinitis pigmentosa GTPase regulator interacting protein (RPGRIP1) (LCA5) and Crumbs homologue 1 (CRB1). Additional genes involved in this group of diseases remain to be identified.

RetGC-1 is involved in recovery of the dark-adapted state following bleaching of rhodopsin, the guanylate cyclase activating proteins (GCAPs) being stimulated in the presence of low Ca^{2+} to activate RetGC-1, catalysing the conversion of GTP to cGMP and restoring high resting intracellular cGMP concentrations which regulate the opening of cGMP-gated cation channels in the photoreceptor outer segment (OS) plasma membrane. Based on the current understanding of molecular mechanisms of

[1]Laboratoire de Physiopathologie Cellulaire et Moléculaire de la Rétine, France; [2]Service de Génétique Médicale et Unité de Recherches sur les Handicaps Génétiques de l'Enfant, France; [3]Centro Oftalmológico de Minas Gerais e Hospital São Geraldo da Universidade Federal de Minas Gerais, Brasil. Corresponding author: F. B. O. Porto, E-mail: fernandaporto@hrfree.com.br

phototransduction, the *GUCY2D* mutations observed in LCA1 would lead to markedly reduced or abolished production of cGMP in photoreceptor cells (Perrault et al., 1996), leading to a situation equivalent to permanent light exposure (Perrault et al., 1996). As a consequence, the excitation process of rod and cone photoreceptors might be markedly impaired due to persistent closure of cation channels with ensuing hyperpolarization of the plasma membrane.

In contrast to other genes whose defects have been implicated in degenerative retinopathies, RPE65 is the first disease gene in this group of inherited disorders expressed exclusively in the RPE, and plays an important role in the visual cycle (Redmond et al., 1998; Gu et al., 1997). RPE65 acts within the visual cycle pathway, representing either the isomerase itself or a closely associated protein, catalyzing the conversion of all-trans retinyl ester to 11-cis retinol within the RPE. Active 11-cis retinal is ultimately transported to the apposing photoreceptors where it becomes covalently linked to lysine 296, forming the photoactivatable holoprotein rhodopsin. Thus RPE65 mutations would impede formation of rhodopsin, leading to a situation equivalent to a retina kept in a constant dark state.

Animal models for both LCA1 and LCA2 have been described. The *retinal degeneration* (*rd*) chicken exhibits many of the features of LCA, and was found to possess mutations in RetGC-1 (Semple-Rowland, et al., 1998). Homozygous RetGC1$^{-/-}$ mice have also been produced, and shown to have virtually non-recordable cone ERG's shortly after birth (Yang et al., 1999). In both strains, histological analyses demonstrated normal retinal architecture at birth, with numbers and morphology of rod and cone cells indistinguishable from age-matched controls. Time of onset of visible degeneration occurs at different ages for the different strains, approximately one week post-hatch for the *rd* chicken (Semple-Rowland, et al., 1998) and one month for RetGC-1$^{-/-}$ mice (Yang et al., 1999), with cone degeneration commencing well before that of rods, which retained apparently normal morphologies and numbers for extended times. Homozygous RPE65$^{-/-}$ mice also show normal retinal histology at birth, although by contrast with RetGC-1$^{-/-}$ mice the rod ERG's are undetectable while cone ERG's appear relatively normal (Redmond et al., 1998). The retina undergoes a slow degeneration, exhibiting a normal architecture for many months although disc organisation within the photoreceptor OS appears slightly perturbed. These animal model data, combined with the normal fundus appearance of newborn LCA patients, have led to the impression that structural breakdown occurs well after functional deficits in this pathology. We show here that contrary to this idea, histological and immunohistochemical examination of foetal retinas obtained from both LCA1 and LCA2 human embryos reveal profound retinal, especially photoreceptor, degeneration by at least the start of the third trimester of gestation, and may hence participate in the severe functional losses observed in these conditions.

2. MATERIALS AND METHODS

2.1. Clinical History

Consanguineous parents originating from North Africa referred for prenatal diagnosis of LCA at 26 weeks pregnancy. In this family, the proband displayed typical clinical features of LCA1 (Perrault et al., 1999) and was found to be homozygous for two

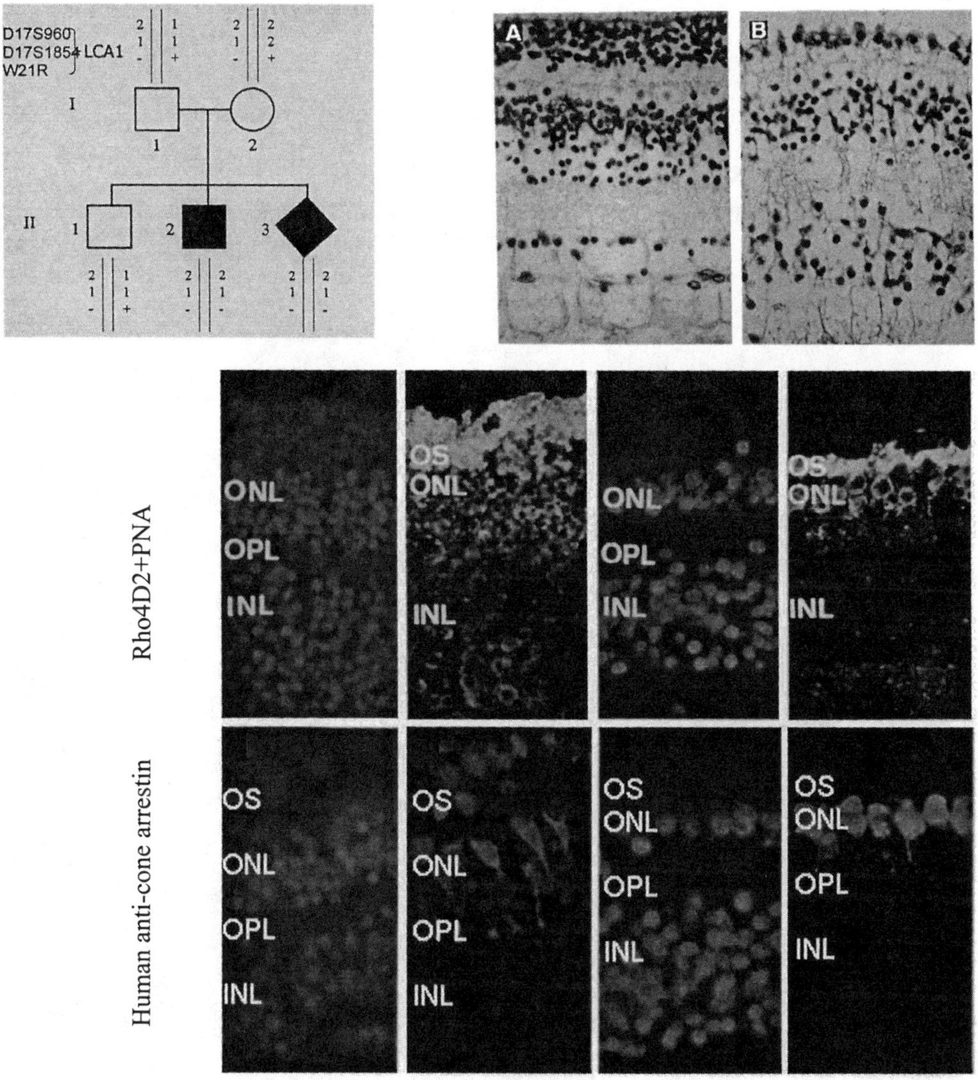

Figure 1. Genetics and histology of LCA1 embryo eyes. **Top left:** Pedigree of LCA1 family originating from North Africa. Numbers designate markers at the LCA1 locus (D17S960 and D17S1854). The patient carried a homozygous rare intragenic variant inherited from each parent while his healthy brother was heterozygote (W21R). **Top right:** 26 week LCA1 stained retina sections of the posterior pole (B) exhibit severe abnormalities. The outer nuclear layer is reduced in thickness compared to controls (A). **Bottom:** Abnormal immunolabeling in LCA1. Anti-rhodopsin antibody immunolabeling was intense in the surface membranes of rod inner segments and outer segments (OS), cell bodies and synapses of normal retina. In contrast, LCA1 foetal rod photoreceptors had short OS and restricted rhodopsin immunolabeling, the immunopositive rod population being restricted to a single cell row at the posterior pole. Labelling with PNA and anti-cone arrestin in normal and mutant 26 week retina was intense in cone inner segments and OS, but showed OS shortening and outer nuclear layer (ONL) thinning in the mutant.

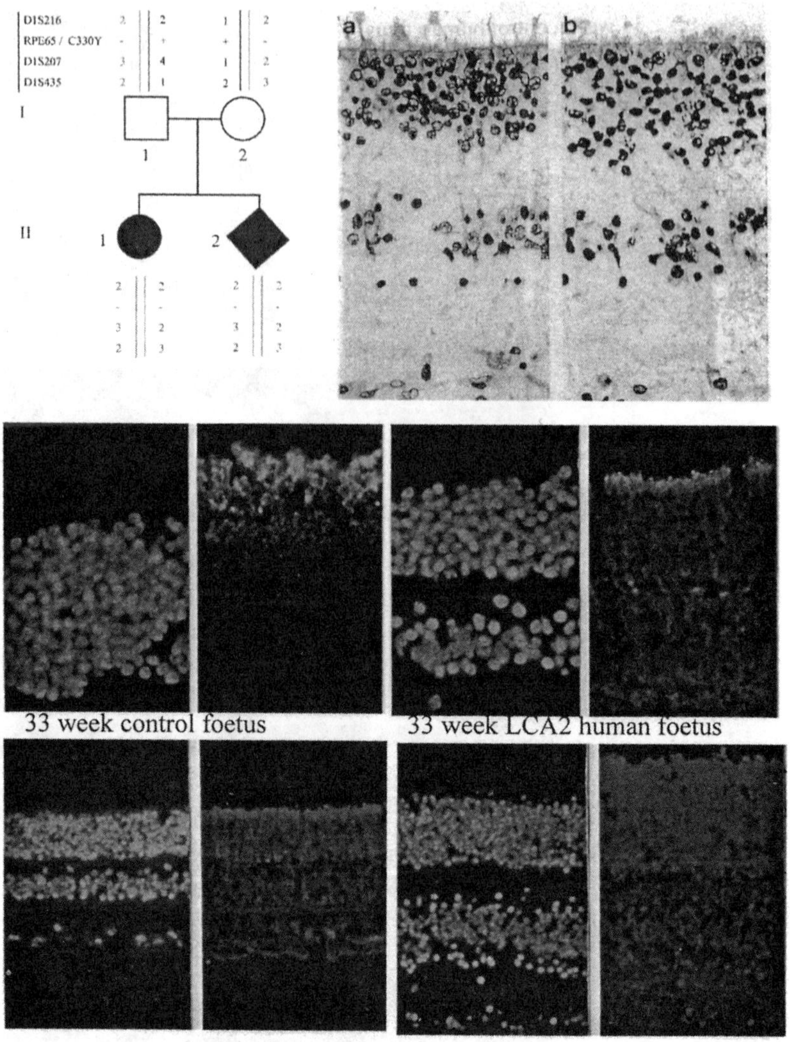

Figure 2. Genetics and histology of LCA2 embryo eyes. **Top left:** Pedigree of LCA2 family from North Africa. Numbers designate alleles at the D1S216, D1S207 and D1S435 loci. C330Y is the mutation identified in exon 9 of the RPE65 gene (+ wild type allele, - mutation). **Top right:** LCA2 human embryonic eyes display abnormal histology compared to age-matched normal tissue. Normal 33 week retina (botton left); inner plexiform layers respectively containing developing synaptic elements; B: By comparison the 33 week LCA2 retina exhibits many dense profiles within the ONL and also INL, and an overall reduction in cell numbers **Bottom:** Photoreceptor immunoreactivity is aberrant in LCA2 retina. DAPI staining of age-matched control human retina shows numerous nuclei in the three cellular layers. Immunolabeling of same field with anti-opsin antibody and PNA respectively, shows intense staining of the broad OS and ONL DAPI staining of 33 week LCA2 retina showing considerable cell loss in layers ONL and especially INL; the same section shows reduced opsin immunoreactivity in shortened OS.

markers at the LCA1 locus (D17S960 and D17S1854) (Perrault et al., 1996) while his healthy brother was heterozygote (Figure 1). The GUCY2D mutation responsible for the disease in this family has not been yet identified but the patient carried a homozygous rare intragenic variant inherited from each parent while his healthy brother was heterozygote (W21R; Figure 1). Taken together, these data gave strong evidence for linkage of the disease to LCA1. The prenatal diagnosis performed in this family revealed that the foetus was haplo-identical to his affected brother at the LCA1 locus. The parents decided to interrupt the pregnancy and gave written permission for a histopathological study of the foetus to be carried out. The lod-score value calculated after the histological findings gave a maximum of $Z=1.88$ at $\Theta = 0$. Linkage analysis was performed in the LCA1 region as described elsewhere (Porto et al., 2002). Lod-score values were calculated using the M-LINK option of the 5.1 version of the LINKAGE program (Wrigstad et al., 1994).

For the second consanguineous family originating from North Africa, the LCA patient was found to carry a homozygous missense mutation in the RPE65 gene, C330Y (figure 2). The parents referred for a prenatal diagnosis while the mother was already 30 weeks pregnant. The prenatal diagnosis performed in this family revealed that the fetus was haplo-identical to his affected sister at the LCA2 locus. The parents decided to interrupt the pregnancy at 33 weeks and gave written permission for a histopathologic study of the foetus to be performed.

2.2. Histology and immunohistochemistry

The eyes were rapidly enucleated, fixed in 10% formalin for 24 to 72 hours, and stored in phosphate buffered saline (PBS) at 4°C. Each globe was sectioned in a horizontal plane comprising the optic nerve and cornea, and embedded in paraffin wax. A globe obtained from a voluntarily aborted normal human embryo of the same gestational age was treated in the same way. Representative sections of the central and peripheral posterior pole were collected and stained with haematoxylin/eosin. Additional tissue sections were dewaxed, rehydrated and labeled with the following antibodies: rods were identified with anti-rhodopsin mouse monoclonal antibody rho4D2 (10 μg/ml). Cone cells were labeled with peanut agglutinin (PNA) lectin (25 μg/ml) (Sigma) and rabbit polyclonal anti-PNA (10 μg/ml) (Sigma), or human polyclonal anti-cone arrestin (generous gift of Dr. Cheryl Craft, Doheny Eye Inst., USC). Synaptic vesicles were identified with mouse monoclonal anti-synaptophysin (SVP-38) (10 μg/ml) (Sigma). Anti-RPE65 monoclonal antibody was generously provided by Dr. Debra Thompson (Univ. Michigan, Ann Arbor, Michigan), and used at a dilution of 1:250. Sections were incubated with primary antibodies overnight, washed thoroughly and incubated with goat anti-mouse IgG conjugated with Texas Red (gαm-TR) and/or goat anti-rabbit IgG conjugated with Alexa 488 (gαr-A488) (both 10 μg/ml) for 1 hour, with 4',6-diamidino-2-phenylindole dihydrochloride (DAPI) added for nuclear labeling. For RPE65 studies, sections were blanched with 0.25% $KMnO_4$ for 30 min. followed by 5% oxalic acid for 5 min. to remove melanin pigmentation from the RPE prior to incubation with the primary antibody. Sections were then incubated in biotinylated anti-mouse IgG-horseradish peroxidase-avidin (Vectastain ABC kit), washed and immunoreactive product detected by treating with NovaRED (Vector Laboratories, Burlingame, California) in TRIS buffer for 10 min. Fragments of control and LCA2 ocular tissue were re-embedded for electron microscopical analyses, and both semi- and ultra-thin sections were obtained using conventional techniques. Ultrastructural observations were performed using a Hitachi H-7500 TEM.

3. RESULTS

3.1. Histopathology

Stained sections of the posterior pole of 26 week normal foetal retinas exhibited a typical stratified appearance with prominent outer, inner nuclear and ganglion cell layers (ONL, INL and GCL respectively), separated by well-defined outer and inner plexiform layers (OPL and IPL respectively). By comparison, the 26 week LCA1 retina carried severe abnormalities. The ONL was reduced to 2-5 rows of nuclei in thickness compared with 8-14 rows in controls, and OS were shorter and the inner segments (IS) disorganized (Figure 1). Quantification of photoreceptor nuclei showed control retina contained seven fold more cells than mutant LCA1 retina in the same posterior field. Within the mutant tissue, the outer limiting membrane and photoreceptor synaptic extensions were not well formed, and the OPL appeared anarchic and sparse. There were also significantly less cells in the INL, although the GCL and nerve fibres had a normal appearance. No apparent abnormality was detected in other ocular tissues, including the RPE.

33 week embryonic normal retina exhibited a typical stratified appearance, with prominent outer and inner nuclear layer, while that of the LCA2 foetus revealed cell loss within the central posterior field, the ONL containing many condensed cellular profiles indicating cell death. The overall thickness of LCA2 retina was reduced compared to age-matched controls, due especially to thinning of the OPL and INL, the latter showing considerable cell loss. The RPE showed numerous inclusions, probably lipid droplets containing all-trans-retinyl esters (Figure 2). Electron microscopic examination of mutant retina showed that the outer limiting layer was formed but photoreceptor IS were disorganized and OS were stunted. Choriocapillaris in normal 33 week fetal eye tissue revealed small vessel profiles distributed close to Bruch's membrane, embedded in loose connective tissue (Porto et al., 2002). The same region in pathological tissue exhibited abundant distended vessels extending into deeper layers.

3.2. Immunohistochemistry

Comparison of DAPI staining in 26 week normal and LCA1 retinas confirmed extensive ONL cell loss. Immunolabeling with anti-rhodopsin antibody showed a typical central to peripheral gradient of rod differentiation in the posterior pole. Immunolabeling was intense in the surface membranes of rod IS and OS, cell bodies and synapses of normal retina (Figure 1). In contrast, LCA1 foetal rod photoreceptors had short OS and restricted rhodopsin immunolabeling, the immunopositive rod population being restricted to a single cell row at the posterior pole. Labeling with PNA and anti-cone arrestin in normal and mutant 26 week retina was intense in cone IS and OS, but showed OS shortening and ONL thinning in the mutant (Figure 1). Cone arrestin immunolabeling stained the somata, axonal extensions and synaptic terminals in normal 26 week foetal retina, while in the LCA1 retina synaptic terminal labeling was completely absent, axonal extensions were small and thin, and cone IS were short or absent. Total arrestin-immunopositive cone numbers were reduced by >50% in the posterior pole in the LCA2 retina. In normal 26 week foetal retina, SVP-38 labeled numerous synaptic terminals of rods and cones within the OPL. The OPL in LCA1 retina was only sparsely labeled by SVP-38. In the normal 33 week foetal retina, immunolabeling with anti-rhodopsin antibody was intense in the surface membranes of developing rod IS and OS, cell bodies and synapses. In contrast, rod photoreceptors of the 33 week LCA2 foetus displayed short OS and restricted rhodopsin labeling (Figure 2). Labeling with anti-PNA in normal and mutant 33 week retina was intense in cone IS and OS and revealed OS shortening in the mutant. Cone arrestin immunolabeling did not show any differences between mutant and control embryonic retinas. Total arrestin-immunopositive cone numbers in the LCA2 retina were only slightly less than controls.

4. DISCUSSION

The principal findings of this study are the marked retinal and photoreceptor degeneration already present in LCA retinas during gestation, observations which were not predicted from existing animal models. In contrast to developing retina in the normal foetus, widespread cell damage and loss was observed in the ONL and INL in both LCA1 and LCA2 mutations. LCA1 showed severe dropout or lack of appearance of photoreceptors by 26 weeks. In cat, monkey, and human retinas, which have a significant number of cones, RetGC-1-like immunoreactivity appears greater in cone than in rod OS (Dizhoor et al., 1994; Liu et al., 1994; Cooper et al., 1995), correlating with the non-recordable ERG and visual fields in patients with autosomal recessive GUCY2D mutations (Perrault et al., 1999). The lack of GUCY2D does not seem to affect the initial differentiation or survival of photoreceptors in RetGC-$1^{-/-}$ mice, since they have morphologically normal rods at least up to one year of age, and cones, initially present in normal numbers, disappear by five weeks, based on ERG and histology (Yang et al., 1999). In *rd* chicken, absence of GUCY2D prevents phototransduction and affects post-hatching survival of rods and cones but does not interfere with normal photoreceptor development (Semple-Rowland et al., 1998). Thus transgenic and spontaneous animal models in which the GUCY2D gene has been inactivated do not display the same pathological phenotype as humans with LCA1 mutations. One possible factor affecting onset of cell damage could be statement of the photoreceptor cGMP-gated cation channel. Immunochemical detection of this channel can not be demonstrated prior to 6 days after birth in the rat, and hence embryonic pertubations in cGMP levels may not induce membrane channel opening and be without visible effect on initial development. It is still possible that the widespread absence at birth of functional retinal photoreceptor cells may be responsible for the subsequent reactive changes observed in animal models and the associated progressive fundoscopic and histopathologic changes observed in patients with advanced stages of LCA.

In the 26 week GUCY2D foetus, changes in the OPL and INL indicate that RetGC-1 may also be linked to retinal synaptic activity. This conclusion is also supported by the immunocytochemical presence of RetGC-1 outside rod OS in the photoreceptor synaptic terminals (Liu et al., 1994) and its ability to be stimulated by micromolar concentrations of Ca+2 (Duda et al., 1998). In contrast, Duda et al. (1998) found that the GUCY2D mutation severely damaged the intrinsic cyclase activity of RetGC-1 and inactivated its CRM1 switch, but did not damage the Ca+2-binding protein-mediated activation of the enzyme. Our data suggest that an early disruption of the normal sequential development of the synaptic organization of the OPL caused by the GUCY2D mutation takes place in the LCA1 retina. The interaction between photoreceptors and bipolar cells required for normal synaptogenesis may also be impaired in LCA1 retina. These findings correlate with a histopathologic study in developing and adult mouse retinas, where another gene responsible for LCA, CRX, is expressed not only in the outer nuclear layer, but also in the bipolar and horizontal cells.

The RPE65 gene is specifically expressed in the RPE and encodes for a highly evolutionary conserved protein (RPE65) present at high levels *in vivo* and essential for the regeneration of rhodopsin after bleaching. We suggest that the disruption of the RPE-based metabolism of vitamin A that appears to underlie the LCA2 phenotype induces severe rod and cone photoreceptor abnormalities in human foetal retina. The ONL showed reduced numbers of nuclei, and rod and cone photoreceptors present disorganized and shorter outer segments in the 33 week LCA2 foetal retina. RPE65-deficient ($RPE65^{-/}$) mice (Redmond et al., 1998), a model of LCA2, exhibit a slow rod degeneration with loss of nuclei and abolished rod function as measured by electroretinography, with disorganized rod photoreceptor OS discs appearing by 7 months. Interestingly cone function remains in these transgenic animals, contrary to what is observed in the human phenotype. Human LCA2 patients have recordable but severely disrupted cone function, compatible with the shorter cone outer segments observed in the 33 week fetus. These differences between cone degeneration in human LCA2 and the mouse model may be due to species-specific differences in cone statement and distribution. The foveal cones have an extreme importance to human central vision, unlike the widely scattered small cones of the mouse. The highly active human foveal cones might be as dependent as rods on RPE for supply of chromophore. Furthermore, in vertebrate eye, the major site of production of 11-cis-retinoids is the RPE, while in some animals production of 11-cis-retinoids is also suspected to occur outside the RPE (Redmond et al., 1998). Importantly, recent animal trials involving either 9-cis retinal oral supplementation in $RPE65^{-/-}$ mice (Van Hooser et al., 2000) or gene therapy trials in the Briard dog (Acland, 2001) were successfully able to restore visual function. Such trials have raised hopes that similar approaches could rescue visual function in LCA2 patients (Gu et al., 1997; Dejneka and Bennett, 2001). Even if the human cones could be regenerated from another available chromophore source besides the RPE, the severe abnormality in rods during photoreceptor development and differentiation may impair the coordinated interaction between rods and cones and induce cone degeneration in human LCA2. The progressive concentric visual field reduction and cone ERG deterioration following massive rod death in LCA2 patients corroborates the conclusion that cone degeneration may occur secondarily, but the current histopathological findings indicate breakdown already starts during gestation. This secondary event might represent another illustration of the dependence on rods of cone survival and of the interaction between rods and cones during photoreceptor development and differentiation. In spite of these differences in cone degeneration, the RPE65 knockout mice present lipid droplets as a hallmark of the disease phenotype and a consequence of the blocked transport of retinoids; similarly the LCA2 foetus presented RPE inclusions which probably correspond to this abnormality, and ultrastructural analyses exhibits empty vesicular structures (Porto et al., 2002). The Briard dog, a model for congenital stationary night blindness due to a functional null allele of canine RPE65 gene, also has lipid vacuoles in RPE cells (Wrigstad et al., 1994).

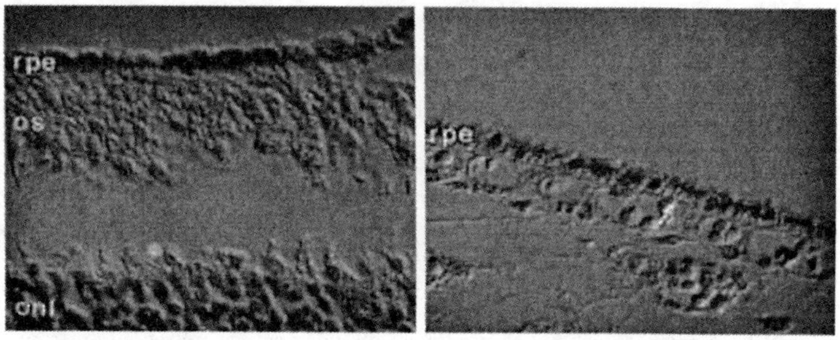

Figure 3. Left: Control fetus normal RPE. **Right:** The RPE present lipid droplets, probably the hallmark of the disease phenotype and a consequence of the blocked transport of retinoids; similarly the RPE65-/- mice also presented the RPE vesicles in the RPE cells.

Embryonic retinal degeneration in LCA2 mutations suggests chromophore addition occurs from the earliest time points and is critical for photoreceptor development and survival. Several lines of evidence have pointed that there is a critical period in photoreceptor development in which rhodopsin exerts an essential structural role. Drosophila rhodopsin null mutants show a catastrophic failure of rhabdomere morphogenesis, resulting in deep, irregular membrane intrusions (Kumar and Ready, 1995). Mice homozygous for a targeted rhodopsin gene disruption are deficient in both opsin and rhodopsin, lacking rod OS 27 days postnatally and losing their photoreceptors over 3 months (Humphries et al., 1997). Whereas both $Rho^{-/-}$ and $RPE65^{-/-}$ mutant mouse rod photoreceptors are non-functional, the OS-lacking photoreceptors in the former degenerate far faster than OS-bearing photoreceptors.

In conclusion, our findings are relevant to improve the current understanding of disease pathobiology and to establish clinicopathologic relations. LCA1 and LCA2 are associated with abnormal synaptogenesis and survival of photoreceptors. RetGC-1 might be essential not only for normal maintenance of photoreceptors (Porto et al., 2002), as demonstrated by mutations causing autosomal dominant cone-rod dystrophy, but also possibly essential for early photoreceptor development and connectivity. RPE65 also appears functionally critical for normal photoreceptor development. These data conflict with the normal histology of currently available avian and rodent models, and caution against directly extrapolating experimental observations obtained from animal studies towards the human disease state.

5. ACKNOWLEDGEMENTS

We gratefully thank Manuel Simonutti and Noëlle Hanoteau for expert technical assistance, Serge Picaud, Hélène Dollfus and Ghislaine Hirth for their helpful discussions and advice, and Fréderic Stoeckel and Dalila Grampp for the photographic work.

6. REFERENCES

Acland G.M., 2001, Gene therapy restores vision in a canine model of childhood blindness. *Nature Genet.* **28**:92-95.

Camuzat A. *et al.*, 1996, Evidence of genetic heterogeneity of Leber's congenital amaurosis (LCA) and mapping of LCA1 to chromosome 17p13. *Hum Genet.* **97**:798-801.

Cooper N, Liu L, Yoshida A, Pozdnyakov N, Margulis A, Sitaramayya A, 1995, The bovine rod outer segment guanylate cyclase, ROS-GC, is present in both outer segment and synaptic layers of the retina. *J Mol Neurosci.* **6**:211-222.

Dejneka, NS & Bennett, J, 2001, Gene therapy and retinitis pigmentosa: advances and future challenges. *Bioessays* **23**:662-668.

Dizhoor AM, Lowe DG, Olshevskaya EV, Laura RP, Hurley JB, The human photoreceptor membrane guanylyl cyclase, RetGC, is present in outer segments and is regulated by calcium and a soluble activator. *Neuron* **12**:1345-1352.

Duda T, Venkataraman V, Goraczniak R, Lange C, Koch KW, Sharma RK, 1998, Functional consequences of a rod outer segment membrane guanylate cyclase (ROS-GC1) gene mutation linked with Leber's congenital amaurosis. *Bioch.* **38**:509-515.

Gu S.M. *et al.*, 1997, Mutations in RPE65 cause autosomal recessive childhood-onset severe retinal dystrophy. *Nature Genet.* **17**:194-197.

Humphries M.M. *et al.*, 1997, Retinopathy induced in mice by targeted disruption of the rhodopsin gene. *Nat Genet* **15**:216-219.

Kaplan J, Bonneau D, Frezal J, Munnich A, Dufier JL, 1990, Clinical and genetic heterogeneity in retinitis pigmentosa. *Hum Genet.* **85**:635-642.

Kumar JP and Ready, DF, 1995, Rhodopsin plays an essential structural role in *Drosophila* photoreceptor development. *Development* **121**:4359-4370.

Lathrop GM et al., 1985, Multilocus linkage analysis in humans: detection of linkage and estimation of recombination. Am J Hum Genet. **37**:482-498.

Liu X, Seno K, Nishizawa Y, Hayashi F, Yamazaki A, Matsumoto H, Wakabayashi T, Usukura J, 1994, Ultrastructural localization of retinal guanylate cyclase in human and monkey retinas. *Exp. Eye Res.* **59**:761-768.

Perrault, I *et al.*, 1996, Retinal-specific guanylate cyclase gene mutations in Leber congenital amaurosis. *Nature Genet.* **14**:461-464.

Perrault I, Rozet JM, Ghazi I, Leowski C, Bonnemaison M, Gerber S, Ducroq D, Cabot A, Souied E, Dufier JL, Munnich A, Kaplan J, 1999, Different functional outcome of RetGC1 and RPE65 gene mutations in Leber congenital amaurosis. *Am J Hum Genet.* **64**:1225-1228.

Porto FBO, Perrault I, Hicks D *et al.*, 2002, Prenatal human ocular degeneration occurs in Leber's congenital amaurosis (LCA2). *J Gene Med.* **4**,390-396.

Redmond, TM *et al.*, 1998, Rpe65 is necessary for production of 11-cis-vitamin A in the retinal visual cycle. *Nature Genet.* **20**:344-351.

Semple-Rowland SL, Lee NR, Van Hooser JP, Paazewski K and Bachr W, 1998, A null mutation in the photoreceptor guanylate cyclase gene causes the *retinal degeneration* chicken phenotype. *Proc Natl Acad Sci USA.* **95:**1271-1276.

Van Hooser J.P. *et al.*, 2000, Rapid restoration of visual pigment and function with oral retinoid in a mouse model of childhood blindness. *Proc Natl Acad Sci USA.* **97**:8623-8628.

Wrigstad A. *et al.*, 1994, Slowly progressive changes of the retina and retinal pigment epithelium in briard dogs with hereditary retinal dystrophy. A morphological study. *Doc Ophthalmol.* **87**:337-354.

Yang RB, Robinson SW, Xiong WH, Yan KW, Birch DG, Garbers DL, 1999, Disruption of a retinal guanylyl cyclase gene leads to cone-specific dystrophy and paradoxical rod behavior. *J Neurosci.* **19**:5889-97.

LEBER CONGENITAL AMAUROSIS - GENOTYPING REQUIRED FOR POSSIBLE INCLUSION IN A CLINICAL TRIAL

Isabelle Perrault[1], Sylvie Gerber[1], Sylvain Hanein[1], Serge Picaud[2], Jean-Michel Rozet[1], Jean-Louis Dufier[3], Arnold Munnich[1], José Sahel[2], and Josseline Kaplan[1]

1. INTRODUCTION

The genetic heterogeneity of LCA no longer needs to be demonstrated. Indeed, the six already known disease-causing genes and the two additional LCA loci only account for 45% of all patients. To this tremendous genetic heterogeneity corresponds an almost equivalent physiopathological heterogeneity and to a minor extent a clinical variability. Therefore, the identification of LCA-causing genes and subsequent genotyping is an obligatory prerequisite to classify patients and include them in clinical trials. For this reason, we have collected several consanguineous multiplex families affected with LCA and performed a genome-wide search for homozygosity. Three distinct chromosomal regions, different from the 8 already known LCA loci, have been identified for three unrelated families. The candidate genes lying in these regions are currently analyzed. On the other hand, very recently, encouraging therapeutic trials have been performed in order to propose gene therapy to the RPE65 subgroup. Nevertheless, to date in Europe, the major LCA gene is GUCY2D. It is why it seems important to us to develop a therapeutic approach for this group. A murine retinal explant model of pharmacological-induced GUCY2D deficiency is currently developed in our group. This model will be of primary importance to screen different drugs for their ability to slow, stop or reverse the phenotype.

2. GENETIC HETEROGENEITY OF LCA

Since 1996, six genes involved in LCA have been identified: the *guanylate cyclase 2D* (*GUCY2D*, *RETGC1*, *Perrault et al., 1996*), *RPE65* (*Marlhens et al., 1997*), *the*

[1] Inserm U393, Hôpital Necker Enfants Malades, Paris, France, [2] Inserm U592, Hôpital Saint Antoine, Paris , France, [3] Service d'ophthalmologie, Hôpital Necker, Paris France. Corresponding author J Kaplan, E-mail: kaplan@necker.fr

Retinal Degenerations: Mechanisms and Experimental Therapy
Edited by LaVail *et al.*, Kluwer Academic/Plenum Publishers, 2003

cone–rod homeobox (*CRX, Freund et al., 1998*), the *aryl hydrocarbon receptor-interacting protein-like 1* (*AIPL1,Sohocki et al., 2000*), the *Crumbs homolog 1* (*CRB1, den Hollander et al., 2001; Lotery et al., 2001*), and the *retinitis pigmentosa GTPase regulator-interacting protein 1* (*RPGRIP1*, Dryja et al., 2001; Gerber et al., 2001). In addition, linkage studies have revealed LCA loci on 14q24 (LCA3, Stockton et al., 1998) and 6q11-q16 (LCA5, Dharmaraj et al., 2000).

These latter are all involved in different physiological pathways . GUCY2D is essential for the recovery of the dark state after the excitation process of photoreceptor cells by light stimulation (Perrault et al., 1999). RPE65 has a crucial role in the metabolism of vitamin A in the visual cycle (Redmond et al., 1998). CRX is a homeobox transcription factor that plays a crucial role in the differentiation and maintenance of photoreceptor cells (Furukawa et al., 1998). AIPL1 may be involved in retinal protein folding or trafficking (van der Spuy et al., 2002).Crumbs is required to maintain the integrity of the ZA in photoreceptor cells (Izaddoost et al.; Pellikka et al.). RPGRIP1 is an interactor of the retinitis pigmentosa GTPase regulator (RPGR, Roepman et al., 2000; Hong et al., 2001).

We performed a screening for mutations in the 6 cloned LCA-genes using DHPLC followed by direct sequencing in 175 unrelated LCA patients hailing from countries worldwide (see figure 1, Perrault et al., 2000). We found mutations in GUCY2D 20%, RPE65 6%, CRX less than 1%, AIPL1 5%, RPGRIP1 7%, CRB1 7%, respectively.
The total represents only 45% of all cases in our series.

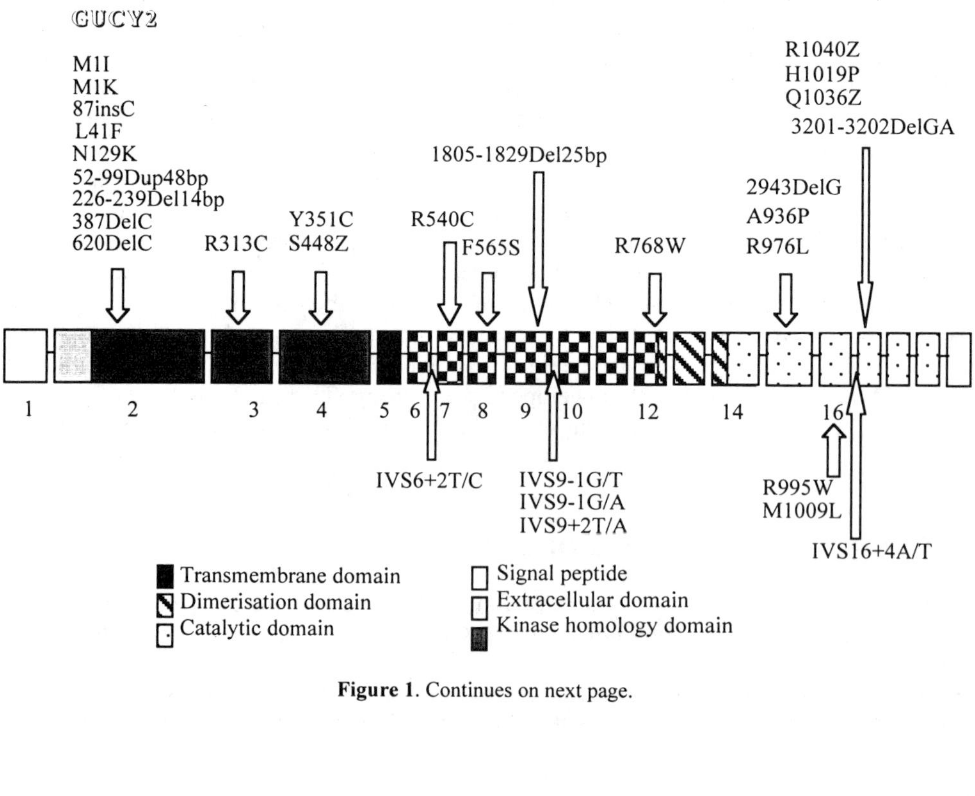

Figure 1. Continues on next page.

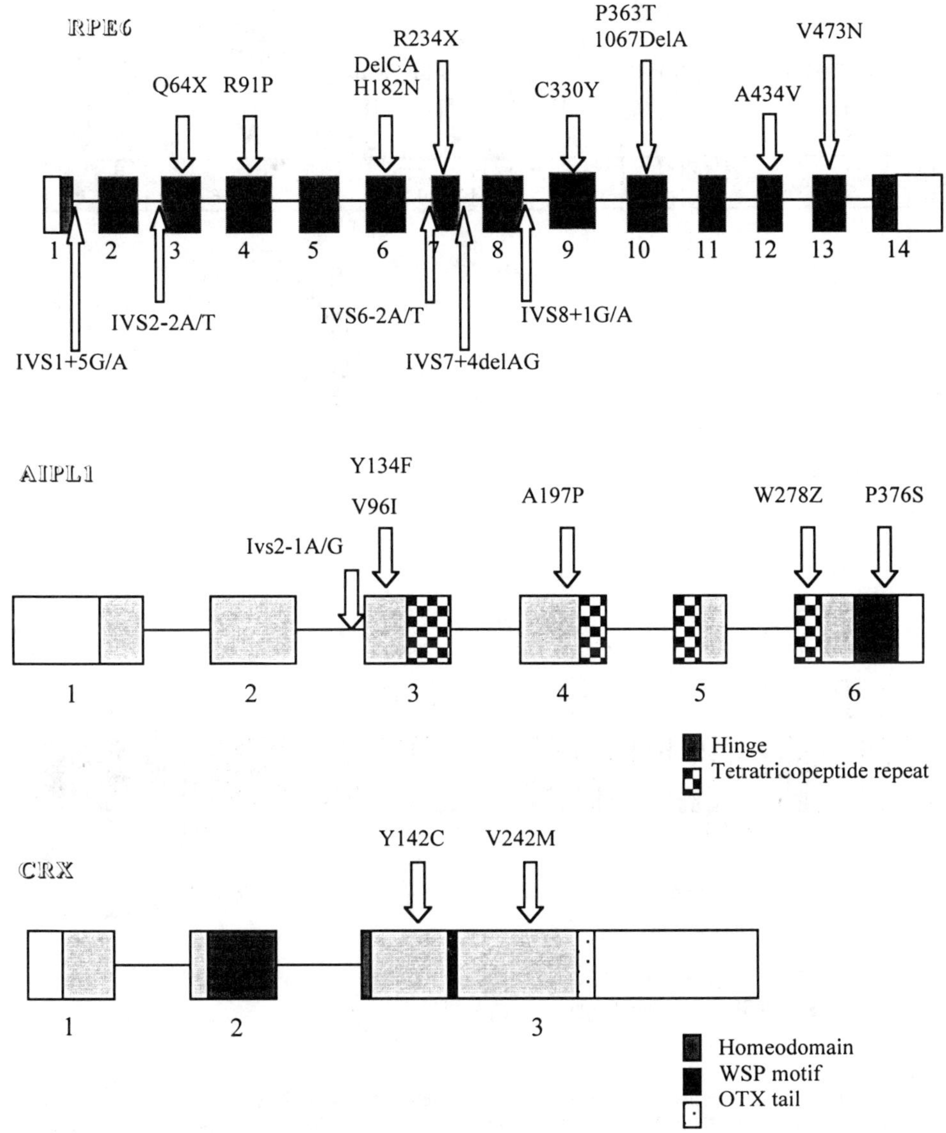

Figure 1. Continues on next page.

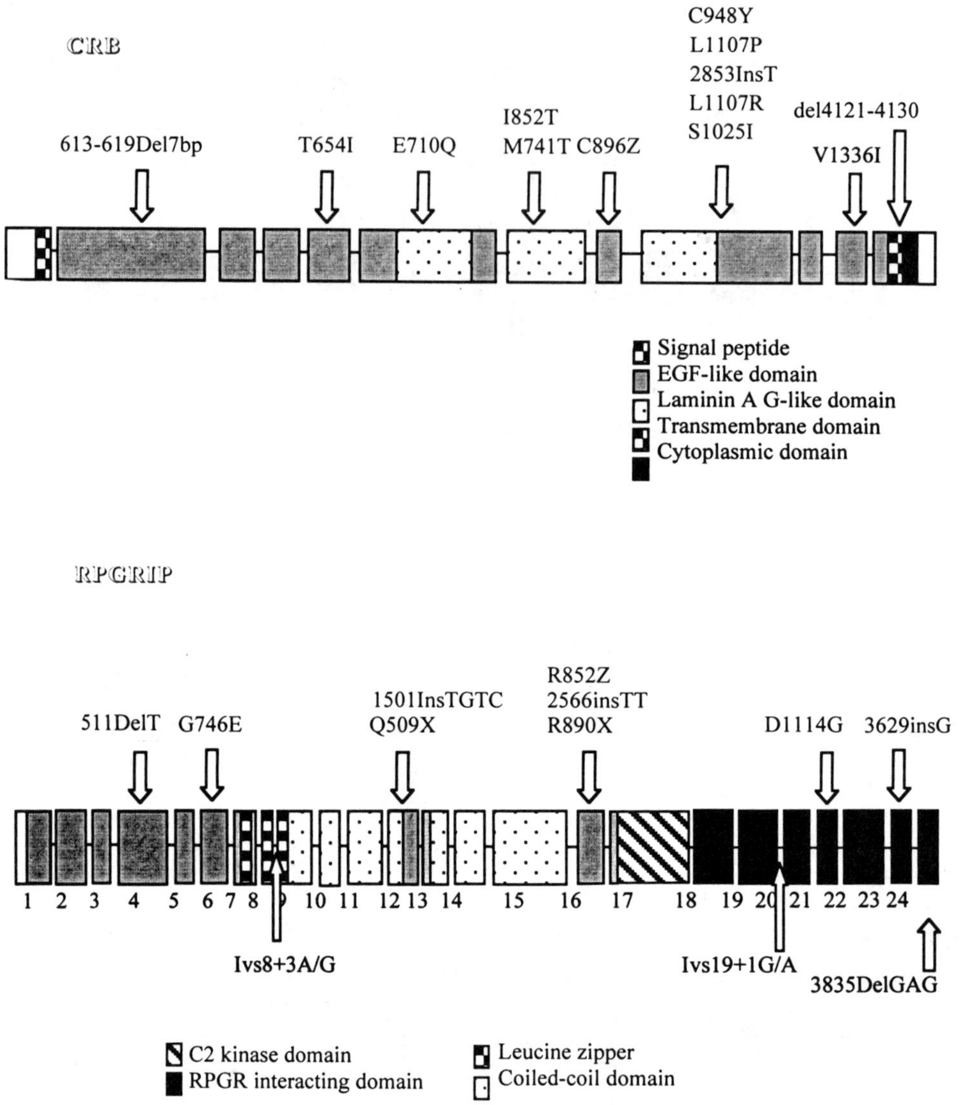

Figure 1. Schematic representation of the intron-exon structure of the six LCA genes and associated mutations in our series. (Parts of Figure 1 for GUCY2, RPE6, AIPL1 and CRX are located on the two preceding pages.)

Table 1. For each LCA gene, the number of LCA families harbouring mutations is indicated.

Gene	Families Compound heterozygote	Families Single heterozygote or homozygote	Total
Gucy2d	32	3	35
Rpe65	9	3	12
Crb1	9	1	10
Rpgrip1	4	5	9
Crx	0	2	2
Aipl1	4	3	7
TOTAL	58	**17**	75

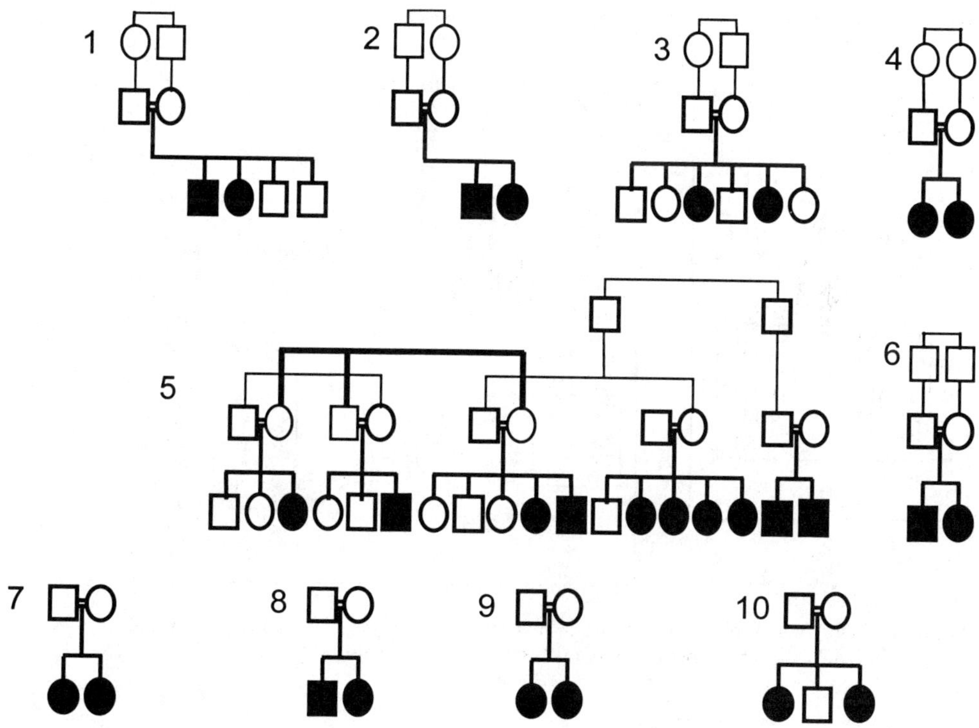

Figure 2. Consanguineous and multiplex LCA families in which a whole genome-wide search for homozygosity has been performed.

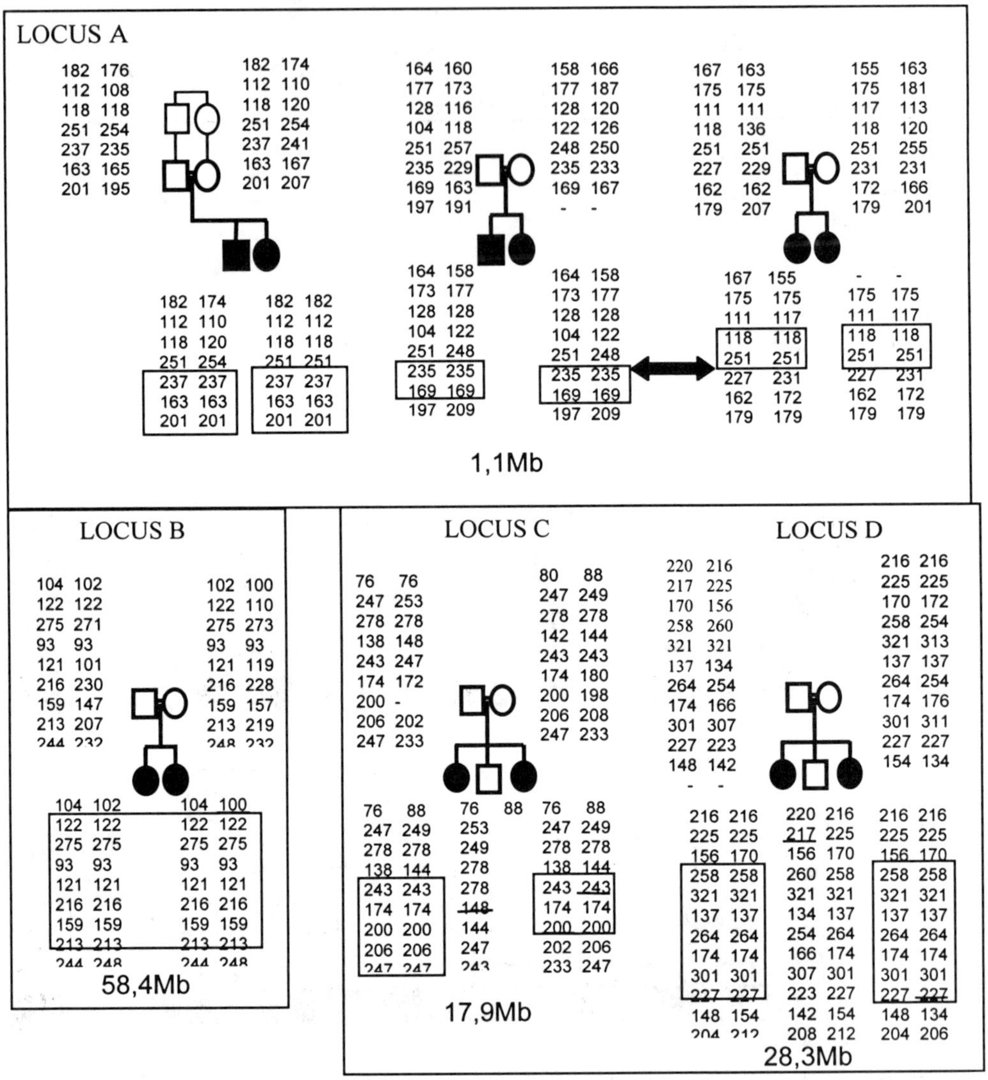

Figure 3. Homozygote haplotypes identified in five unrelated LCA families unlinked to any known LCA genes. Families 2, 7, 8, 9, 10 point to four distinct chromosomal regions (loci A, B, C/D, respectively).

Of the 75 unrelated patients found to carry mutations in one of the 6 cloned LCA genes, 17 were found to be single heterozygote, although the whole coding sequence as well as the intron-exon boundaries (30bp up to the junction) were sequenced in all cases.

The dominant inheritance is unlikely (all mutations being inherited from a healthly parent in all cases)

The unidentified mutations might lie in unexplored regions : non described exons, promoter, introns, or close vicinity of the gene (*eg* already described for aniridia).

It is also possible that a large deletion can exist, undetectable when using PCR-based methodologies.

3. IDENTIFICATION OF NEW LCA LOCI

In 2000 : To identify new LCA loci, a genome wide search was undertaken in 10 LCA families multiplex and consanguineous (Figure 2).

Families 1 and 4 allowed us to identify RPGRIP1 as LCA-causing gene. Families 3 and 5 allowed us to identify CRB1 as LCA-causing gene (Gerber et al., 2002; Figure 2).

The candidate genes lying in the A (Families 2, 7,8), B (Family 9), C and D (Family 10) chromosomal regions are currently analysed.

4. DISCUSSION AND CONCLUSION

The results of the study reported here have three main interests :

1- They confirm the outstanding genetic heterogeneity of the disease of which three main considerations emerged from:

-RETGC1 is unambiguously the major gene in our series : 20 % of all cases (35 families world-wide). It is probably why it has been identified first.

-The respective frequencies of the implication of the 5 other LCA genes are almost equivalent and equal to 7%.

-All together the six LCA genes (2002) only account for 46% of all cases

For the remaining families, the search for homozygosity in consanguineous families failed to detect a major locus. By the way, our recent results showed that only one or two loci were compatible for linkage in only one or two families suggesting that a lot of LCA causing genes remain to be identified.

2- The study of about 200 unrelated patients allowed to give a new definition of Leber congenital amaurosis. Indeed, Theodore Leber described a stationary congenital blindness; This description still matches patients harbouring mutations in retGC1 and possibly in RPGRIP1 and AIPL1.

On the other hand, with the passing years, childhood-onset severe retinal dystrophies (CSRD) with onset in the infancy, have been included in the so –called Leber congenital amaurosis. It is particularly the case of patients harbouring mutations in RPE65, CRX or CRB1.

These findings led to reconsider LCA as the extreme end of a spectrum of severity of RP.

3- The identification of the molecular bases of LCA as well as the new clinical definition of the disease allow in most cases to

- anticipate the final outcome in a blind infant.
- to direct genetic studies in older patients.

5. DEVELOPMENT OF A CELLULAR MODEL OF GUCY2D DEFICIENCY IN MOUSE

Considering the high frequency of retGC1 mutations in our series of patients hailing from countries world-wide, it seems essential, in our point of view, to develop a therapeutic approach for this group of patients.

To study the physiopathology of photoreceptors carrying mutation in retGC1, we will block the visual transduction cascade using pharmacological agents.

We are currently developing a cellular model of pharmalogically-induced GUCY2D deficiency in mouse. The effect of the treatment of the mouse retinal explants will be investigated by studying the morphology of the retina and by evaluating the cell death in the tissue. The morphology will be analysed histologically using sections of the treated retinas and by a stereotopic counting of the different retinal cells. The cell death will be investigated by searching DNA fragmentation (Tunel) and by studying the cascade of caspases. In addition functionnal analyses will be performed after treatment by micro ERG. This approach is currently in development in the laboratory of Professor Sahel (Frasson et al., 1999).

This model will be of primary importance to screen different drugs for their ability to slow, stop or reverse the phenotype.

6. ACKNOWLEDGEMENTS

This work was supported by Association Retina France and the Foundation Fighting Blindness. We thank the National Eye Institute and Foundation Fighting Blindness for the travel award provided to IP to attend this meeting.

7. REFERENCES

den Hollander, A.I., Heckenlively, J.R., van den Born, L.I., de Kok, Y.J.M., van der Velde-Visser, S.D., Kellner, U., Jurklies, B., van Schooneveld, M.J., Blankenagel, A., Rohrschneider, K. *et al.*, 2001, Leber congenital amaurosis and retinitis pigmentosa with Coats-like exudative vasculopathy are associated with mutations in the Crumbs homologue 1 (*CRB1*) gene. *Am. J. Hum. Genet.* **69**:198–203.

Dharmaraj, S., Li, Y., Robitaille, J.M., Silva, E., Zhu, D., Mitchell, T.N., Maltby, L.P., Baffoe-Bonnie, A.B. and Maumenee, I.H., 2000, A novel locus for Leber congenital amaurosis maps to chromosome 6q. *Am. J. Hum. Genet.* **66**:319–326.

Dryja, T.P., Adams, S.M., Grimsby, J.L., McGee, T.L., Hong, D.-H., Li, T., Andréasson, S. and Berson, E.L., 2001, Null *RPGRIP1* alleles in patients with Leber congenital amaurosis. *Am. J. Hum. Genet.* **68**:1295–1298.

Frasson, M., Sahel, J.A., Fabre, M., Simonutti, M., Dreyfus, H. and Picaud, S., 1999, Retinitis pigmentosa: rod photoreceptor rescue by a calcium-channel blocker in the rd mouse. *Nat. Med.*, **5**,1183–1187

Freund, C.L., Wang, Q.-L., Chen, S., Muskat, B.L., Wiles, C.D., Sheffield, V.C., Jacobson, S.G., McInnes, R.R., Zack, D.J. and Stone, E.M., 1998, *De novo* mutations in the *CRX* homeobox gene associated with Leber congenital amaurosis. *Nat. Genet.*, **18**:311–312.

Furukawa, T., Morrow, E.M. and Cepko, C.L., 1997, Crx, a novel otx-like homeobox gene, shows photoreceptor-specific expression and regulates photoreceptor differentiation. *Cell* **91**:531–541

Gerber, S., Perrault, I., Hanein, S., Barbet, F., Ducroq, D., Ghazi, I., Martin-Coignard, D., Leowski, C., Homfray, T., Dufier, J.L. *et al.*, 2001, Complete exon-intron structure of the RPGR-interacting protein (*RPGRIP1*) gene allows the identification of mutations underlying Leber congenital amaurosis. *Eur. J. Hum. Genet.* **9**:561–571

Gerber S, Perrault I, Hanein S, Shalev S, Zlotogora J, Barbet F, Ducroq D, Dufier JL, Munnich A, Rozet JM, Kaplan J., 2002, A novel mutation disrupting the cytoplasmic domain of CRB1 in a large consanguineous family of Palestinian origin affected with Leber congenital amaurosis. *Ophthal. Genet.* **22**:3.

Hong, D.H., Yue, G., Adamian, M. and Li, T., 2001, Retinitis pigmentosa GTPase regulator (RPGR)-interacting protein is stably associated with the photoreceptor ciliary axoneme and anchors RPGR to the connecting cilium. *J. Biol. Chem.* **276**:12091–12099

Izaddoost, S., Nam, S.-C., Bhat, M.A., Bellen, H.J. and Choi, K.-W.,2002, *Drosophila* Crumbs is a positional cue in photoreceptor adherens junctions and rhabdomeres. *Nature* **416**:178–183.

Lotery, A.J., Jacobson, S.G., Fishman, G.A., Weleber, R.G., Fulton, A.B., Namperumalsamy, P., Heon, E., Levin, A.V., Grover, S., Rosenow, J.R. *et al.*, 2001, Mutations in the *CRB1* gene cause Leber congenital amaurosis. *Arch. Ophthalmol.* **119**:415–420.

Marlhens, F., Bareil, C., Griffoin, J.-M., Zrenner, E., Amalric, P., Eliaou, C., Liu, S.-Y., Harris, E., Redmond, T.M., Arnaud, B. *et al.*, 1997, Mutations in *RPE65* cause Leber's congenital amaurosis. *Nat. Genet.*,**17**:139–141.

Pellikka, M., Tanentzapf, G., Pinto, M., Smith, C., McGlade,CJ., Ready, D.F. and Tepass, U., 2002, Crumbs, the *Drosophila* homologue of human CRB1/RP12, is essential for photoreceptor morphogenesis. *Nature* **416**:143–149.

Perrault, I., Rozet, J.M., Calvas, P., Gerber, S., Camuzat, A., Dollfus, H., Châtelin, S., Souied, E., Ghazi, I., Leowski, C. *et al.*, 1996, Retinal-specific guanylate cyclase gene mutations in Leber's congenital amaurosis. *Nat. Genet.* **14**:461–464.

Perrault, I., Rozet, J.-M., Gerber, S., Ghazi, I., Leowski, C., Ducroq, D., Souied, E., Dufier, J.-L., Munnich, A. and Kaplan, J., 1999, Leber congenital amaurosis. *Mol. Genet. Metab.* **68**:200–208

Perrault, I., Rozet, J.M., Gerber, S., Ghazi, I., Ducroq, D., Souied, E., Leowski, C., Bonnemaison, M., Dufier, J.L., Munnich, A. *et al.*, 2000, Spectrum of *retGC1* mutations in Leber's congenital amaurosis. *Eur. J. Hum. Genet.* **8**:578–582.

Redmond, T.M., Yu, S., Lee, E., Bok, D., Hamasaki, D., Chen, N., Goletz, P., Ma, J.-X., Crouch, R.K. and Pfeifer, K., 1998, *Rpe65* is necessary for production of 11-*cis*-vitamin A in the retinal visual cycle. *Nat. Genet.* **20**:344–351.

Roepman, R., Bernoud-Hubac, N., Schick, D., Maugeri, A., Berger, W., Ropers, H.-H., Cremers, F.P.M. and Ferreira, P.A., 2000, The Retinitis Pigmentosa GTPase Regulator (RPGR) interacts with novel transport-like proteins in the outer segments of rod photoreceptors. *Hum. Mol. Genet.* **9**: 2095–2105.

Sohocki, M.M., Bowne, S.J., Sullivan, L.S., Blackshaw, S., Cepko, C.L., Payne, A.M., Bhattacharya, S.S., Khaliq, S., Mehdi, S.Q., Birch, D.G. *et al.*, 2000, Mutations in a new photoreceptor-pineal gene on 17p cause Leber congenital amaurosis. *Nat. Genet.* **24**:79–83

Stockton, D.W., Lewis, R.A., Abboud, E.B., Al Rajhi, A., Jabak, M., Anderson, K.L. and Lupski, J.R., 1998, A novel locus for Leber congenital amaurosis on chromosome 14q24. *Hum. Genet.* **103**:328–333.

van der Spuy J, Chapple JP, Clark BJ, Luthert PJ, Sethi CS, Cheetham ME., 2002, The Leber congenital amaurosis gene product AIPL1 is localized exclusively in rod photoreceptors of the adult human retina. *Hum Mol Genet*. **11**:823-31

TREATMENT OF CYSTOID MACULAR EDEMA RELATED TO RETINITIS PIGMENTOSA WITH INTRAVITREAL TRIAMCINOLONE ACETONIDE: CASE REPORT

Juliana M. F. Sallum, Michel E. Farah, and Vinicius S. Saraiva[1]

1. INTRODUCTION

Cystoid macular edema is an important cause of central visual loss in patients suffering from retinitis pigmentosa (Weleber et al., 2001). A few treatment modalities have been described, including the use of carbonic anhydrase inhibitors and grid laser photocoagulation (Fetkenhour et al., 1977).

Although intravitreal triamcinolone acetonide has been successfully employed to treat macular edema related to several diseases, no previous reports have described its use in retinitis pigmentosa complicated by cystoid macular edema.

2. PURPOSE

To describe a case of cystoid macular edema related to retinitis pigmentosa treated with intravitreal injection of triamcinolone acetonide.

3. CASE REPORT

The patient was a thirty-year old, white male with retinitis pigmentosa, complaining of progressive visual loss during the last 12 months.

Best corrected visual acuity was 20/40 in the right eye and 20/80 in the left eye. Biomicroscopy of the anterior segment was unremarkable.

Fundus findings included temporal optic disc pallor, arteriolar attenuation, equatorial and peripheral hypopigmentation and cystoid macular edema (confirmed by fluorescein angiography and optic coherence tomography) in both eyes..

After being treated with a course of oral acetazolamide (250 mg, bid, for 3 weeks) without clinical improvement, intravitreal injection of 0.1 ml of triamcinolone acetonide

[1]Department of Ophthalmology, Federal University of São Paulo. R. Botucatu 822, São Paulo, SP, Brazil 04023-062. Corresponding author: J. M. F. Sallum, E-mail: juliana@pobox.com

(40 mg/ml solution) was performed bilaterally. The left eye was treated five months prior to the right eye.

4. RESULTS

In the left eye, 18 days after injection, visual acuity improved to 20/50. Optic coherence tomography, performed on the 40th day, showed no macular edema and re-establishment of foveal depression (Figure 1. A). The results were stable up to the sixth month post-injection, when examination disclosed 20/200 visual acuity and recurrence of cystoid macular edema.

In the right eye, 30 days after injection, cystoid macular edema resorbed, but visual acuity was unchanged (Figure 1. B). No recurrence was detected up to the third month of follow-up.

Ocular pressure increase and cataract formation were not observed in either eye.

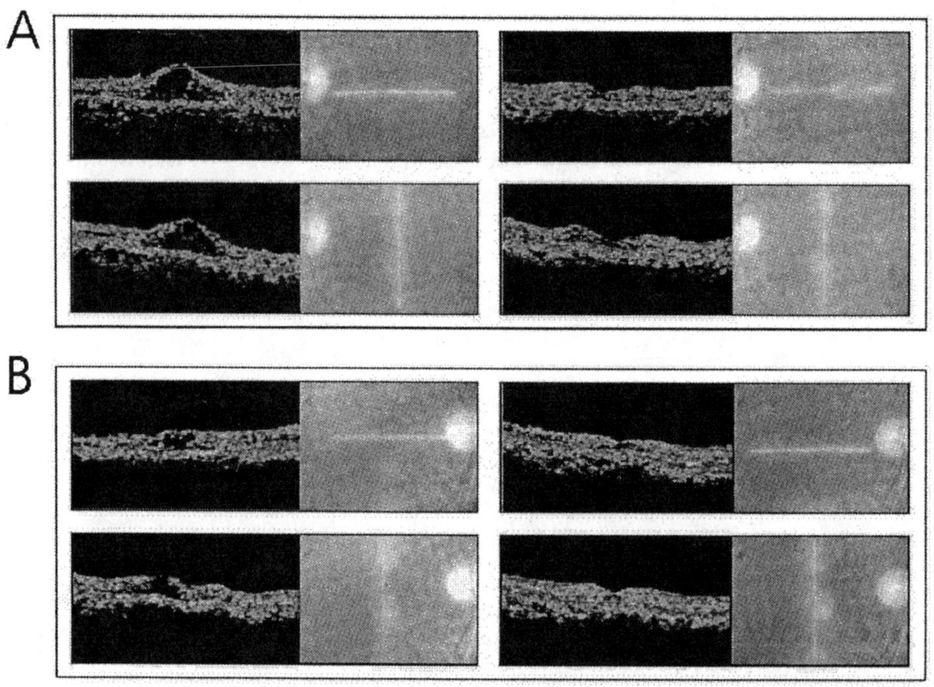

FIGURE 1
A. Optical coherence tomography of the left eye. Horizontal scan (top left) and vertical scan(bottom left) before injection. Horizontal scan (top right) and vertical scan(bottom right) 40 days after intravitreal injection of triamcinolone acetonide. **B.** Optical coherence tomography of the right eye. Horizontal scan (top left) and vertical scan(bottom left) before injection. Horizontal scan (top right) and vertical scan(bottom right) 30 days after intravitreal injection of triamcinolone acetonide.

5. DISCUSSION

Cystoid macular edema in retinitis pigmentosa is a consequence of the blood-retinal barrier and retinal pigment epithelium dysfunction with increased vascular permeability and fluid leakage (Cunha-Vaz and Travassos: 1984; Vinores et al., 1995; Vinores, et al., 1999). The high prevalence of antiretinal auto-antibodies in this subset of patients indicates an inflammatory, autoimmune process playing a role in the pathogenesis of such cases (Heckenlively et al., 1999).

Triamcinolone acetonide is a potent, long-acting steroid drug, capable of inhibiting inflammation, improving blood-retinal barrier status and decreasing vascular permeability and leakage. Intravitreal delivery enhances its performance and decreases systemic side effects. The main ocular complications with this route of administration are secondary glaucoma and cataract.

The left eye was treated first because of poorer visual acuity. Initially successful results encouraged treatment of the fellow eye, but the effect of intravitreal triamcinolone acetonide was temporary and edema recurred 6 months after injection.

In the right eye, despite good anatomical results, visual acuity improvement was not observed. Up to the third month of follow-up, no recurrence was detected.

Ocular complications were absent in this case during the follow-up period.

Additional studies are necessary to evaluate efficacy, indications, repeated injections for recurrences and complications.

6. CONCLUSION

Intravitreal treatment with triamcinolone acetonide may be useful for selected cases of cystoid macular edema related to retinitis pigmentosa.

7. REFERENCES

Cunha-Vaz JG, Travassos A, 1984, Breakdown of the blood-retinal barriers and cystoid macular edema. *Surv Ophthalmol.* **28**:485-92.

Fetkenhour CL, Choromokos E, Weinstein J et al., 1977, Cystoid macular edema in retinitis pigmentosa. *Trans Am Acad Ophthalmol Otolaryngol.* **83**:515-21.

Heckenlively JR, Jordan BL, Aptsiauri N, 1999, Association of antiretinal antibodies and cystoid macular edema in patients with retinitis pigmentosa. Am J Ophthalmol. **127**:565-73.

Vinores SA, Kuchle M, Derevjanik NL, Henderer JD, Mahlow J, Green WR, Campochiaro PA., 1995, Blood-retinal barrier breakdown in retinitis pigmentosa: light and electron microscopic immunolocalization. *Histol Histopathol.* **10**:913-23.

Vinores SA, Derevjanik NL, Ozaki H, Okamoto N, Campochiaro PA, 1999, Cellular mechanisms of blood-retinal barrier dysfunction in macular edema. *Doc Ophthalmol.* **97**:217-28.

Weleber RG, Gregory-Evans K, Retinitis Pigmentosa and Allied Disorders, 2001, In Ryan, SJ: Retina, 3rd Edition, Volume 1, St Louis, Mosby, 362-460.

PROTEOMIC APPROACHES TO UNDERSTANDING AGE-RELATED MACULAR DEGENERATION

Joe G. Hollyfield[1], Robert G. Salomon[2], and John W. Crabb[1]

1. SUMMARY

Microdissection methods have been developed for isolating drusen and Bruch's membrane from human eyes. Comparative proteomic studies of these isolates from normal and AMD donors were pursued for clues to the biochemical pathways involved in the pathogenesis of AMD. A total of 129 potential drusen proteins were identified by LC MS/MS and immunocytochemical analyses have confirmed drusen localization for ~16% of the proteins. The most common drusen proteins appear to be TIMP-3, clusterin, vitronectin and serum albumin. Western blot analysis suggests that carboxyethyl pyrrole-protein adducts derived from docosahexaenoate-containing lipids are more abundant in AMD than in normal tissues. Abnormal protein cross-links and advanced glycation end products were also observed in drusen and Bruch's membrane. Lipid oxidation products and oxidative protein modifications may be causally involved in drusen formation and Bruch's membrane thickening.

2. INTRODUCTION

Drusen are extracellular deposits that accumulate on Bruch's membrane below the retinal pigment epithelium during aging. These deposits can be observed during a routine eye exam and are referred to as either "soft drusen" or "hard drusen", depending on their relative size and shape. Hard drusen are nodular, with distinct borders while soft drusen are typically larger, more globular and irregular in shape. Normal eyes may have maculas free of drusen yet have abundant drusen in the retinal periphery. The presence of numerous and/or confluent, soft drusen in the macula is considered a major risk factor for developing age-related macular degeneration (AMD) and the loss of central vision. Late stage AMD, known as "wet" AMD, is accompanied with new blood vessel growth from the choroid into the retina (choroidal neovascularization) while the earlier stage, "dry"

[1]Ophthalmic Research, Cole Eye Institute, Cleveland Clinic Foundation, Cleveland, OH 44195, USA; [2]Department of Chemistry, Case Western Reserve University, Cleveland, OH 44106, USA. Corresponding author: J. G. Hollyfield, E-mail: hollyfj@ccf.org

AMD is characterized by progressively more macular drusen and geographic atrophy. Drusen formation appears to be associated with the progression of AMD, therefore a better understanding of drusen biogenesis may lead to the development of therapeutic agents that slow or prevent AMD. To this end we have initiated efforts to determine the protein and lipid composition of drusen isolated from normal and AMD donor eyes (Crabb et al., 2002).

3. MATERIALS AND METHODS

Normal eyes were obtained through the Cleveland Eye Bank, (Cleveland, OH), or the National Disease Research Interchange (Philadelphia, PA). AMD donor eyes were from clinically diagnosed patients registered through the Eye Donor Program of the Foundation Fighting Blindness, Inc. Using normal and AMD eyes from donors aged 70 years and older, we developed a procedure for microdissecting drusen directly from the surface of Bruch's membrane with fine forceps.

After lipid extraction with chloroform/methanol, the aqueous fraction of the drusen preparations were dried, resuspended in 8 M urea, and cysteine residues were reduced and alkylated. The preparation was then digested overnight with trypsin (100 ng) at 37°C (Crabb et al., 1998). Soluble tryptic peptides were analyzed by LC MS/MS using a CapLC system (Micromass) and a quadrupole time-of-flight mass spectrometer (QTOF2, Micromass, Beverly, MA) (Miyagi et al., 2002). Protein identifications from MS/MS data utilized ProteinLynx™ Global Server and MassLynx™, software (Micromass), and the Swiss-Protein and NCBI protein sequence databases. Identifications from peptide mass mapping utilized a Voyager DE Pro MALDI-TOF mass spectrometer (PE Biosystems, Framingham, MA) (West et al., 2001).

Immunocytochemical and Western analyses of Bruch's membrane preparations to confirm drusen protein localization and to probe for oxidative protein modifications were performed as detailed elsewhere (Crabb et al., 2002).

4. RESULTS

Drusen Isolation. Frozen eyes used for drusen isolation were thawed overnight at 4°C. After dissecting the retina from the globe, drusen appear as bump-like elevations of the RPE when viewed under a dissecting microscope. The RPE was removed from the interior surface of the globe with gentle brushing, revealing the drusen as opaque deposits attached to Bruch's membrane. The choroid-Bruch's membrane complex was removed from the eyecup, the macular region isolated with a trephine, and the remainder cut into smaller fragments that lay flat in a petri dish. Working under a dissecting microscope, most of the choroid was then removed, leaving only the choriocapillaris/Bruch's membrane complex. Drusen were isolated using fine forceps. In tissues with heavy drusen deposits, drusen could be scraped from Bruch's membrane with blunt forceps or a small blade. Based on Coomassie blue staining after SDS-PAGE of the most abundant drusen sample isolated, the average protein yield per eye was in the low microgram range (<10 μg).

Drusen Proteins. Liquid chromatography tandem mass spectrometry (LC MS/MS) analyses of drusen preparations from 18 normal donors and 5 AMD donors identified 129 proteins. (Crabb et al., 2002). The identified proteins are normal components of several tissues including the RPE, blood and photoreceptors. These identifications include 107 proteins from the normal donors and 62 proteins from AMD donors. TIMP3, clusterin, vitronectin and serum albumin appear to be common drusen proteins, detectable in up to 80% of the drusen from the normal donors and 60% of drusen preparations from the AMD donors. Crystallins were detected more frequently in AMD donor drusen.

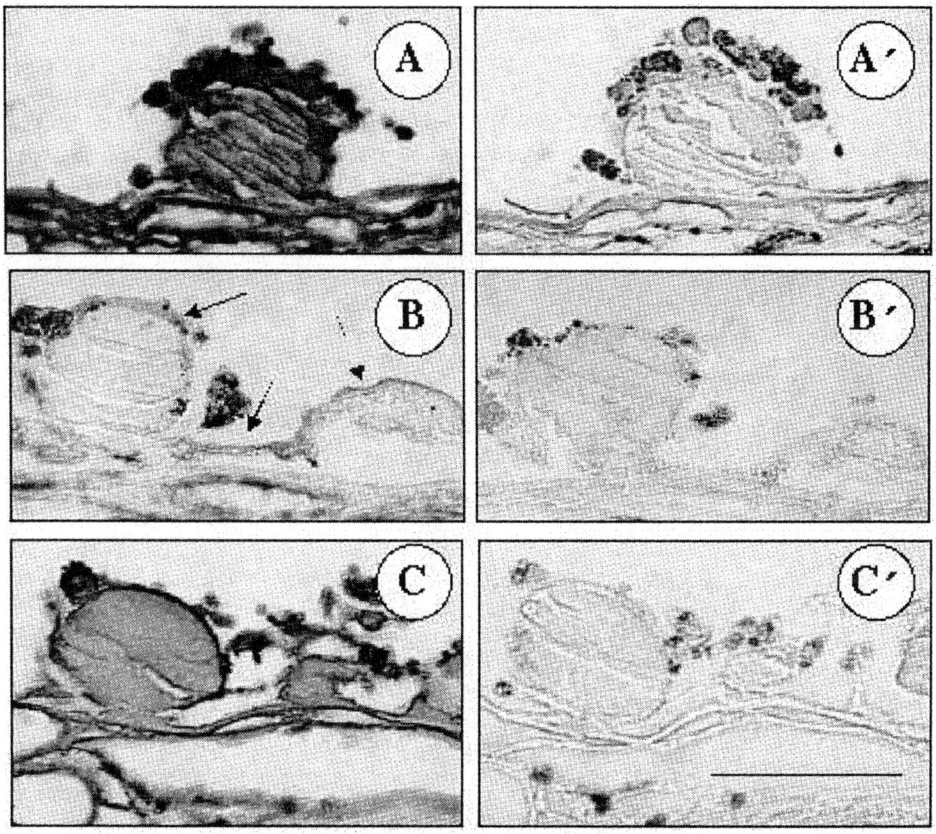

Figure 1. Immunocytochemical localization of annexins in drusen. (A) Annexin-I localization in drusen and Bruch's membrane using a polyclonal anti-Annexin antibody. (A') Control incubation with non-immune rabbit serum. (B) Annexin-II is localized to the surface of drusen and the region between drusen (arrows) and not to the interior of drusen with a mouse monoclonal antibody. (C) Annexin-VI is localized to both the surface of drusen and the interior, along with widespread distribution in the choroid. (B' and C') Control incubations with non-immune mouse serum. All micrographs are at the same magnification. Bar length in C' represents 50

We expect the majority of the identified proteins are components of drusen, however the possibility exists that some may be soluble contaminating proteins. As illustrated in Figure 1, immunocytochemical studies are in progress to verify drusen localization, and to date 21 of the identified proteins (16%) have been confirmed as drusen components in our laboratories and elsewhere (Hageman et al., 2001).

Drusen Oxidative Protein Modifications. A drusen sample with particularly abundant protein was subjected to one-dimensional SDS-PAGE, bands were excised, protein digested *in situ* with trypsin and the digest analyzed by LC MS/MS. Proteins identified were the more common components of normal donor drusen, including TIMP3 and vitronectin which migrated in multiple mass ranges, from the top to the bottom of the gel (Crabb et al., 2002). The multiple, higher mass electrophoretic components containing these proteins suggest the presence of covalent crosslinks, protein modifications that can be caused by reactive lipid and carbohydrate oxidation products (Friguet et al., 1994) (Elgawish et al., 1996).

Immunocytochemical analysis of Bruch's membrane/choroid tissue (Figure 2) demonstrated carboxyethyl pyrrole (CEP) and carboxymethyllysine immunoreactivity in drusen and Bruch's membrane (Crabb et al., 2002). CEP protein modifications are uniquely generated by oxidation products of docosahexaenoate-containing lipids (Hollyfield et al., 2002). In complimentary experiments, Bruch's membrane/choroid/RPE from AMD and age-and-gender-matched normal tissue donors were screened by Western analysis for CEP modifications. Western results revealed more frequent CEP immunoreactivity among the AMD samples than in the normal controls (Crabb et al., 2002). Specifically, 9 of the 11 AMD samples were immunopositive while only 2 of the 11 age matched controls were CEP immunoreactive (Crabb et al., 2002). Statistically the

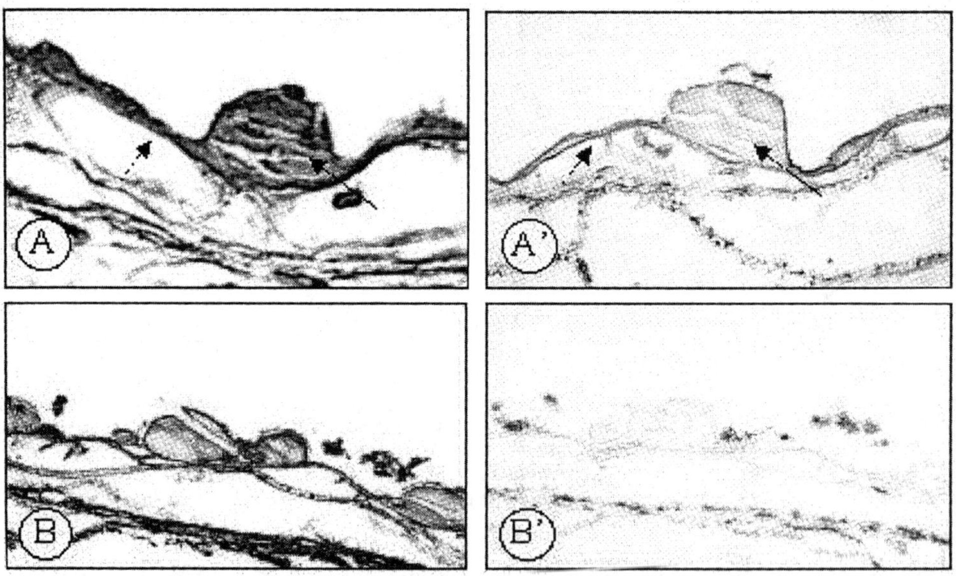

Figure 2. Carboxyethyl pyrrole (CEP) and Advanced Glycation End Product (AGE) Immunoreactivity in Human Drusen and Bruch's Membrane. (A) Left panel: section probed with anti-CEP mAb. Right panel: control section, probed with preimmune IgG. (Arrows indicate Bruch's membrane and drusen; bar = 50 μm). **(B)** Drusen was probed with a monoclonal antibody to carboxymethyllysine and a control consecutive section treated with normal mouse IgG. This figure demonstrates CEP and AGEs immunoreactivity associated with drusen and Bruch's membrane. (bar = 50 μm).

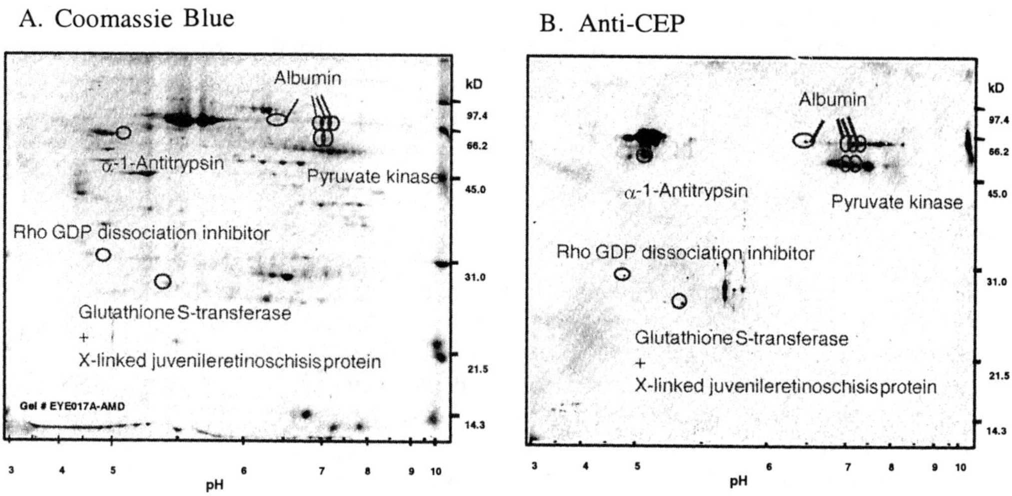

Protein	Peptide Matches	Sequence Coverage	On Gel MW	On Gel pI	Theoretical MW	Theoretical PI
alpha-1-Antitrypsin	8/16	33%	70 kD	5.1	46736	5.40
Albumin	10/16	15%	80	6.7	66472	5.67
Pyruvate kinase, M1 or M2	6/13	18%	70	7.0	57746	7.57
Pyruvate kinase, M1 or M2	8/29	22%	70	7.0	57746	7.57
Rho GDP dissociation inhibitor	5/16	33%	31	5.0	23207	5.03
Glutathione S-transferase	5/18	41%	27	5.4	23382	5.43
X-linked juvenile retinoschisis protein	4/18	26%	27	5.4	23038	5.47

Figure 3. Western 2D Gel Analysis and Protein Identification from AMD Bruch's Membrane/RPE/Choroid Tissue. About 200 μg protein from Bruch's Membrane/RPE/Choroid from a 84 year old female AMD donor was isoelectric focused on an 18 cm IPG strip (88, 402 Vh) then separated by SDS-PAGE on a large gel (23.5 x 18 x 0.1 cm). Following electrophoresis, protein was partially transferred to PVDF membrane: (A) the gel stained with colloidal Coomassie Blue and (B) the membrane probed by Western analysis with rabbit polyclonal anti-CEP antibody. Immunoreactive spots were excised from the gel, digested with trypsin and the indicated proteins identified by peptide mass mapping using MALDI TOF MS as described in Methods.

results of CEP immunoreactivity between AMD and normal donors are significantly different (single-sided Fischer Exact Test, p-value = 0.004).

Western 2D gel analyses of Bruch's membrane/RPE/choroid from an AMD donor revealed several CEP immunoreactive proteins (Figure 3), nine of which were identified by MALDI TOF mass spectrometry. Among the CEP immunopositive proteins were albumin and α1-antitrypsin, both of which were also detected in drusen by LC MS/MS

(Crabb et al., 2002) and by immunocytochemistry (Hageman et al., 2001). X-linked juvenile retinoschisis protein (XLRS) was also identified with apparent CEP immunoreactivity. Oxidative modification of XLRS may be another mechanism for causing retinal degenerations (Sauer et al., 1997) and/or play a role in the pathogenesis of AMD.

5. DISCUSSION

Drusen are considered a major risk factor for developing AMD, however little is known about their composition or mechanism of formation. Our recent results show that the protein content of drusen from AMD and normal donor eyes are similar and that the components originate from both the retina and the systemic circulation (blood). The effect of oxidative damage during aging and has long been suspected of contributing to the pathogenesis of AMD (Beatty et al., 2000). Indirect evidence that oxidative damage plays a role in AMD comes from epidemiological studies showing that smoking significantly increases the risk of AMD (Seddon et al., 1996) and that antioxidant vitamins and zinc can slow the progression of the disease for select individuals (Group, 2001). Our findings establish a direct molecular link between oxidative damage and AMD. We have found protein modifications in drusen and Bruch's membrane that can be generated from the oxidation of lipids and carbohydrates. Furthermore, CEP protein adducts originating from the oxidative fragmentation of docosahexaenoate-containing lipids was found to be more abundant in tissue from AMD donors than from normal eye donors. Notably, docosahexaenoic acid is abundant in photoreceptor cell outer segments and is the most oxidizable fatty acid in humans. The high photooxidative stress conditions normally prevailing in the retina suggests that other oxidative protein modification are likely to be found in drusen and Bruch's membrane.

A recent hypothesis for drusen formation suggests that immune mediated, complement activation events, triggered by unknown signals from the RPE, are causally involved in the process (Hageman et al., 2001) (Johnson et al., 2000). Others have associated immune mediated events with the pathogenesis of AMD (Penfold et al., 2001), but it is not yet clear whether this influence is a causative or a secondary contributory event. Reactive oxidation products from lipids and carbohydrate and their resulting protein modifications such as crosslinks, advanced glycation end products (AGEs), and CEP adducts may be the primary catalysts for drusen formation. AGEs stimulate angiogenesis *in vivo* (Okamoto et al., 2002), suggesting choroidal neovascularization in AMD may also be associated with such modifications. As genetically programmed oxidative stress defense mechanisms deteriorate with age, oxidation products and oxidative modifications may gradually lock proteins to Bruch's membrane with crosslinks, preventing normal turnover and initiating drusen development. The photooxidative environment in the retina and the lipid-rich photoreceptor outer segments provide an excellent source of reactive oxygen species and oxidation products. Waste products from the RPE and blood components from the choriocapillaris provide a ready source of extracellular material for oxidative modification and drusen formation. Over time, oxidation products and oxidative modifications may stimulate immune mediated events, causing the expansion of drusen on Bruch's membrane. Efforts in our laboratories are under way to test the hypothesis that oxidation products and oxidative protein modifications are the very early events that initiate drusen formation.

6. ACKNOWLEDGEMENTS

This work was supported in part by NIH grants EY06603, EY02362, GM21249, EY014239, EY014240, a Research Center grant from The Foundation Fighting Blindness, a grant from Merck, Inc., and funds from the Wolf Family Foundation, and The Cleveland Clinic Foundation.

7. REFERENCES

Beatty, S., H. Koh, M. Phil, D. Henson, and M. Boulton, 2000, The role of oxidative stress in the pathogenesis of age-related macular degeneration, *Surv. Ophthalmol.* **45**:115-134.

Crabb, J., M. Miyagi, X. Gu, K. Shadrach, K. West, H. Sakaguchi, M. Kamei, A. Hasan, L. Yan, M. Rayborn, R. Salomon, and J. Hollyfield, 2002, Drusen proteome analysis: an approach to the etiology of age-related macular degeneration, *Proc. Natl. Acad. Sci. USA.* **99**:14682-14687.

Crabb, J., Z. Nie, Y. Chen, J. Hulmes, K. West, J. Kapron, S. Ruuska, N. Noy, and J. Saari, 1998, Cellular retinaldehyde-binding protein ligand interactions; Gln^{210} and Lys^{221} are in the retinoid-binding pocket, *J. Biol. Chem.* **273**:20712-20720.

Elgawish, A., M. Glomb, M. Friedlander, V.M.R. Monnier, 1996, Involvement of hydrogen peroxide in collagen cross-linking by high glucose in vitro and in vivo, *J. Biol. Chem.* **271**:12964-12971.

Friguet, B., E.R. Stadtman and L.I. Szweda, 1994, Modification of glucose-6-phosphate dehydrogenase by 4-hydroxy-2-nonenal. Formation of cross-linked protein that inhibits the multicatalytic protease, *J. Biol. Chem.* **269**:21639-21643.

Group, A.R., 2001, A randomized, placebo-controlled, clinical trial of high-dose supplementation with vitamins C and E and beta carotene for age-related cataract and vision loss: AREDS report no. 9, *Arch. Ophthalmol.* **119**:1439-1452.

Hageman, G.S., P.J. Luthert, N.H.V. Chong, L.V. Johnson, D.H. Anderson and R.F. Mullins, 2001, An integrated hypothesis that considers drusen as biomarkers of immune-mediated processes at the RPE-Bruch's membrane interface in aging and age-related macular degeneration, *Prog. Ret. Eye Res.* **20**: 705-732.

Hollyfield, J.G., R.G. Salomon, J.W. Crabb and X. Gu, 2002, Patent Application 10/135, 196.

Johnson, L., S. Ozaki, M. Staples, P. Erickson, and D. Anderson, 2000, A potential role for immune complex pathogenesis in drusen formation, *Exp. Eye Res..* **70**:441-449.

Miyagi, M., H. Sakaguchi, R. Darrow, L. Yan, K. West, K. Aulak, D. Stuehr, J. Hollyfield, D. Organisciak, and J. Crabb, 2002, Evidence that light modulates protein nitration in rat retina, *Mol. Cell Proteomics.* **1**:293-303.

Okamoto, T., S. tanaka, A.C. Stan, T. Koike, M. Kase, Z. Makita, E. Sawa and K. Nagashima, 2002, Advanced glycation end products induce angiogenesis in vivo, *Microvasc. Res.* **63**:186-195.

Penfold, P., M. Madigan, M. Gillies, and J. Provis, 2001, Immunological and aetiological aspects of macular degeneration, *Prog. Ret. Eye Res..* **20**:385-414.

Sauer, C.G., A. Gehrig, R. Warneke-Wittstock, A. Marquardt, C.C. Ewing, A. Gibson, B. Lorenz, B. Jurklies and B.H.F. Weber, 1997, Positional cloning of the gene associated with X-linked juvenile retinoschisis, *Nat. Genet.* **17**:164-170.

Seddon, J.M., W.C. Willett, F.E. Speizer, and S.E. Hankinson, 1996, A prospective study of cigarette smoking and age-related macular degeneration in women, *J. Am. Med. Assoc.* **276**:1141-1146.

West, K., L. Yan, M. Miyagi, J. Crabb, A. Marmorstein, L. Marmorstein, and J. Crabb, 2001, Proteome survey of proliferating and differentiating rat RPE-J cells, *Exp. Eye Res.* **73**:479-491.

PROGRESSIVE PATHWAYS IN AGE-RELATED MACULAR DEGENERATION

Kimberly A. Howes[1], Jeanne M. Frederick[1], Alexander Marks[2], and Wolfgang Baehr[1]

1. INTRODUCTION

Age-related macular degeneration (AMD), the leading cause of blindness in the elderly, is characterized by progressive damage to supportive layers and photoreceptor cell loss. No current therapy halts vision loss. Numerous genetic and/or environmental factors likely contribute to the initiating events leading to AMD. Progressive events from cellular dysfunction and tissue destruction result in the histopathological features of AMD including drusen formation, basal laminar deposits, accumulation of lipofuscin in the retinal pigmented epithelium (RPE), thickening of Bruch's membrane, choriovasculature and RPE cell loss, and photoreceptor cell death. Studies on the biochemical and physiological perturbations seen in the RPE, Bruch's membrane and the choroid correlate well with AMD, however, progressive events leading to these characteristic features are not well understood. Resolving these issues is essential for successful therapeutic intervention of AMD. We propose that receptor-mediated cellular activation by advanced glycation end (AGE) products, a major contributor to progressive tissue damage in other aging disorders, may also contribute significantly to progressive damage in AMD. In this paper, we review and compare pathological features of AMD to AGE/RAGE- mediated progressive events incurred in other aging disorders.

2. THE RPE IN AMD

It is generally believed that perturbations brought about by aging of the RPE and its affects on surrounding tissues may be inductive to events leading to the characteristic features of AMD (Winkler et al., 1999; Boulton and Dayhaw-Barker, 2001). RPE cells must phagocytose daily shed photoreceptor outer segments, and any other cellular debris that might accumulate in this region, and thoroughly degrade and remove these products.

[1]Moran Eye Center, University of Utah Health Science Center, Salt Lake City, UT 84112, USA; [2]Banting and Best Department of Medical Research, University of Toronto, Toronto, Ontario, Canada, M5G 1L6.
Corresponding author: K. A. Howes, E-mail: kim.howes@hsc.utah.edu

Retinal Degenerations: Mechanisms and Experimental Therapy
Edited by LaVail *et al.*, Kluwer Academic/Plenum Publishers, 2003

If RPE cells are metabolically challenged or functionally defective, these deposits will not be digested efficiently and can accumulate intracellularly and extracellularly thereby contributing to lipofuscin, basal laminar and drusen formation (Sarks et al., 1994). Rakoczy et al., (2002) recently described a transgenic mouse line specifically showing that impaired degradation of photoreceptor outer segments by the RPE is sufficient for inducing a retinal pathology closely resembling AMD. A mutant (inactivated) cathepsin $D^{mcd/mcd}$ transgene impairs lysosomal degradation of photorecepter outer segments in RPE cells and ultimately gives rise to a retinal pathology showing most of the hallmark features of AMD. The RPE cell layer, lying adjacent to AMD progressive pathological features is phenotypically and functionally altered in AMD (Alizadeh et al., 2001; Guidry et al., 2002). RPE cells transdifferentiate into a myofibroblast-like cell type in AMD, presumably through the ability of RPE cells to respond to environmental perturbations by initiating the altered expression of genes - downregulating differentiation specific genes and upregulating genes involved with migration, proliferation and matrix repair. Aged RPE cells lose antioxidant capabilities, and it has been proposed that enhanced oxidant stress may signal RPE cells to transdifferentiate in AMD (Alizadeh et al., 2001).

3. OXIDANT STRESS IN AGING

A disruption of redox homeostasis through sustained increases in reactive oxygen species has been linked to a variety of aging and aging-like disorders including diabetes, atherosclerosis, arthritis, cancer, and neurodegenerative diseases (reviewed in Droge, 2002). Alterations in redox potential have also been reported for AMD. An excessive cellular imbalance of oxidants and free radicals versus antioxidant capabilities can result in an oxidatively damaging retinal environment. The retinal pigmented epithelium (RPE) and photoreceptor cell layers are particularly susceptible to oxidant stress due to, sunlight exposure, a high oxygen tension environment, enriched polyunsaturated fatty acid (PUFA) content of outer segment membranes, and progressive accumulation of photooxidatively damaging Lipofuscin (Winkler et al., 1999). Lipofuscin is composed of photoxidative fluorophores, the primary components of which are green-emitting retinol and retinyl esters. A particularly stable and photoactive fluorophore, A2E, a diretinal adduct of all-trans-retinal and phosphatidylethanolamine, can accumulate in the RPE and cause photooxidative toxicity in RPE cells (Sun and Nathans, 2001). This has been effectively shown in ABCR knockout mice, a mouse model for Stargardt's disease, and in cell culture in which subsequent A2E formation leads to photoxidation and RPE cell death Mata et al., 2001; Sun and Nathans, 2001).

4. ADVANCED GLYCATION END (AGE) PRODUCT FORMATION IN AGING

Regions of oxidant stress, as occurs in aging tissues, favor advanced glycation end (AGE) product formation (Glomb and Pfahler, 2001; Singh et al., 2001). AGE products are formed by reducing sugars reacting nonenzymatically with free amino groups on proteins to form Schiff base and Amadori products. During aging, nonenzymatic glycation of proteins and lipids followed by oxidation results in stable advanced glycoxidation endproducts (Kislinger et al., 1999), incurring toxic damage through crosslinking proteins (particularly in extracellular matrices) and activating cellular receptors (Fig. 1). Similar to excessive ROS production, enhanced AGE formation and

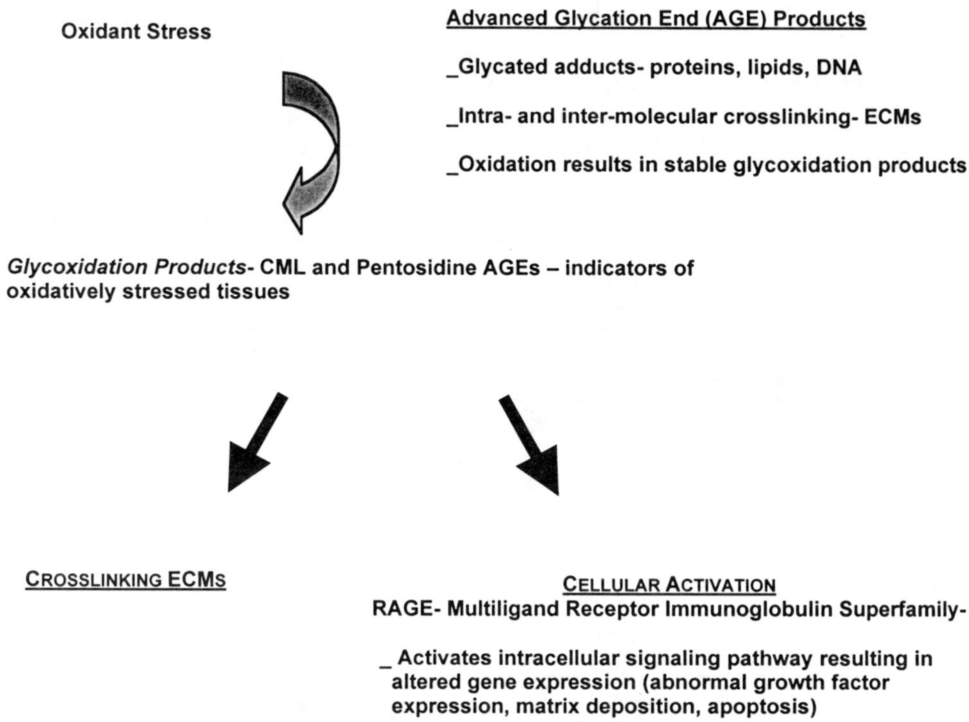

Figure 1. Oxidant stress leads to AGE formation inducing tissue toxicity through crosslinking ECMs and receptor-mediated cellular activation.

deposition has been well-documented in age-related disorders such as diabetic pathologies, Alzheimer's disease, amyloidosis, elastosis, pulmonary fibrosis, and atherosclerosis (reviewed in Vitek et al., 1994; Yan et al., 1996; Park et al., 1998; Schmidt et al., 1999; Jenkins et al., 2002; Stern et al., 2002). CML-AGE deposition, a glycoxidation marker for oxidatively stressed tissues (Farboud et al., 1999), has been reported in extracellular supportive layers in AMD including subfoveal membranes, Bruch's membrane, and drusen deposits (Ishibashi et al., 1998; Farboud et al., 1999; Stitt, 2001). Similar to other aging tissues, AGE formation in these extracellular matrices is thought to impede normal barrier/filtration functions across this essential blood-retinal barrier. However, further cellular activating events are induced by AGE products. Several reports have described AGE induced RPE alterations in expression of genes governing differentiation, growth factors (including PDGF and VEGF), and apoptosis (Honda et al., 2001; Treins et al., 2001).

5. CELLULAR ACTIVATION THROUGH RAGE

While AGE deposition in itself is damaging to matrices, further damage is incurred through chronic AGE receptor (RAGE)-mediated cellular activation. Cellular damage induced by the sustained activation of RAGE is central to many chronic aging disorders

Thornalley, 1998; Schmidt et al., 1999). RAGE is a transmembrane protein of the immunoglobulin superfamily, binding a class of molecules that form nondegradable aggregates in aging disorders. Examples include AGE products as found in diabetic tissues, amyloid in brain neuropathies, S100/calgranulins in chronic inflammatory disorders, and amphoterin in tumors (Hofmannet al., 1999; Schmidt et al., 1999; Yan et al., 2000; Schmidt et al., 2001). Sustained availability of ligand results in repeated

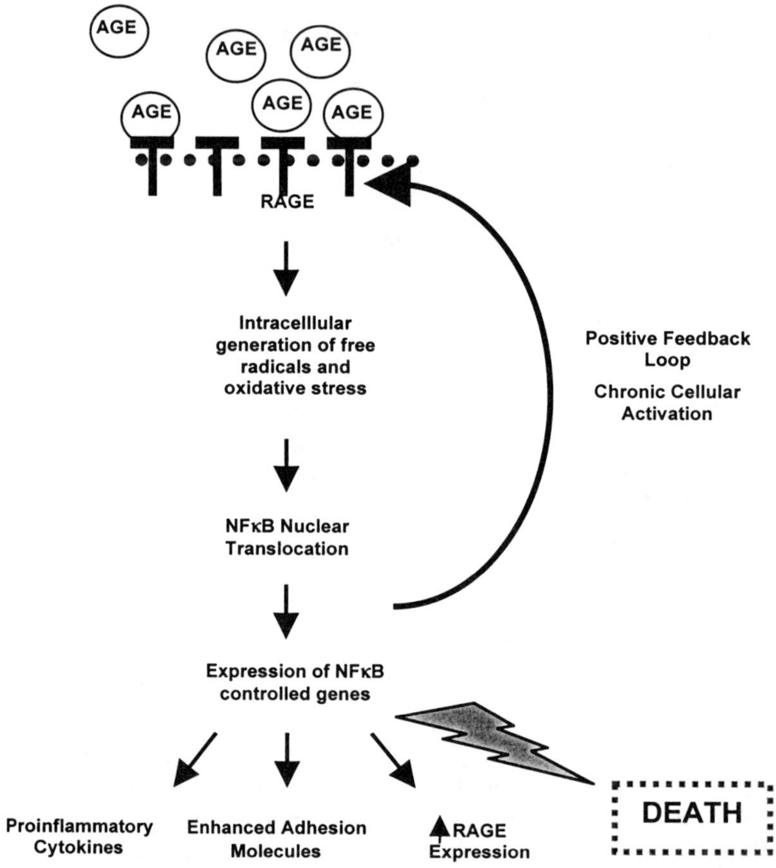

Figure 2. Model for sustained AGE/RAGE mediated cellular activation leading to abnormal gene expression and cell death.

cellular insult and ultimately cell death through the uncontrolled upregulation of RAGE expression. The cytotoxic effects induced through chronic activity of ligand/RAGE-mediated events are an exciting topic of research in studies of age-related disorders. Indeed, AGE/RAGE has also been implicated in the progression of other ocular aging diseases including diabetic retinopathy, glaucoma, and cataracts (reviewed in Stitt, 2001).

Cellular activation elicited through RAGE alters growth factor expression, matrix protein deposition, abnormal growth characteristics, vascular tissue perturbations, and apoptosis (Schmidt et al., 1999) (Fig. 2).

6. FUTURE DIRECTIONS IN AMD RESEARCH

We are currently addressing the involvement of RAGE-mediated cellular activation in AMD tissues, RPE cell culture and transgenic mice. Our preliminary results suggest that his pathway may very well play a role in progressive damage leading to AMD.
Based on results from other aging disorders, a more encompassing approach to immediate therapeutic intervention may be to target mediators of chronic cellular activation leading to progressive tissue damage and AMD. This has proven to be a very promising approach for diabetes, amyloidoses, inflammatory disorders and tumor biologies (reviewed in Schmidt et al., 2001) (Table 1). Animal models for these disorders show significant reduction in progressive tissue damage by intervention of ligand/RAGE mediated cellular activation. A significant number of AGE and RAGE inhibitors are currently being developed and tested *in vivo* for therapeutic value in these disorders. We propose that this pathway should be thoroughly addressed in future studies of AMD, since this mechanism would provide a more immediately feasible target for delaying the progression of AMD in the elderly thereby improving quality of life.

Table 1. Ligand/RAGE mediated involvement in other aging pathologies. Significant inhibition of progressive tissue damage through treatments in animal models for these disorders.*

Chronic ligand/RAGE-mediated cellular activation

*Diabetes- progressive tissue complications (wound healing, retinopathy, nephropathy, periodontal disease)- AGEs, S100/calgranulins

*Amyloidosis- Alzheimer Disease- β-sheet fibrils, S100/calgranulins

*Chronic inflammation (rheumatoid arthritis)- AGEs, S100/calgranulins, amphoterin

*Tumor biologies- amphoterin

Animal Models for these Diseases _ Treatments- sRAGE, anti-RAGE, anti-S100s, and AGE inhibitors_ Results- SIGNIFICANT SUPPRESSION OF PATHOLOGIES

*reviewed in Schmidt et al., 2001, J. Clin. Invest.

7. REFERENCES

Alizadeh, M., Wada, M., Gelfman, C. M., Handa, J. T., and Hjelmeland, L. M., 2001, Downregulation of differentiation specific gene expression by oxidative stress in ARPE-19 cells. *Invest Ophthalmol. Vis. Sci.* **42**:2706.

Boulton, M.and Dayhaw-Barker, P., 2001, The role of the retinal pigment epithelium: topographical variation and ageing change, *Eye* **15**:384.

Droge, W., 2002, Free radicals in the physiological control of cell function. *Physiol. Rev.* **82**:47.

Farboud, B., Aotaki-Keen, A., Miyata, T., Hjelmeland, L. M., and Handa, J. T., 1999, Development of a polyclonal antibody with broad epitope specificity for advanced glycation endproducts and localization of these epitopes in Bruch's membrane of the aging eye. *Mol. Vis.* **5**:11.

Glomb, M. A. and Pfahler, C., 2001, Amides are novel protein modifications formed by physiological sugars. *J. Biol. Chem.* **276**:41638.

Guidry, C., Medeiros, N. E., and Curcio, C. A., 2002, Phenotypic variation of retinal pigment epithelium in age-related macular degeneration. *Invest Ophthalmol. Vis. Sci.* **43**:267.

Hofmann, M. A., Drury, S., Fu, C., Qu, W., Taguchi, A., Lu, Y., Avila, C., Kambham, N., Bierhaus, A., Nawroth, P., Neurath, M. F., Slattery, T., Beach, D., McClary, J., Nagashima, M., Morser, J., Stern, D., and Schmidt, A. M., 1999, RAGE mediates a novel proinflammatory axis: a central cell surface receptor for S100/calgranulin polypeptides. *Cell* **97**:889.

Honda, S., Farboud, B., Hjelmeland, L. M., and Handa, J. T., 2001, Induction of an aging mRNA retinal pigment epithelial cell phenotype by matrix-containing advanced glycation end products in vitro. *Invest Ophthalmol. Vis. Sci.* **42**:2419.

Ishibashi, T., Murata, T., Hangai, M., Nagai, R., Horiuchi, S., Lopez, P. F., Hinton, D. R., and Ryan, S. J., 1998, Advanced glycation end products in age-related macular degeneration. *Arch. Ophthalmol.* **116**:1629.

Jenkins, A. J., Cooper, M. E., and Stitt, A. W., 2002, Advanced glycation end products and diabetic complications. *Expert. Opin. Investig. Drugs* **11**:1205.

Kislinger, T., Fu, C., Huber, B., Qu, W., Taguchi, A., Du, Y.S., Hofmann, M., Yan, S. F., Pischetsrieder, M., Stern, D., and Schmidt, A. M., 1999, N(epsilon)-(carboxymethyl)lysine adducts of proteins are ligands for receptor for advanced glycation end products that activate cell signaling pathways and modulate gene expression. *J. Biol. Chem.* **274**:31740.

Mata, N. L., Tzekov, R. T., Liu, X., Weng ,J., Birch, D. G., and Travis, G. H., 2001, Delayed dark-adaptation and lipofuscin accumulation in abcr+/- mice: implications for involvement of ABCR in age-related macular degeneration. *Invest Ophthalmol. Vis. Sci.* **42**:1685.

Park, L., Raman, K. G., Lee, K. J., Lu, Y., Ferran, L. J. Jr., Chow, W. S., Stern, D., and Schmidt, A. M., 1998, Suppression of accelerated diabetic atherosclerosis by the soluble receptor for advanced glycation endproducts. *Nat. Med.* **4**:1025.

Rakoczy, P. E., Zhang, D., Robertson, T., Barnett, N. L., Papadimitriou, J., Constable, I. J., and Lai, C. M., 2002, Progressive age-related changes similar to age-related macular degeneration in a transgenic mouse model. *Am. J. Pathol.* **161**:1515.

Sarks, J. P., Sarks, S. H., and Killingsworth, M.C., 1994, Evolution of soft drusen in age-related macular degeneration. *Eye* **8**:269.

Singh, R., Barden, A., Mori, T., and Beilin, L., 2001, Advanced glycation end-products: a review. *Diabetologia* **44**:129.

Schmidt, A. M., Yan, S. D., Wautier, J. L., and Stern, D., 1999, Activation of receptor for advanced glycation end products: a mechanism for chronic vascular dysfunction in diabetic vasculopathy and atherosclerosis. *Circ. Res.* **84**: 489.

Schmidt, A. M., Yan, S. D., Yan, S. F., and Stern, D. M., 2001, The multiligand receptor RAGE as a progression factor amplifying immune and inflammatory responses. *J. Clin. Invest* **108**:949.

Stern, D. M., Yan, S. D., Yan, S. F., and Schmidt, A. M., 2002, Receptor for advanced glycation endproducts (RAGE) and the complications of diabetes. *Ageing Res. Rev.* **1**:1.

Stitt, A. W. 2001, Advanced glycation: an important pathological event in diabetic and age related ocular disease. *Br. J Ophthalmol.* **85**:746.

Sun, H. and Nathans, J., 2001, ABCR, the ATP-binding cassette transporter responsible for Stargardt macular dystrophy, is an efficient target of all-trans-retinal-mediated photooxidative damage in vitro. Implications for retinal disease. *J Biol. Chem.* **276**:11766.

Thornalley, P. J. 1998, Cell activation by glycated proteins. AGE receptors, receptor recognition factors and functional classification of AGEs. *Cell Mol. Biol. (Noisy. -le-grand)* **44**:1013.

Treins, C., Giorgetti-Peraldi, S., Murdaca, J., and Van Obberghen, E., 2001, Regulation of vascular endothelial growth factor expression by advanced glycation end products. *J. Biol. Chem.* **276**:43836.

Vitek, M. P., Bhattacharya, K., Glendening, J. M., Stopa, E., Vlassara, H., Bucala, R., Manogue, K., and Cerami, A., 1994, Advanced glycation end products contribute to amyloidosis in Alzheimer disease. *Proc. Natl. Acad. Sci. U. S. A* **91**:4766.

Winkler, B. S., Boulton, M. E., Gottsch, J. D., and Sternberg, P., 1999, Oxidative damage and age-related macular degeneration. *Mol. Vis.* **5**:32.

Yan, S. D., Chen, X., Fu, J., Chen, M., Zhu, H., Roher, A., Slattery, T., Zhao, L., Nagashima, M., Morser, J., Migheli, A., Nawroth, P., Stern, D., and Schmidt, A., M., 1996, RAGE and amyloid-beta peptide neurotoxicity in Alzheimer's disease. *Nature* **382**: 685.

Yan, S. D., Zhu, H., Zhu, A., Golabek, A., Du, H., Roher, A., Yu, J., Soto, C., Schmidt, A. M., Stern, D., and Kindy, M., 2000, Receptor-dependent cell stress and amyloid accumulation in systemic amyloidosis. *Nat. Med.* **6**:643.

TISSUE INHIBITOR OF METALLOPROTEINASES-3 AND SORSBY FUNDUS DYSTROPHY

Jian Hua Qi[1], Quteba Ebrahem[1], and Bela Anand-Apte[1,2]

1. INTRODUCTION

Some of the difficulties in studying age-related macular degeneration (ARMD) are its late onset, complex genetics and probable strong environmental component (Zack et al., 1999). Thus it has been proposed, that to gain a better understanding of this disease, it might be useful to study related clinical diseases such as Sorsby fundus dystrophy, Stargardt disease and Best disease, that demonstrate early onset forms of macular degeneration with simple Mendelian inheritance patterns. Sorsby fundus dystrophy (SFD) is a fully penetrant, autosomal dominant retinal dystrophy disease that was first described by Sorsby in 1949 (Sorsby and Joll Mason, 1949). Clinically, patients first complain of night blindness, which is usually followed by central visual loss from subretinal neovascularization and haemorrhage with a late loss of peripheral vision (Capon et al., 1988; (Hamilton et al., 1989; Hoskin et al., 1981; Polkinghorne et al., 1989a; Sorsby and Joll Mason, 1949; Wu et al., 1991). SFD is a relatively rare disease, but commands considerable interest because of its striking similarity to age-related macular degeneration (ARMD), the leading cause of blindness in individuals over the age of 50 years in the western world.

SFD is characterized by accumulation of extracellular deposits (drusen) in Bruch's membrane, the five-layered sheet of connective tissue that separates the retinal pigment epithelium (RPE) from the choriocapillaris (Capon et al., 1989). The predominant histopathological feature in the eyes of a 63 year old patient with SFD was a confluent 30μm thick, lipid containing amorphous deposit found between the basement membrane of the RPE and the inner collagenous layer of Bruch's membrane (Capon et al., 1989). This is distinct from the basal linear deposits seen in age-related macular degeneration,

[1]Cole Eye Institute and [2]Department Cell Biology, Lerner Research Institute, Cleveland Clinic Foundation, Cleveland, OH 44106, USA. Corresponding author: B. Anand-Apte, E-mail: anandab@ccf.org

Retinal Degenerations: Mechanisms and Experimental Therapy
Edited by LaVail *et al.*, Kluwer Academic/Plenum Publishers, 2003

which consists of filamentous fine granular material and may represent a thickened basement membrane of the RPE (Capon et al., 1989). A serious complication of SFD is the invasion of the thickened Bruch's membrane by newly formed, thin walled vessels derived from the choriocapillaris. These vessels grow into the subretinal space, causing exudative detachment of the RPE and loss of photoreceptors (Polkinghorne et al., 1989b). The symptom of night blindness in SFD can be reversed by mega doses of vitamin A, supporting the notion that nutritional deprivation of photoreceptors, due to the thickened Bruch's membrane may be a part of the pathophysiology of the disease (Jacobson et al., 1995). Based on similar clinical and histopathological features, in particular at the level of Bruch's membrane, SFD is thought to provide a genetic model for the more common macular degenerations (Bird, 1992; Bird, 1993; Capon et al., 1989; Wu et al., 1991).

2. TIMP-3 MUTATIONS CAUSE SORSBY FUNDUS DYSTROPHY (SFD)

SFD has been linked with mutations in the Tissue Inhibitor of Metalloproteinase –3 SFD has been linked with mutations in the Tissue Inhibitor of Metalloproteinase –3 (TIMP-3) gene (Langton et al., 2000; Tabata et al., 1998; Weber et al., 1994) with five different missense mutations and a splice site mutation having been identified to date. Interestingly, all mutations identified so far occur in the COOH-terminal portion of the mature TIMP-3 protein with the introduction of a new cysteine. These additional thiol groups are located close to the last two conserved cysteine residues which are believed to participate in normal intrachain disulphide bond rearrangements (Williamson et al., 1990). Structurally, TIMPs have a two-domain structure with each domain folded into three loops held together by three disulfide bonds (Woessner, 2001). It has been proposed that the C terminal domains may participate in the modulation of the TIMP/MMP interactions by increasing the low affinity binding characteristics between the two molecules (Nguyen et al., 1994; (Willenbrock et al., 1993). It still remains to be determined whether the introduction of a new cysteine in SFD-TIMP-3 affects its binding to and/or activation of MMPs. Varied reports have been published suggesting that SFD mutations have reduced association rate constants for MMPs (Yeow et al., 2001) as well as reduced MMP inhibitory activity when expressed in human RPE cells (Qi et al., 2002). However these reports are in contrast to that reported previously with expression in COS-7 and BHK (baby hamster kidney) cells, in which no loss of MMP inhibition by SFD protein was observed (Langton et al., 1998; Langton et al., 2000; Yeow et al., 2001). Dimerization of the mutant protein has also been demonstrated in some systems (Langton et al., 1998; Langton et al., 2000; Weber et al., 2002; Yeow et al., 2001). Based on the dominant nature of the disease, its similarity in unrelated families, it is possible that the new extra cysteine may participate in abnormal interchain bonds with another ECM present specifically in the retina. This might explain the presence of an exclusive ocular phenotype and sparing of other organs in SFD.

3. TIMP-3 EXPRESSION IN THE EYE

In *situ* hybridization studies have demonstrated the presence of TIMP-3 mRNA in retinal pigment epithelial (RPE) cells and choroid (Della et al., 1996; Ruiz et al., 1996a; Ruiz et al., 1996b; Vranka et al., 1996; Zucker, 1996) . TIMP-3 protein is present in low levels around blood vessels in human and non-human primate retinas, Bruch's membrane and drusen (Fariss et al., 1997; Kamei and Hollyfield, 1999). Accumulation of TIMP-3 has been observed in Bruch's membrane of aging eyes (Kamei and Hollyfield, 1999) as well as in SFD (Chong et al., 2000; Fariss et al., 1998) and ARMD (Fariss et al., 1997; Kamei et al., 1997). Based on our results of inhibition of angiogenesis by TIMP-3 (Anand-Apte et al., 1997), we hypothesize that TIMP-3 in Bruch's membrane may be required to maintain a controlled and localized degradation of the matrix around the blood vessels in the retina.

4. CHOROIDAL NEOVASCULARIZATION (CNV)

It has been suggested that a balance between local stimulators (angiogenic factors) and inhibitors regulate the growth of the microvasculature. It is becoming increasingly clear that the mechanisms involved in the distribution and maintenance of retinal and choroid vessels during development, and throughout life, are also probably involved in pathological ocular neovascularization (D'Amore, 1994; Gariano et al., 1996). The pathogenesis and molecular signals involved in CNV are poorly understood. There is some evidence that choroidal blood flow may be altered in patients with AMD which may lead to hypoxia (Grunwald et al., 1998; Ross et al., 1998). Another hypothesis that has been put forward is that diffuse thickening of Bruch's membrane decreases the diffusion of oxygen from the choroid to the RPE and retina and results in hypoxia which can induce upregulation of angiogenic factors such as VEGF. Additional evidence suggests that an inflammatory response around Bruch's membrane may be able to induce an angiogenic response (Dastgheib and Green, 1994; Killingsworth et al., 1990). There has been longstanding evidence that vessel formation by choroidal endothelial cells is modulated by retinal pigment epithelial cells (Sakamoto et al., 1995), which suggests that these cells may play a key role in regulating the angiogenic state of the choroid. Angiogenic factors VEGF and bFGF have been shown to play a critical role in choroidal neovascularization (Amin et al., 1994; Frank et al., 1996; Kvanta et al., 1996; Lopez et al., 1996; Ogata et al., 1996; Soubrane et al., 1994; Yi et al., 1997). The most consistent pathologic finding in the wet form of AMD, is the presence of diffuse thickening of Bruch's membrane (presumably due to accumulation of abnormal ECM) (Green and Enger, 1993). This, together with the focal areas of thickening resulting from drusen, suggests that a disrupted ECM turnover may play a critical role in this process. Altered matrix metalloproteinase metabolism may lead to a phenotype like that seen in AMD and may even be able to induce choroidal neovascularization (Capon et al., 1989; Capon et al., 1988; Hoskin et al., 1981; Weber et al., 1994). We have demonstrated that Tissue

Inhibitor of Metalloproteinase-3 (TIMP-3) is a potent angiogenesis inhibitor and may play an important role in the regulation of choroidal neovascularization (Anand-Apte et al., 1997; Qi et al., 2002).

4.1. Matrix Metalloproteinases and Neovascularization

Recent studies have indicated that members of the MMP family play an important role in angiogenesis (Hiraoka et al., 1998; (Stetler-Stevenson, 2001; Stetler-Stevenson and Yu, 2001; Werb et al., 1999). MMP-2 is the most widely distributed MMP and is localized on the surface of angiogenic blood vessels (Brooks, 1996). Mice deficient in MMP-2 and MMP-9 exhibit reduced angiogenesis in vivo (Itoh et al., 1998; Vu et al., 1998). The matrix metalloproteinases (MMPs, e.g. collagenases, gelatinases (A, 72kDa and B, 92 kDa) and stromelysins) are a family of zinc binding, Ca^{2+} dependent neutral endopeptidases that can act together or in concert with other enzymes to degrade most components of the ECM (Matrisian, 1992). These enzymes have been implicated strongly in invasive cell behavior. MMPs need to be activated before they can exert their biochemical functions. MMP activity is tightly regulated in response to the potential consequences of disrupted ECM integrity (Alexander and Werb, 1992). The complex regulation of these enzymes occurs at three levels; transcriptional regulation of the genes, activation of secreted proenzymes, and via specific inhibitors, the Tissue Inhibitors of Metalloproteinases (TIMPs) which bind to activated MMPs with 1:1 molar stoichiometry. **TIMPs,** constitute a family of distinct MMP inhibitors whose members are TIMP-1, TIMP-2, TIMP-3 and TIMP-4. The TIMPs show 35% to 40% identity at the amino acid level and are capable of inhibiting the MMP family of enzymes with equal efficacy in in- vitro assays (Apte et al., 1995). TIMP-1, 2 and 4 are "soluble" proteins and are present in numerous body fluids (Murphy, 1991). TIMP-3 is distinct in that it is sequestered in the ECM.

4.2. Inhibition of Angiogenesis by TIMP-3

Our previous work has shown that TIMP-3 inhibits some of the endothelial cell responses to physiological angiogenic agents such as VEGF and bFGF. TIMP-3 inhibits migration and prevents invasion and tube formation by endothelial cells in a three-dimensional collagen matrix. That this inhibition of activity was relevant to angiogenesis *in vivo* was confirmed in a Chick Chorio-Allantoic Membrane assay (CAM) assay (Anand-Apte et al., 1997). As in SFD, since TIMP-3 accumulates in Bruch's membrane with ARMD, understanding the molecular mechanisms of its actions will help in our understanding of this complex process.

5. DIVERGENT FUNCTIONS OF TIMP-3

Besides its role in MMP inhibition and angiogenesis, TIMP-3 has been shown to induce apoptosis in both normal and cancer lines (Ahonen et al., 2002; Ahonen et al., 1998a; Ahonen et al., 1998b; Baker et al., 1999; Baker et al., 1998; Bond et al., 2000; George et al., 2000). However TIMP-3 has also been proposed to be a cell survival factor during mammary gland involution (Fata et al., 2001). Thus the divergent effects of TIMP-3 on cell survival, and other potential functions may depend on the cell type, microenvironment and amount of TIMP-3 in the microenvironment (Fata et al., 2001).

6. PATHOGENESIS OF SORSBY FUNDUS DYSTROPHY

Sub-RPE TIMP-3 immunoreactivity has been observed to be more extensive in SFD than in control subjects. The co-localization of TIMP-3 and elastin in the elastic layer of Bruch's membrane, in the elastic layer of choroidal vessels and in the basement membrane of the non pigmented ciliary epithelium, suggests that TIMP-3 might play a role in the turnover of elastic fibres (Chong et al., 2000). This might also invite conjecture on the possibility of a link between SFD mutant TIMP-3 and altered elastin processing (Chong et al., 2000). We have reported a constitutive activation of gelatinase A in the conditioned medium and extracellular matrix of RPE cells expressing S156C TIMP-3 (Qi et al., 2002). Unscheduled MMP-2 activation has also been observed in the lungs and mammary glands of homozygous TIMP-3 null mice (Fata et al., 2001; Leco et al., 2001). It can be postulated, based on the known anti-angiogenic activity of TIMP-3, that WT-TIMP-3 deposited by RPE cells into Bruch's membrane efficiently inhibits neovascular invasion from the choriocapillaris. This might occur by inhibiting MMP activity, resulting in an intact matrix, and/or binding directly to the surface of endothelial cells to inhibit endothelial cell responses. Conditioned medium from RPE cells expressing SFD mutant TIMP-3 could induce an increased angiogenic response in chicken chorioallantoic membranes (Qi et al., 2002). There are several possible mechanisms by which mutant TIMP-3 can induce this effect. Intact matrix might provide a physical barrier that limits endothelial cell migration and invasion into the surrounding ECM. Breakdown of the basement membrane allows the endothelial cells to initiate the process of sprouting. Alternatively, proteolysis of ECM might release sequestered angiogenic factors such as VEGF and/or bFGF, which promote angiogenesis. Finally, a third possibility that has been recently proposed is the exposure of an angiogenic cryptic site upon proteolytic cleavage of ECM molecules such as collagen type IV (Xu et al., 2001). Accumulation of TIMP-3 in both ARMD and SFD, with similar clinical pictures might also suggest that the disease phenotype may be entirely due to accumulation of TIMP-3 protein, and the SFD mutation somehow increases the accumulation of TIMP-3 in the sub RPE, leading to early onset disease. Thus, understanding the mechanism by which CNV occurs in these pathologies will allow the design of strategies aimed at combating these devastating diseases.

7. ACKNOWLEDGEMENTS

This work was supported by the following grants to B.A-A, 1 R29EY12109-01 (NIH), module grant (Foundation Fighting Blindness) and Seed grant (Cleveland Clinic Foundation). We thank the National Eye Institute for the travel award provided to BAA to attend this meeting.

8. REFERENCES

Ahonen, M., R. Ala-Aho, A.H. Baker, S.J. George, R. Grenman, U. Saarialho-Kere, and V.M. Kahari. 2002. Antitumor activity and bystander effect of adenovirally delivered tissue inhibitor of metalloproteinases-3. *Mol Ther.* 5:705-15.

Ahonen, M., A.H. Baker, and V.M. Kahari. 1998a. Adenovirus-mediated gene delivery of tissue inhibitor of metalloproteinases-3 inhibits invasion and induces apoptosis in melanoma cells. *Cancer Res.* 58:2310-5.

Ahonen, M., A.H. Baker, and V.M. Kahari. 1998b. High level expression of tissue inhibitors of metalloproteinases-1,-2 and -3 in melanoma cells achieved by adenovirus mediated gene transfer. *Adv Exp Med Biol.* 451:69-72.

Alexander, C.M., and Z. Werb. 1992. Targeted disruption of the tissue inhibitor of metalloproteinases gene increases the invasive behavior of primitive mesenchymal cells derived from embryonic stem cells in vitro. *J Cell Biol.* 118:727-39.

Amin, R., J.E. Puklin, and R.N. Frank. 1994. Growth factor localization in choroidal neovascular membranes of age-related macular degeneration. *Invest Ophthalmol Vis Sci.* 35:3178-88.

Anand-Apte, B., M.S. Pepper, E. Voest, R. Montesano, B. Olsen, G. Murphy, S.S. Apte, and B. Zetter. 1997. Inhibition of Angiogenesis by Tissue Inhibitor of Metalloproteinase-3. *Invest. Ophthal. Vis. Sci.* 38:817-823.

Apte, S.S., B. Olsen, and G. Murphy. 1995. The gene structure of tissue inhibitor of metalloproteinases (TIMP)-3 and its inhibitory activities define the distinct TIMP gene family. *J. Biol. Chem.* 270:14313-14318.

Baker, A.H., S.J. George, A.B. Zaltsman, G. Murphy, and A.C. Newby. 1999. Inhibition of invasion and induction of apoptotic cell death of cancer cell lines by overexpression of TIMP-3. *Br J Cancer.* 79:1347-55.

Baker, A.H., A.B. Zaltsman, S.J. George, and A.C. Newby. 1998. Divergent effects of tissue inhibitor of metalloproteinase-1, -2, or -3 overexpression on rat vascular smooth muscle cell invasion, proliferation, and death in vitro. TIMP-3 promotes apoptosis. *J Clin Invest.* 101:1478-87.

Bird, A.C. 1992. Bruch's membrane change with age. *Br J Ophthalmol.* 76:166-8.

Bird, A.C. 1993. Choroidal neovascularisation in age-related macular disease [editorial]. *Br J Ophthalmol.* 77:614-5.

Bond, M., G. Murphy, M.R. Bennett, A. Amour, V. Knauper, A.C. Newby, and A.H. Baker. 2000. Localization of the death domain of tissue inhibitor of metalloproteinase-3 to the N terminus. Metalloproteinase inhibition is associated with proapoptotic activity. *J Biol Chem.* 275:41358-63.

Brooks, P.C. 1996. Cell adhesion molecules in angiogenesis. *Cancer Metastasis Rev.* 15:187-94.

Capon, M.R., J. Marshall, J.I. Krafft, R.A. Alexander, P.S. Hiscott, and A.C. Bird. 1989. Sorsby's fundus dystrophy. A light and electron microscopic study. *Ophthalmology.* 96:1769-77.

Capon, M.R., P.J. Polkinghorne, F.W. Fitzke, and A.C. Bird. 1988. Sorsby's pseudoinflammatory macula dystrophy--Sorsby's fundus dystrophies. *Eye.* 2:114-22.

Chong, N.H., R.A. Alexander, T. Gin, A.C. Bird, and P.J. Luthert. 2000. TIMP-3, collagen, and elastin immunohistochemistry and histopathology of Sorsby's fundus dystrophy. *Invest Ophthalmol Vis Sci.* 41:898-902.

D'Amore, P.A. 1994. Mechanisms of retinal and choroidal neovascularization. *Invest Ophthalmol Vis Sci.* 35:3974-9.

Dastgheib, K., and W.R. Green. 1994. Granulomatous reaction to Bruch's membrane in age-related macular degeneration. *Arch Ophthalmol.* 112:813-8.

Della, N.G., P.A. Campochiaro, and D.J. Zack. 1996. Localization of TIMP-3 mRNA expression to the retinal pigment epithelium. *Invest Ophthalmol Vis Sci.* 37:1921-4.

Fariss, R.N., S.S. Apte, P.J. Luthert, A.C. Bird, and A.H. Milam. 1998. Accumulation of tissue inhibitor of metalloproteinases-3 in human eyes with Sorsby's fundus dystrophy or retinitis pigmentosa. *Br J Ophthalmol.* 82:1329-34.

Fariss, R.N., S.S. Apte, B.R. Olsen, K. Iwata, and A.H. Milam. 1997. Tissue inhibitor of metalloproteinases-3 is a component of Bruch's membrane of the eye. *Am J Pathol.* 150:323-8.

Fata, J.E., K.J. Leco, E.B. Voura, H.Y. Yu, P. Waterhouse, G. Murphy, R.A. Moorehead, and R. Khokha. 2001. Accelerated apoptosis in the Timp-3-deficient mammary gland. *J Clin Invest.* 108:831-41.

Frank, R.N., R.H. Amin, D. Eliott, J.E. Puklin, and G.W. Abrams. 1996. Basic fibroblast growth factor and vascular endothelial growth factor are present in epiretinal and choroidal neovascular membranes. *Am J Ophthalmol.* 122:393-403.

Gariano, R.F., R.E. Kalina, and A.E. Hendrickson. 1996. Normal and pathological mechanisms in retinal vascular development. *Surv Ophthalmol.* 40:481-90.

George, S.J., C.T. Lloyd, G.D. Angelini, A.C. Newby, and A.H. Baker. 2000. Inhibition of late vein graft neointima formation in human and porcine models by adenovirus-mediated overexpression of tissue inhibitor of metalloproteinase-3. *Circulation.* 101:296-304.

Green, W.R., and C. Enger. 1993. Age-related macular degeneration histopathologic studies. The 1992 Lorenz E. Zimmerman Lecture. *Ophthalmology.* 100:1519-35.

Grunwald, J.E., S.M. Hariprasad, J. DuPont, M.G. Maguire, S.L. Fine, A.J. Brucker, A.M. Maguire, and A.C. Ho. 1998. Foveolar choroidal blood flow in age-related macular degeneration. *Invest Ophthalmol Vis Sci.* 39:385-90.

Hamilton, W.K., C.C. Ewing, E.J. Ives, and J.D. Carruthers. 1989. Sorsby's fundus dystrophy. *Ophthalmology.* 96:1755-62.

Hiraoka, N., E. Allen, I.J. Apel, M.R. Gyetko, and S.J. Weiss. 1998. Matrix metalloproteinases regulate neovascularization by acting as pericellular fibrinolysins. *Cell.* 95:365-77.

Hoskin, A., K. Sehmi, and A.C. Bird. 1981. Sorsby's pseudoinflammatory macular dystrophy. *Br J Ophthalmol.* 65:859-65.

Itoh, T., M. Tanioka, H. Yoshida, T. Yoshioka, H. Nishimoto, S. Itohara, P. Hofman, B.C. van Blijswijk, P.J. Gaillard, G.F. Vrensen, and R.O. Schlingemann. 1998. Reduced angiogenesis and tumor progression in gelatinase A-deficient mice. *Cancer Res.* 58:1048-51.

Jacobson, S.G., A.V. Cideciyan, G. Regunath, F.J. Rodriguez, K. Vandenburgh, V.C. Sheffield, and E.M. Stone. 1995. Night blindness in Sorsby's fundus dystrophy reversed by vitamin A. *Nat Genet.* 11:27-32.

Kamei, M., S.S. Apte, M.E. Rayborn, H. Lewis, and J.G. Hollyfield. 1997. TIMP-3 accumulation in drusen and Bruch's membrane in eyes from donors with age-related macular degeneration (1997). *In* Degenerative Diseases of the Retina. M.M. Luvail, R.E. Anderson, and J.G. Hollyfield, editors. Plenum Press, New York. 11-15.

Kamei, M., and J. Hollyfield. 1999. TIMP-3 in Bruch's membrane: changes during aging and in age-related macular degeneration. *Invest Ophthalmol Vis Sci.* 40: p2367-75.

Killingsworth, M.C., J.P. Sarks, and S.H. Sarks. 1990. Macrophages related to Bruch's membrane in age-related macular degeneration. *Eye.* 4:613-21.

Kvanta, A., P.V. Algvere, L. Berglin, and S. Seregard. 1996. Subfoveal fibrovascular membranes in age-related macular degeneration express vascular endothelial growth factor. *Invest Ophthalmol Vis Sci.* 37:1929-34.

Langton, K.P., M.D. Barker, and N. McKie. 1998. Localization of the functional domains of human tissue inhibitor of metalloproteinases-3 and the effects of a Sorsby's fundus dystrophy mutation. *J Biol Chem.* 273:16778-81.

Langton, K.P., N. McKie, A. Curtis, J.A. Goodship, P.M. Bond, M.D. Barker, and M. Clarke. 2000. A novel tissue inhibitor of metalloproteinases-3 mutation reveals a common molecular phenotype in Sorsby's fundus dystrophy. *J Biol Chem.* 275:27027-31.

Leco, K.J., P. Waterhouse, O.H. Sanchez, K.L. Gowing, A.R. Poole, A. Wakeham, T.W. Mak, and R. Khokha. 2001. Spontaneous air space enlargement in the lungs of mice lacking tissue inhibitor of metalloproteinases-3 (TIMP-3). *J Clin Invest.* 108:817-829.

Lopez, P.F., B.D. Sippy, H.M. Lambert, A.B. Thach, and D.R. Hinton. 1996. Transdifferentiated retinal pigment epithelial cells are immunoreactive for vascular endothelial growth factor in surgically excised

age-related macular degeneration-related choroidal neovascular membranes. *Invest Ophthalmol Vis Sci.* 37:855-68.
Matrisian, L.M. 1992. The matrix-degrading metalloproteinases. *Bioessays.* 14:455-63.
Murphy, G. 1991. The regulation of connective tissue metalloproteinases by natural inhibitors. *In* AAS 35: Progress in Inflammation Research and Therapy. Birkhauser, Basel. 69-76.
Nguyen, Q., F. Willenbrock, M.I. Cockett, M. O'Shea, A.J. Docherty, and G. Murphy. 1994. Different domain interactions are involved in the binding of tissue inhibitors of metalloproteinases to stromelysin-1 and gelatinase A. *Biochemistry.* 33:2089-95.
Ogata, N., M. Matsushima, Y. Takada, T. Tobe, K. Takahashi, X. Yi, C. Yamamoto, H. Yamada, and M. Uyama. 1996. Expression of basic fibroblast growth factor mRNA in developing choroidal neovascularization. *Curr Eye Res.* 15:1008-18.
Polkinghorne, P.J., M.R. Capon, T. Berninger, A.L. Lyness, K. Sehmi, and A.C. Bird. 1989a. Sorsby's fundus dystrophy. A clinical study. *Ophthalmology.* 96:1763-8.
Polkinghorne, P.J., M.R.C. Capon, T. Berninger, A.L. Lyness, K. Sehmi, and A.C. Bird. 1989b. Sorsby's fundus dystrophy. A clinical study. *Ophthalmol.* 96:1763-1768.
Qi, J.H., Q. Ebrahem, K. Yeow, D.R. Edwards, P.L. Fox, and B. Anand-Apte. 2002. Expression of Sorsby's fundus dystrophy mutations in human retinal pigment Epithelial cells reduces matrix metalloproteinase inhibition and may promote angiogenesis. *J Biol Chem.* 30:30.
Ross, R.D., J.M. Barofsky, G. Cohen, W.B. Baber, S.W. Palao, and K.A. Gitter. 1998. Presumed macular choroidal watershed vascular filling, choroidal neovascularization, and systemic vascular disease in patients with age-related macular degeneration. *Am J Ophthalmol.* 125:71-80.
Ruiz, A., P. Brett, and D. Bok. 1996a. TIMP-3 is expressed in the human retinal pigment epithelium. *Biochem Biophys Res Commun.* 226: p467-74.
Ruiz, A., B. Peterson, and D. Bok. 1996b. Localization and quantification of TIMP-3 mRNA in the human retina. *Invest. Ophthalmol. Vis, Sci.* 37:S1143.
Sakamoto, T., H. Sakamoto, T.L. Murphy, C. Spee, D. Soriano, T. Ishibashi, D.R. Hinton, and S.J. Ryan. 1995. Vessel formation by choroidal endothelial cells in vitro is modulated by retinal pigment epithelial cells. *Arch Ophthalmol.* 113:512-20.
Sorsby, A., and M.E. Joll Mason. 1949. A fundus dystrophy with unusual features. *British journal of Ophthalmology*:67-97.
Soubrane, G., S.Y. Cohen, T. Delayre, J. Tassin, M.P. Hartmann, G.J. Coscas, Y. Courtois, and J.C. Jeanny. 1994. Basic fibroblast growth factor experimentally induced choroidal angiogenesis in the minipig. *Curr Eye Res.* 13:183-95.
Stetler-Stevenson, W.G. 2001. The role of matrix metalloproteinases in tumor invasion, metastasis, and angiogenesis. *Surg Oncol Clin N Am.* 10:383-92, x.
Stetler-Stevenson, W.G., and A.E. Yu. 2001. Proteases in invasion: matrix metalloproteinases. *Semin Cancer Biol.* 11:143-52.
Tabata, Y., Y. Isashiki, K. Kamimura, K. Nakao, and N. Ohba. 1998. A novel splice site mutation in the tissue inhibitor of the metalloproteinases-3 gene in Sorsby's fundus dystrophy with unusual clinical features. *Hum Genet.* 103:179-82.
Vranka, J.A., A. Shepardson, E.C. johnson, X.-H. Zhu, J.P. Alexander, and J.M.B. Bradley. 1996. TIMP3 expression in the human retina and choroid. *Invest. Ophthalmol. Vis. Sci.* 15:S273.
Vu, T.H., J.M. Shipley, G. Bergers, J.E. Berger, J.A. Helms, D. Hanahan, S.D. Shapiro, R.M. Senior, and Z. Werb. 1998. MMP-9/gelatinase B is a key regulator of growth plate angiogenesis and apoptosis of hypertrophic chondrocytes. *Cell.* 93:411-22.
Weber, B.H., B. Lin, K. White, K. Kohler, G. Soboleva, S. Herterich, M.W. Seeliger, G.B. Jaissle, C. Grimm, C. Reme, A. Wenzel, E. Asan, and H. Schrewe. 2002. A mouse model for Sorsby fundus dystrophy. *Invest Ophthalmol Vis Sci.* 43:2732-40.
Weber, B.H., G. Vogt, R.C. Pruett, H. Stohr, and U. Felbor. 1994. Mutations in the tissue inhibitor of metalloproteinases-3 (TIMP3) in patients with Sorsby's fundus dystrophy. *Nat Genet.* 8:352-6.
Werb, Z., T.H. Vu, J.L. Rinkenberger, and L.M. Coussens. 1999. Matrix-degrading proteases and angiogenesis during development and tumor formation. *Apmis.* 107:11-8.
Willenbrock, F., T. Crabbe, P.M. Slocombe, C.W. Sutton, A.J. Docherty, M.I. Cockett, M. O'Shea, K. Brocklehurst, I.R. Phillips, and G. Murphy. 1993. The activity of the tissue inhibitors of metalloproteinases is regulated by C-terminal domain interactions: a kinetic analysis of the inhibition of gelatinase A. *Biochemistry.* 32:4330-7.

Williamson, R.A., F.A. Marston, S. Angal, P. Koklitis, M. Panico, H.R. Morris, A.F. Carne, B.J. Smith, T.J. Harris, and R.B. Freedman. 1990. Disulphide bond assignment in human tissue inhibitor of metalloproteinases (TIMP). *Biochem J.* 268:267-74.

Woessner, J.F., Jr. 2001. That impish TIMP: the tissue inhibitor of metalloproteinases-3. *J Clin Invest.* 108:799-800.

Wu, G., R.C. Pruett, J. Baldinger, and T. Hirose. 1991. Hereditary hemorrhagic macular dystrophy. *Am J Ophthalmol.* 111:294-301.

Xu, J., D. Rodriguez, E. Petitclerc, J.J. Kim, M. Hangai, S.M. Yuen, G.E. Davis, and P.C. Brooks. 2001. Proteolytic exposure of a cryptic site within collagen type IV is required for angiogenesis and tumor growth in vivo. *J Cell Biol.* 154:1069-80.

Yeow, K.M., N.S. Kishnani, M. Hutton, S.P. Hawkes, G. Murphy, and D.R. Edwards. 2001. Sorsby's fundus dystrophy Tissue Inhibitor of Metalloproteinases-3 (TIMP-3) mutants have unimpaired matrix metalloproteinase inhibitory activities but affect cell adhesion to the extracellular matrix. *Matrix Biology.* In Press.

Yi, X., N. Ogata, M. Komada, C. Yamamoto, K. Takahashi, K. Omori, and M. Uyama. 1997. Vascular endothelial growth factor expression in choroidal neovascularization in rats. *Graefes Arch Clin Exp Ophthalmol.* 235:313-9.

Zack, D.J., M. Dean, R.S. Molday, J. Nathans, T.M. Redmond, E.M. Stone, A. Swaroop, D. Valle, and B.H. Weber. 1999. What can we learn about age-related macular degeneration from other retinal diseases? *Mol Vis.* 5:30.

Zucker, S. 1996. Correspondence re: N-B. Liabakk et al., matrix metalloprotease 2 (MMP-2) and matrix metalloprotease 9 (MMP-9) type IV collagenases in colorectal cancer. Cancer Res., 56: 190-196, 1966 [letter]. *Cancer Res.* 56:3610.

INVESTIGATIONS OF RPE CELLS OF CHORIODAL NEOVASCULAR MEMBRANES FROM PATIENTS WITH AGE-RELATED MACULA DEGENERATION

Rita Rosenthal[1] and Olaf Strauß[2]

1. INTRODUCTION

Age-related macular degeneration (AMD) is the most frequent cause for blindness in older age. The most severe complication is the development of choroidal neovascularisation (CNV). The retinal pigment epithelium (RPE) plays an important role in the induction of angiogenesis in AMD (Glaser *et al.*, 1985; Young, 1987; Morse *et al.*, 1989; Bird, 1991) by secreting the majority of angiogenic factors (Kliffen *et al.*, 1997). The most important angiogenic factor is VEGF (vascular endothelial growth factor). CNV membranes contain high amounts of VEGF and it is known that overexpression of VEGF in the RPE leads to choroidal neovascularisation (Schwesinger *et al.*, 2001). *In vitro* and *in situ* investigations revealed several extracellular stimuli inducing secretion of angiogenic factors by the RPE. A major part takes a hypoxia-comparable situation in the area of soft drusen (Eagle, 1984). In consequence to this, VEGF and factors inducing the VEGF production are secreted by the RPE and the neuronal retina (D' Amore, 1994; Shima *et al.*, 1995; Stone *et al.*, 1995; Frank *et al.*, 1996; Kliffen *et al.*, 1997; Tanihara *et al.*, 1997). VEGF secretion is stimulated by advanced-glycation endproducts (AGE) accumulating in CNV membranes, and also by vitronectin, the major extracellular matrix component in drusen (Hammes *et al.*, 1996; Hageman *et al.*, 1999). In addition, insulin-like growth factor-1 (IGF-1) secreted by the neuronal retina and tumour necrosis factor-α (TNFα) secreted by macrophages which migrate towards the retina during the development of the first blood vessels are known to induce VEGF secretion by RPE cells (Punglia *et al.*, 1997; Oh *et al.*, 1999).

However, in contrast to many extracellular factors, the corresponding changes in intracellular signalling leading to the changed secretory activity of the RPE are not known. Purpose of this study is the cell physiological characterization of freshly isolated RPE

[1]Department of Clinical Physiology, Benjamin Franklin University Hospital, Free University Berlin, 12200 Berlin, Germany; [2]Experimental Ophthalmology, Eye Hospital, University, Hospital Hamburg-Eppendorf, 20246 Hamburg, Germany. Corresponding author: Olaf Strauss, E-mail: o.strauss@uke.uni-hamburg.de

cells from CNV membranes to get new insights into mechanisms leading to a changed secretory activity.

2. METHODS

Cell physiological characterization was performed using freshly isolated and cultured RPE cells of CNV membranes. For cell isolation the surgical removed CNV membranes were dissolved into single cell suspension by incubation in trypsin/protease and then dissociated by gentle pipetting. The cells were plated on glasscover slips and used for electrophysiological and immunocytochemical investigations. Freshly isolated RPE cells were identified by pigmentation and staining for classIII beta-tubulin (Figure 1).

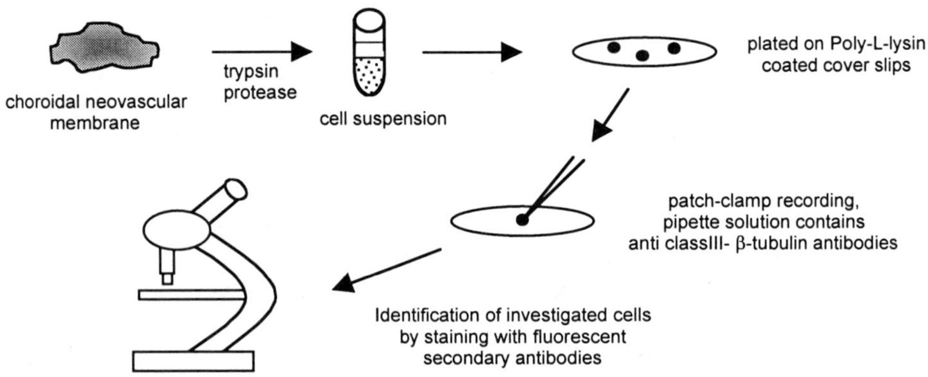

Figure1. Method for isolation and investigation of RPE cells from CNV membranes

For culture of CNV RPE cells the membranes were placed to microporous aluminium filter membranes mounted in culture inserts. To stimulate growth 20 % fetal calf serum (FCS) was applied for the first two weeks, thereafter 10 % FCS was used.

Changes in the properties of CNV RPE cells will be detected by comparison with identical experiments on human RPE cells from healthy donors without CNV. To culture these cells the RPE was removed from a donor eye and suspended into a single cell suspension which was plated in culture dishes.

Ion channels were investigated using the perforated-patch configuration of the patch-clamp technique.

3. RESULTS

Cell culture: Under our culture conditions, only the RPE cells in CNV membranes were able to proliferate. After a period of time RPE cells migrate out of the membrane and start to proliferate. Purity of RPE cell culture was confirmed by immunostaining with epithelial cell markers cytokeratin18 and classIII-β-tubulin and the mesenchymal cell marker vimentin and in addition by RT-PCR specific for the RPE marker RPE65.

Patch-Clamp: The examination of freshly isolated RPE cells of CNV membranes using the perforated-patch configuration of the patch-clamp technique demonstrated the presence of voltage-dependent ion currents. Depolarisation of RPE cells from a holding potential of –45 mV led to activation of outwardly rectifying currents with characteristics of the delayed rectifier potassium current (Figure 2). These currents could also be observed in cultured CNV RPE cells as well as in cultured human RPE cells without CNV. No differences in delayed rectifier currents were found between CNV RPE cells and control RPE cells of donor eyes without CNV.

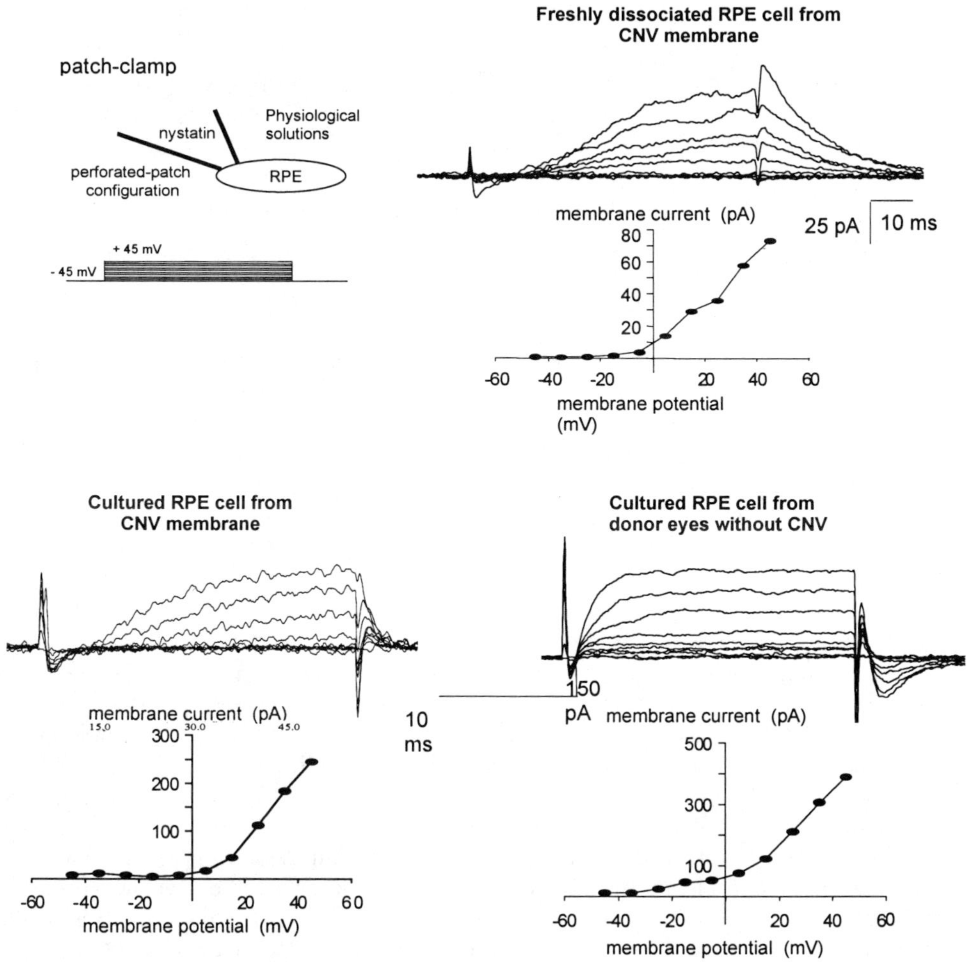

Figure 2. Outward currents with characteristics of the delayed rectifier K^+ current in freshly isolated and cultured RPE cells of CNV membranes and cultured human RPE cells.

However, when changing the holding potential to –70 mV, which is normally used for investigation of L-type Ca^{2+} channels, CNV RPE cells displayed depolarisation-induced inward currents. These currents showed an activation threshold of –30 mV and were reversibly activated by the dihydropyridine derivative BAYK8644, a specific L-type channel activator (Figure 3). This indicates the presence of L-type Ca^{2+} channels in freshly isolated RPE cells of CNV membranes. In human RPE cells inward currents could never be observed under the same conditions. These conditions comprise physiological extracellular Ca^{2+} concentration (1mM) and presence of physiological K^+ concentrations. This data were the first indication of an increased activity of L-type Ca^{2+} channels in RPE cells of CNV membranes.

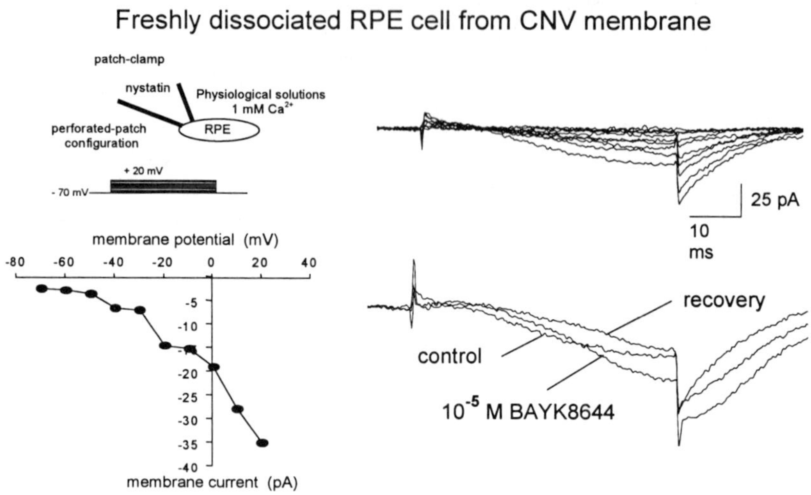

Figure 3. Inward currents with characteristics of L-type Ca^{2+} currents in RPE cells of CNV membranes.

4. DISCUSSION

This is the first cell physiological study of freshly isolated RPE cells from CNV membranes. Until now, RPE cells in CNV membranes are regarded as atrophic cells without function because the development of CNV membranes results into the formation of a subretinal fibrovascular scar. This study demonstrated that CNV RPE cells are highly active cells having the ability to migrate out of the surrounding tissue and to proliferate. CNV RPE cells could not be cultured as single cells in contrast to RPE cells from healthy donors. This observation suggests that CNV cells need the environmental factors of the membrane, perhaps stimuli from surrounding cells to start migration and proliferation.

Patch-clamp experiments revealed the presence of an outward current with characteristics of the delayed rectifier potassium current in freshly isolated and cultured CNV RPE. There seem to be no differences in the properties of this current between CNV RPE cells and RPE cells of donors without CNV. Patch-clamp studies documented several types of potassium channels in the RPE: M-type channels (Takahira and Hughes, 1997), Ca^{2+}-dependent potassium channels (Tao and Kelly, 1995), inward rectifier channels (Strauß *et al.*, 1993; Tao *et al.*, 1994; Hughes and Takahira, 1996) and delayed rectifier channels (Hughes and Steinberg, 1990; Strauß *et al.*, 1993; Strauß *et al.*, 1994; Tao *et al.*,

1994; Hughes *et al.*, 1995). Immunocytochemical and PCR-based investigations demonstrated the expression of outwardly rectifying potassium channel subunits Kv1.2, Kv1.3, Kv1.4, Kv2.1, and Kv4.2 (Pinto and Klumpp, 1998). Western blot analysis and immunocytochemical staining revealed presence of the Kv1.3 subtype of the delayed rectifier in RPE cells (Strauss *et al.*, 2002). The role of the delayed rectifier channels in the RPE is not fully understood. In many tissues K^+ channels are functional antagonists of voltage-dependent Ca^{2+} channels to control the intracellular Ca^{2+} homeostasis (Nelson *et al.*, 1990). Both channels are regulated by tyrosine kinase of the src-subtype (Holmes *et al.*, 1996; Strauß *et al.*, 1997). Tyrosine kinase activation leads to a simultaneous stimulation of delayed rectifier and L-type Ca^{2+} channels. This limits the depolarisation-induced activation of L-type channels and controls Ca^{2+} influx into the cell.

The first investigations of ion currents in CNV RPE indicated an increased density of L-type Ca^{2+} currents. Former studies demonstrated that L-type channels in RPE cells belong to the neuroendocrine subtype which corresponds to the expression of the α_{1D} (Cav1.3) subunits (Strauss *et al.*, 2000). Neuroendocrine L-type channels are expressed in tissues displaying secretory activity but also in neurons, receptor cells and myocardial cells (Catterall, 1998; Catterall, 2000). These L-type channels participate in the regulation of secretory activity (Catterall, 1998) and also in transmitters release by receptor cells (Platzer *et al.*, 2000). In neurons these Ca^{2+} channels are located in the cell body and dendrites (Hell *et al.*, 1993), and seem to mediate activity-dependent modulation of gene expression (Murphy *et al.*, 1991; Hell *et al.*, 1993; Bito *et al.*, 1996; Hell *et al.*, 1996; Deisseroth *et al.*, 1998).

Thus, the function of L-type channels in the RPE could be both regulation of secretion and modulation of gene expression and therefore differentiation. The latter function might result in the direct interaction of the growth factor receptor FGFR2 and the L-type channel α_{1D} subunit (Rosenthal *et al.*, 2001). An increased activity of L-type channels corresponds with increased secretory activity.

By means of their regulation, L-type channels in the RPE are ideal candidates to transduce the different extracellular stimuli which have been described for AMD into increased secretory activity. These Ca^{2+} channels are activated by tyrosine kinase (Wijetunge and Hughes, 1995; Strauß *et al.*, 1997), protein kinase C (Strauß *et al.*, 1997; Summers *et al.*, 2000), activation of growth factor receptors (Koike *et al.*, 1993; Puro *et al.*, 1996; Bence-Hanulec *et al.*, 2000; Rosenthal *et al.*, 2001) and binding of ligands to integrins (Wu *et al.*, 1998). In addition, in carotid body glomus cells L-type channel activity is augmented by hypoxia (Premkumar *et al.*, 2000a; Premkumar *et al.*, 2000b). Increased activity of L-type channels in CNV RPE cells may result from altered regulation by extracellular stimuli or intracellular modulators. If L-type channels in the RPE are involved in the regulation of secretion, the increased activity of L-type channels in CNV membranes seems to be of importance for the angiogenesis in AMD.

A second consequence of an increased L-type channel activity may be an influence on the gene expression in these cells. RPE cells in CNV membranes are known to show many different phenotypes. In addition, an increased secretory activity must include changes in gene expression for the production of the corresponding angiogenic factors.

Thus, changes in the activity of the L-type Ca^{2+} channel in CNV RPE cells may concern many different cellular functions and led to the phenotype of RPE cells which was found in AMD.

5. ACKNOWLEDGEMENTS

This study was supported by the DFG grants Str 480/3-1 and Str 480/8-1, by Pro Retina Germany and The Foundation Fighting Blindness. We thank the National Eye Institute for the travel award provided to RR to attend this meeting.

6. REFERENCES

Bence-Hanulec, K.K., Marshall, J. and Blair, L.A., 2000, Potentiation of neuronal L calcium channels by IGF-1 requires phosphorylation of the alpha1 subunit on a specific tyrosine residue, *Neuron* **27**:121-31.

Bird, A.C., 1991, Doyne Lecture. Pathogenesis of retinal pigment epithelial detachment in the elderly; the relevance of Bruch's membrane change, *Eye* **5**:1-12.

Bito, H., Deisseroth, K. and Tsien, R.W., 1996, CREB phosphorylation and dephosphorylation: a Ca^{2+}- and stimulus duration-dependent switch for hippocampal gene expression, *Cell* **87**:1203-14.

Catterall, W.A., 1998, Structure and function of neuronal Ca^{2+} channels and their role in neurotransmitter release, *Cell Calcium* **24**:307-23.

Catterall, W.A., 2000, Structure and regulation of voltage-gated Ca2+ channels, *Annu Rev Cell Dev Biol* **16**:521-55.

D' Amore, P.A., 1994, Mechanisms of retinal and choroidal neovascularization, *Invest Ophthalmol Vis Sci* **35**:3974-9.

Deisseroth, K., Heist, E.K. and Tsien, R.W., 1998, Translocation of calmodulin to the nucleus supports CREB phosphorylation in hippocampal neurons, *Nature* **392**:198-202.

Eagle, R.C., Jr., 1984, Mechanisms of maculopathy, *Ophthalmology* **91**:613-25.

Frank, R.N., Amin, R., Eliott, D., Puklin, J.E. and Abrams, G.W., 1996, Basic fibroblast growth factor and vascular endothelial growth factor present in epiretinal and choroidal neovascular membranes., *Am J Ophthalmology* **122**:393-403.

Glaser, B.M., Campochiaro, P.A., Davis, J.L., Jr. and Sato, M., 1985, Retinal pigment epithelial cells release an inhibitor of neovascularization, *Arch Ophthalmol* **103**:1870-5.

Hageman, G.S., Mullins, R.F., Russell, S.R., Johnson, L.V. and Anderson, D.H., 1999, Vitronectin is a constituent of ocular drusen and the vitronectin gene is expressed in human retinal pigmented epithelial cells, *FASEB J* **13**:477-84.

Hammes, H.P., Brownlee, M., Jonczyk, A., Sutter, A. and Preissner, K.T., 1996, Subcutaneous injection of a cyclic peptide antagonist of vitronectin receptor-type integrins inhibits retinal neovascularization, *Nat Med* **2**:529-33.

Hell, J.W., Westenbroek, R.E., Breeze, L.J., Wang, K.K.W., Chavkin, C. and Catterall, W.A., 1996, *N*-Methyl-D-aspartate receptor-induced proteolytic conversion of postsynaptic class C L-type calcium channels in hippocampal neurons., *Proc Natl Acad Sci* **93**:3362-67.

Hell, J.W., Westenbroek, R.E., Warner, C., Ahlijanian, M.K., Prystay, W., Gilbert, M.M., Snutch, T.P. and Catterall, W.A., 1993, Identification and differential subcellular localization of the neuronal class C and class D L-type calcium channel alpha 1 subunits, *J Cell Biol* **123**:949-62.

Holmes, T.C., Fadool, D.A. and Levitan, I.B., 1996, Tyrosine phosphorylation of the Kv 1.3 potassium channel., *J Neurosci* **16**:1581-90.

Hughes, B.A. and Steinberg, R.H., 1990, Voltage-dependent currents in isolated cells of the frog retinal pigment epithelium., *J Physiol* **428**:273-97.

Hughes, B.A. and Takahira, M., 1996, Inwardly rectifying K^+ currents in isolated human retinal pigment epithelial cells., *Invest Ophthalmol Vis Sci* **37**:1125-1139.

Hughes, B.A., Takahira, M. and Segawa, Y., 1995, An outwardly rectifying K^+ current active near resting potential in human retinal pigment epithelial cells., *Am J Physiol* **269**:179-187.

Kliffen, M., Sharma, H.S., Mooy, C.M., Kerkvliet, S. and de Jong, P.T., 1997, Increased expression of angiogenic growth factors in age-related maculopathy, *Brit J Ophthalmol* **81**:154-62.

Koike, H., Saito, H. and Matsuki, N., 1993, Effect of fibroblast growth factors on calcium currents in acutely isolated neuronal cells from rat ventromedial hypothalamus, *Neurosci Lett* **150**:57-60.

Morse, L.S., Terrell, J. and Sidikaro, Y., 1989, Bovine retinal pigment epithelium promotes proliferation of choroidal endothelium in vitro, *Arch Ophthalmol* **107**:1659-63.

Murphy, T.H., Worley, P.F. and Baraban, J.M., 1991, L-type voltage-sensitive calcium channels mediate synaptic activation of immediate early genes, *Neuron* **7**:625-35.

Nelson, M.T., Patlak, J.B., Worley, J.F. and Standen, N.B., 1990, Calcium channels, potassium channels, and voltage dependence of arterial smooth muscle tone., *Am J Physiol* **259**:3-18.

Oh, H., Takagi, H., Takagi, C., Suzuma, K., Otani, A., Ishida, K., Matsumura, M., Ogura, Y. and Honda, Y., 1999, The potential angiogenic role of macrophages in the formation of choroidal neovascular membranes, *Invest Ophthalmol Vis Sci* **40**:1891-8.

Pinto, L.H. and Klumpp, D.J., 1998, Localization of potassium channels in the retina, *Prog Retinal Eye Res* **17**:207-30.

Platzer, J., Engel, J., Schrott-Fischer, A., Stephan, K., Bova, S., Chen, H., Zheng, H. and Striessnig, J., 2000, Congenital deafness and sinoatrial node dysfunction in mice lacking class D L-type Ca2+ channels, *Cell* **102**:89-97.

Premkumar, D.R., Adhikary, G., Overholt, J.L., Simonson, M.S., Cherniack, N.S. and Prabhakar, N.R., 2000a, Intracellular pathways linking hypoxia to activation of c-fos and AP-1, *Adv Exp Med Biol* **475**:101-9.

Premkumar, D.R., Mishra, R.R., Overholt, J.L., Simonson, M.S., Cherniack, N.S. and Prabhakar, N.R., 2000b, L-type Ca(2+) channel activation regulates induction of c-fos transcription by hypoxia, *J Appl Physiol* **88**:1898-906.

Punglia, R.S., Lu, M., Hsu, J., Kuroki, M., Tolentino, M.J., Keough, K., Levy, A.P., Levy, N.S., Goldberg, M.A., RJ, D.A. and Adamis, A.P., 1997, Regulation of vascular endothelial growth factor expression by insulin-like growth factor I, *Diabetes* **46**:1619-26.

Puro, D.G., Hwang, J.-J., Kwon, O.-J. and Chin, H., 1996, Characterization of an L-type calcium channel expressed by human retinal Müller (glial) cells., *Mol Brain Res* **37**:41-48.

Rosenthal, R., Thieme, H. and Strauss, O., 2001, Fibroblast growth factor receptor 2 (FGFR2) in brain neurons and retinal pigment epithelial cells act via stimulation of neuroendocrine L-type channels (Ca_v1.3)., *FASEB J* **15**:970-97.

Schwesinger, C., Yee, C., Rohan, R.M., Joussen, A.M., Fernandez, A., Meyer, T.N., Poulaki, V., Ma, J.J., Redmond, T.M., Liu, S., Adamis, A.P. and RJ, D.A., 2001, Intrachoroidal neovascularization in transgenic mice overexpressing vascular endothelial growth factor in the retinal pigment epithelium, *Am J Pathol* **158**:1161-72.

Shima, D.T., Adamis, A.P., Ferrara, N., Yeo, K.-T., Yeo, T.-K., Allende, R., Folkman, J. and D'Amore, P.A., 1995, Hypoxic induction of endothelial cell growth factors in retinal cells: identification and characterization of vascular endothelial growth factor (VEGF) as the mitogen., *Mol Med* **1**:182-93.

Stone, J., Itin, A., Alon, T., Pe'er, J., Gnessin, H., Chan-Ling, T. and Keshet, E., 1995, Development of retinal vasculature is mediated by hypoxia-induced vascular endothelial growth factor (VEGF) expression by neuroglia, *J Neurosci* **15**:4738-47.

Strauss, O., Buss, F., Rosenthal, R., Fischer, D., Mergler, S., Stumpff, F. and Thieme, H., 2000, Activation of neuroendocrine L-type channels (alpha1D subunits) in retinal pigment epithelial cells and brain neurons by $pp60^{c\text{-}src}$, *Biochem Biophys Res Commun* **270**:806-10.

Strauß, O., Mergler, S. and Wiederholt, M., 1997, Regulation of L-type calcium channels by tyrosine kinase and protein kinase C in cultured rat and human retinal pigment epithelial cells., *FASEB J* **11**:859-867.

Strauß, O., Richard, G. and Wienrich, M., 1993, Voltage-dependent potassium currents in cultured human retinal pigment epithelial cells., *Biochem Biophys Res Commun* **191**:775-81.

Strauss, O., Rosenthal, R., Dey, D., Beninde, J., Wollmann, G., Thieme, H. and Wiederholt, M., 2002, Protein kinase C determines the effect of tyrosine kinase on delayed rectifier K^+ channels in rat RPE cells, *Invest Ophthalmol Vis Sci* **43**:1645-54.

Strauß, O., Weiser, T. and Wienrich, M., 1994, Potassium currents in cultured cells of the rat retinal pigment epithelium., *Comp Biochem Physiol A Physiol* **109A**:975-83.

Summers, B.A., Overholt, J.L. and Prabhakar, N.R., 2000, Augmentation of L-type calcium current by hypoxia in rabbit carotid body glomus cells: evidence for a PKC-sensitive pathway, *J Neurophysiol* **84**:1636-44.

Takahira, M. and Hughes, B.A., 1997, Isolated bovine retinal pigment epithelial cells express delayed rectifier type and M-type K^+ currents., *Am J Physiol* **273**:790-803.

Tanihara, H., Inatani, M. and Honda, Y., 1997, Growth factors and their receptors in the retina and pigment epithelium, *Prog Retinal Eye Res* **16**:271-301.

Tao, Q. and Kelly, M.E.M., 1995, Calcium-activated potassium current in cultured rabbit retinal pigment epithelial cells., *Curr Eye Res* **15**:237-46.

Tao, Q., Rafuse, P.E. and Kelly, M.E.M., 1994, Potassium currents in cultured rabbit retinal pigment epithelial cells., *J Membrane Biol* **141**:123-38.

Wijetunge, S. and Hughes, A.D., 1995, pp60c-src increases voltage-operated calcium channel currents in vascular smooth muscle cells., *Biochem Biophys Res Commun* **217**:1039-44.

Wu, X., Mogford, J.E., Platts, S.H., Davis, G.E., Meininger, G.A. and Davis, M.J., 1998, Modulation of calcium current in arteriolar smooth muscle by alphav beta3 and alpha5 beta1 integrin ligands, *J Cell Biol* **143**:241-52.

Young, R.W., 1987, Pathophysiology of age-related macular degeneration, *Surv Ophthalmol* **31**:291-306.

RETINAL AND CHOROIDAL ALTERATIONS FOLLOWING PHOTODYNAMIC THERAPY

Yusuf K. Durlu[1]

1. INTRODUCTION

Photodynamic therapy (PDT) has been reported as an effective method of treatment in patients with subfoveal choroidal neovascularisation (SFCNV) caused by age-related macular degeneration (AMD), presumed ocular histoplasmosis syndrome (POHS) and myopic maculopathy (No Authors, 1999; Bressler, 2001; No Authors, 2001). The basic mechanism of PDT is the light activation of photosensitizing agent based on its absorption profile. PDT involves two crucial steps;

I.The intravenous injection of photosensitizing drug (verteporfin)
II.The delivery of 689 nm. wavelength laser to the target tissue which is SFCNV

In Step I, verteporfin is injected intravenously and binds to low-density lipoprotein (LDL) that acts as a carrier protein. At the posterior segment eye tissues, verteporfin binds to the choroidal vascular endothelium, retinal pigment epithelium (RPE) and photoreceptor outer segments. It has been postulated that, the binding of verteporfin in SFCNV is highest as proliferating cells of choroidal neovascularized tissue exhibit strong expression of LDL receptors (Schmidt-Erfurth et al., 2000). Following Step II or after light activation, the photosensitizer is converted into two forms at the cellular level; a-excited state, b-lower energy triplet state which then leads to two distinct reaction types. Type I reaction involves the production of hydroxyl radicals, peroxides and superoxides, whereas a direct interaction to produce singlet oxygen is the main mechanism of Type II reaction which is believed to play a primary role in PDT-mediated phototoxicity (Schmidt-Erfurth et al., 2000). Local inflammatory response is initiated with the production of prostaglandin E2 and some cytokines including interleukin-2, interleukin-1β, and tumor necrosis factor-α (Schmidt-Erfurth et al., 2000; Hunt, et al., 1998). The occlusion of abnormal vessels at SFCNV occurs with thrombus formation. The permanent closure of SFCNV is the main goal in PDT, however this result may be

[1]Dünya Göz Hastanesi, World Eye Hospital, Nispetiye caddesi Aydın sk. No:1, 1.Levent 80620, Istanbul, TURKEY. E-mail: laser.lens@superonline.com

achieved after multipl PDT sessions repeated at three months interval if required as decided by fluorescein fundus angiography (FFA) results.

The aim of this retrospectively designed study is to disclose retinal and choroidal alterations by digital FFA, digital indocyanin green angiography (ICGA) and optical coherence tomography (OCT) recordings following PDT.

2. MATERIALS AND METHODS

Between August 2000 and September 2002, 119 patients with SFCNV were treated by PDT in Dünya Eye Hospital Retina Section. Complete ophthalmological examinations of all patients were done including visual acuity measurements by Snellen chart, biomicroscopy, tonometry and binocular fundus examinations. Digital stereo color, FFA and ICGA recordings were made by using Topcon Retinal Camera TRC-50IX with Sony DXXC-950P 3CCD and Hitachi CCD cameras. Fluorescein and indocyanin green dyes were supplied from Novartis and Akorn, respectively. OCT recordings were performed by Model 3000 OCT Zeiss Optical Systems at the same day and during post-PDT period as indicated. The greatest linear dimension (GLD) measurements and subtraction analysis were done with a software incorporated into Topcon Imagenet program. A total of 192 PDT applications were performed according to the previously published protocol in 65 male (55%) and 54 female (45%) patients. Verteporfin (commercially known as Visudyne supplied from Novartis) at a dose of 5 mg/m (Bressler, 2001) body surface, was injected from antecubital vein of the patient wih a pump (Braun Melsungen AG/Perfusor TM) infusing at a rate of 180 ml/hr for ten minutes. After five minutes, laser at a wavelength of 689 nm (Coherent Opal Photoactivator) was delivered at an intensity of 50 J/cm (Bressler, 2001) or 600 mW/cm (Bressler, 2001) over 83 seconds. PDT lens (X1.6) was used for the delivery of laser. The spot size of the laser is determined by adding 1000 μ to GLD.

The indications were as follows; age-related macular degeneration (93 cases-78%), myopic maculopathy (11 cases-9%), angioid streaks (6 cases-5%), post-laser (5 cases-4%), idiopathic (2 cases-2%), presumed ocular histoplasmosis syndrome (1 case-1%) and multifocal choroiditis (1 case-1%). The age range of the patients was 29-94 yrs. 71 cases (59.6%) received single PDT application, whereas 32 cases (26.9%) received two PDT applications, 9 cases (7.6%) received three PDT applications, 5 cases (4.2%) received four PDT applications and 2 cases (1.7%) received five PDT applications. After each consecutive PDT session, the patients were strongly advised to avoid direct sunlight and bright halogen indoor light and to wear dark protective sunglasses for 2 days. The follow-up period was 1-24 months.

3. RESULTS

The systemic side effects observed in our PDT patients include backpain during the infusion of verteporfin (4 cases (3.4%); 3 cases with AMD and a case with angioid streaks) and pain at the injection site (7 cases (5.9%); 6 cases with AMD and a case with myopic maculopathy). We did not observe any extravasation, inflammation and necrosis in none of the cases.

Chorio-retinal side effects seen after PDTwere shown at Table 1.

Table 1. Chorio-retinal side effects after PDT

SUBMACULAR FIBROSIS: 11 CASES (10.1 %)
- 9 CASES: AMD
- 2 CASES: ANGIOID STREAKS
- 1 CASE: SFCNV AFTER ARGON LASER TREATED BY PDT

SUBMACULAR HEMORRHAGE: 5 CASES (4.2 %)
- 4 CASES: AMD
- 1 CASE: ANGIOID STREAKS

EXUDATIVE MACULAR DETACHMENT: 3 CASES (2.5%)
- 2 CASES: AMD
- 1 CASE: PRESUMED OCULAR HISTOPLASMOSIS SYNDROME

CHOROIDAL HYPOPERFUSION: 3 CASES (2.5 %): AMD

RPE ATROPHY: 2 CASES (1.6%): AMD

RPE TEAR: 1 CASE (0.8 %): AMD

3.1. Case 1

75 yrs-old female complained of decreased vision in her right eye two weeks ago, and in her left eye for the last five years. She had stable systemic hypertension under medical control. Her ophthalmic examination disclosed best corrected visual acuity (BCVA) as right eye (OD):0.1 and left eye (OS):0.05. Bilateral nuclear sclerosis was noticed at anterior segment examination. Ocular tensions were OD:16 mmHg and OS:14 mmHg, respectively. SFCNV was seen at her right eye and submacular scarring at her left eye during fundus examination. AMD was diagnosed. The signs of bilateral temporal conus and and hypertensive retinopathy at Stage II (Keith-Wegener Classification) were also found in both eyes. FFA disclosed a delay in early venous circulation time and predominantly classic SFCNV (Fig. 1).The GLD of the lesion (SFCNV) was 2300 μ .

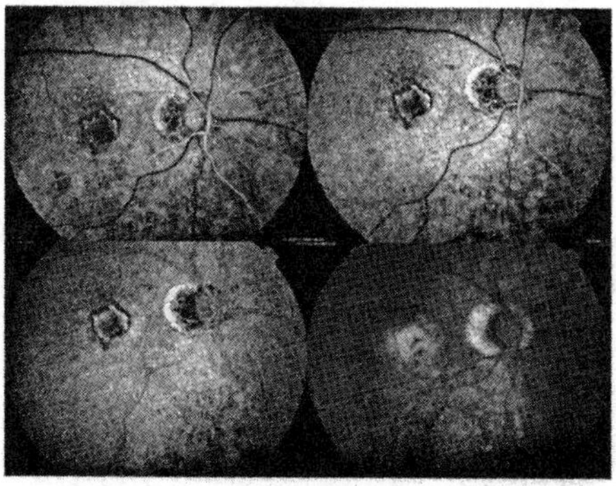

Figure 1. Case 1: FFA, early (19-22-28 sec.; upper left, upper right and lower left) and late (6:52 min.; lower right) frames showing a delay in early venous circulation time and predominantly classic SFCNV of the right eye.

Figure 2. Case 1: FFA, three months after PDT (upper left and upper right) showing subfoveal scar tissue, and fifteen months after PDT (lower left and lower right) showing marked peri-lesional RPE/choriocapillaris atrophy and subfoveal scar tissue at the right eye.

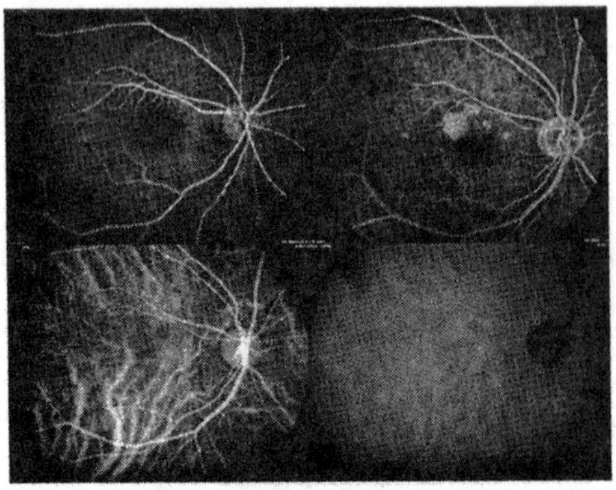

Figure 3. Case 2: FFA (upper left and upper right) disclosing extrafoveal fibrovascular pigment epithelial detachment and ICGA (lower left and lower right) disclosing no hot spot at the right eye.

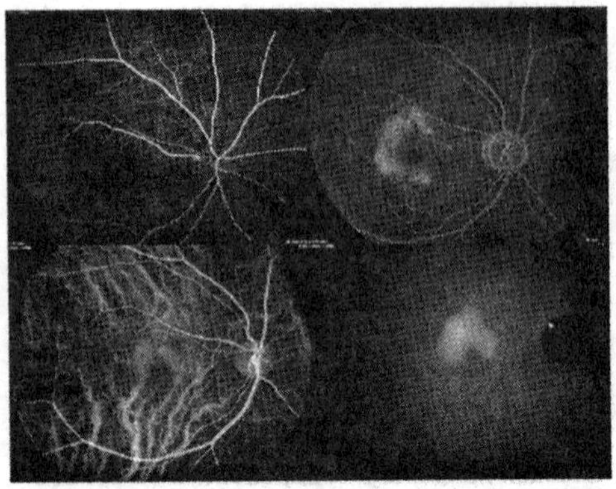

Figure 4. Case 2: FFA (upper left and upper right) and ICGA (lower left and lower right) showing signs of occult SFCNV at the right eye before PDT.

After infusion of verteporfin [5 mg/m (Bressler, 2001) body surface], 3300 μ spot size delivering 51 mw laser was applied over SFCNV at her right eye. Three months after PDT, her fundus examination showed subfoveal scarring at her right eye. The other ophthalmic findings were same as before PDT. The FFA was repeated and the cessation of the leakage of the SFCNV was found (Fig. 2). The patient was re-examined fifteen months after PDT, and perilesional RPE atrophy was noticed in addition to subfoveal scarring of her right eye. The FFA disclosed RPE and choriocapillaris atrophy at the same region corresponding to the laser applied area with 3300 μ diameter size (Fig. 2).

3.2. Case 2

72 yrs-old male was examined because of distorted images in his right eye for the last ten days. He had decreased vision in his left eye for four years. His BCVA was 1.0 at his right eye and 0.05 at his left eye. Bilateral nuclear sclerosis was noticed at anterior segment examination. Ocular tensions were OD:15 mmHg and OS:18 mmHg, respectively. Shallow RPE detachment was noticed at the macula of his right eye and submacular scarring at his left eye during fundus examination. His FFA disclosed the RPE detachment which was fibrovascular in nature, whereas no *hot spot* was seen at ICGA (Fig. 3). The patient was advised for follow-up at monthly interval by Amsler test. Four weeks after his initial eye examination, the patient complained of a sudden decrease of vision at his right eye. The BCVA was dropped to 0.1 with the loss of macular reflex at his right eye. The FFA and ICGA disclosed signs of occult SFCNV at his right eye which was due to AMD (Fig. 4). The GLD of the lesion was 4100 μ, therefore 5100 μ spot size delivering 121 mw laser was applied over macula at his right eye, after infusion of verteporfin [5 mg/m (Bressler, 2001) body surface]. Three weeks after PDT, the visual acuity of the right eye was reduced to counting fingers. His fundus examination, FFA and ICGA of the right eye showed giant RPE tear involving the infero-temporal region of the

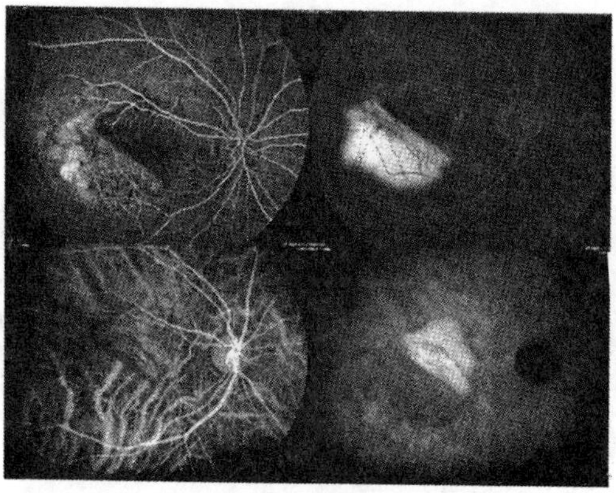

Figure 5. Case 2: FFA (upper left and upper right) and ICGA (lower left and lower right) showing signs of giant RPE tear located at the infero-temporal region of the macula and the redundant RPE locating subfoveally at the right eye three weeks following PDT.

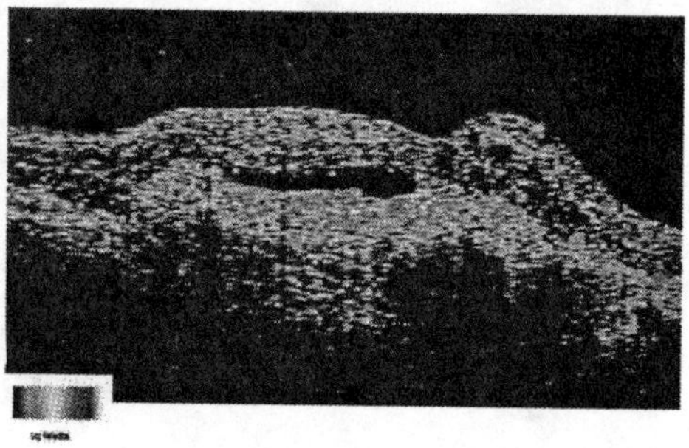

Figure 6. Case 2: OCT of the macular region of the right eye, three months after second session of PDT. Exudative macular detachment and macular edema were seen with a relatively preserved foveal contour.

macula (Fig. 5). Three months after PDT, the visual acuity of the right eye was increased at 0.2. The FFA and ICGA of the right eye disclosed RPE tear at the infero-temporal macular region with choroidal neovascularisation and redundant RPE involving the foveal region. The patient was advised for regular follow-up at three months intervals. Three months later, his visual acuity was dropped to 0.05 at his right eye. The fundus examination, FFA and ICGA showed the growth of SFCNV lesion at the same eye and the second session of PDT was performed with the 6400 μ spot size delivering 196 mw laser at his right eye, after infusion of verteporfin [5 mg/m (Bressler, 2001) body surface]. Three months after second session of PDT, the visual acuity remained same as before second session of PDT. During fundus examination, submacular and subretinal hemorrhage was noticed at his right eye. FFA and ICGA showed an increase of the SFCNV lesion at his right eye. The OCT disclosed localized macular detachment at his right eye (Fig. 6). The patient was again advised for regular follow-up at three months intervals. Six months after second session of PDT, submacular hemorrhage was diminished with a mild increase in BCVA to 0.2 (Fig. 7), and the foveal contour was remarkable with reduced localized macular detachment at the same time at his right eye (Fig. 8). The retinal thickness change analysis (RTCA) compared as three months versus six months after second PDT session showed the resolution of the edema at the center of fovea (umbo) at his right eye (Fig. 9).

4. DISCUSSION

In this study, it was shown that the potential complications seen after PDT might decrease the quality of vision in AMD patients. It was known that all of the complications observed during post-PDT period including choroidal hypoperfusion and RPE/choriocapillaris atrophy noticed at the laser applied area in this study, could also be

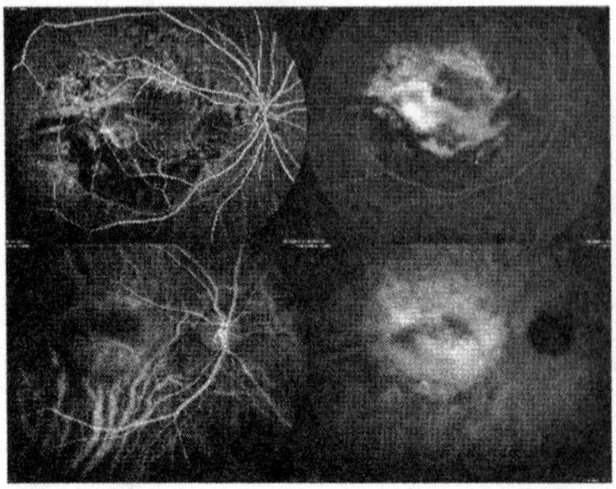

Figure 7. Case 2: FFA (upper left and upper right) and ICGA (lower left and lower right) disclosing submacular hemorrhage and extension of submacular fibrovascular tissue at the right eye six months after second session of PDT

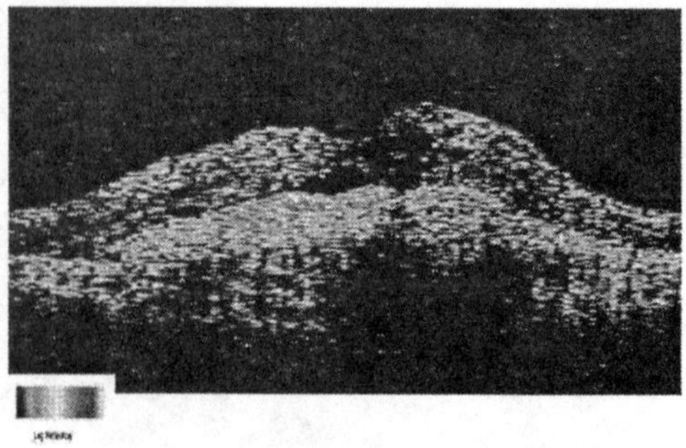

Figure 8. Case 2: OCT of the macular region of the right eye, six months after second session of PDT. Exudative macular detachment and macular edema was resolved with highly preserved foveal contour.

seen during the course of AMD. However, it could also be possible that PDT might effect the occurence and prognosis of those complications.

It has been reported that verteporfin binds to choroidal neovascular/vascular endothelial cells, RPE and photoreceptor outer segments (Schmidt-Erfurth et al., 2000).

In Case 1 with AMD in this study, a delay in early venous circulation time of FFA might explain a disturbance of retinal blood circulation (Fig. 1). PDT application might have an effect on the RPE and choriocapillaris endothelial cells seen in this case (Fig. 2), as the size of the annular-shaped RPE and choriocapillaris atrophy was almost the same diameter of the spot size of the laser. Michels et. al. indicated that perilesional hypofluorescence was noted by ICGA in their cases and speculated that the hypofluorescence reflected choroidal hypoperfusion which reached its maximum size at 12 months then gradually recovered by 24 months (Michels et al., 2000; Michels et al., 2002). The RPE atrophy might lead to choriocapillaris atrophy as known from the experimental and clinical studies (Korte et al., 1984; Gass, 1997). Experimentally, RPE atrophy could be developed by sodium iodate in the rabbits which was followed by choriocapillaris atrophy, therefore, it was commented that RPE might modulate choriocapillaris structure and function (Korte et al., 1984). Clinically, the atrophy of RPE and choriocapillaris could be seen together in AMD cases (Gass, 1997). In choroideremia, choriocapillaris atrophy coincides with the RPE atrophy (Gass, 1997).

In Case 2, a giant RPE tear was noticed three weeks after PDT in this study (Fig. 5). Although, this complication might be due to the natural course of the AMD, PDT might fascilitate the development of RPE tear in occult AMD cases. After PDT, exudative macular detachment could be seen in 2.5% of the cases in this study. The exudation might be related with the inflammatory response of the tissue. The swollen tissue might stretch the underlying RPE which could result to a giant RPE tear in this case. Several cases of RPE tear following PDT have been reported (Gelisken et al., 2001; Pece, et al., 2001). It has been speculated that the destructive effect of PDT on RPE may lead to RPE tear which was seen four weeks after PDT in AMD with classic SFCNV (Gelisken et al., 2001). RPE tears could also been observed as a final outcome of pigment epithelial

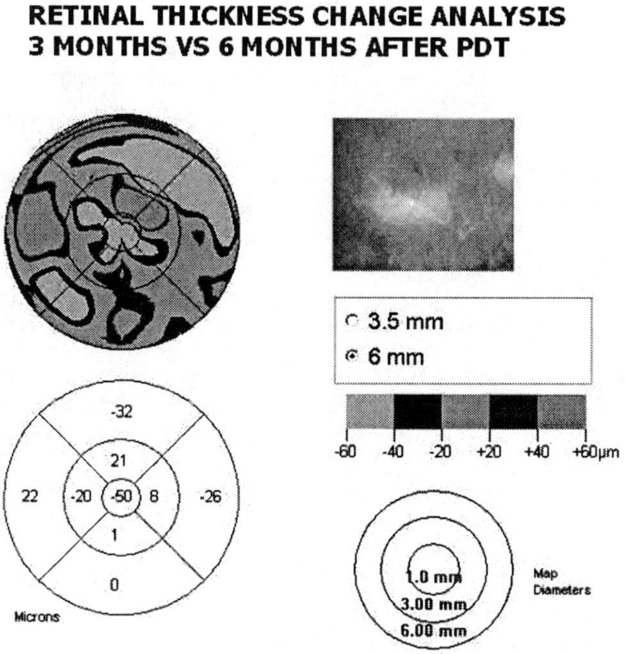

Figure 9. Case 2: Comparative Retinal Thickness Change Analysis of Figs. 6 and 8 disclosed reduction of macular edema which was more prominent at the subfoveal location.

detachments during the natural course of AMD (Gass, 1997). Pece et. al. reported four cases of RPE tears that were developed at 2 to 15 days after PDT in occult AMD with pigment epithelial detachment (Pece et al., 2001). There was no direct evidence indicating that PDT causes RPE tear in the previous studies. OCT findings in this case disclosed that micron based changes in the fovea after PDT could be useful to monitor the recovery of the patient (Fig. 9) (Rogers et al., 2002).

5. CONCLUSION

The most of the undesirable ocular side effects following PDT may be the result of the natural course of AMD. The rate of the retinal and choroidal alterations seen after PDT in this study is in accordance with the previous reports. The choroidal hypoperfusion and exudative macular detachment noticed during the early post-PDT period may be limited to an individual basis. The application of macular OCT after PDT may give additional clues, only if evaluated together with FFA and ICGA. In conclusion, PDT is found as a safe procedure in this study.

6. REFERENCES

Bressler, N.M., 2001, Photodynamic therapy of subfoveal choroidal neovascularization in age-related macular degeneration with verteporfin: two-year results of 2 randomized clinical trials-tap report 2., *Arch. Ophthalmol.*, Feb; **119**(2):198-207.

Gass, J.D.M., 1997, *Stereoscopic Atlas of Macular Diseases Diagnosis and Treatment*, Mosby, St. Louis.

Gelisken, F., Inhoffen, W., Partsch, M., Schneider, U., and Kreissig, I., 2001, Retinal pigment epithelial tear after photodynamic therapy for choroidal neovascularization, *Am. J. Ophthalmol.*, Apr; **131**(4):518-20.

Hunt, D.W.C., Chan, A.H., and Levy, J.G., 1998, Immunological aspects of photodynamic therapy, *Photodynamics News* **1**(3):2-4.

Korte, G.E., Reppucci, V., and Henkind, P., 1984, RPE destruction causes choriocapillary atrophy, *Invest. Ophthalmol. Vis. Sci.*, Oct; **25**(10):1135-45.

Michels, S., Barbazetto, I., and Schmidt-Erfurth, U., 2000, Choroidal changes after photodynamic therapy (PDT). A two-year follow-up study of 38 patients[Article in German], *Klin. Monatsbl. Augenheilkd.*, Aug; **217**(2):94-9.

Michels, S., Barbazetto, I., and Schmidt-Erfurth, U., 2002, Changes in neovascular membranes and normal choroid blood vessels after multiple photodynamic therapy treatments [Article in German], *Ophthalmologe,* Feb; **99**(2):96-100.

No Authors Listed, 1999, Photodynamic therapy of subfoveal choroidal neovascularization in age-related macular degeneration with verteporfin: one-year results of 2 randomized clinical trials--TAP report. Treatment of age-related macular degeneration with photodynamic therapy (TAP) Study Group, *Arch. Ophthalmol.*, Oct; **117**(10):1329-45.

No Authors Listed, 2001, Photodynamic therapy of subfoveal choroidal neovascularization in pathologic myopia with verteporfin. 1-year results of a randomized clinical trial--VIP report no. 1., *Ophthalmology,* May; **108**(5):841-52.

Pece, A., Introini, U., Bottoni, F., and Brancato, R., 2001, Acute retinal pigment epithelial tear after photodynamic therapy, *Retina.*, **21**(6):661-5.

Rogers, A.H., Martidis, A., Greenberg, P.B., and Puliafito, C.A., 2002, Optical coherence tomography findings following photodynamic therapy of choroidal neovascularization, *Am. J. Ophthalmol.*, Oct; **134**(4):566-76.

Schmidt-Erfurth, U., and Hasan, T., 2000, Mechanisms of action of photodynamic therapy with verteporfin for the treatment of age-related macular degeneration, *Surv. Ophthalmol.*, Nov-Dec; **45**(3):195-214.

USHER SYNDROME: CORRELATION BETWEEN VISUAL FIELD SIZE AND MAXIMAL ERG RESPONSE B-WAVE AMPLITUDE

Alessandro Iannaccone[1]

1. INTRODUCTION

The evaluation of a patient affected with hereditary retinal degenerations is based on the combination of psychophysical (e.g., visual acuity, visual field) and objective measurements [(e.g., the electroretinogram (ERG)]. The ERG provides essential information on the amount and quality of functioning retina left. The a-wave of the full-field flash maximal ERG response reflects the combined activity of rod and cone photoreceptors, whereas the b-wave reflects the activity of both the rod- and the cone-driven postreceptoral pathways. Vision begins at the level of photoreceptors but is subsequently fine-tuned via a series of gain and filtering steps taking place both inside the retina and, of course, at the brain level. Therefore, among the ERG parameters, the b-wave of the full-field flash maximal ERG response offers, however imperfect it may be, the best approximation of the full visual potential of any given retina, as it already encompasses the gain of the visual response performed intraretinally by the bipolar cells. In addition, since both the rod- and cone-driven pathways can contribute to the detection of the test targets used conventionally to delimit visual fields with the Goldmann kinetic technique, the maximal ERG response b-wave is well suited also for correlation analysis between ERG and visual fields obtained under these conditions.

A relationship between GVF size and ERG sensitivity (i.e., the reciprocal of the energy required to elicit a criterion ERG response of 1μV) in patients with retinitis pigmentosa (RP) was first noted by Gouras et al.(Gouras, et al., 1964) Despite these premises, the relationship between ERG amplitude and Goldmann visual fields (GVFs) has been object of significant controversy until recently. In a number of subsequent reports (reviewed in Iannaccone, et al., 1995), the maximal ERG response was frequently noted to underestimate GVF area in patients with RP, being often reported to be barely

[1]Department of Ophthalmology, University of Tennessee Health Science Center, 956 Court Avenue, Suite D228, Memphis, TN 38163, USA. E-mail: iannacca@utmem.edu

Retinal Degenerations: Mechanisms and Experimental Therapy
Edited by LaVail *et al.*, Kluwer Academic/Plenum Publishers, 2003

in patients with relatively well-maintained GVF outer limits. This raised doubts about the concordance of these two measures of visual function. However, more recent investigations on large RP patient populations have demonstrated that, if adequate ERG recording techniques are used (Rispoli et al., 1994) and accurate GVF area measurements are made, the maximal ERG response b-wave is in fact closely correlated to GVF area in most instances (Iannaccone et al., 1995). Independent confirmation of the close correlation between the ERG and the area of the visual field has been subsequently provided by others (Sandberg et al., 1996). Therefore, the ERG is nowadays considered to be an accurate and meaningful measure of residual retinal function in non-syndromic RP that can be reasonably used for studies of natural history of retinal disease and as an outcome measure for treatment trials.

Table 1. Clinical classification of Usher syndrome

Clinical variables	Usher syndrome Type 1	Usher syndrome Type 2	Usher syndrome Type 3
Retina	RP	RP	RP
Hearing loss:			
Onset	From birth	From birth	Late-onset
Severity	Moderate to profound	Mild to moderate	Mild to moderate
Progressive	No	No	Yes
Speech abnormality	Moderate to severe	Mild to moderate	Absent
Vestibular dysfunction	Present	Absent or minimal	Variable

Usher syndrome is a condition characterized by a retinal degeneration of the RP family and hearing loss. Three major clinical Usher syndrome subtypes can be distinguished on clinical grounds (Table 1). Retinal degeneration in Usher syndrome is as of yet clinically undistinguishable from non-syndromic RP, and there are presently no established criteria to differentiate retinal degeneration across Usher syndrome types. Despite these similarities, Usher syndrome is genetically distinct from RP. To date, 11 distinct Usher syndrome chromosomal loci have been identified (USH1A-G, USH2A-C, and USH3A). Many of the causal genes at these loci have been cloned: *USH1B*, encoding the unconventional myosin VIIa (Weil et al., 1995), *USH1C*, encoding the PDZ-containing domain protein harmonin (Bitner-Glindzicz et al., 2000), *USH1D*, encoding cadherin 23 (Bolz et al., 2001; Bork et al., 2001; Di Palma et al., 2001), *USH1F*, encoding protocadherin 15 (Ahmed et al., 2001; Alagramam et al., 2001), *USH2A*, encoding for the basement membrane protein usherin (Eudy et al., 1998; Bhattacharya et al., 2002), and *USH3A*, encoding the transmembrane protein clarin-1 (Joensuu et al., 2001). Of these, mutations in *USH1B* and *USH2A* account for most cases of Usher syndrome type 1 and 2, respectively (Astuto et al., 2000; Leroy et al., 2001; Najera et al., 2002; Weston et al., 1996).

With the exception of the Cys759Phe missense mutation in the *USH2A* gene (Rivolta et al., 2000), no other change in any of the Usher's genes has yet been implicated in causing RP without hearing loss. The mechanism(s) leading to retinal degeneration in Usher syndrome are not yet understood and may be different from those underlying non-syndromic RP. Therefore, the extrapolation of data on non-syndromic RP to Usher syndrome requires caution, because the relationship between functional phenotypic variables may also be different. For example, it has been noted that, different from classical RP, ERGs are non-recordable in the majority of patients with Bardet-Biedl syndrome, even when good visual field expanses can be documented (Iannaccone et al., 1996). In addition, although previous studies have reported on visual functional outcomes in Usher syndrome (Fishman et al., 1995; Edwards et al., 1998; Seeliger et al., 1999), the relationship between ERG and GVF area in Usher syndrome in unknown. It is also unknown whether this relationship may be different between Usher syndrome type 1 and type 2 patients. Therefore, the purpose of this investigation was to determine if a correlation between the size of the residual visual field and the amplitude of the b-wave of the maximal ERG response could be identified also in a cohort of affected patients with Usher syndrome.

2. METHODS

Twenty-three patients with an established diagnosis of Usher syndrome (Usher syndrome-1: n=9; Usher syndrome-2: n=14) were included in this study (age range: 13-56 years; M=15; F=8). Some of the patients had serial measurements over the years, obtained several years apart. The average areas of a total of 70 Goldmann visual fields (GVFs) were calculated for the I4e and the V4e target from the mean radius of the field, determined in degrees at 12 meridians. Maximal ERG responses were elicited with a 0.5-Hz full-field white flash of about 2 cd·s /m^2 meeting the ISCEV standards, and the amplitude of the b-wave was measured from trough to peak. Signal averaging was performed in all cases as needed, up to 100 iterations (Rispoli et al., 1994).

Data were tested for normal distribution with the Kolmogorov-Smirnov test. Since the assumption of normality was violated, data were normalized by log transformation to be suitable for linear regression analysis. Correlation coefficients R and associated p values were determined for the sample as a whole. The same analyses were applied also to subgroups based on: clinical diagnosis (Type 1 Usher syndrome GVFs: n=28; Type 2 Usher syndrome GVFs: n=40); ERG b-wave size ($\leq$10μV: n=39); and GVF size (radius $\leq$10 degrees to the I4e target: n=51).

For ease of interpretation of the log-transformed data: for ERG b-wave amplitude, 1 log unit (logU) equals to 10 μV and 2 logU corresponds to 100 μV; whereas for GVF areas, 2 logU indicate a GVF of 5.5 degrees in average radius, 3 logU a GVF 18.5 degrees in average radius, and 4 logU a GVF of 57 degrees in average radius.

3. RESULTS

The results of this analysis are summarized in Table 2 and are illustrated in part in Figures 1 through 4. A reasonably good correlation between VF area and ERG b-wave amplitude was observed for both the I4e and V4e targets in the entire cohort of Usher syndrome patients (Figures 1 and 2). However, when compared to results for the I4e target, a small subset of patients appeared to differ from the others by having non-recordable ERGs despite fairly well preserved GVFs to the V4e target (2.75 to 3.1 logU, approximately equivalent to 15 to 30 degrees of radius, and a possible outlier at 3.9 logU). Most of these patients had Usher syndrome type 1 with a I4e target area that was either quite small or not measurable, which is more consistent with the increased likelihood of non-recordable ERGs.).

Table 2. Summary of the results of correlation analysis between visual field area (targets I4e and V4e) and maximal ERG b-wave amplitude

	b-wave *vs.* I4e		b-wave *vs.* V4e	
Subgroups	*R value*	*p value*	*R value*	*p value*
Usher syndrome (n=70)	0.767	0.0001	0.798	0.0001
Usher syndrome Type 1 (n=28)	0.865	0.0001	0.831	0.0001
Usher syndrome Type 2 (n=40)	0.698	0.0001	0.784	0.0001
Usher syndrome with ERG amplitude ≤10 μV (n=39)	0.696	0.0001	0.494	0.0014
Usher syndrome with GVF radius ≤10 degs (n=51)	0.665	0.0001	–	–

No significant difference in the correlation coefficients was observed between Usher syndrome type 1 (Figure 3) and type 2 (Figure 4) patients, indicating that, for the most part, the correlation between the two measures held across different types of the disease. Despite persistent statistical significance, The only notable exception to this overall behavior was a sizable drop in the correlation coefficient for the V4e areas in type 1 patients. This was accounted for by the cases of non-recordable ERG despite relatively well preserved field expanses discussed above.

When the analysis was limited to ERG b-waves ≤10 μV in amplitude, a significant correlation was still appreciable for both the I4e target and the V4e target (Table 2). This is an amplitude range in which ERG responses recorded without signal averaging and adequate low-noise recording techniques (Rispoli et al., 1994) are indistinguishable from noise in most laboratories. Likewise, when limiting the correlation analysis to patients with

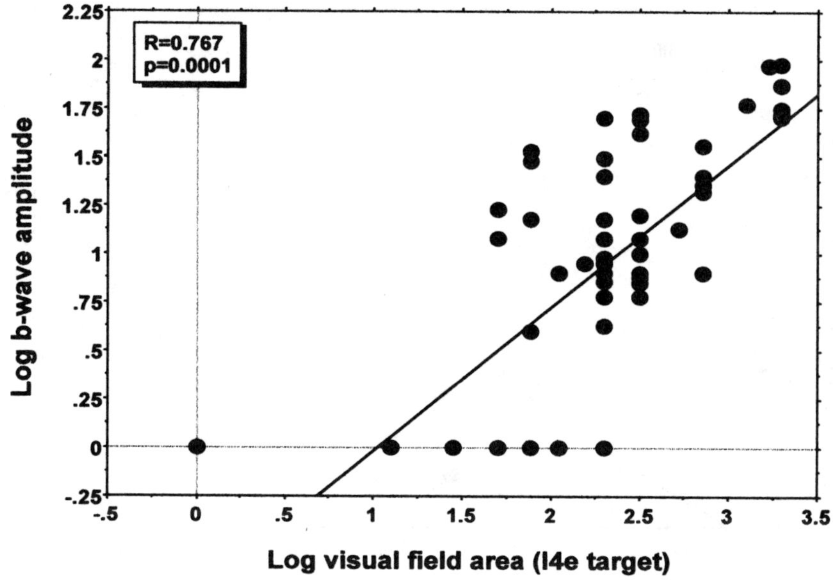

Figure 1. Correlation between ERG (log b-wave amplitude) and visual field size (log of the area delimited with the I4e target, in square degrees)

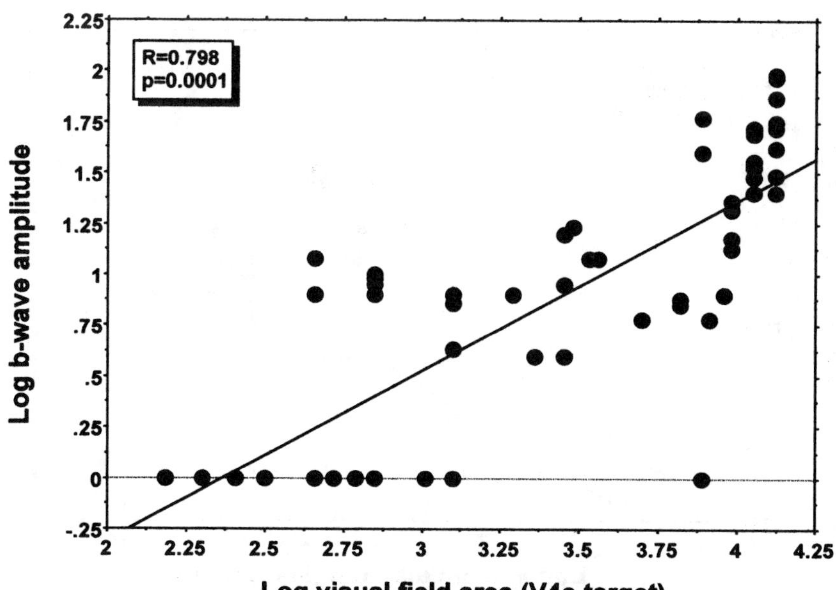

Figure 2. Correlation between ERG (log b-wave amplitude) and visual field size (log of the area delimited with the V4e target, in square degrees)

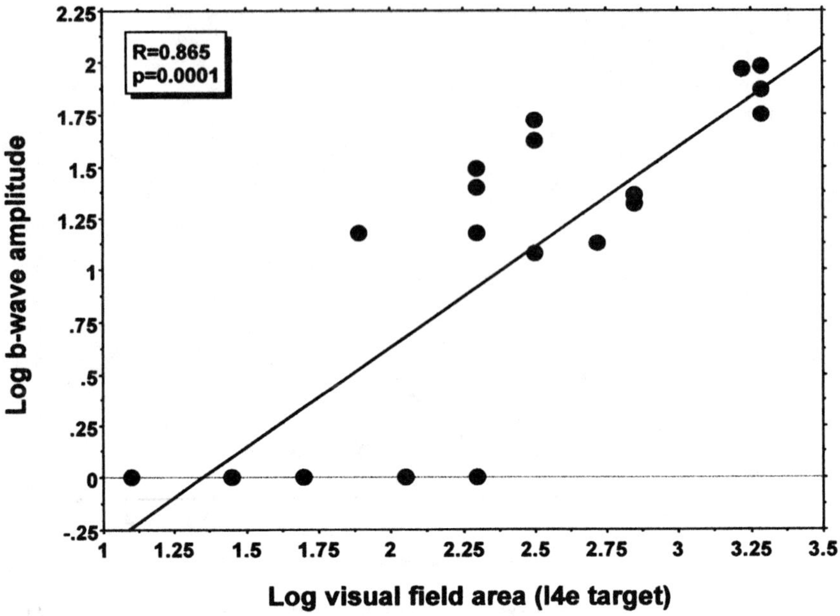

Figure 3. Correlation between ERG (log b-wave amplitude) and visual field size (log of the area delimited with the I4e target, in square degrees) in patients with Usher syndrome type 1.

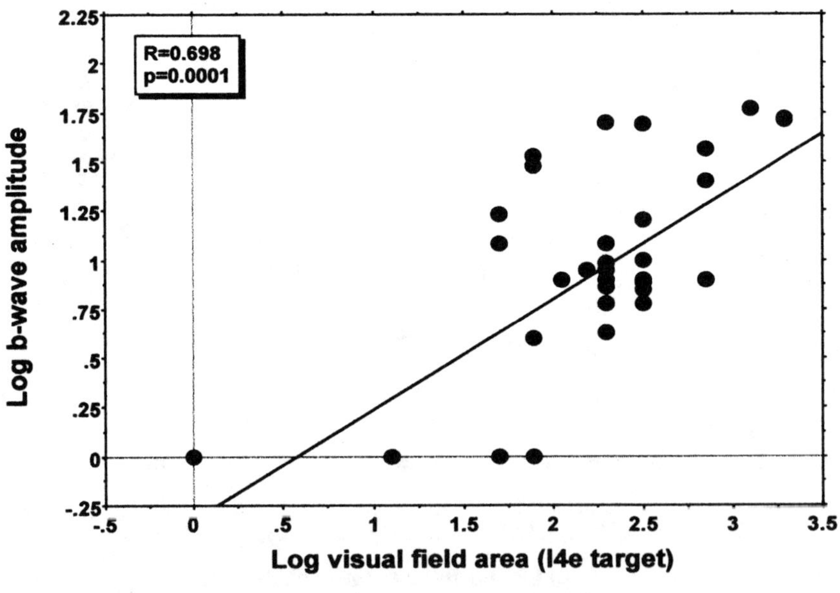

Figure 4. Correlation between ERG (log b-wave amplitude) and visual field size (log of the area delimited with the I4e target, in square degrees) in patients with Usher syndrome type 2.

VF constricted to ≤ 10 degrees around the fixation point, correlations held to the same extent as when using ERG amplitude as a limiting factor. GVFs ≤ 10 degrees with the V-4e target were not numerous enough to allow for statistical analysis. Mid- to late-stage patients often have ERGs and GVFs in this narrow size range. Therefore, the results of these analyses show that the correlations observed across the entire sample are not driven exclusively by larger ERGs or larger visual fields characterizing patients at the early stage of their conditions, but rather that they hold fairly well across the entire spectrum of disease severity.

4. DISCUSSION

Consistent with the most recent reports and with our previous findings in non-syndromic RP, psychophysical and objective ERG functional variables appear to be closely correlated also in patients with Usher syndrome. The relationship held irrespective of the syndrome type, suggesting that the correlation between ERG b-wave amplitude and GVF residual area may largely depend on physiological and functional mechanisms that are apart from genetic differences intrinsic to Usher syndrome.

With appropriate recording techniques, on average, ERGs and GVFs are well correlated also for ERG amplitudes ≤10 μV and for GVFs constricted ≤ 10 degrees around the fixation point. Some persisting cases of marked discrepancy between ERG and GVF were observed, primarily in a subset of patients with Usher syndrome type 1. This may be simply due to inaccuracy of GVF measurements in type 1 patients, whose profound dual sensory impairment may preclude them from performing optimally to GVF testing. Alternatively, these outliers may represent a genetically distinct subset of patients, whose disease mechanisms differ from that of the others.

While there is no one-to-one relationship between ERGs and GVFs either in RP (Iannaccone et al., 1995; Sandberg et al., 1996) or in Usher syndrome, this study confirms that, on average, the ERG is an accurate measure of retinal function also in Usher syndrome, and can be used reliably as a disease-monitoring tool in natural history studies and as a meaningful outcome measure in forthcoming treatment trials of Usher syndrome. However, for a subset of patients this relationship may not hold and other techniques [such as focal flash ERG, multifocal ERG, pattern ERG, or submicrovolt ERG recording techniques (Sieving et al., 1998)] may be better indicators of residual retinal function. It will be interesting to see if genetic characterization of our patient population, which is presently in progress, will facilitate the understanding of the reasons for these discrepancies.

Over the forthcoming years, the unveiling of the molecular mechanisms underlying the various forms of Usher syndrome will likely lead to a better understanding of the disease mechanism(s). Meanwhile, an improved understanding of the value of clinical measures of retinal function in Usher syndrome can allow researchers and clinicians alike to study and manage this severely disabling condition better. Further studies are currently in progress to characterize in greater detail the relationship between GVF and ERG in Usher syndrome.

5. ACKNOWLEDGMENTS

This investigation was generously supported by grants to AI by: Fight for Sight, Schaumburg, IL (Grant-in-Aid No. GA99056); the University of Tennessee Medical Group Research Grants Program, Memphis, TN; the Mid-South Lions Sight and Hearing Service, Memphis, TN; the National Institute on Deafness and Other Communication Disorders (subcontract to Program Project Grant No. P01 DC01813-07 from Boys Town National Research Hospital), NIH, Bethesda, MD; and by an unrestricted grant to the University of Tennessee HSC Department of Ophthalmology from Research to Prevent Blindness, Inc., New York, NY.

6. REFERENCES

Ahmed, Z. M., Riazuddin, S., Bernstein, S. L., Ahmed, Z., Khan, S., Griffith, A. J., Morell, R. J., Friedman, T. B., Riazuddin, S. and Wilcox, E. R., 2001, Mutations of the protocadherin gene PCDH15 cause Usher syndrome type 1F, *Am J Hum Genet.* **69**:25-34.

Alagramam, K. N., Yuan, H., Kuehn, M. H., Murcia, C. L., Wayne, S., Srisailpathy, C. R., Lowry, R. B., Knaus, R., Van Laer, L., Bernier, F. P., Schwartz, S., Lee, C., Morton, C. C., Mullins, R. F., Ramesh, A., Van Camp, G., Hageman, G. S., Woychik, R. P. and Smith, R. J., 2001, Mutations in the novel protocadherin PCDH15 cause Usher syndrome type 1F, *Hum Mol Genet.* **10**:1709-1718.

Astuto, L. M., Weston, M. D., Carney, C. A., Hoover, D. M., Cremers, C. W., Wagenaar, M., Moller, C., Smith, R. J., Pieke-Dahl, S., Greenberg, J., Ramesar, R., Jacobson, S. G., Ayuso, C., Heckenlively, J. R., Tamayo, M., Gorin, M. B., Reardon, W. and Kimberling, W. J., 2000, Genetic heterogeneity of Usher syndrome: analysis of 151 families with Usher type I, *Am J Hum Genet.* **67**:1569-1574.

Bhattacharya, G., Miller, C., Kimberling, W. J., Jablonski, M. M. and Cosgrove, D., 2002, Localization and expression of usherin: a novel basement membrane protein defective in people with Usher's syndrome type IIa, *Hear Res.* **163**:1-11.

Bitner-Glindzicz, M., Lindley, K. J., Rutland, P., Blaydon, D., Smith, V. V., Milla, P. J., Hussain, K., Furth-Lavi, J., Cosgrove, K. E., Shepherd, R. M., Barnes, P. D., O'Brien, R. E., Farndon, P. A., Sowden, J., Liu, X. Z., Scanlan, M. J., Malcolm, S., Dunne, M. J., Aynsley-Green, A. and Glaser, B., 2000, A recessive contiguous gene deletion causing infantile hyperinsulinism, enteropathy and deafness identifies the Usher type 1C gene, *Nat Genet.* **26**:56-60.

Bolz, H., von Brederlow, B., Ramirez, A., Bryda, E. C., Kutsche, K., Nothwang, H. G., Seeliger, M., del C-Salcedo Cabrera, M., Vila, M. C., Molina, O. P., Gal, A. and Kubisch, C., 2001, Mutation of CDH23, encoding a new member of the cadherin gene family, causes Usher syndrome type 1D, *Nat Genet.* **27**:108-112.

Bork, J. M., Peters, L. M., Riazuddin, S., Bernstein, S. L., Ahmed, Z. M., Ness, S. L., Polomeno, R., Ramesh, A., Schloss, M., Srisailpathy, C. R., Wayne, S., Bellman, S., Desmukh, D., Ahmed, Z., Khan, S. N., Kaloustian, V. M., Li, X. C., Lalwani, A., Riazuddin, S., Bitner-Glindzicz, M., Nance, W. E., Liu, X. Z., Wistow, G., Smith, R. J., Griffith, A. J., Wilcox, E. R., Friedman, T. B. and Morell, R. J., 2001, Usher syndrome 1D and nonsyndromic autosomal recessive deafness DFNB12 are caused by allelic mutations of the novel cadherin-like gene CDH23, *Am J Hum Genet.* **68**:26-37.

Di Palma, F., Holme, R. H., Bryda, E. C., Belyantseva, I. A., Pellegrino, R., Kachar, B., Steel, K. P. and Noben-Trauth, K., 2001, Mutations in Cdh23, encoding a new type of cadherin, cause stereocilia disorganization in waltzer, the mouse model for Usher syndrome type 1D, *Nat Genet.* **27**:103-107.

Edwards, A., Fishman, G. A., Anderson, R. J., Grover, S. and Derlacki, D. J., 1998, Visual acuity and visual field impairment in Usher syndrome, *Arch Ophthalmol.* **116**:165-168.

Eudy, J. D., Weston, M. D., Yao, S., Hoover, D. M., Rehm, H. L., Ma-Edmonds, M., Yan, D., Ahmad, I., Cheng, J. J., Ayuso, C., Cremers, C., Davenport, S., Moller, C., Talmadge, C. B., Beisel, K. W., Tamayo, M., Morton, C. C., Swaroop, A., Kimberling, W. J. and Sumegi, J., 1998, Mutation of a gene encoding a protein with extracellular matrix motifs in Usher syndrome type IIa, *Science.* **280**:1753-1757.

Fishman, G. A., Anderson, R. J., Lam, B. L. and Derlacki, D. J., 1995, Prevalence of foveal lesions in type 1 and type 2 Usher's syndrome, *Arch Ophthalmol.* **113**:770-773.

Gouras, P., Armington, J. C., Kropfl, W. J. and Gunkel, R. D., 1964, Electronic computation of human retinal and brain responses to light stimulation, *NY Acad Sci.* **115**:763-775.

Iannaccone, A., Rispoli, E., Vingolo, E. M., Onori, P., Steindl, K., Rispoli, D. and Pannarale, M. R., 1995, Correlation between Goldmann perimetry and maximal electroretinogram response in retinitis pigmentosa, *Doc Ophthalmol.* **90**:129-142.

Iannaccone, A., Vingolo, E. M., Rispoli, E., De Propris, G., Tanzilli, P. and Pannarale, M. R., 1996, Electroretinographic alterations in the Laurence-Moon-Bardet-Biedl phenotype, *Acta Ophthalmol Scand.* **74**:8-13.

Joensuu, T., Hamalainen, R., Yuan, B., Johnson, C., Tegelberg, S., Gasparini, P., Zelante, L., Pirvola, U., Pakarinen, L., Lehesjoki, A. E., de la Chapelle, A. and Sankila, E.-M., 2001, Mutations in a novel gene with transmembrane domains underlie Usher syndrome type 3, *Am J Hum Genet*. **69**:673-684.

Leroy, B. P., Aragon-Martin, J. A., Weston, M. D., Bessant, D. A., Willis, C., Webster, A. R., Bird, A. C., Kimberling, W. J., Payne, A. M. and Bhattacharya, S. S., 2001, Spectrum of mutations in USH2A in British patients with Usher syndrome type II, *Exp Eye Res.* **72**:503-509.

Najera, C., Beneyto, M., Blanca, J., Aller, E., Fontcuberta, A., Millan, J. M. and Ayuso, C., 2002, Mutations in myosin VIIA (MYO7A) and usherin (USH2A) in Spanish patients with Usher syndrome types I and II, respectively, *Hum Mutat.* **20**:76-77.

Rispoli, E., Iannaccone, A. and Vingolo, E. M., 1994, Low-noise electroretinogram recording techniques in retinitis pigmentosa, *Doc Ophthalmol.* **88**:27-37.

Rivolta, C., Sweklo, E. A., Berson, E. L. and Dryja, T. P., 2000, Missense mutation in the USH2A gene: association with recessive retinitis pigmentosa without hearing loss, *Am J Hum Genet.* **66**:1975-1978.

Sandberg, M. A., Weigel-DiFranco, C., Rosner, B. and Berson, E. L., 1996, The relationship between visual field size and electroretinogram amplitude in retinitis pigmentosa, *Invest Ophthalmol Vis Sci.* **37**:1693-1698.

Seeliger, M., Pfister, M., Gendo, K., Paasch, S., Apfelstedt-Sylla, E., Plinkert, P., Zenner, H. P. and Zrenner, E., 1999, Comparative study of visual, auditory, and olfactory function in Usher syndrome, *Graefes Arch Clin Exp Ophthalmol.* **237**:301-307.

Sieving, P. A., Arnold, E. B., Jamison, J., Liepa, A. and Coats, C., 1998, Submicrovolt flicker electroretinogram: cycle-by-cycle recording of multiple harmonics with statistical estimation of measurement uncertainty, *Invest Ophthalmol Vis Sci.* **39**:1462-1469.

Weil, D., Blanchard, S., Kaplan, J., Guilford, P., Gibson, F., Walsh, J., Mburu, P., Varela, A., Levilliers, J., Weston, M. D. and et, a., 1995, Defective myosin VIIA gene responsible for Usher syndrome type 1B, *Nature.* **374**:60-61.

Weston, M. D., Kelley, P. M., Overbeck, L. D., Wagenaar, M., Orten, D. J., Hasson, T., Chen, Z. Y., Corey, D., Mooseker, M., Sumegi, J., Cremers, C., Moller, C., Jacobson, S. G., Gorin, M. B. and Kimberling, W. J., 1996, Myosin VIIA mutation screening in 189 Usher syndrome type 1 patients, *Am J Hum Genet.* **59**:1074-1083.

THE CELLULAR FUNCTION OF THE USHER GENE PRODUCT MYOSIN VIIA IS SPECIFIED BY ITS LIGANDS

Uwe Wolfrum[1]

1. INTRODUCTION

Defects in myosin VIIa are responsible for Usher Syndrome 1B (Weil et al., 1995). Human Usher syndrome (USH), named after the British ophthalmologist Charles Usher (Usher, 1914), is the most common hereditary form of combined blind- and deafness (≈ 50 % of cases in the developed countries). USH designates a group of clinically and genetically heterogeneous disorders with hearing loss and retinitis pigmentosa (RP). Three different USH types (USH1, 2 and 3; see Table 1) can be distinguished according to the degree of clinical symptomes. USH1 is the most severe subtype, characterized by severe to profound congenital sensorineuronal deafness, constant vestibular dysfunction (balance deficiency) and prepubertal onset of retinitis pigmentosa. USH1 is genetically heterogeneous. Out of at least seven distinct genetic loci (*USH1A-G*) four corresponding genes have been identified, namely *USH1B*, *C*, *D*, and *F* (Mustapha et al., 2002; Petit, 2001; Table 1). *USH1B* encodes for the molecular motor protein myosin VIIa (Weil et al., 1995), *USH1C* encodes harmonin (Bitner-Glindzicz et al., 2000; Verpy et al., 2000), a scaffold protein containing PDZ-domains, motifs which are known to organize supramolecular protein complexes (Sheng and Sala, 2001). Mutations in the cadherin-related proteins, cadherin 23 (Cdh23) and protocadherin 15 (PCdh15) underline USH1D and USH1F, respectively (Bolz et al., 2001, 2002; Bork et al., 2001; Ahmed et al., 2001; Alagramam et al., 2001a, b).

As in other hereditary human diseases the knowledge of the cellular function of the USH gene products is a necessary prerequisite for the development of founded therapeutic strategies for the treatment of the disease. Essential insights into the cellular function of the *USH1B* gene product myosin VIIa were gathered by the identification and characterization of ligands to myosin VIIa. Recently obtained data further identify myosin VIIa as a component of a supramolecular USH1-protein complex composed of

[1]Johannes Gutenberg-Universität, Institut für Zoologie, 55099 Mainz, Germany. E-mail: wolfrum@mail.uni-mainz.de

several USH gene products (Boëda et al., 2002; Reiners et al., 2003). The purpose of this chapter is to integrate the present knowledge on the function of myosin VIIa and its ligands in the cellular environment of the retina.

Table 1. Usher syndrome (USH) subtypes (changed after Petit, 2001); fall 2002 update.

Type	Gene locus	Gene	Protein (cellular function)	Mouse model
1A	14q32			
1B	11q13.5	*MYO7A*	Myosin VIIa (molecular motor)	*shaker-1*
1C	11q15.1	*USH1C*	Harmonin (PDZ scaffold protein)	in prep.
1D	10q21-q22	*CDH23*	Cadherin 23 (cell-cell adhesion)	*waltzer*
1E	21q21			
1F	10q11.2-q21	*PCDH15*	Protocadherin 15 (cell-cell adhesion)	*ames waltzer*
1G	17q24-25			
2A	1q41	*USH2A*	Usherin (extracellular matrix)	in prep.
2B	3p23-24.2			
2C	5q14.3-21.3			
3A	3q21-25	*USH3A*	Clarin-1 (cell-cell adhesion)	in prep.

2. MYOSIN VIIA DOMAIN STRUCTURE

The myosin VIIa gene (*USH1B*) was the first USH gene identified (Weil et al., 1995). It encodes myosin VIIa, an unconventional myosin composed of several functional domains (Figure in Table 2): a motor domain is linked by a neck region with the long tail. The motor or head domain bears an ATP-binding site and an actin filament-binding site common for myosins (Sellers, 2000). Recent studies demonstrated that myosin VIIa functions indeed as an actin filament-based molecular motor protein (Udovichenko et al., 2002). The neck region of myosin VIIa contains 5 IQ (isoleucine/glutamine) motifs expected to bind myosin light-chains regulating the motile function in other myosins. The Ca^{2+}-binding EF-hand protein calmodulin binds to one or more IQ-motifs and may serve as a myosin light-chain (Todorov et al., 2001; Udovichenko et al., 2002). Among the increasing number of myosin family members, the tail region is the most variable protein region indicating that the tail may be responsible for the specific cellular function of a particular myosin. The long myosin VIIa tail contains several domains. It begins with a short coiled-coil domain that has been shown to form homodimers (Weil et al., 1997). The coiled-coil domain is followed by two tandem repeats, each containing a MyTH4 (myosin tail homology 4) and a FERM (4.1 (four point one) protein, ezrin, radixin, moeosin)-like domain which are separated by a poor conserved SH3 (src homology type 3) domain (Chen et al., 1996; Mermall et al., 1998; Oliver et al., 1999). Since this tandem of MyTH4 and FERM domains also is present in other myosins, a functional significance has been suggested (Mermall et al., 1998; Oliver et al., 1999). Nevertheless, isolated FERM-domains have been shown to be responsible for protein attachment to the plasma membrane either via direct binding to phospholipids or through the interaction with specific transmembrane adaptor proteins (Chishti et al., 1998). This tail diversity in unconventional myosins is thought to specify their cellular function and it is commonly accepted that proteins interacting with the tail domain can determine the target (e.g. a vesicle or an organelle) to which the force generated by the motor domain is applied (Oliver et al., 1999). Therefore, it does not surprise that recent identification of proteins interacting with the tail of myosin VIIa provided essential insights into the function of myosin VIIa in retinal cells.

3. MYOSIN VIIA EXPRESSION AND SUBCELLULAR LOCALIZATION IN THE MAMMALIAN RETINA

Myosin VII is expressed as a 230 kDa protein in a variety of tissues as an alternatively spliced truncated variant lacking portions of the c-terminal tail (Weil et al., 1995; Hasson et al., 1995; Self et al., 1998; Wolfrum et al., 1998). A comparative analysis of different tissues demonstrated myosin VIIa as a common component of the transition zone of motile and in-motile cilia and microvilli (Wolfrum et al., 1998).

In the retina, myosin VIIa contains its tail and is expressed in the cells of the retinal pigment epithelium and in both cone and rod photoreceptor cells (Hasson et al.,1995; El-Amraoui et al., 1996; Liu et al., 1997). In the cells of the retinal pigment epithelium, myosin VIIa is concentrated in the apical microvilli-like processes (Hasson et al.,1995; Liu et al., 1997). Since these extensions are involved in the phagocytosis of the distal ends of photoreceptor outer segments, it has been suggested that the retinitis pigmentosa in USH1B patients may result from defects in the phagocytotic ability of the retinal pigment epithelium (Hasson et al., 1995).

In retinal photoreceptor cells, myosin VIIa is predominantly localized at the membrane of the connecting cilium at the joint between the inner and the outer segment. In the ciliary membrane, the actin-based and membrane associated molecular motor myosin VIIa is associated with actin filaments and co-localizes with opsin (Wolfrum and Schmitt, 1999, 2000). Further, evidence for participation of myosin VIIa in the ciliary transport of opsin is coming from studies on myosin VIIa deficient *shaker-1* mice which revealed an abnormal opsin accumulation at the membrane of the connecting cilium (Liu et al., 1998; Wolfrum and Schmitt, 1999, 2000). Beside the ciliary localization, myosin VIIa was also detected in the synaptic terminals of photoreceptor cells (El-Amraoui et al., 1996; Reiners et al., 2003).

4. MYOSIN VIIa CELLULAR FUNCTION IS DETERMINED BY INTERACTING PROTEINS

Unconventional myosins play critical roles in the morphogenesis and maintenance of sensory cells (Hasson and Mooseker, 1997). A variety of publications shows that myosin VIIa is expressed in different subcellular compartments of cells (e.g. Wolfrum et al., 1998). This indicates that the motor activity of myosin VIIa serves in several distinct motile processes in the various cell types studied. As described above, in the retinal pigment epithelium, myosin VIIa contributes to the migration of melanosomes (Liu et al., 1998; El-Amraoui et al., 2002), whereas in the mechanosensory hair cells and photoreceptor cells it is suggested to participate in the transport of receptors and opsin (hair cells: Richardson et al., 1997; 1999; opsin transport in photoreceptors: Liu et al., 1999; Wolfrum and Schmitt, 1999; 2000). In *Dictyostelium,* myosin VII has been suggested to act as an important mediator of cell adhesion (Tuxworth et al., 2001) which is consistent with association of myosin VIIa with cell-cell adhesions in sensory hair cells (Küssel-Andermann et al., 2000b). The tail domains of unconventional myosins are believed to be largely responsible for their specific function (Oliver et al., 1999). Recent identifications of proteins interacting with the domains encompassed in the tail of myosin VIIa indicate that these proteins specify the cellular function of myosin VIIa.

Table 2. Scheme of myosin VIIa domain structure and binding sites of myosin VIIa binding proteins.

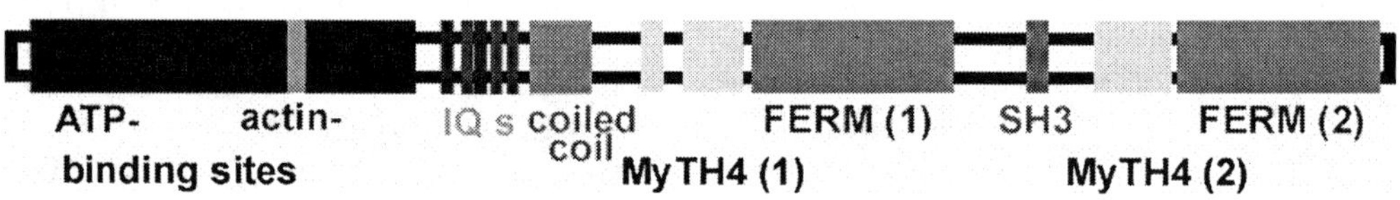

Interacting protein	Myosin VIIa target domain	Reference
Calmodulin (Ca^{2+}-binding protein)	IQ-motifs	Todorov et al., 2001 Udovichenko et al., 2002
MAP2B (microtubule association)	SH3, MyTH4 (2), FERM (2)	Todorov et al., 2001
Ri_ of PKA (regulation)	MyTH4 (2), FERM (2)	Küssel-A. et al., 2000a
Keap1 (actin-associated protein)	SH3	Velichkova et al., 2002
Vezatin (transmembrane protein)	FERM (2)	Küssel-A. et al., 2000b
MyRIP (Rap-interacting protein)	MyTH4 (2), FERM (2)	El-Amraoui et al., 2002
Harmonin (scaffold protein)	FERM (2)	Boëda et al., 2002

The proteins identified as ligands to myosin VIIa are summarized in Table 2. As mentioned above, calmodulin binds to one or more IQ-motifs in the neck region of myosin VIIa and probably regulates the motor function of myosin VIIa (Todorov et al., 2001; Udovichenko et al., 2002). In biochemical approaches, Todorov et al. (2001) also found the microtubuli associated protein 2 (MAP2) to interact with the C-terminus portion of the tail domain of myosin VIIa. In neurons, MAP2 plays an important role in the integrative signaling between the actin and the microtubule cytoskeleton anchoring protein kinase A (PKA) in subcellular compartments (Harada et al., 2002). Interestingly, the regulatory subunit of the PKA, RIα, was identified as a ligand to the myosin VIIa-tail in yeast two hybrid screens (Küssel-Andermann et al. 2000a). It is worth to speculate that myosin VIIa brings the MAP2-PKA complex and the RIα in spatial vicinity at the myosin VIIa tail for the regulation of kinase activity.

In yeast two hybrid screens, further molecules were identified as myosin VIIa binding proteins: ***i)*** the actin-associated protein Keap1 specifically interacts with the SH3-domain in the myosin VIIa tail which was used as the bait in the screen (Velichkova et al., 2002). Keap1 is the human ortholog to *Drosophila* kelch-protein which is associated with filamentous actin at ring channels of fruit fly larvae. Mammalian Keap1 has been localized at the endoplasmic junctions which are complex cell-cell adhesion specializations at the contacts between Sertoli cells and the differentiating sperm heads in testes. Keap1 is also expressed in the retina. However, neither the subcellular distribution nor the cellular function of Keap1 in the retina has yet been studied. ***ii.)*** Vezatin (derived from Slowenian *vezati* = bind, connect) was identified as a myosin VIIa ligand by yeast two hybrid screens using the C-terminal region of myosin VIIa, corresponding to the second tandem of MyTH4 and FERM domains as bait (Küssel–Andermann et al., 2000b). Vezatin is an ubiquitous transmembrane protein of adherens cell-cell junctions, where it interacts with myosin VIIa and the cadherin/catenin complex. ***iii.)*** Applying the same bait for screening a human retinal cDNA library, an additional myosin VIIa-interacting protein MyRIP, named after its ability to interact with myosin VIIa and small Rab GTPases, was identified (El-Amraoui et al., 2002). ***iv.)*** More recently, we demonstrated that myosin VIIa specifically interacts via its second FERM domain with harmonin (Boëda et al., 2002). Harmonin is a scaffold protein previously identified as the *USH1C* gene product (Verpy et al., 2000). It contains three PDZ-domains (derived from the first described proteins containing this domain: post synaptic density protein PSD-95, *Drosophila* septated junction protein Disc large, and tight junction protein ZO-1) known to interact with other proteins (Fanning and Anderson, 1996; Sheng and Sala, 2001). Our studies revealed that harmonin PDZ1 binds to myosin VIIa (USH1B) whereas the PDZ2 specifically interacts with the cell-cell adhesion protein cadherin 23 (USH1D) (Boëda et al., 2002). In photoreceptors, the USH1 protein complex may contribute to specialized cortical cytoskeletal matrices of these synapses which are thought to play a fundamental role in the organization, structure and function of the synaptic junction (Garner et al., 2000).

4.1. Vezatin Recruits Myosin VIIa To Specific Sites Of The Plasma Membrane

Vezatin is a transmembrane protein ubiquitously localized at cell-cell adhesion junctions associated with the cadherin/catenin complex (Küssel-Andermann et al., 2000b). Vezatin may anchor myosin VIIa to the adhesion junction and is a component of fibrillar ankle links that interconnect stereovilli (= stereocilia) of mechanosensitive hair cells (Küssel-Andermann et al., 2000b). In the retina, vezatin is expressed in the cells of the retinal pigment epithelium and the photoreceptor cells at sites where it co-localizes with myosin VIIa (Schäfer et al., in prep.). Its localization indicates that vezatin recruits myosin VIIa to the microvilli-like extensions of the retinal pigment cells and to the complex membrane contacts at the ribbon synapses of the photoreceptor cells. In addition, vezatin may anchor myosin VIIa at the membrane of the connecting cilium of photoreceptor cells supporting the transport of opsin through the cilium (Wolfrum et al., 2001).

4.2. MyRIP Bridges Myosin VIIa To Its Cargo, Namely RPE Melanosomes And Synaptic Vesicles

Myosin VIIa also specifically binds MyRIP (MyosinVIIa- and Rab-Interacting Protein), a novel Rab effector protein that directly interacts with the small GTPase Rab27 in a GTP dependent manner (El-Amraoui et al., 2002). In cells of the retinal pigment epithelium, Rab27A, MyRIP, and myosin VIIa are found to be associated with melanosomes. This molecular motor complex bridges the melanosomes to actin filaments and thereby mediates the trafficking of these organelles. Defects in this complex are likely to account for the melanosome mislocalization in the retinal pigment epithelium of myosin VIIa defective *shaker-1* mice previously described by Liu et al. (1998). However, we also identified the molecular complex of MyRIP-myosin VIIa at the synapses of photoreceptor cells. It is well known that the targeting of synaptic vesicles is controlled by Rab proteins (e.g. Rab3) in crosstalk with Rab effectors (Rabphilin or RIM) (e.g. Wang et al., 1997). Although MyRIP does not interact with Rab3 nor with Rab5, our data suggested that the MyRIP–myosin VIIa complex is associated with the membrane of vesicles at the synapse of photoreceptor cells.

4.3. Myosin VIIa Is A Component Of A Supramolecular USH1 Complex Integrated Via Harmonin

Recently, two independent studies showed that USH1-proteins can interact with each other. Siemens et al. (2002) demonstrated that the USH1C protein harmonin binds cadherin 23 (USH1D). Moreover, our investigations provided striking evidence that harmonin interacts also with myosin VIIa (USH1B) (Boëda et al., 2002). In the USH1 complex, the scaffold protein harmonin binds the myosin VIIa tail via PDZ-domain 1 and cadherin 23 via PDZ-domain 2. In the mechanosensitive hair cells, this supramolecular USH1 complex is necessary for the proper differentiation of their stereovilli (Boëda et al., 2002). In our studies on the retina, all three components of the USH1 complex were found to be co-localized at the ribbon synapse of mature photoreceptor cells (Reiners et al., 2003). However, single components of the supramolecular USH1 protein complex are also localized in compartments of the retina separate from the other components (Reiners et al., 2003). Biochemical experiments with fractionated photoreceptor compartments in combination with cytochemical analysis strongly indicate that harmonin isoforms serve as scaffold proteins in the light sensitive outer segment. Nevertheless, at the photoreceptor synapse the USH1 protein complex may contribute to specialized cortical cytoskeletal matrices of the pre- and postsynaptic regions which are thought to play a fundamental role in the organization, structure and function of the synaptic junction (Garner et al., 2000). As other PDZ-domain proteins, based on its modular

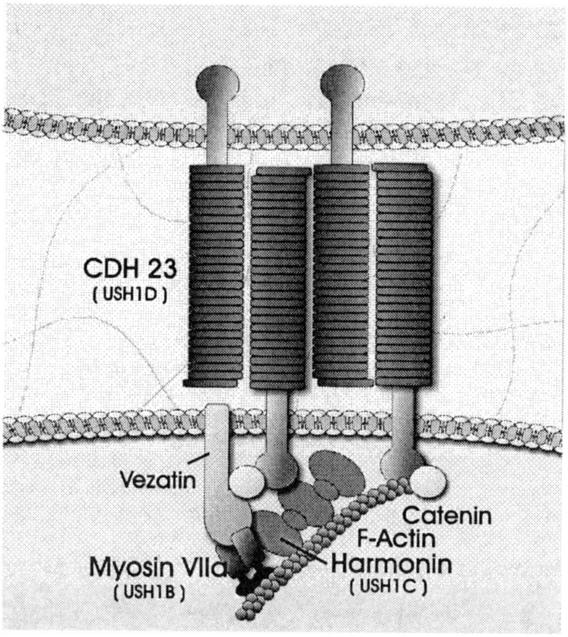

Figure 1. Scheme of the supramolecular USH1 protein complex at cell-cell adhesions. Description see text.

character and its ability to form multivalent interactions, harmonin is ideally suited to act as a potent organizer for a specialized protein complex at ribbon synapses of sensory cells. It physically and functionally bridges the motor activity of myosin VIIa probably involved in endo- or exocytotic processes with the cell-cell adhesion complex. The identified adhesion complex The transmembrane proteins of this complex, cadherin 23 and vezatin, may contribute to holding the synaptic cleft in close register. The theoretical characteristics of the cadherin-like protocadherin 15 and the 4-tansmembrane protein clarin-1, the gene products of two additional USH genes, namely *USH1F* and *USH3A* (Alagramam et al., 2001a, b; Ahmed et al., 2001; Adato et al., 2002), respectively, indicate that both proteins may also be present in a membrane associated protein complex.

Although the defects in USH1 complex components should lead to abnormalities in retina function in USH1 animal models, all available USH1-mouse models (see Table 1) do not show progressive retinal degeneration. However, the minor reduction of a and b waves in electroretinograms (ERG) of mice with mutations in *Myo7a* (*shaker-1 mice*) may reflect defects at photoreceptor synapses introduced by the absence or dysfunction of the USH1 complex component, myosin VIIa (Libby and Steel, 2001). In patients, dysfunction of one USH1 complex partner protein may cause retinal dysfunction and the clinical Usher syndrome type 1 phenotype found in patients with USH1.

5. ACKNOWLEDGEMENTS

This work was supported by the FAUN-Stiftung, Nürnberg, Germany, Forschung contra Blindheit – Initative Usher Syndrom e.V., and the Deutsche Forschungsgemeinschaft Wo548/3 and Wo548/4. I would like to thank B. Reidel, J. Reiners, and R. Schäfer providing me with figures for this chapter and L. Clermont-Wocker for her help preparing the manuscript.

6. REFERENCES

Adato, A., Vreugde, S., Joensuu, T., Avidan, N., Hamalainen, R., Belenskiy, O., Olender, T., Bonne-Tamir, B., Ben-Asher, E., Espinos, C., Millan, J. M., Lehesjoki, A. E., Flannery, J. G., Avraham, K. B., Pietrokovski, S., Sankila, E. M., Beckmann, J. S., Lancet, D., 2002, USH3A transcripts encode clarin-1, a four-transmembrane-domain protein with a possible role in sensory synapses, *Eur. J. Hum. Genet.* **10**: 339-350.

Ahmed, Z., Khan, S. N., Der Kaloustian, V. M., Li, X. C., Lalwani, A., Bitner-Glindzicz, M., Nance, W. E., Liu, X.-Z., Wistow, G., Smith, R. J. H., Griffith, A. J., Wilcox, E. R., Friedman, T. B., Morell, R. J., 2001, Usher syndrome 1D and nonsyndromic autosomal recessive deafness DFNB12 are caused by allelic mutations of the novel cadherin-like gene *CDH23*, *Am. J. Hum. Genet.* **68**:26-37.

Alagramam, K. N., Murcia, C. L., Kwon, H. Y., Pawlowski, K. S., Wright, C. G., Woychik, R. P., 2001a, The mouse Ames waltzer hearing-loss mutant is caused by mutation of *Pcdh15*, a novel protocadherin gene, *Nat. Genet.* **27**:99-102.

Alagramam, K. N., Yuan, H. J., Kuehn, M. H., Murcia, C. L., Wayne, S., Srisailpathy, C. R. S., Lowry, R. B., Knaus, R., Van Laer, L., Bernier, F. P., Schwartz, S., Lee, C., Morton, C. C., Mullins, R. F., Ramesh, A., Van Camp, G., Hagemen, G. S., Woychik, R. P., Smith, R. J. H., 2001b, Mutations in the novel protocadherin PCDH15 cause Usher syndrome type 1F, *Hum. Mol. Genet.* **10**:1709-1718.

Bitner-Glindzicz, M., Lindley, K. J., Rutland, P., Blaydon, D., Smith, V. V., Milla, P. J., Hussain, K., Furth-Lavi, J., Cosgrove, K. E., Shepherd, R. M., Barnes, P. D., O'Brien, R. A., Farndon, P. A., Sowden, J., Liu, X.-Z., Scanlan, M. J., Malcolm, S., Dunne, M. J., Aynsley-Green, A., Glaser, B., 2000, A recessive contiguous gene deletion causing infantile hyperinsulinism, enteropathy and deafness identifies the Usher type 1C gene, *Nat. Genet.* **26:**56-60.

Boëda, B., El-Amraoui, A., Bahloul, A., Goodyear, R., Daviet, L., Blanchard, S., Perfettini, I., Fath, K. R., Shorte, S., Reiners, J., Houdusse, A., Legrain, P., Wolfrum, U., Richardson, G., Petit, C., 2002, Myosin VIIa, harmonin and cadherin 23, three Usher 1 gene products that cooperate to shape the sensory hair cell bundle, *EMBO J.***21**:6689-6699.

Bolz, H., von Brederlow, B., Ramirez, A., Bryda, E. C., Kutsche, K., Nothwang, H. G., Seeliger, M., Salcedo Cabrera, M. D., Caballero Vila, M., Pelaez Molina, O., Gal, A., Kubisch, C, 2001, Mutation of *CDH23*, encoding a new member of the cadherin gene family, causes Usher syndrome type 1D, *Nat. Genet.* **27**:108-112.

Bolz, H., Reiners, J., Wolfrum, U., Gal, A., 2002 The role of cadherins in Ca^{2+}-mediated cell adhesion and inherited photoreceptor degeneration, in: *Photoreceptors and Calcium*, W. Baehr, K. Palczewski, eds, Ronald G. Landes (Landes Bioscience), Georgetown, in press

Bork, J. M., Peters, L. M., Riazuddin, S., Bernstein, S. L., Ahmed, Z. M., Ness, S. L., Polomeno, R., Ramesh, A., Schloss, M., Srikumari, C. R. S., Wayne, S., Bellman, S., Desmukh, D., Ahmed, Z., Khan, S. N., Der Kaloustian, V. M., Li, X. C., Lalwani, A., Bitner-Glindzicz, M., Nance, W. E., Liu, X.-Z., Wistow, G., Smith, R. J. H., Griffith, A. J., Wilcox, E. R., Friedman, T. B., Morell, R. J., 2001, Usher syndrome 1D and nonsyndromic autosomal recessive deafness DFNB12 are caused by allelic mutations of the novel cadherin-like gene *CDH23*, *Am. J. Hum. Genet.* **68**:26-37.

Chen, Z. Y., Hasson, T., Kelley, P. M., Schwender, B. J., Schwartz, M. F., Ramakrishnan, M., Kimberling, W. J., Mooseker, M. S., Corey, D. P., 1996, Molecular cloning and domain structure of human myosin- VIIa. The gene product defective in usher syndrome 1B, *Genomics* **36**:440-448.

Chishti, A. H., Kim, A. C., Marfatia, S. M., Lutchman, M., Hanspal, M., Jindal, H., Liu, S. C., Low, P. S., Rouleau, G. A., Mohandas, N., Chasis, J. A., Conboy, J. G., Gascard, P., Takakuwa, Y., Huang, S. C., Benz, E. J., Bretscher, A., Fehon, R. G., Gusella, A. F., Ramesh, V., Solomon, F., Marchesi, V. T., Tsukita, S., Arpin, M., Louvard, D., Tonks, N. K., Anderson, J. M., Fanning, A. S., Bryant, P. J., Woods, D. F., Hoover, K. B., 1998, The FERM domain: a unique module involved in the linkage of cytoplasmic proteins to the membrane, *Trends Biochem. Sci.* **23**:281-282.

El-Amraoui, A., Sahly, I., Picaud, S., Sahel, J., Abitbol, M., Petit, C., 1996, Human Usher 1B mouse shaker-1: The retinal phenotype discrepancy explained by the presence/absence of myosin VIIA in the photoreceptor cells, *Hum. Mol. Genet.* **5**:1171-1178.

El-Amraoui, A., Schonn, J.-S., Küssel-Andermann, P., Blanchard, S., Desnos, C., Henry J.-P., Wolfrum, U., Darchen, F., Petit, C., 2002, MyRIP, a novel Rab effector, enables myosin VIIa recruitement to melanosomes and synaptic vesicles, *EMBO Rep.* **3**:463-470.

Fanning, A. S., Anderson, J. M., 1996, Protein-protein interactions: PDZ domain networks, *Curr. Biol.* **6**:1385-1388.

Garner, C. C., Nash, J., Huganir, R. L., 2000, PDZ domains in synapse assembly and signalling, *Trends Cell Biol.* **10**:274-280.

Harada, A., Teng, J., Takei, Y., Oguchi, H., Hirokawa, N., 2002, MAP2 is required for dendrite elongation, PKA anchoring in dendrites, and proper PKA signal transduction, *J. Cell Biol.* **158**:541-549.

Hasson, T., Heintzelman, M. B., Santos-Sacchi, J., Corey, D. P., Mooseker, M. S., 1995, Expression in cochlea and retina of myosin VIIa, the gene product defective in Usher syndrome type 1B, *Proc. Natl. Acad. Sci. USA* **92**:9815-9819.

Hasson, T., Mooseker, M. S., 1997, The growing family of myosin motors and their role in neurons and sensory cells, *Cur.r Opin. Neurobiol.* **7**:615-623.

Küssel-Andermann, P., El-Amraoui, A., Safieddine, S., Nouaille, S., Perfettini, I., Lecuit, M., Cossart, P., Wolfrum, U., Petit, C., 2000a Vezatin, a novel transmembrane protein, bridges myosin VIIa to the cadherin/catenin complex, *EMBO J.* **19**:6020-6029.

Küssel-Andermann, P., El-Amraoui, A., Safieddine, S., Hardelin, J. P., Nouaille, S., Camonis, J., Petit, C., 2000b, Unconventional myosin VIIA is a novel A-kinase anchoring protein, *J. Biol. Chem.* **275**:29654-29659.

Libby, R. T., Steel, K. P., 2001, Electroretinographic anomalies in mice with mutations in Myo7a, the gene involved in human Usher syndrome type 1B, *Invest. Ophthalmol. Visual Sci.* **42**:770-778.

Liu, X. R., Vansant, G., Udovichenko, I. P., Wolfrum, U., Williams, D. S., 1997, Myosin VIIa, the product of the Usher 1B syndrome gene, is concentrated in the connecting cilia of photoreceptor cells, Cell Motil. Cytoskeleton **37**:240-252.

Liu, X. R., Ondek, B., Williams, D. S., 1998, Mutant myosin VIIa causes defective melanosome distribution in the RPE of shaker-1 mice, *Nat. Genet.* **19**:117-118.

Liu, X., Udovichenko, I. P., Brown, S. D., Steel, K. P., Williams, D. S., 1999, Myosin VIIa participates in opsin transport through the photoreceptor cilium, *J. Neurosci.* **19**:6267-6274.

Mermall, V., Post, P. L., Mooseker, M. S., 1998, Unconventional myosins in cell movement, membrane traffic, and signal transduction, *Science* **279**:527-533.

Mustapha, M., Chouery, E., Torchard-Pagnez, D., Nouaille, S., Khrais, A., Sayegh, F. N., Mégarbané, A., Loiselet, J., Lathrop, M., Petit, C., Weil, D. 2002, A novel locus for Usher syndrome type I, USH1G, maps to chromosome 17q24-25, *Hum. Genet.* **110**:348-350.

Oliver, T. N., Berg, J. S., Cheney, R. E., 1999, Tails of unconventional myosins, *Cell Mol. Life Sci.* **56**:243-257.

Richardson, G. P., Forge, A., Kros, C. J., Fleming, J., Brown, S. D. M., Steel, K. P., 1997, Myosin VIIA is required for aminoglycoside accumulation in cochlear hair cells. *J. Neurosci.* **17**:9506-9519.

Richardson, G. P., Forge, A., Kros, C. J., Marcotti, W., Becker, D., Williams, D. S., Thorpe, J. R., Fleming, J., Brown, S. D., Steel, K. P., 1999, A missense mutation in myosin VIIA prevents aminoglycoside accumulation in early postnatal cochlear hair cells, *Ann. N. Y. Acad. Sci.* **884**:110-124.

Petit, C., 2001, Usher syndrome: from genetics to pathogenesis, Annu. Rev. Genomics Hum. Genet. **2**:271-297.

Reiners, J., Reidel, B., El-Amraoui, A., Boëda, B., Huber, I., Petit, C., Wolfrum, U., 2003, Spatial colocalization of the PDZ-protein harmonin with the other Usher I syndrome gene products Cadherin 23, and myosin VIIa at ribbon synapses, *submitted.*

Self, T., Mahony, M., Fleming, J., Walsh, J., Brown, S. D. M., Steel, K. P. 1998, Shaker-1 mutations reveal roles for myosin VIIa in both development and function of cochlear hair cells, *Development* **125**:557-566.

Sellers, J. R., 2000, Myosins: a diverse superfamily, *Biochim. Biophys. Acta* **1496**:3-22.

Sheng, M., Sala, C., 2001, PDZ domains and the organization of supramolecular complexes, *Ann. Rev. Neurosci.* **24**:1-29.

Siemens, J. Kazmierczak, P., Reynolds, A., Sticker, M., Littlewood Evans, A., Müller, U., 2002, The Usher syndrome proteins cadherin 23 and harmonin form a complex by means of PDZ-domain interactions, *Proc. Natl. Acad. Sci. USA* **99**: 14946-14951.

Todorov, P. T., Hardisty, R. E., Brown, S. D. M., 2001, Myosin VIIA is specifically associated with calmodulin and microtubule-associated protein-2B (MAP-2B), *Biochem. J.* **354**:267-274.

Tuxworth, R. I., Weber, I., Wessels, D., Addicks, G. C., Soll, D. R. , Gerisch, G., Titus, M. A. 2001, A role for myosin VII in dynamic cell adhesion, *Curr. Biol.* **11**:318-329.

Udovichenko, I. P., Gibbs, D., Williams, D. S., 2002, Actin-based motor properties of native myosin VIIa, *J. Cell Sci.* **115**:445-450.

Usher, C. H., 1914, On the inheritance of Retinitis pigmentosa, with notes of cases, *R. Lond. Ophthalmol. Hosp. Rep.* **19**:130-236.

Velichkova, M., Guttman, J., Warren, C., Eng, L., Kline, K., Vogl, A. W., Hasson, T., 2002, A human homologue of *Drosophila* kelch associates with myosin-VIIa in specialized adhesion junctions, *Cell Motil. Cytoskeleton* **51**:147-164.

Verpy, E., Leibovici, M., Zwaenepoel, I., Liu, X.-Z., Gal, A., Salem, N., Mansour, A., Blanchard, S., Kobayashi, I., Keats, B. J. B., Slim, R., Petit, C. 2000, A defect in harmonin, a PDZ domain-containing protein expressed in the inner ear sensory hair cells, underlies Usher syndrome type 1C, Nature Genet. **26**:51-55.

Wang, J., Powers, N. L., Hofstetter, P., Trautwein, P., Ding, D., Salvi, R. 1997, Effects of selective inner hair cell loss on auditory nerve fiber threshold, tuning and spontaneous and driven discharge rate, *Hearing Res.* **107**:67-82.

Weil, D., Blanchard, S., Kaplan, J., Guilford, P., Gibson, F., Walsh, J., Mburu, P., Varela, A., Levilliers, J., Weston, M. D., Kelley, P. M., Kimberling, W. J., Wagenaar, M., Levi-Acobas, F., Larget-Piet, D., Munnich, A., Steel, K. P., Brown, S. D. M., Petit, C. 1995, Defective myosin VIIA gene responsible for Usher syndrome type 1B, Nature **374**:60-61.

Weil, D., Küssel, P., Blanchard, S., Levy, G., Leviacobas, F., Drira, M., Ayadi, H., Petit, C., 1997, The autosomal recessive isolated deafness, DFNB2, and the Usher 1B syndrome are allelic defects of the myosin-VIIA gene, *Nat. Genet.* **16**:191-193.

Wolfrum, U., Liu, X. R., Schmitt, A., Udovichenko, I. P., Williams, D. S., 1998, Myosin VIIa as a common component of cilia and microvilli, *Cell Motil. Cytoskeleton* **40**:261-271.

Wolfrum, U., Schmitt, A., 1999, Evidence for myosin VIIa driven transport of rhodopsin in the plasma membrane of the photoreceptor-connecting cilium, in: *Retinal degeneration diseases and experimental therapy*, J. G. Hollyfield, R. E. Andersson, M. LaVail, eds, Kluwer Academic/Plenum Publishers, New York, pp. 3-14.

Wolfrum, U., Schmitt, A. 2000, Rhodopsin transport in the membrane of the connecting cilium of mammalian photoreceptor cells, *Cell Motil. Cytoskeleton.* **46**:95-107.

Wolfrum, U., El-Amraoui, A., Fath, K. R., Petit, C., Schmitt, A., 2001, Identification and characterization of the membrane-associated actin cytoskeleton in the photoreceptor cilium, *Invest. Ophthalmol. Vis. Sci.* **42**:296

MOUSE MODELS FOR USHER SYNDROME 1B

Concepcion Lillo[1], Junko Kitamoto[1], Xinran Liu[1], Elizabeth Quint[2], Karen P. Steel[2], and David S. Williams[1]

1. INTRODUCTION

Patients with Usher syndrome (USH) suffer from inherited deafness coupled with blindness (Usher, 1914). Three types of USH have been described, with USH1 being the most severe. Children with USH1 are born profoundly deaf, and then begin to lose their sight as a result of retinitis pigmentosa usually in their second decade of life. Mutations in at least seven different genes can cause USH1. Four of these have now been identified. The most common form of USH1 is USH1B, which accounts for at least 50% of USH1 cases (Astuto et al., 2000). USH1B is caused by mutations in human myosin VIIa (*MYO7A*) (Weil et al., 1995).

Shaker1 mice have been shown to possess mutations in *Myo7a* (Gibson et al., 1995; Mburu et al., 1997). These mice are deaf and have vestibular dysfunction that results in hyperactivity and circling and head-tossing behavior (Lord and Gates, 1929). The retinas of some alleles of shaker1 mice have been examined and were originally noted to be normal at the light microscope level (Hasson et al., 1997). However, later studies noted defects in both the RPE and the photoreceptor cells, consistent with the localization of the protein in both these cell types (Hasson et al., 1995; Liu et al., 1997). In the mutant RPE, melanosomes are absent from the apical processes (Liu et al., 1998), and a defect in RPE phagocytosis has been observed (Gibbs et al., submitted). In mutant photoreceptor cells, there is an abnormal accumulation of opsin in the connecting cilium and a slight retardation in the rate of outer segment disk membrane renewal (Liu et al., 1999). ERG a and b wave amplitudes show a small but significant reduction in amplitude in several of the *Myo7a* mutant alleles (Libby and Steel, 2001). Nevertheless, so far, there have been no reports of retinal degeneration in any of the alleles of shaker1 mice.

We have undertaken a study of retinas from shaker1 mice in order to determine if mutations in *Myo7a* can cause retinal degeneration. In this report we describe our results from examining all ten known alleles of shaker1, as well as compound heterozygotes from two of the alleles, plus shaker 1 mice that have been crossed with waltzer mice.

[1]Departments of Pharmacology and Neurosciences, UCSD School of Medicine, La Jolla, CA 92093-0983, USA; [2]MRC Institute of Hearing Research, University Park, Nottingham NG7 2RD, UK. Corresponding author: D. S. Williams, E-mail: dswilliams@ucsd.edu

Waltzer mice have mutations in their cadherin23 (*Cdh23*) gene (Di Palma et al., 2001), which is the homologue of the USH1D gene (Bolz et al., 2001).

2. MATERIALS AND METHODS

The following alleles of shaker1 were bred from stocks maintained in Nottingham, UK: $Myo7a^{sh1}$ (the original allele, R502P), $Myo7a^{816SB}$, $Myo7a^{26SB}$, $Myo7a^{3336SB}$, $Myo7a^{4494SB}$, $Myo7a^{4626SB}$, and $Myo7a^{6J}$. Details of genetic background have been described (Libby and Steel, 2001). The $Myo7a^{7J}$, $Myo7a^{8J}$, and $Myo7a^{9J}$ alleles were obtained from Jackson Labs, and bred in La Jolla, CA. The $Myo7a^{7J}$ and $Myo7a^{8J}$ mice were bred directly from the animals received. Because the $Myo7a^{9J}$ allele was on the C3H background, which has the rd1 mutation, this allele was backcrossed on to the CBA/Ca background. Backcrossed mice were checked for the absence of the rd1 mutation by PCR genotyping (Pittler and Baehr, 1991). These $Myo7a^{9J}$ mice were crossed with $Myo7a^{4626SB}$ mice, also on a CBA/Ca background, to generate compound heterozygotes.

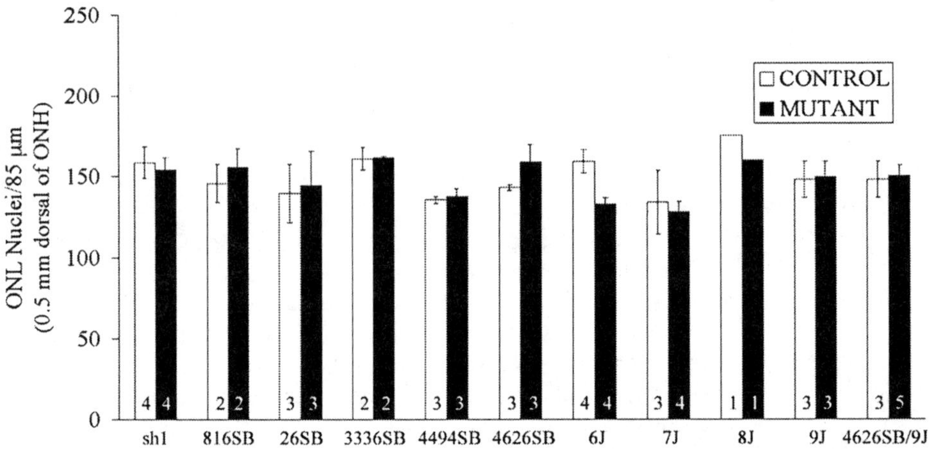

Figure 1. Histogram of numbers of photoreceptor cells counted in the dorsal retinas (0.5 mm from the optic nerve head in dorso-ventral sections) of shaker1 mice carrying different alleles of *Myo7a*. Control (open bars) mice were littermates of the mutants (filled bars) and included both wildtype and heterozygous mice. The number of different animals used for the counts is indicated at the bottom of each bar. Ages of the animals were as follows: sh1, 3-4 months; 816SB, 3-6 months; 26SB, 12-16 months; 3336SB, 5 months; 4494SB, 5-6 months; 4626SB, 3 months; 6J, 4-9 months; 7J, 3-5 months; 8J, 11 months; 9J, 11-16 months; 4626SB/9J, 12-14 months.

Double homozygotes of waltzer and shaker1 mice and a range of control genotypes were generated by crossing $Cdh23^{v}$ and $Myo7a^{4626SB}$ mutant stocks, followed by molecular genotyping. Details of genetic backgrounds and genotyping have been described (Holme and Steel, 2002). Mice from this cross ranged from 350 to 400 days old.

Eyes were fixed with a mixture of buffered gluteraldehyde and formaldehyde by cardiac perfusion of the animal or immersion of the enucleated eye. Before removal of an eye from an animal, its dorsal region was identified by a mark from a hot needle. Eyecups were processed and embedded in Epon. Semithin sections were cut along the dorso-ventral axis, passing through the optic nerve head, and with the photoreceptor cells aligned along their long axes. The number of photoreceptor cells were determined by counting the number of nuclei in an 85-µm length of retina at 0.5-mm intervals from the optic nerve head. In all cases, control and mutants came from the same litters, although data were obtained typically from more than one litter. Control mice were represented by either wildtype or heterozygous animals, or a mixture of both. Where appropriate, a one-tailed, Student's t test was applied to test the probability of no significant difference.

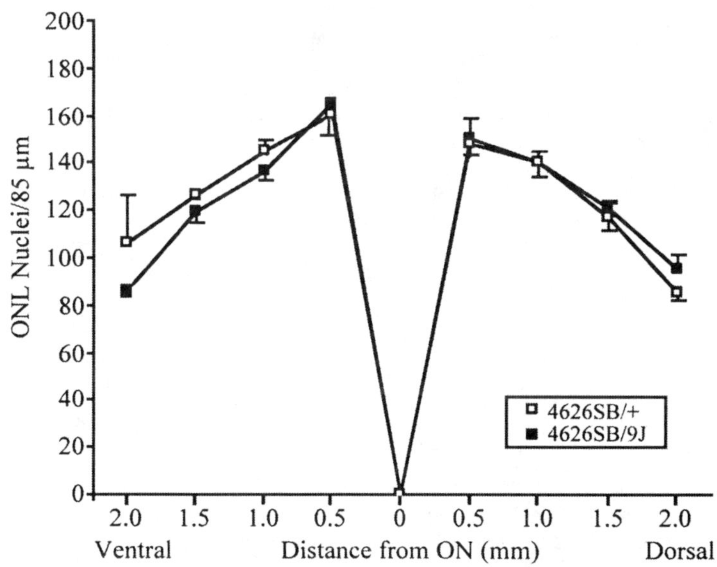

Figure 2. Graph showing numbers of photoreceptor cells counted in different parts of dorso-ventral retinal sections from control ($Myo7a^{4626SB/+}$) (n = 3) and mutant ($Myo7a^{4626SB/9J}$) (n = 5) mice from a compound heterozygote breeding. Counts were made at intervals of 0.5 mm from the optic nerve head (indicated by 0).

3. RESULTS

3.1. Different Alleles of Shaker1 Mice

The mutations responsible for seven different alleles of shaker1 mice have been described (Gibson et al., 1995; Mburu et al., 1997). In addition, three other alleles, $Myo7a^{7J}$, $Myo7a^{8J}$, and $Myo7a^{9J}$, had been identified at Jackson Labs; the mutations responsible for these alleles are unknown. Where possible photoreceptor cells were counted in animals that were at least 5 months old. However, the $Myo7a^{7J}$ allele was found to be on a genetic background that resulted in a slow photoreceptor cell loss in both mutants and controls, beginning after ~5 months, so that the age that counts were taken from these mice had to be restricted to less than 5 months old.

We detected no significant difference in the number of photoreceptor cells between control and mutant retinas, indicating a lack of photoreceptor degeneration induced by mutant *Myo7a*, in any of the alleles. Counts of photoreceptor cells in the dorsal retina, 0.5 mm from the optic nerve head, are illustrated in Fig. 1, for retinas of all alleles. We also examined a cross between two different alleles, $Myo7a^{4626SB}$ and $Myo7a^{9J}$. Mice carrying one copy of each of these two mutant alleles also had no photoreceptor cell degeneration (Figs. 1 and 2).

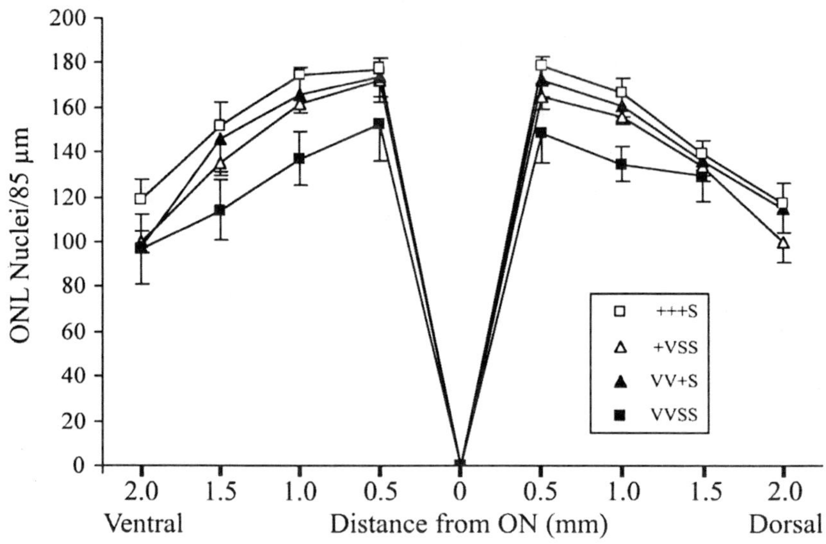

Figure 3. Graph showing numbers of photoreceptor cells counted in different parts of dorso-ventral retinal sections for different genotypes bred from a cross between shaker1 ($Myo7a^{4626SB}$) and waltzer mice ($Cdh23^{v}$). Counts were made at intervals of 0.5 mm from the optic nerve head (indicated by 0). v indicates a mutant $Cdh23^{v}$ allele (v+, heterozygote; vv, mutant homozygote) and s indicates a mutant $Myo7a^{4626SB}$ allele. For +++s, n = 3; vv+s, n = 5; +vss, n=5; vvss, n=8.

3.2. Double Shaker1 and Waltzer Mutants

Although the retinal distribution of CDH23, or any other USH1 gene product besides MYO7A, is unknown, it is plausible that some of the seven known USH1 gene products might interact. $Myo7a^{4626SB}$ mice have been shown to be null (Liu et al., 1999) and $Cdh23^{v}$ mice are presumed to be null (Di Palma et al., 2001). We have previously examined crosses of $Myo7a^{4626SB}$ and $Cdh23^{v}$ mice that were 100 days old, and detected no retinal degeneration (Libby et al., submitted). Here, we examined retinas of mice from this cross that were around one year old. Although no pronounced photoreceptor cell death is evident, the double homozygous mutant retinas have significantly fewer photoreceptor cells in the central retina (Figs. 3 and 4). In the dorsal retina, 1 mm from the optic nerve head, there are at least 15% fewer photoreceptor cells in the double mutant compared with any of the other genotypes ($p < 0.01$). With respect to *Myo7a*, a pertinent comparison is between the double homozygous mutants and the *Cdh23*$^{v/v}$ *Myo7a*$^{4626SB/+}$ mice. The difference between these genotypes indicates that on the appropriate genetic background (in this case, containing *Cdh23*$^{v/v}$), a recessive *Myo7a* mutation can induce photoreceptor cell loss in the mouse retina. (Our data also show that a recessive *Cdh23* mutation can induce photoreceptor cell loss on a *Myo7a*$^{4626SB/4626SB}$ background.)

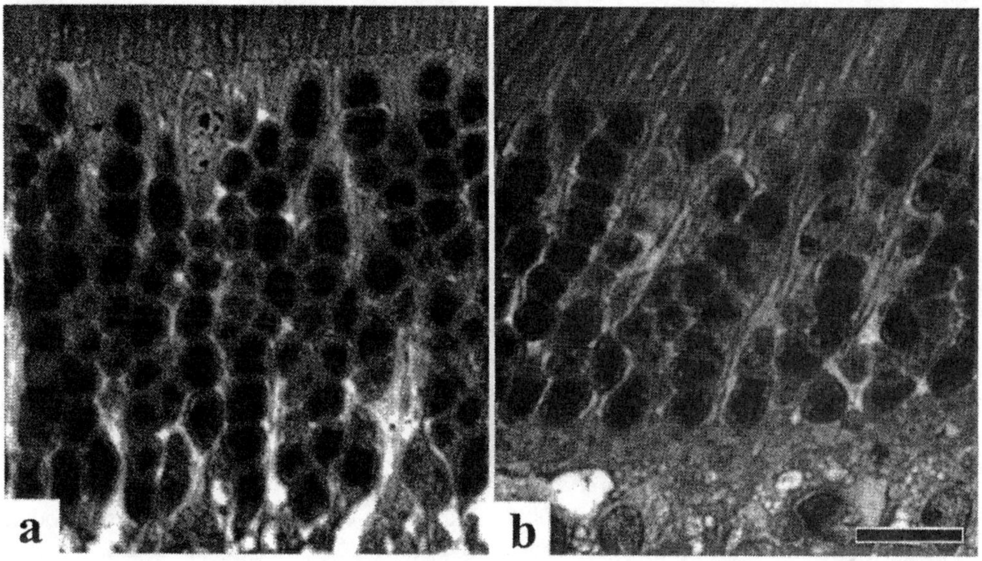

Figure 4. Light micrographs of the photoreceptor cell nuclei (outer nuclear layer) of the dorsal retinas (1 mm from the optic nerve head) of *Cdh23*$^{v/v}$ *Myo7a*$^{4626SB/+}$ (a) and *Cdh23*$^{v/v}$ *Myo7a*$^{4626SB/4626SB}$ (b) littermates, 388 days old. Scale bar = 10 μm.

4. DISCUSSION

Our studies found no photoreceptor cell loss in any of the shaker1 alleles, in the absence of mutant *Cdh23*. These alleles represented a variety of mutations, including a known null mutation in the *Myo7a*^{4626SB} allele (Liu et al., 1999). While none of the specific shaker1 mutations have been found in a homozygous state in USH1B patients, there have been two reports of presumptive homozygous null mutations (Cuevas et al., 1998; Weston et al., 1996). Interestingly, this lack of photoreceptor cell degeneration is also characteristic of the other known USH1 mouse models. Homozygous mutant waltzer mice, represented by three different alleles, including a presumptive null, also had a normal number of photoreceptor cells, even at one year old (Libby et al., submitted). A third mouse model of USH1 is the Ames waltzer, which has a mutant protocadherin15 gene (*Pcdh15*) (Alagramam et al., 2001), the homologue of the USH1F gene (Bolz et al., 2001; Bork et al., 2001). Retinas of Ames waltzer mice have yet to be examined in detail, however, ERG recordings failed to detect any abnormalities and thick-section histology showed no gross anatomical abnormalities (Ball et al., 2002).

Although we examined all known alleles of shaker1 mice, we have not explored thoroughly the effects of different genetic backgrounds on a given mutation. An indication that mutant *Myo7a* might elicit photoreceptor cell degeneration on an appropriate genetic background comes from our observation of reduced photoreceptor cell numbers in mice that were homozygous mutant for both *Myo7a* and *Cdh23*. These results suggest that there is a genetic interaction between *Myo7a* and *Cdh23*. Importantly, they demonstrate photoreceptor cell loss that results from the presence of homozygous mutant *Myo7a* ($Myo7a^{4626SB/+}$ mice on the same background had no photoreceptor cell loss). This observation suggests that an USH1B mouse model that includes substantial retinal degeneration within a practicable period of time might be feasible. It will be interesting to test the effects of mutant Myo7a on a genetic background that is not only compromised by the presence of mutant *Cdh23/Cdh23,* but also by the presence of, say, mutant *Pcdh15/Pcdh15*. Given the existence of Ames waltzer mice, this experiment is feasible, albeit rather arduous to set up.

Mouse models that mimic the mutant phenotypes of human diseases are especially important in the development of therapies. Although there are other mammals that are genetically and physiologically more like humans, mice offer the distinct advantage over other mammals of being suitable for relatively short experiments with large numbers of congenic individuals. In the field of photoreceptor degeneration, we appear to be fortunate with many good animal models. A number of different mouse lines, with mutant genes that are orthologous to those found in retinitis pigmentosa (RP) patients, undergo photoreceptor cell degeneration. By comparison, mouse models for other CNS degenerative disorders appear to be less satisfactory. Transgenic mice have been generated as models for neurodegenerative diseases, such as Alzheimer's disease (AD) and amyotrophic lateral sclerosis (ALS). However, even the best of these models recapitulate only some of the features of the human condition (Wong et al., 2002).

Upon a closer examination of the RP mouse models, we find, perhaps not surprisingly, that they are less than perfect. When a mouse carrying an RP mutation does not undergo photoreceptor cell degeneration, as in the USH1 models, the imperfection is more obvious. However, perhaps a less-than-perfect temporal correlation in RP mouse models is more common than generally thought. A number of popular retinal degeneration models have very rapid photoreceptor cell degeneration – which is perhaps too rapid to be an exact model. For example, in the rd1 mouse, cell death begins before the photoreceptors are even fully developed, and nearly all rod photoreceptor cells are gone within the first month of postnatal life (Portera-Cailliau et al., 1994; Sanyal and Hawkins, 1981). By contrast, ERG responses emanating from rod photoreceptors have been detected in adult RP patients, carrying similar types of mutations in the β-phosphodiesterase (β-PDE) gene (McLaughlin et al., 1995). It seems possible that the cellular basis for eliciting the cell death might be quite different between rd1 mice and β-PDE RP patients.

Nevertheless, it is clear that rod photoreceptor cell death is caused by mutations in the β-PDE gene in both the rd1 mouse and the β-PDE RP patient (Bowes et al., 1990; McLaughlin et al., 1995; Pittler and Baehr, 1991). Find a way to correct the β-PDE gene in either (before the photoreceptor cells die) and sight will not be lost. Therefore the rd1 mouse is a suitable model to test the efficacy of β-PDE gene therapy. If the therapy is successful, no photoreceptor cell death will ensue. However, by the same argument, correction of any mutant phenotype can be used as an indicator of successful gene therapy. In this respect, the shaker1 mouse is still a satisfactory model for USH1B. Several mutant phenotypes have been described in the photoreceptor and RPE cells (Gibbs et al., submitted; Libby and Steel, 2001; Liu et al., 1998; Liu et al., 1999), with one, the misplaced pigment granules in the RPE, providing a particularly convenient assay. Correction of these mutant phenotypes would demonstrate successful therapy of the USH1B gene.

5. SUMMARY

Photoreceptor cell degeneration was not detected in any of the shaker1 alleles, except for a small but significant loss of photoreceptor cells found in *Myo7a*$^{4626SB/4626SB}$ mice that were also homozygous mutant for *Cdh23*^{v}. Perhaps greater and/or faster photoreceptor cell loss that is dependent on mutant *Myo7a* can be effected by having additional mutant USH1 genes in the genetic background. In any case, it is argued that shaker1 mice are a useful model for testing USH1B gene therapy, due to the presence of mutant phenotypes other than photoreceptor cell death.

6. ACKNOWLEDGMENTS

This research was supported by NIH R01 grant EY07042, NIH core grant EY EY12598, and a grant from the Foundation Fighting Blindness (to DSW), and the MRC, Defeating Deafness, and EC contract QLG2-CT-1999-00988 (to KPS).

7. REFERENCES

Alagramam, K. N., Murcia, C. L., Kwon, H. Y., Pawlowski, K. S., Wright, C. G., and Woychik, R. P., 2001, The mouse Ames waltzer hearing-loss mutant is caused by mutation of Pcdh15, a novel protocadherin gene. *Nat. Genet.* **27**:99-102.

Astuto, L. M., Weston, M. D., Carney, C. A., Hoover, D. M., Cremers, C., Wagenaar, M., Moller, C., Smith, R. J. H., Pieke-Dahl, S., Greenberg, J., Ramesar, R., Jacobson, S. G., Ayuso, C., Heckenlively, J. R., Tamayo, M., Gorin, M. B., Reardon, W., and Kimerling, W. J., 2000, Genetic heterogeneity of Usher syndrome: Analysis of 151 families with Usher type I. *Amer. J. Hum. Genet.* **67**:1569-1574.

Ball, S. L., Bardenstein, D., and Alagramam, K., 2002, Characterization of retinal function in the Ames Waltzer mouse. *Invest Ophthalmol Vis Sci ARVO abstracts* **43**:1778.

Bolz, H., von Brederlow, B., Ramirez, A., Bryda, E. C., Kutsche, K., Nothwang, H. G., Seeliger, M., del C-Salcedo Cabrera, M., Vila, M. C., Molina, O. P., Gal, A., and Kubisch, C., 2001, Mutation of CDH23, encoding a new member of the cadherin gene family, causes Usher syndrome type 1D. *Nat. Genet.* **27**:108-112.

Bork, J. M., Peters, L. M., Riazuddin, S., Bernstein, S. L., Ahmed, Z. M., Ness, S. L., Polomeno, R., Ramesh, A., Schloss, M., Srisailpathy, C. R., Wayne, S., Bellman, S., Desmukh, D., Ahmed, Z., Khan, S. N., Kaloustian, V. M., Li, X. C., Lalwani, A., Bitner-Glindzicz, M., Nance, W. E., Liu, X. Z., Wistow, G., Smith, R. J., Griffith, A. J., Wilcox, E. R., Friedman, T. B., and Morell, R. J., 2001, Usher syndrome 1D and nonsyndromic autosomal recessive deafness DFNB12 are caused by allelic mutations of the novel cadherin-like gene CDH23. *Am. J. Hum. Genet.* **68**:26-37.

Bowes, C., Li, T., Danciger, M., Baxter, L. C., Applebury, M. L., and Farber, D. B., 1990, Retinal degeneration in the rd mouse is caused by a defect in the beta subunit of rod cGMP-phosphodiesterase. *Nature* **347**:677-680.

Cuevas, J. M., Espinos, C., Millan, J. M., Sanchez, F., Trujillo, M. J., Garcia-Sandoval, B., Ayuso, C., Najera, C., and Beneyto, M., 1998, Detection of a novel Cys628STOP mutation of the myosin VIIA gene in Usher syndrome type Ib. *Mol. Cell. Probes* **12**:417-420.

Di Palma, F., Holme, R. H., Bryda, E. C., Belyantseva, I. A., Pellegrino, R., Kachar, B., Steel, K. P., and Noben-Trauth, K., 2001, Mutations in Cdh23, encoding a new type of cadherin, cause stereocilia disorganization in waltzer, the mouse model for Usher syndrome type 1D. *Nat. Genet.* **27**:103-107.

Gibbs, D., Kitamoto, J., and Williams, D. S., submitted, Myosin VIIa in RPE phagocytosis and a potential cause of blindness in Usher syndrome 1B.

Gibson, F., Walsh, J., Mburu, P., Varela, A., Brown, K. A., Antonio, M., Beisel, K. W., Steel, K. P., and Brown, S. D. M., 1995, A type VII myosin encoded by mouse deafness gene shaker-1. *Nature* **374**:62-64.

Hasson, T., Heintzelman, M. B., Santos-Sacchi, J., Corey, D. P., and Mooseker, M. S., 1995, Expression in cochlea and retina of myosin VIIa, the gene product defective in Usher syndrome type 1B. *Proc. Natl. Acad. Sci. USA* **92**:9815-9819.

Hasson, T., Walsh, J., Cable, J., Mooseker, M. S., Brown, S. D. M., and Steel, K. P., 1997, Effects of shaker-1 mutations on myosin-VIIa protein and mRNA expression. *Cell Motil. Cytoskel.* **37**:127-138.

Holme, R. H., and Steel, K. P., 2002, Stereocilia defects in waltzer (Cdh23), shaker1 (Myo7a) and double waltzer/shaker1 mutant mice. *Hear. Res.* **169**:13-23.

Libby, R. T., Kitamoto, J., Holme, R. H., Williams, D. S., and Steel, K. P., submitted, *Cdh23* mutations in the mouse are associated with retinal dysfunction but not retinal degeneration.

Libby, R. T., and Steel, K. P., 2001, Electroretinographic anomalies in mice with mutations in Myo7a, the gene involved in human Usher syndrome type 1B. *Invest. Ophthalmol. Vis. Sci.* **42**:770-778.

Liu, X., Ondek, B., and Williams, D. S., 1998, Mutant myosin VIIa causes defective melanosome distribution in the RPE of shaker-1 mice. *Nat. Genet.* **19**:117-118.

Liu, X., Udovichenko, I. P., Brown, S. D. M., Steel, K. P., and Williams, D. S., 1999, Myosin VIIa participates in opsin transport through the photoreceptor cilium. *J. Neurosci.* **19**:6267-6274.

Liu, X., Vansant, G., Udovichenko, I. P., Wolfrum, U., and Williams, D. S., 1997, Myosin VIIa, the product of the Usher 1B syndrome gene, is concentrated in the connecting cilia of photoreceptor cells. *Cell Motil. Cytoskel.* **37**:240-252.

Lord, E. M., and Gates, W. H., 1929, Shaker, a new mutation of the house mouse (*Mus musculus*). *Am. Nat.* **63**:435-442.

Mburu, P., Liu, X. Z., Walsh, J., Saw, D., Jamie, M., Cope, T. V., Gibson, F., Kendrick-Jones, J., Steel, K. P., and Brown, S. D. M., 1997, Mutation analysis of the mouse myosin VIIA deafness gene. *Genes Funct.* **1**:191-203.

McLaughlin, M. E., Ehrhart, T. L., Berson, E. L., and Dryja, T. P., 1995, Mutation spectrum of the gene encoding the beta subunit of rod phosphodiesterase among patients with autosomal recessive retinitis pigmentosa. *Proc. Natl. Acad. Sci. U S A* **92**:3249-3253.

Pittler, S. J., and Baehr, W., 1991, Identification of a nonsense mutation in the rod photoreceptor cGMP phosphodiesterase beta-subunit gene of the rd mouse. *Proc. Natl. Acad. Sci. U S A* **88**:8322-6.

Portera-Cailliau, C., Sung, C. H., Nathans, J., and Adler, R., 1994, Apoptotic photoreceptor cell death in mouse models of retinitis pigmentosa. *Proc. Natl. Acad. Sci. U S A* **91**:974-978.

Sanyal, S., and Hawkins, R. K., 1981, Genetic interaction in the retinal degeneration of mice. *Exp. Eye Res.* **33**:213-22.

Usher, C., 1914, On the inheritance of retinitis pigmentosa with notes of cases. *R. Lond. Ophthalmol. Hosp. Rep.* **19**:130-236.

Weil, D., Blanchard, S., Kaplan, J., Guilford, P., Gibson, F., Walsh, J., Mburu, P., Varela, A., Levilliers, J., Weston, M. D., Kelley, P. M., Kimberling, W. J., Wagenaar, M., Levi-Acobas, F., Larget-Piet, D., Munnich, A., Steel, K. P., Brown, S. D. M., and Petit, C., 1995, Defective myosin VIIA gene responsible for Usher syndrome type 1B. *Nature* **374**:60-61.

Weston, M. D., Kelley, P. M., Overbeck, L. D., Wagenaar, M., Orten, D. J., Hasson, T., Chen, Z. Y., Corey, D., Mooseker, M., Sumegi, J., Cremers, C., Moller, C., Jacobson, S. G., Gorin, M. B., and Kimberling, W. J., 1996, Myosin VIIa mutation screening in 189 Usher syndrome type 1 patients. *Amer. J. Hum. Genet.* **59**:1074-1083.

Wong, P. C., Cai, H., Borchelt, D. R., and Price, D. L., 2002, Genetically engineered mouse models of neurodegenerative diseases. *Nat. Neurosci.* **5**:633-639.

SCREEN FOR USHER SYNDROME 1B MUTATIONS IN THE OVINE MYOSIN VIIA GENE

Tania Slatter[1], Sassan M. Azarian[2], Scott Tebbutt[1], Marion Maw[1], and David S. Williams[2]

1. INTRODUCTION

Recessive mutations in human myosin VIIa (*MYO7A*) cause either nonsyndromic deafness or, in most cases, Usher syndrome 1B (USH1B), which includes retinitis pigmentosa as well as deafness (Liu et al., 1997a; Liu et al., 1997b; Weil et al., 1995). Mutations in myosin VIIa have also been reported in shaker1 mice (Gibson et al., 1995) and zebrafish (Ernest et al., 2000). However, retinal degeneration has not been observed in any of the mutant mice or zebrafish. In this respect, ten different alleles of shaker1 mice have been studied (Lillo et al., this volume).

For future studies on therapies for USH1B, it would be helpful to have an animal model in which mutant myosin VIIa results in photoreceptor degeneration. Moreover, it maybe useful to have a mammalian model with an eye of similar size to that of a human. For these reasons, we took advantage of the availability of DNA samples from 1350 sheep, and thus screened the myosin VIIa gene in these samples to see if we could identify any carriers of USH1B-like mutations.

2. MATERIALS AND METHODS

2.1. Exonic and Flanking Intronic Sequences of the Ovine Myosin VIIa Gene

Primers for amplifying and sequencing ovine myosin VIIa exons were based on nucleotide sequences that were identical between mouse (accession number U81453) and human myosin VIIa cDNA (accession numbers U39226 and U55208). Initially, primers were designed to amplify the region between exons 6 and 7, since the corresponding intron was reported to be relatively small (Kelley et al., 1997; Levy et al., 1997). Primer

[1]Department of Biochemistry, University of Otago, Dunedin, New Zealand; [2]Departments of Pharmacology and Neurosciences, UCSD School of Medicine, La Jolla, CA 92093-0983, USA. Corresponding author: D. S. Williams, E-mail: dswilliams@ucsd.edu

sequences were: 5'-TTCCTGGCAGCCATCAGTGG (forward) and 5'-CAGACACGTGACTTTTCCAGCAGG (reverse). Ovine genomic DNA was amplified using HotStarTaq DNA polymerase kit (Qiagen) per manufacturer's instructions, with 40 cycles of 20 sec at 95°C, 40 sec at 53°C, 1 min at 72°C. The resulting PCR product was sequenced for confirmation. Using the same primers, a library of ovine BAC clones was screened by PCR at the BAC-YAC Resource Center of the Animal Genetics Department of the INRA (Institut National De La Recherche Agronomique). The BAC clones had an average insert size of 123 kb in the pBeloBAC11 vector (Vaiman et al., 1999). Three PCR-positive BAC clones were obtained: clones 104C8, 206A11, and 196A2. Sequencing-grade BAC DNA was prepared by Amplicon Express (Pullman, WA). Primers to various exons were designed as above. BAC clones were sequenced by the DNA Sequencing Shared Resource, UCSD Cancer Center (funded in part by NCI Cancer Center Support Grant # 2 P30 CA23100-18). Both exonic and flanking intronic sequences were obtained. Exonic sequences were identified by alignment with cDNA sequence using either the Sequencher (Gene Codes) or SeqMan (Lasergene) software. All sequences were obtained from clone 104C8, which contained at least exons 3 through 49, and probably the entire myosin VIIa gene.

2.2. Ovine DNA Samples

Sheep DNA samples were taken from 1200 lambs that had just died on a number of farms in the South Island of New Zealand. In addition, 150 sheep DNA samples from Australia were available as a result of a previous study that sought a sheep model for cystic fibrosis and other pediatric diseases (Tebbutt et al., 1996).

2.3. SSCP Analysis of Ovine Myosin VIIa Exons

PCR for single-strand conformation polymorphism (SSCP) analysis was performed for exons 5-8, 16,17, 21, 28, and 29 of the ovine myosin VIIa gene from each of the 1350 samples. Each PCR reaction was optimized for minimal nonspecific amplification. Each forward primer was labeled using [γ-^{33}P]-ATP. The labeled PCR product was mixed with an equal volume of fomamide dye, denatured for 5 min at 95°C, and then placed on ice for 5 min. Samples were applied to non-denaturing polyacrylamide gels. Electrophoresis was conducted in 1x TBE buffer and 1x TME buffer. All electrophoresis was carried out at 40 W at 18°C. Any PCR product that was associated with a band shift was sequenced.

3. RESULTS

MYO7A has 49 exons (Kelley et al., 1997; Levy et al., 1997). Rather than screen all 49, we focused on seven exons that contained the most reported USH1B mutations. These were exons 5, 7, 16,17, 21, 28, and 29 (Fig. 1). To these seven, we added exons 6 and 8 because they could be obtained in a single PCR reaction with exons 5 and 7 due to small intervening introns.

For SSCP and sequencing analyses, a pair of intronic primers was designed for each exon studied (Table 1). A total of nine single-nucleotide substitutions were identified: four intronic and five exonic. None of these variations was considered likely to be pathogenic as splice site consensus sequences were conserved and the amino acids encoded were unaltered (Table 2).

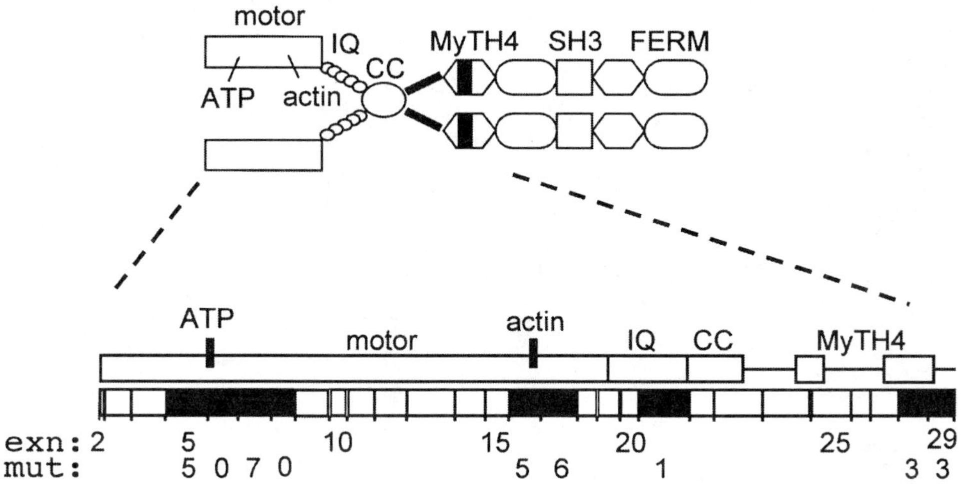

Figure 1. Ovine myosin VIIa exons that were screened. A schematic diagram of the myosin VIIa protein and the corresponding exons of the gene is shown. The filled-in exons were screened. They included those which contained the most reported USH1B mutations, which tend to be concentrated in exons that encode critical functional regions, such as ATP and actin binding regions in the motor domain. IQ, light chain binding sites; CC, coiled coil; MyTH4, myosin tail homology four, (which is split); FERM, band 4.1, ezrin, radixin, moesin homology; exn, exon number; mut, number of USH1B mutations reported in the exons we screened.

Table 1. Sequence of primers used for mutational analysis

Exon	Sequence
5	5'-CCCTACATAGGAAGGGATG
	5'-CTAAGCCTGGCGCTCCATG
6	5'-CTGATGAGGAGCTGACTGG
	5'-CCCTGCTCCTCCCTGAGTG
7	5'-CTGTGACCCATGACCCGTG
	5'-TCCTGTGCAGGGGCTCAGG
8	5'-GTTCCCCAACACCTTGTGC
	5'-GTGCCCACATGGGCCTCAC
16	5'-CCTCCCATCGCTGTGGTCCG
	5'-CCCATCCCTCCAGGGCTG
17	5'-CAGCACTGACGCCCTTGGCTCC
	5'-GCCTGGCCCAGGGCTGGTACC
21	5'-CATCAGGGCCGTGGTTCTG
	5'-CCAGGGGTTCTCGTGGACTGGG
28	5'-CTTGTATCCCACTCTGCTTAG
	5'-CATTGACACCAAAGACTGAG
29	5'-CTGAGTTGGTCCTCTGTCCG
	5'-CCAGGGCACTGGAGTGTATG

Table 2. Ovine myosin VIIa variations

Location	Type of variation	Predicted amino acid	Equivalent amino acid position in human MYO7A
Intron 5	IVS5-8C>T		
Intron 5	IVS5-13G>A		
Intron 5	IVS5-37A>G		
Intron 5	IVS5-43A>G		
Exon 6	CTG-TTG	L-L	L170
Exon 7	GGT-GGC	G-G	G229
Exon 7	ATC-ATT	I-I	I232
Exon 16	CCC-CCT	P-P	P608
Exon 16	GAG-GAA	E-E	N639

4. DISCUSSION

The frequency of Usher syndrome in the U.S.A. has been estimated at 4.4 per 100,000 (Boughman et al., 1983). Estimates are slightly lower from Scandanavia at 3.5 per 100,000, and as high as 8 per 100,000 from Germany (Otterstedde et al., 2001). About 25% of all USH patients have USH1B (Astuto et al., 2000), so that the frequency of USH1B affected individuals is around 1.1 per 100,000 in the U.S.A. It follows that about 1 person out of 150 is a carrier for USH1B. If this frequency were the same in the sheep population from which we obtained our samples, the probability that one of our 1350 samples would contain a pathological mutation would have been quite high. We focused on only nine exons out of a total of 49 exons in myosin VIIa. Nevertheless, seven of the nine exons were those that had been reported to contain the most USH1B mutations. Domesticated New Zealand and Australian sheep, from which our samples were obtained, may not represent such a genetically diverse population as the human population. Alternatively, pathological mutations in myosin VIIa may occur at a higher frequency in human than in sheep.

5. ACKNOWLEDGMENTS

This research was funded by the Foundation Fighting Blindness and Ted and Elaine Welp.

6. REFERENCES

Astuto, L.M., M.D. Weston, C.A. Carney, D.M. Hoover, C. Cremers, M. Wagenaar, C. Moller, R.J.H. Smith, S. Pieke-Dahl, J. Greenberg, R. Ramesar, S.G. Jacobson, C. Ayuso, J.R. Heckenlively, M. Tamayo, M.B. Gorin, W. Reardon, and W.J. Kimerling. 2000. Genetic heterogeneity of Usher syndrome: Analysis of 151 families with Usher type I. *Amer. J. Hum. Genet.* **67**:1569-1574.

Boughman, J., M. Vernon, and K. Shaver. 1983. Usher syndrome: definition and estimate of prevalence from two high-risk populations. *J. Chronic Dis.* **36**:595-603.

Ernest, S., G.J. Rauch, P. Haffter, R. Geisler, C. Petit, and T. Nicolson. 2000. Mariner is defective in myosin VIIA: a zebrafish model for human hereditary deafness. *Hum. Mol. Genet.* **9**:2189-2196.

Gibson, F., J. Walsh, P. Mburu, A. Varela, K.A. Brown, M. Antonio, K.W. Beisel, K.P. Steel, and S.D.M. Brown. 1995. A type VII myosin encoded by mouse deafness gene shaker-1. *Nature* **374**:62-64.

Kelley, P.M., M.D. Weston, Z.Y. Chen, D.J. Orten, T. Hasson, L.D. Overbeck, J. Pinnt, C.B. Talmadge, P. Ing, M.S. Mooseker, D. Corey, J. Sumegi, and W.J. Kimberling. 1997. The genomic structure of the gene defective in Usher syndrome type Ib (MYO7A). *Genomics* **40**:73-79.

Levy, G., F. Levi Acobas, S. Blanchard, S. Gerber, D. Larget Piet, V. Chenal, X.Z. Liu, V. Newton, K.P. Steel, S.D. Brown, A. Munnich, J. Kaplan, C. Petit, and D. Weil. 1997. Myosin VIIA gene: heterogeneity of the mutations responsible for Usher syndrome type IB. *Hum. Mol. Genet.* **6**:111-116.

Liu, X.-Z., J. Walsh, P. Mburu, J. KendrickJones, M.J.T.V. Cope, K.P. Steel, and S.D.M. Brown. 1997a. Mutations in the myosin VIIA gene cause non-syndromic recessive deafness. *Nat. Genet.* **16**:188-190.

Liu, X.Z., J. Walsh, Y. Tamagawa, K. Kitamura, M. Nishizawa, K.P. Steel, and S.D. Brown. 1997b. Autosomal dominant non-syndromic deafness caused by a mutation in the myosin VIIA gene. *Nat. Genet.* **17**:268-289.

Otterstedde, C.R., U. Spandau, A. Blankenagel, W.J. Kimberling, and C. Reisser. 2001. A new clinical classification for Usher's syndrome based on a new subtype of Usher's syndrome type I. *Laryngoscope* **111**:84-86.

Tebbutt, S.J., A. Harris, and D.F. Hill. 1996. An ovine CFTR variant as a putative cystic fibrosis causing mutation. *J. Med. Genet.*. **33**:623-4.

Vaiman, D., A. Billault, K. Tabet-Aoul, L. Schibler, D. Vilette, A. Oustry-Vaiman, C. Soravito, and E.P. Cribiu. 1999. Construction and characterization of a sheep BAC library of three genome equivalents. *Mamm Genome* **10**:585-7.

Weil, D., S. Blanchard, J. Kaplan, P. Guilford, F. Gibson, J. Walsh, P. Mburu, A. Varela, J. Levilliers, M.D. Weston, P.M. Kelley, W.J. Kimberling, M. Wagenaar, F. Levi-Acobas, D. Larget-Piet, A. Munnich, K.P. Steel, S.D.M. Brown, and C. Petit. 1995. Defective myosin VIIA gene responsible for Usher syndrome type 1B. *Nature* **374**:60-61.

PHOTORECEPTOR INTERSEGMENTAL TRANSPORT AND RETINAL DEGENERATION

A conserved pathway common to motile and sensory cilia

Joseph C. Besharse[1], Sheila A. Baker[1], Katherine Luby-Phelps[1], and Gregory J. Pazour[2]

1. INTRODUCTION

During early development of rod and cone photoreceptors, the outer segment (OS) assembles on the apical cell surface by an elaborate process of membrane and protein transport followed by membrane reorganization to form the highly organized stack of membrane disks found in the OS of mature cells. The process of membrane and protein transport continues in mature photoreceptors as the molecular components of the OS turnover at a prodigious rate for the life of the cell (Young, 1967). Intersegmental transport between the inner segment (IS) and the OS is a major logistical problem as all OS macromolecules are synthesized in the IS and must be transported to the OS. Elucidating the molecular pathways involved in intersegmental transport will ultimately require a refined understanding of the photoreceptor connecting cilium as this structure is the only stable connecting link between the IS and the OS and is the likely transport corridor (reviewed in Besharse and Horst, 1990). In addition, understanding the details of these pathways is likely to provide an improved understanding of photoreceptor degeneration diseases that involve trafficking between the two segments.

Like all other cilia, photoreceptor cilia have a microtubule based cytoskeleton called the axoneme, which is templated from a basal body (Horst, et al., 1990). In the region of the connecting cilium, the axoneme has a 9+0 configuration of microtubules that corresponds anatomically to the transition zone of other cilia and flagella (Rohlich, 1975). Beyond the connecting cilium, the axoneme loses it stereotypical 9+0 organization. However the microtubules extend for some distance into the outer segment.

[1]Department of Cell Biology, Neurobiology and Anatomy, Medical College of Wisconsin, 8701 Watertown Plank Road, Milwaukee, WI 53226, USA; [2]Program in Molecular Medicine, University of Massachusetts Medical School, Worcester, MA 01605, USA. Corresponding author: J. C. Besharse, E-mail: jbeshars@mcw.edu

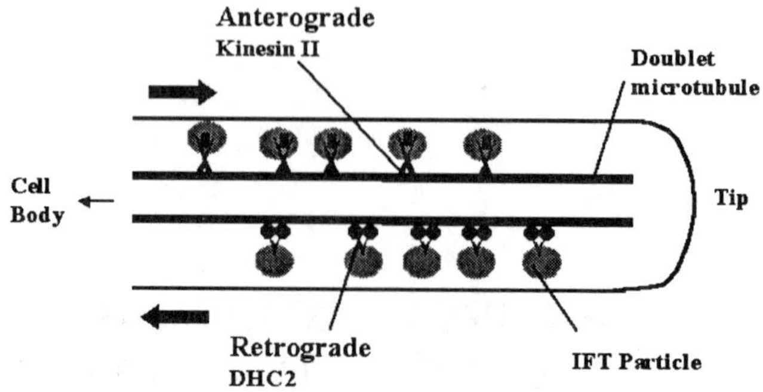

Figure 1. Diagram illustrating IFT components in cilia and flagella.

The conserved structure of eukaryotic cilia led to the idea that a conserved process may underlie assembly and turnover of all sensory and motile cilia (Besharse and Horst, 1990; Cole, et al., 1998; Horst, et al., 1990; Rosenbaum, et al., 1999). Currently, intraflagellar transport (IFT) is the best understood ciliary transport system. This system was first elucidated in the motile flagella of the green alga, *Chlamydomonas*. Analysis of mutations in components of the IFT pathway in *Chlamydomonas*, *C. elegans,* and mouse indicates that IFT is likely to be required for assembly of all motile and sensory cilia (reviewed in Rosenbaum and Witman, 2002). In IFT (Figure 1), protein and membrane cargo necessary for cilium assembly are carried distally by kinesin-II (Cole, et al., 1998; Piperno and Mead, 1997; Piperno, et al., 1996; Signor, et al., 1999b; Walther, et al., 1994) for incorporation into a growing cilium or proximally by cytoplasmic dynein-2 containing the DHC2/Dhc1b heavy chain (Pazour, et al., 1999; Pazour, et al., 1998; Porter, et al., 1999; Signor, et al., 1999a) for exchange with a precursor pool within the cell body. This transport is associated with movement of a large IFT particle consisting of at least 15 different IFT polypeptides (Cole, et al., 1998; Piperno and Mead, 1997).

2. KINESIN II AND DYNEIN 2 IN PHOTORECEPTORS

Kinesin II, the anterograde IFT motor, is a heterotrimeric complex containing two motor subunits (KIF3A and KIF3B) and an accessory subunit (KAP). Both motor subunits have been localized to photoreceptor basal bodies and connecting cilia (Beech, et al., 1996; Whitehead, et al., 1999). Furthermore, photoreceptor specific deletion of the KIF3A motor subunit in mice results in accumulation of opsin and arrestin within the inner segment and eventual apoptotic cell death in photoreceptors (Marszalek, et al., 2000). These observations suggest that kinesin II is directly involved in transport to the outer segment and are consistent with the idea that kinesin II is the anterograde motor for photoreceptor IFT (Pazour, et al., 2002).

Dyneins are large protein complexes containing heavy chain motor subunits and an assortment of intermediate, light intermediate, and light chains that are likely involved in motor-cargo interactions and regulatory functions. Dynein-2, the predicted motor for retrograde IFT, is thought to function as a large protein complex but the only known subunits are a heavy chain (Gibbons, et al., 1994) and a light-intermediate chain called

D2LIC (Grissom, et al., 2002) or LIC3 (Mikami, et al., 2002). In photoreceptors both the dynein-2 heavy chain and D2LIC/LIC3 are associated with the ciliary axoneme and the basal body (Mikami, et al., 2002).

Photoreceptor discs and the associated transduction machinery are formed proximally and shed distally (Young, 1967; Young and Bok, 1969) . Consequently, there is an obvious need for anterograde transport that would deliver components from their site of synthesis in the IS to the OS. However, the relative importance of a dynein-2 motor and retrograde transport in photoreceptor cilia has not been defined. Several lines of data indicating that components of the transduction machinery shuttle between the segments lead us to suggest that a robust retrograde transport system in photoreceptor OS is likely. The alpha and beta subunits of transducin and phosducin appear to move from outer to inner segment upon light exposure (Sokolov, et al., 2002; Whelan and McGinnis, 1988). In addition, translocation of rod arrestin from outer to inner segment reportedly occurs in darkness (Philp, et al., 1987). These studies have all been interpreted as reflecting a bidirectional, light-regulated transport pathway in outer segments and the connecting cilium.

Both kinesin-II and cytoplasmic dynein-2 motors have additional functions outside of the cilium. Kinesin-II is localized to ribbon synapses in photoreceptors (Muresan, et al., 1999) and is thought to be involved in cytoplasmic membrane trafficking, neuronal transport and melanosome movement (Marszalek and Goldstein, 2000). Cytoplasmic dynein-2 is localized to inner segment domains outside of the cilium in photoreceptors (Mikami, et al., 2002), and to the Golgi complex in tissue culture cells (Grissom, et al., 2002). The multi-functional role of these motors means that they are not unique signatures of IFT. For example, mis-localization of outer segment proteins seen in photoreceptors with mutated KIF3A (Marszalek, et al., 2000) could reflect defective IFT, defective trafficking pathways within the inner segment, or both.

3. IFT PROTEINS IN PHOTORECEPTORS

A 17S IFT particle with a unique assemblage of at least 15 different proteins is transported during IFT in *Chlamydomonas* (Figure 1). As an initial test of the hypothesis that IFT supports intersegmental transport in photoreceptors, we studied the expression of 4 mammalian proteins identified by their sequence homology to *Chlamydomonas* IFT proteins (Pazour, et al., 2002). We found that IFT88, IFT57, IFT52, and IFT20* all fractionate with a detergent extracted photoreceptor cytoskeleton from bovine rods, and localize to the inner segment of rods and cones. Within photoreceptor inner segments, distinct concentration of all 4 IFT proteins was found at the basal body and in association with the ciliary axoneme. A similar pattern of localization of both endogenous protein and transgenically over-expressed IFT57 and 52 in *Xenopus* rod photoreceptors has also been detected (Luby-Phelps and Besharse, unpublished data). In the very long *Xenopus* rods IFT proteins are arranged at the basal body and in a punctate pattern along the length of the axoneme. The localization of IFT proteins in photoreceptors (Pazour, et al., 2002) is remarkably similar to that of IFT proteins, kinesin-II and dynein-2 in *Chlamydomonas* (Cole, et al., 1998; Deane, et al., 2001; Iomini, et al., 2001; Pazour, et al., 1999).

* IFT numbers refer to the apparent M_r of the protein in *Chlamydomonas* (Cole, et al., 1998); mammalian homologues are similar but not identical in M_r to those of *Chlamydomonas*.

4. A 17S IFT PARTICLE IN PHOTORECEPTORS

Although the localization of IFT proteins in photoreceptor cilia implies a conserved function, the ~17S IFT protein complex has been identified only in association with the motile flagella of *Chlamydomonas* (Cole, et al., 1998; Piperno and Mead, 1997) and in mouse testis (Pazour, et al., 2002). Furthermore, IFT particle-like structures are not routinely seen in photoreceptors examined by EM. Thus, an important question remains as to whether a ~17S protein complex is involved in IFT in photoreceptors where the IFT cargo is potentially very different from that in motile flagella. To address this question, we have used sucrose density sedimentation and immunoprecipitation to analyze extracts of purified bovine photoreceptor outer segments and compared them directly to similar fractions derived from mouse testis (Baker, Luby-Phelps, Pazour and Besharse, unpublished data). IFT88, IFT57, IFT52, and IFT20 co-sedimented at ~17S and co-immunoprecipitated. This indicates that in photoreceptors, like in testis, the IFT proteins are organized in a large macromolecular complex. Furthermore, a pair-wise yeast 2-hydrid analysis of IFT88, 57, 52 and 20 shows that IFT57 and IFT20 uniquely interact via coiled-coil domains; neither hetero- nor homodimeric interactions can be detected among the remaining combinations. The interaction of IFT20 and IFT57 is the first direct interaction identified for any IFT proteins and is likely to form a basis for understanding how the IFT complex is assembled.

5. PHOTORECEPTOR DEGENERATION IN TG737ORPK MICE

Our recent characterization of progressive photoreceptor degeneration in $Tg737^{orpk}$ mice carrying a mutation in the gene encoding IFT88 (also called Polaris) shows that IFT protein function is critical in photoreceptor assembly and maintenance (Pazour, et al., 2002). The hypomorphic $Tg737^{orpk}$ mutation was generated by a random integration of a small insert into an intron. This does not affect the coding potential of the gene but appears to cause improper processing of the message and its destabilization (Moyer, et al., 1994). Targeted deletion of IFT88/polaris in mice results in an embryonic lethal phenotype in which embryos exhibit randomization of the left-right symmetry of organs (Murcia, et al., 2000). Polycystic kidneys in $Tg737^{orpk}$ mice (Pazour, et al., 2000) and embryonic lethality in mice with the targeted deletion (Murcia, et al., 2000) are associated with poorly formed primary cilia on kidney epithelial cells and un-developed nodal cilia in embryos respectively. The striking ciliary defects in mice with mutations in IFT88 along with finding that mutations in the *Chlamydomonas* (Pazour, et al., 2000) and *C. elegans* (Haycraft, et al., 2001; Qin, et al., 2001) homologues of IFT88 result in failure of flagellar or sensory cilium assembly indicates that IFT88 is critical for assembly of cilia and flagella in multiple organisms.

Disruption of outer segment assembly is the earliest manifestation of the $Tg737^{orpk}$ mutation (Pazour, et al., 2000). $Tg737^{orpk}$ homozygous mutants form a full complement of rod and cone photoreceptors and have normal retinal architecture during early development. At p10 (postnatal day 10), however, the mutant photoreceptor outer segments are shorter than littermate controls and have poorly formed and vesiculated discs. By p21 the disrupted disc structure is more apparent and rod photoreceptors exhibit mis-localization of opsin to the inner segment. At p21 the interphotoreceptor space accumulates massive numbers of extracellular vesicles that are immuno-positive for opsin. The disrupted structure of outer segment discs and the accumulation of extracellular vesicles is remarkably similar to the phenotype described for the targeted deletion of the Tulp1 gene (Hagstrom, et al., 1999) , which may function in intracellular trafficking of rhodopsin (Hagstrom, et al., 2001). By p45 a major loss of nuclei from the outer nuclear layer is evident and further loss of photoreceptors is seen though the postnatal week eleven. In general, abnormal outer segment assembly followed by cell death characterizes retinal degeneration in $Tg737^{orpk}$ mutants. This phenotype is consistent with a role for IFT88 in trafficking of critical "cargo" into the outer segment.

6. WHAT IS THE IFT CARGO IN PHOTORECEPTORS?

Figure 2 illustrates the basic components of IFT in photoreceptors. Doublet microtubules of the axoneme serve as tracks for outward and inward transport using the kinesin-II and dynein-2 motors respectively. The IFT protein complex moves in association with the motors and is thought to serve as a loading scaffold for "cargo". Although a clear role for IFT in the assembly and maintenance of photoreceptor outer segments has been demonstrated, the mechanistic details and a direct link between IFT and a specific "cargo" has not been established. Nonetheless, a large number of known macromolecules pass bi-directionally through the connecting cilium making photoreceptors an ideal system for identification of IFT cargo.

First, components of the axoneme itself are likely cargo. The axoneme is a highly complex structure made up of α and β tubulin and a multitude of other proteins. By analogy to flagella, the tubulin subunits of the axoneme are predicted to turnover with replacement occurring at the distal (plus end) of the axoneme (Marshall and Rosenbaum, 2001). In *Chlamydomonas* the abundance of IFT proteins is closely correlated with the rate of axoneme turnover suggesting that axonemal components are cargo (Marshall and Rosenbaum, 2001). A similar relationship could pertain in Tg737 mice; the shortened and disrupted OS in these mice could be related to a deficit in distal axoneme assembly.

Second, mis-localization of opsin (Marszalek, et al., 2000; Pazour, et al., 2002) in mice carrying mutations in either the kinesin-II motor or IFT88 suggests that it may be moved into the outer segment by IFT. Post-Golgi vesicles containing nascent rhodopsin are targeted to the basal body, possibly through a direct interaction with cytoplasmic dynein-1 (Tai, et al., 1999). At the basal body, dynein-1 may be exchanged for kinesin-II or an IFT subunit. However, a large proportion of rhodopsin is correctly targeted to the outer segments in mice with a mutation in IFT88 (Pazour, et al., 2002), and mis-localization of opsin is a general feature of numerous photoreceptor degenerations (Hagstrom, et al., 1999; Nir and Papermaster, 1986; Nir, et al., 1987), suggesting that the link to IFT may be indirect.

Third, the aberrant regulation of OS disc size and the formation of extracellular vesicles in IFT88 mutants (Pazour, et al., 2002) suggests that proteins responsible for disc morphogenesis, such as peripherin/rds and ROM1 are potential IFT cargo. Mutations in those proteins lead to aberrant disc morphogenesis and release of extracellular vesicles (reviewed in Boesze-Battaglia and Goldberg, 2002).

Fourth, IFT may facilitate the dramatic light-regulated shuttling of transducin, phosducin, and arrestin between the segments (Philp, et al., 1987; Sokolov, et al., 2002; Whelan and McGinnis, 1988), a suggestion consistent with the mis-localization of

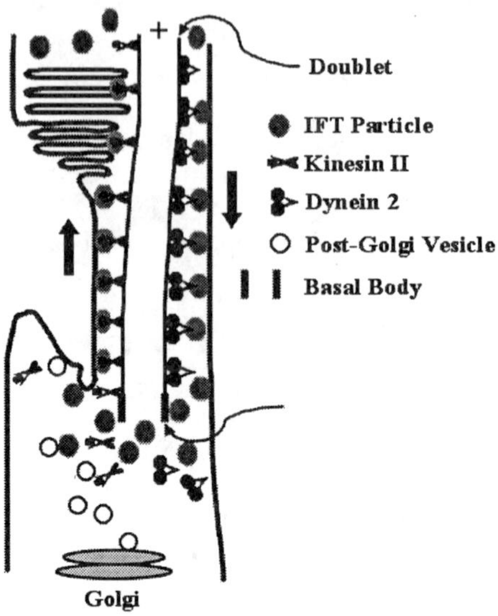

Figure 2. Model for intraflagellar transport in photoreceptors.

arrestin in photoreceptors with disrupted kinesin-II (Marszalek, et al., 2000). In particular, IFT may be necessary to support the rapid translocation in the retrograde direction of transducin at light onset.

7. PERSPECTIVE

The IFT model for intersegmental transport arose through exploitation of the idea that ancient mechanisms conserved in all cilia and flagella play a critical role in the assembly of photoreceptor outer segments (Besharse and Horst, 1990; Rosenbaum, et al., 1999). At this point, the study of IFT is in its infancy and there is much to learn about the assembly and regulation of IFT complexes and their associated molecular motors and cargo. There is a striking lack of biochemical data supporting a direct link between IFT and any of the proposed cargo and this gap needs to be filled by molecular analysis. In addition, it seems likely that IFT is but one of multiple transport pathways that underlie intersegmental transport. For example, there is evidence that myosin VIIa in connecting cilia supports transport as well (Liu, et al., 1999) and the relationship between myosyn VIIa and IFT remains unexplored.

It is clear that disruption of IFT through mutations in either its molecular motors or the IFT protein complex can lead to photoreceptor degeneration in mice. Thus, the genes encoding IFT components are potential candidates for mutations involved in photoreceptor degeneration diseases in man. Because IFT components are found in all cilia and flagella, individuals with mutations in IFT genes are likely to exhibit defects in multiple tissues, and severe deletions may lead to embryonic lethal phenotypes. For example, mice with the Tg737^{orpk} mutation have retinal degeneration and polycystic kidneys and deletion of the gene leads to embryonic lethality. It is therefore of considerable interest that several human syndromes including Senior-Loken syndrome, Jeune syndrome and Bardet-Biedl syndrome are characterized by cystic kidneys and retinal degeneration along with other defects that could be associated with improperly formed cilia (Traboulsi, 1998). IFT components are candidate genes for these and other disorders.

8. REFERENCES

Beech, P.L., K. Pagh-Roehl, Y. Noda, N. Hirokawa, B. Burnside, and J.L. Rosenbaum, 1996, Localization of kinesin superfamily proteins to the connecting cilium of fish photoreceptors, *J Cell Sci.* **109**:889-97.

Besharse, J.C., and C.J. Horst. 1990. The Photoreceptor Connecting Cilium. A Model for the Transition Zone. *In* Ciliary and Flagellar Membranes. R.A. Bloodgood, editor. Plenum Press, New York. 389-417.

Boesze-Battaglia, K., and A.F. Goldberg, 2002, Photoreceptor renewal: a role for peripherin/rds, *Int Rev Cytol.* **217**:183-225.

Cole, D.G., D.R. Diener, A.L. Himelblau, P.L. Beech, J.C. Fuster, and J.L. Rosenbaum, 1998, Chlamydomonas kinesin-II-dependent intraflagellar transport (IFT): IFT particles contain proteins required for ciliary assembly in Caenorhabditis elegans sensory neurons, *J Cell Biol.* **141**:993-1008.

Deane, J.A., D.G. Cole, E.S. Seeley, D.R. Diener, and J.L. Rosenbaum, 2001, Localization of intraflagellar transport protein IFT52 identifies basal body transitional fibers as the docking site for IFT particles, *Curr Biol.* **11**:1586-90.

Gibbons, B.H., D.J. Asai, W.J. Tang, T.S. Hays, and I.R. Gibbons, 1994, Phylogeny and expression of axonemal and cytoplasmic dynein genes in sea urchins, *Mol Biol Cell.* **5**:57-70.

Grissom, P.M., E.A. Vaisberg, and J.R. McIntosh, 2002, Identification of a novel light intermediate chain (D2LIC) for mammalian cytoplasmic dynein 2, *Mol Biol Cell.* **13**:817-29.

Hagstrom, S.A., M. Adamian, M. Scimeca, B.S. Pawlyk, G. Yue, and T. Li, 2001, A role for the Tubby-like protein 1 in rhodopsin transport, *Invest Ophthalmol Vis Sci.* **42**:1955-62.

Hagstrom, S.A., M. Duyao, M.A. North, and T. Li, 1999, Retinal degeneration in tulp1-/- mice: vesicular accumulation in the interphotoreceptor matrix, *Invest Ophthalmol Vis Sci.* **40**:2795-802.

Haycraft, C.J., P. Swoboda, P.D. Taulman, J.H. Thomas, and B.K. Yoder, 2001, The C. elegans homolog of the murine cystic kidney disease gene Tg737 functions in a ciliogenic pathway and is disrupted in osm-5 mutant worms, *Development.* **128**:1493-505.

Horst, C.J., L.V. Johnson, and J.C. Besharse, 1990, Transmembrane assemblage of the photoreceptor connecting cilium and motile cilium transition zone contain a common immunologic epitope, *Cell Motil Cytoskeleton.* **17**:329-44.

Iomini, C., V. Babaev-Khaimov, M. Sassaroli, and G. Piperno, 2001, Protein particles in Chlamydomonas flagella undergo a transport cycle consisting of four phases, *J Cell Biol.* **153**:13-24.

Liu, X., I.P. Udovichenko, S.D. Brown, K.P. Steel, and D.S. Williams, 1999, Myosin VIIa participates in opsin transport through the photoreceptor cilium, *J Neurosci.* **19**:6267-74.

Marshall, W.F., and J.L. Rosenbaum, 2001, Intraflagellar transport balances continuous turnover of outer doublet microtubules: implications for flagellar length control, *J Cell Biol.* **155**:405-14.

Marszalek, J.R., and L.S. Goldstein, 2000, Understanding the functions of kinesin-II, *Biochim Biophys Acta.* **1496**:142-150.

Marszalek, J.R., X. Liu, E.A. Roberts, D. Chui, J.D. Marth, D.S. Williams, and L.S. Goldstein, 2000, Genetic evidence for selective transport of opsin and arrestin by kinesin-II in mammalian photoreceptors [In Process Citation], *Cell.* **102**:175-87.

Mikami, A., S.H. Tynan, T. Hama, K. Luby-Phelps, T. Saito, J.E. Crandall, J.C. Besharse, and R.B. Vallee, 2002, Molecular structure of cytoplasmic dynein 2 and its distribution in neuronal and ciliated cells, *J Cell Sci.* **115**:4801-4808.

Moyer, J.H., M.J. Lee-Tischler, H.Y. Kwon, J.J. Schrick, E.D. Avner, W.E. Sweeney, V.L. Godfrey, N.L. Cacheiro, J.E. Wilkinson, and R.P. Woychik, 1994, Candidate gene associated with a mutation causing recessive polycystic kidney disease in mice, *Science.* **264**:1329-33.

Murcia, N.S., W.G. Richards, B.K. Yoder, M.L. Mucenski, J.R. Dunlap, and R.P. Woychik, 2000, The Oak Ridge Polycystic Kidney (orpk) disease gene is required for left-right axis determination, *Development.* **127**:2347-55.

Muresan, V., A. Lyass, and B.J. Schnapp, 1999, The kinesin motor KIF3A is a component of the presynaptic ribbon in vertebrate photoreceptors, *J Neurosci.* **19**:1027-37.

Nir, I., and D.S. Papermaster, 1986, Immunocytochemical localization of opsin in the inner segment and ciliary plasma membrane of photoreceptors in retinas of rds mutant mice, *Invest Ophthalmol Vis Sci.* **27**:836-40.

Nir, I., G. Sagie, and D.S. Papermaster, 1987, Opsin accumulation in photoreceptor inner segment plasma membranes of dystrophic RCS rats, *Invest Ophthalmol Vis Sci.* **28**:62-9.

Pazour, G.J., S.A. Baker, J.A. Deane, D.G. Cole, B.L. Dickert, J.L. Rosenbaum, G.B. Witman, and J.C. Besharse, 2002, The intraflagellar transport protein, IFT88, is essential for vertebrate photoreceptor assembly and maintenance, *J Cell Biol.* **157**:103-14.

Pazour, G.J., B.L. Dickert, Y. Vucica, E.S. Seeley, J.L. Rosenbaum, G.B. Witman, and D.G. Cole, 2000, Chlamydomonas IFT88 and its mouse homologue, polycystic kidney disease gene tg737, are required for assembly of cilia and flagella, *J Cell Biol.* **151**:709-18.

Pazour, G.J., B.L. Dickert, and G.B. Witman, 1999, The DHC1b (DHC2) isoform of cytoplasmic dynein is required for flagellar assembly, *J Cell Biol.* **144**:473-81.

Pazour, G.J., C.G. Wilkerson, and G.B. Witman, 1998, A dynein light chain is essential for the retrograde particle movement of intraflagellar transport (IFT), *J Cell Biol.* **141**:979-92.

Philp, N.J., W. Chang, and K. Long, 1987, Light-stimulated protein movement in rod photoreceptor cells of the rat retina, *FEBS Lett.* **225**:127-32.

Piperno, G., and K. Mead, 1997, Transport of a novel complex in the cytoplasmic matrix of Chlamydomonas flagella, *Proc Natl Acad Sci U S A.* **94**:4457-62.

Piperno, G., K. Mead, and S. Henderson, 1996, Inner dynein arms but not outer dynein arms require the activity of kinesin homologue protein KHP1(FLA10) to reach the distal part of flagella in Chlamydomonas, *J Cell Biol.* **133**:371-9.

Porter, M.E., R. Bower, J.A. Knott, P. Byrd, and W. Dentler, 1999, Cytoplasmic dynein heavy chain 1b is required for flagellar assembly in Chlamydomonas, *Mol Biol Cell.* **10**:693-712.

Qin, H., J.L. Rosenbaum, and M.M. Barr, 2001, An autosomal recessive polycystic kidney disease gene homolog is involved in intraflagellar transport in C. elegans ciliated sensory neurons, *Curr Biol.* **11**:457-61.

Rohlich, P., 1975, The sensory cilium of retinal rods is analogous to the transitional zone of motile cilia, *Cell Tissue Res.* **161**:421-30.

Roof, D.J., and C.A. Heth, 1988, Expression of transducin in retinal rod photoreceptor outer segments, *Science.* **241**:845-7.

Rosenbaum, J.L., D.G. Cole, and D.R. Diener, 1999, Intraflagellar transport: the eyes have it, *J Cell Biol.* **144**:385-8.

Rosenbaum, J.L., and G.B. Witman, 2002, Intraflagellar transport, *Nat Rev Mol Cell Biol.* **3**:813-25.

Signor, D., K.P. Wedaman, J.T. Orozco, N.D. Dwyer, C.I. Bargmann, L.S. Rose, and J.M. Scholey, 1999a, Role of a class DHC1b dynein in retrograde transport of IFT motors and IFT raft particles along cilia, but not dendrites, in chemosensory neurons of living Caenorhabditis elegans, *J Cell Biol.* **147**:519-30.

Signor, D., K.P. Wedaman, L.S. Rose, and J.M. Scholey, 1999b, Two heteromeric kinesin complexes in chemosensory neurons and sensory cilia of Caenorhabditis elegans, *Mol Biol Cell.* **10**:345-60.

Sokolov, M., A.L. Lyubarsky, K.J. Strissel, A.B. Savchenko, V.I. Govardovskii, E.N. Pugh, and V.Y. Arshavsky, 2002, Massive Light-Driven Translocation of Transducin between the Two Major Compartments of Rod Cells. A Novel Mechanism of Light Adaptation, *Neuron.* **34**:95-106.

Tai, A.W., J.Z. Chuang, C. Bode, U. Wolfrum, and C.H. Sung, 1999, Rhodopsin's carboxy-terminal cytoplasmic tail acts as a membrane receptor for cytoplasmic dynein by binding to the dynein light chain Tctex-1, *Cell.* **97**:877-87.

Traboulsi, E.I., 1998, Ocular malformations and developmental genes, *J Aapos.* **2**:317-23.

Walther, Z., M. Vashishtha, and J.L. Hall, 1994, The Chlamydomonas FLA10 gene encodes a novel kinesin-homologous protein, *J Cell Biol.* **126**:175-88.

Whelan, J.P., and J.F. McGinnis, 1988, Light-dependent subcellular movement of photoreceptor proteins, *J Neurosci Res.* **20**:263-70.

Whitehead, J.L., S.Y. Wang, L. Bost-Usinger, E. Hoang, K.A. Frazer, and B. Burnside, 1999, Photoreceptor localization of the KIF3A and KIF3B subunits of the heterotrimeric microtubule motor kinesin II in vertebrate retina, *Exp Eye Res.* **69**:491-503.

Young, R.W., 1967, The renewal of photoreceptor cell outer segments, *J Cell Biol.* **33**:61-72.

Young, R.W., and D. Bok, 1969, Participation of the retinal pigment epithelium in the rod outer segment renewal process, *J Cell Biol.* **42**:392-403.

INHERITED RETINAL DYSTROPHY IN *MER* KNOCKOUT MICE

Jacque L. Duncan[1], Haidong Yang[1], Douglas Vollrath[2], Douglas Yasumura[1], Michael T. Matthes[1], Nikolaus Trautmann[1], Aimee V. Chappelow[1], Wei Feng[2], H. Shelton Earp[3], Glenn K. Matsushima[3], and Matthew M. LaVail[1]

1. INTRODUCTION

Abnormalities of the retinal pigment epithelial (RPE) cells are seen in several inherited degenerations such as Best disease (Petrukhin et al., 1998; Marmorstein et al., 2000), Stargardt disease (Weng et al., 1999), Sorsby's fundus dystrophy (Steinmetz et al., 1992; Jacobson et al., 1995), and childhood-onset severe retinal dystrophy (Gu et al., 1997) and Leber congenital amaurosis due to RPE65 mutations (Marlhens et al., 1997; Morimura et al., 1998; Lotery et al., 2000; Thompson et al., 2000). RCS rats have been studied extensively as a model of photoreceptor (PR)-RPE cell interactions (Mullen and LaVail, 1976; LaVail, 2001) and retinal degeneration (Strauss et al., 1998; LaVail, 2001). Retinal degeneration in RCS results from failure of the RPE to phagocytize shed rod outer segments (OS) (Herron et al., 1969; Bok and Hall, 1971). The unphagocytized OS membranes form membranous whorls at the RPE surface, and the rod OS grow abnormally long (Dowling and Sidman, 1962; LaVail and Battelle, 1975). Eventually, the OS layer degenerates into a debris zone with subsequent PR cell death (Dowling and Sidman, 1962).

A null mutation in the receptor tyrosine kinase *Mertk* gene was recently found in RCS rats (D'Cruz et al., 2000). The murine gene orthologous to rat and human *Mertk* is designated *Mer,* and is a member of the Axl/Mer/Tyro3 receptor tyrosine kinase subfamily. The growth arrest-specific protein 6 (Gas6) is a ligand for all three members of this subfamily (Varnum et al., 1995; Nagata et al., 1996; Chen et al., 1997). It has been suggested that Gas6 may mediate RPE phagocytosis of OS by binding older OS

[1] University of California, San Francisco, CA 94143, USA; [2]Stanford University, Stanford, CA 94305, USA; [3]University of North Carolina, Chapel Hill, NC 27599, USA. Corresponding author: J. L. Duncan, E-mail: jduncan@itsa.ucsf.edu

membranes and Mer (D'Cruz et al., 2000). Mer localizes to the sites of OS uptake in the RPE and appears to trigger the ingestion step of phagocytosis in cell culture (Feng et al., 2002).

Wild-type (WT) *Mertk* delivery to the RPE of RCS rats resulted in reversal of the RPE phagocytosis defect and PR rescue (Vollrath et al., 2001), conclusively demonstrating that mutation of *Mertk* is the cause of retinal degeneration in RCS rats. Human patients with retinitis pigmentosa have been identified with mutations in the *MERTK* gene (Gal et al., 2000). The role *Mer* plays in RPE cell phagocytosis and retinal degeneration may be clarified by studying other species with *Mer* mutations. Mice carrying a targeted disruption of *Mer* (mer^{kd}) exhibit a generalized defect in phagocytosis of apoptotic thymocytes by macrophages (Scott et al., 2001). Because the gene mutated in these mice is orthologous to the mutant gene responsible for retinal dystrophy in RCS rats, we examined mer^{kd} mice for all of the key retinal phenotypic characteristics of RCS rats.

2. ASSESSMENT OF MER PROTEIN

Mertk protein cannot be detected in RCS rat tissues (Feng et al., 2002), indicating that the retinal dystrophy gene, *rdy*, is a null allele. Protein samples from the retinas and kidneys of WT and mer^{kd} mice were tested for Mer protein using polyclonal antisera directed against either the cytoplasmic tail or the ectodomain of the receptor. Both antibodies detected Mer proteins of the expected sizes in tissues from WT animals. In contrast, Mer protein was not detected in tissues from mer^{kd} mice and there was no

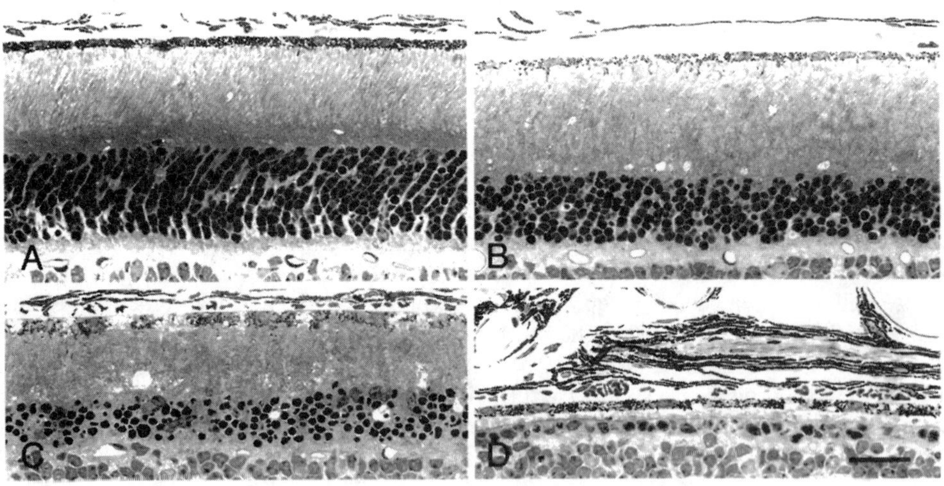

Figure 1. Light micrographs from the posterior retinas of mer^{kd} mice at different ages are shown, including P21 (**A**), P30 (**B**), P39 (**C**) and P45 (**D**). Epon-Araldite, 1 μm; bar, 25 μm. Adapted with permission from Duncan JL, LaVail MM, Yasumura D, Matthes MT, Yang H, Trautmann N, Chappelow AV, Feng W, Earp HS, Matsushima GK, Vollrath D. An RCS-like retinal dystrophy phenotype in *Mer* knockout mice. *Invest Ophthalmol Vis Sci.* **44**:826.

evidence of a truncated protein. These data indicate that mer^{kd} is a null allele, similar to *rdy* in RCS rats.

3. HISTOLOGICAL CHARACTERISTICS OF PR DEGENERATION

The retinal phenotype of mer^{kd} mice displayed a striking similarity to the retinal dystrophy characteristic of RCS rats. As is the case for *rdy* in RCS rats (LaVail, 1981a), the mer^{kd} retinal phenotype is completely recessive because mice heterozygous for the mutation have normal retinal structure and function.

As in RCS rats (LaVail and Battelle, 1975), PR degeneration in mer^{kd} mice led to progressive thinning of the ONL (Fig. 1). The ONL thickness in mer^{kd} was significantly less than in age-matched WT mice at P25 ($P<0.01$) and thereafter. Most mer^{kd} mice had less than 1 complete row of nuclei in the ONL by P45. Some variability in PR cell loss was observed after P45; a few mer^{kd} mice retained a single row of ONL cell nuclei in some parts of the retina up to about P60, while others had less than one row throughout the retina. After P90 at least a few surviving nuclei were found in each section, even in mice older than P250. Although mer^{kd} mice are pigmented animals, the rate of retinal degeneration in the mer^{kd} mouse strain was similar to, but slightly more rapid than the rate of degeneration in pink-eyed RCS rats (LaVail and Battelle, 1975).

With progressive PR degeneration in mer^{kd} mice, the percentage of surviving PRs that was cones increased as rods were lost preferentially. About 50% of the surviving PRs were cones by 2 months, increasing to a maximum of 75% at 6 months of age. RCS rats show a similar pattern of cone preservation relative to rods during retinal degeneration (LaVail, 1981b).

As PRs degenerated, pyknotic nuclei were observed in greater numbers in the ONL of mer^{kd} mice than in WT mice as early as P20. The percentage of nuclei in the ONL that was pyknotic in mer^{kd} mice reached as high as 71% at P40. A similarly high percentage of pyknotic cells in the ONL has been reported in RCS rats (LaVail and Battelle, 1975), but this phenomenon is not typically seen in other animal models of retinal degeneration. Delayed clearance of pyknotic PR cells from the ONL may be a direct consequence of a generalized defect in phagocytosis of postnatal apoptotic cells in animals with mutations in M*er*. Abnormal clearance of dying thymocytes and lymphocytes by macrophages (Scott et al., 2001) is characteristic of mer^{kd} mice, and defective clearance of apoptotic cells results in increased reactivity to self antigens in these mice (Scott et al., 2001). Humans with retinal degenerations due to *MERTK* mutations (Gal et al., 2000) may have as yet unrecognized systemic defects in the clearance of apoptotic cells, which could predispose toward autoimmune disease.

RCS rats show a failure to phagocytize rod OS (Herron et al., 1969; Bok and Hall, 1971; LaVail et al., 1972), which is accompanied by membranous whorls adjacent to the RPE surface, longer than normal OS and increased rhodopsin content in the eye. mer^{kd} mice also demonstrated each of these features. Mice were studied at the peak of circadian rod disk shedding to look for direct evidence of phagocytosis. While numerous phagosomes were observed in the apical RPE cell processes and RPE cell bodies of WT mouse retinas, few, if any, phagosomes were identified in mer^{kd} mice. We identified a gross reduction in the level of rod OS disk shedding and phagocytosis by the RPE cells in mer^{kd} mice, virtually identical to that seen in RCS rats, suggesting that the tyrosine kinase Mer plays an important role in rod OS ingestion by the RPE *in vivo.*

4. CORRELATIONS BETWEEN RETINAL STRUCTURE AND FUNCTION

A rapid and progressive loss of retinal function with age was apparent in *mer*^{kd} mice (Fig. 2A-B). Representative examples of scotopic and photopic ERG recordings from a WT mouse at P30 and from *mer*^{kd} mice of different ages are shown in figures 2A and 2B, respectively. At the youngest age examined (P20), *mer*^{kd} scotopic a- and b-wave amplitudes were significantly lower than normal ($P \leq 0.001$ and $P \leq 0.0005$, respectively). In contrast, photopic b-wave amplitudes were significantly lower than normal ($P \leq 0.04$) only after P30. Compared to WT, scotopic and photopic implicit times after P33 were delayed by approximately 75% ($P < 0.0001$) and 30% ($P < 0.02$), respectively. Interpretation of the scotopic a-wave implicit time was complicated at advanced stages of degeneration (P36 and beyond) due to interference from the cornea-negative STR (Fig. 2C). Scotopic a- and b-wave and photopic b-wave amplitudes declined and implicit times increased with age. No clear ERG a- or b-waves were detectable beyond P40 in most mice, although small amplitude scotopic and photopic b-waves were seen in a few mice at ages P58-65 (Fig. 2D).

As in most retinal degenerations (Dowling and Sidman, 1962; Kedzierski et al., 1997; Nishikawa et al., 1997; Toda et al., 1999; Machida et al., 2000; Sugawara et al., 2000), scotopic ERG amplitudes decreased with age and PR loss in *mer*^{kd} mice. The ONL thickness decreased with scotopic a-wave ($r^2 = 0.83$, $P < 0.0001$) and b-wave ($r^2 = 0.82$, $P < 0.0001$) and photopic b-wave ($r^2 = 0.75$, $P < 0.0001$) amplitudes (Fig. 2D). However, unlike other retinal degenerations (Machida et al., 2000), ERG amplitudes fell significantly earlier than ONL thickness in *mer*^{kd} mice. We considered that the persistence of pyknotic nuclei may have resulted in artificially thick ONL measurements. In an effort to correct for this possibility, we subtracted from the ONL thickness (Fig. 2D) the percentage of pyknotic nuclei at each age. When the estimated ONL thickness including only non-pyknotic cells was compared to scotopic a- and b-wave and photopic b-wave amplitudes, a much stronger positive correlation was evident (Fig. 2D). The corrected ONL thickness, excluding the percentage of pyknotic cells, decreased with scotopic a-wave ($r^2 = 0.96$, $P < 0.005$) and b-wave ($r^2 = 0.93$, $P < 0.01$) and photopic b-wave ($r^2 = 0.82$, $P < 0.05$) amplitudes (Fig. 2D). These correlation coefficients are comparable to those reported by others for light-damaged rats (Sugawara et al., 2000) and transgenic rats with rhodopsin mutations (Machida et al., 2000). Thus, in retinal degenerations where pyknotic nuclei persist, a correction in the ONL thickness should be made to accurately estimate the number of viable PR cells for structural-functional correlations.

Others have found that in RCS and light-damaged rats the scotopic b-wave threshold is more closely related to the number of surviving PR nuclei than is the maximum b-wave amplitude (Bush et al., 1995; Sugawara et al., 2000). In *mer*^{kd} mice, scotopic b-wave thresholds were significantly higher than normal at the youngest ages examined ($P \leq 0.001$), and increased with age in *mer*^{kd} mice until no b-wave was elicited at the highest stimulus intensities after P65 (Fig 2D). In contrast, photopic thresholds were normal in *mer*^{kd} mice until P36, then increased rapidly with age.

Photopic amplitudes and thresholds were better preserved than scotopic responses in *mer*^{kd} mice. Consistent with this, histological analysis showed a preferential preservation of cones.

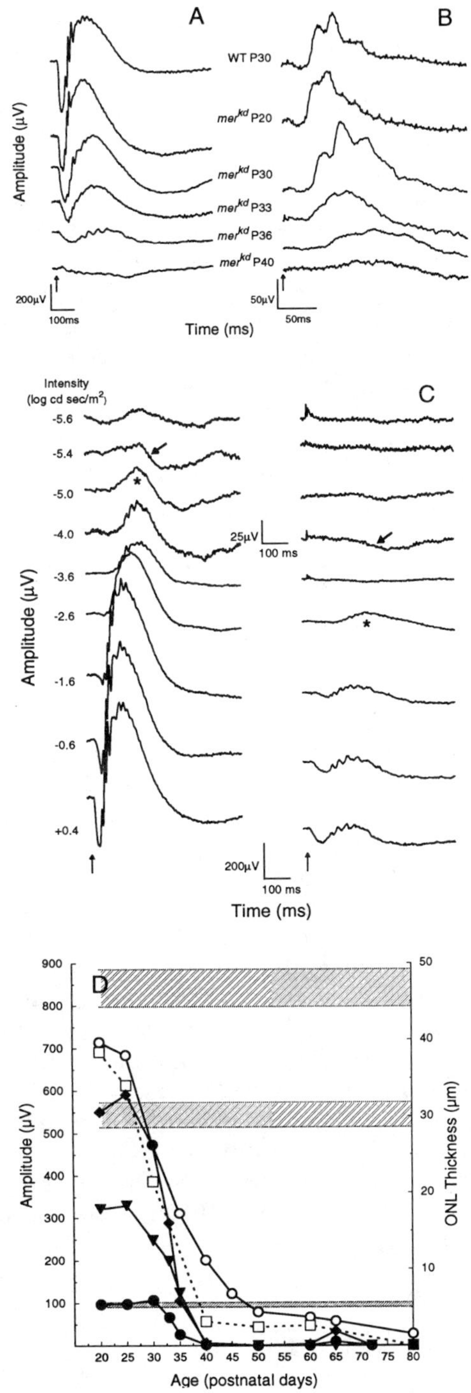

Figure 2. Time course of ERG amplitude changes in mer^{kd} mice. (**A**) Representative scotopic and (**B**) photopic ERG recordings from a WT mouse at P30 (top waveform) and from mer^{kd} mice at different ages are shown. Recordings were made in response to a standard flash of 0.4 log cd sec/m^2. Scotopic a- and b-wave amplitudes are lower than normal at the earliest age tested (P20) and rapidly decline with age. Photopic amplitudes are normal until P30 before rapidly declining with age. (**C**) Representative scotopic ERG waveforms from a WT (left) and a mer^{kd} (right) mouse at P36. The STR (large arrows) and b-wave (asterisks) thresholds are elevated 1.5 and 2.5 log units, respectively, and b- and a-wave amplitudes are reduced in the mer^{kd} mouse. Fine arrows indicate stimulus onset. (**D**) Hatched gray zones illustrate WT mean ± SEM amplitudes for scotopic b-waves (top zone), scotopic a-waves (middle zone) and photopic b-waves (bottom zone). Scotopic b-wave and photopic amplitudes are measured in response to a standard flash of 0.4 log cd sec/m^2; scotopic a-waves are in response to a bright flash of 2.4 log cd sec/m^2. Mean amplitudes for scotopic b-waves (diamonds), scotopic a-waves (triangles) and photopic b-waves (circles) for mer^{kd} mice are shown. While scotopic amplitudes are lower than WT at the youngest ages tested, photopic responses remain normal until P30 before rapidly declining. The ONL thickness declines (open circles) in mer^{kd} mice at a rate slower than the rates of ERG amplitude loss. The modified ONL thickness (open boxes, dashed line) reflects only non-pyknotic nuclei, and declines at a rate comparable to the rates of ERG amplitude loss. Adapted with permission from Duncan JL, LaVail MM, Yasumura D, *et al.* An RCS-like retinal dystrophy phenotype in *Mer* knockout mice. *Invest Ophthalmol Vis Sci.* **44**:826.

4.1. Scotopic Threshold Response

The STR, a cornea-negative waveform elicited at flash intensities below the b-wave threshold, has been observed in a number of animals (Sieving et al., 1986) including RCS rats (Bush et al., 1995) and WT mice (Saszik et al., 2002). We observed the STR at low flash intensities in WT mice at all ages tested (data not shown). In young mer^{kd} mice, cornea-negative waveforms were elicited by low flash intensities similar to those observed in WT mice (Fig. 2C, arrows). This response was markedly attenuated after intravitreal injection with N-methyl-DL-aspartic acid in both WT and mer^{kd} mice (data not shown), consistent with the STR response observed in RCS rats (Bush et al., 1995). At the youngest ages examined, the threshold to elicit the STR was not significantly different between mer^{kd} and WT mice, but at P33 and older, the STR threshold was significantly higher for mer^{kd} mice than for age-matched WT mice (P<0.005). The STR threshold increased with age in mer^{kd} mice (r^2 =0.80, P<0.0001).

Usually after P40, and always after P65, neither scotopic nor photopic b-waves could be recorded in response to the brightest stimulus intensity (+2.4 log cd sec/m^2) in mer^{kd} mice. However, the STR could be elicited even at advanced stages of PR degeneration in some mer^{kd} mice up to P253. Thus, the STR was measurable at bright stimulus intensities when only scattered PR nuclei remained.

5. CONCLUSIONS

The mer^{kd} mouse strain displays a retinal degeneration phenotype almost identical to that of the RCS rat. This mouse model may be particularly useful for studying the role of *Mer* in PR-RPE cell interactions, phagocytosis and retinal degeneration, given the powerful tools available for forward and reverse genetic studies in mice.

6. ACKNOWLEDGMENTS

The authors thank Jose Velarde, Nancy Lawson and Dean Cruz for technical assistance, and Dr. Dean Bok for advice in carrying out the rhodopsin analysis. This work was supported by a Career Development Award from Research to Prevent Blindness (JLD), National Institutes of Health Grants EY01919, EY02162 (MML) and EY00415 (JLD); the Ruth and Milton Steinbach Fund and the Karl Kirchgessner Foundation (DV); the Foundation Fighting Blindness, Inc.; the Macular Vision Research Foundation (MML), a postdoctoral fellowship from the Fight for Sight Division of Prevent Blindness America (WF); and That Man May See, Inc. MML is a Research to Prevent Blindness Senior Scientist Investigator. We thank the National Eye Institute and Foundation Fighting Blindness for the travel award provided to JLD to attend this meeting.

7. REFERENCES

Bok, D., and Hall, M. O., 1971, The role of the pigment epithelium in the etiology of inherited retinal dystrophy in the rat, *J Cell Biol.* **49**:664.

Bush, R. A., Hawks, K. W., and Sieving, P. A., 1995, Preservation of inner retinal responses in the aged Royal College of Surgeons rat. Evidence against glutamate excitotoxicity in photoreceptor degeneration, *Invest Ophthalmol Vis Sci.* **36**:2054.

Chen, J., Carey, K., and Godowski, P. J., 1997, Identification of Gas6 as a ligand for Mer, a neural cell adhesion molecule related receptor tyrosine kinase implicated in cellular transformation, *Oncogene.* **14**:2033.

D'Cruz, P. M., Yasumura, D., Weir, J., Matthes, M. T., Abderrahim, H., LaVail, M. M., and Vollrath, D., 2000, Mutation of the receptor tyrosine kinase gene *Mertk* in the retinal dystrophic RCS rat, *Hum Mol Genet.* **9**:645.

Dowling, J. E., and Sidman, R. L., 1962, Inherited retinal dystrophy in the rat, *J Cell Biol.* **14**:73.

Feng, W., Yasumura, D., Matthes, M. T., LaVail, M. M., and Vollrath, D., 2002, Mertk triggers uptake of photoreceptor outer segments during phagocytosis by cultured retinal pigment epithelial cells, *J Biol Chem.* **277**:17016.

Gal, A., Li, Y., Thompson, D. A., Weir, J., Orth, U., Jacobson, S. G., Apfelstedt-Sylla, E., and Vollrath, D., 2000, Mutations in *MERTK*, the human orthologue of the RCS rat retinal dystrophy gene, cause retinitis pigmentosa, *Nat Genet.* **26**:270.

Gu, S. M., Thompson, D. A., Srikumari, C. R., Lorenz, B., Finckh, U., Nicoletti, A., Murthy, K. R., Rathmann, M., Kumaramanickavel, G., Denton, M. J., and Gal, A., 1997, Mutations in *RPE65* cause autosomal recessive childhood-onset severe retinal dystrophy, *Nat Genet.* **17**:194.

Herron, W. L., Riegel, B. W., Myers, O. E., and Rubin, M. L., 1969, Retinal dystrophy in the rat--a pigment epithelial disease, *Invest Ophthalmol.* **8**:595.

Jacobson, S. G., Cideciyan, A. V., Regunath, G., Rodriguez, F. J., Vandenburgh, K., Sheffield, V. C., and Stone, E. M., 1995, Night blindness in Sorsby's fundus dystrophy reversed by vitamin A, *Nat Genet.* **11**:27.

Kedzierski, W., Lloyd, M., Birch, D. G., Bok, D., and Travis, G. H., 1997, Generation and analysis of transgenic mice expressing P216L-substituted rds/peripherin in rod photoreceptors, *Invest Ophthalmol Vis Sci.* **38**:498.

LaVail, M. M., Sidman, R. L., and O'Neil, D., 1972, Photoreceptor-pigment epithelial cell relationships in rats with inherited retinal degeneration. Radioautographic and electron microscope evidence for a dual source of extra lamellar material, *J Cell Biol.* **53**:185.

LaVail, M. M., and Battelle, B. A., 1975, Influence of eye pigmentation and light deprivation on inherited retinal dystrophy in the rat, *Exp Eye Res.* **21**:167.

LaVail, M. M., 1981a, Photoreceptor characteristics in congenic strains of RCS rats, *Invest Ophthalmol Vis Sci.* **20**:671.

LaVail, M. M., 1981b, Analysis of neurological mutants with inherited retinal degeneration. Friedenwald lecture, *Invest Ophthalmol Vis Sci.* **21**:638.

LaVail, M. M., 2001, Legacy of the RCS rat: impact of a seminal study on retinal cell biology and retinal degenerative diseases, in: *Progress in Brain Research*, H. Kolb, H. Ripps, and S. Wu, ed., Elsevier Science, Paris. pp. 617.

Lotery, A. J., Namperumalsamy, P., Jacobson, S. G., Weleber, R. G., Fishman, G. A., Musarella, M. A., Hoyt, C. S., Heon, E., Levin, A., Jan, J., Lam, B., Carr, R. E., Franklin, A., Radha, S., Andorf, J. L., Sheffield, V. C., and Stone, E. M., 2000, Mutation analysis of 3 genes in patients with Leber congenital amaurosis, *Arch Ophthalmol.* **118**:538.

Machida, S., Kondo, M., Jamison, J. A., Khan, N. W., Kononen, L. T., Sugawara, T., Bush, R. A., and Sieving, P. A., 2000, P23H rhodopsin transgenic rat: correlation of retinal function with histopathology, *Invest Ophthalmol Vis Sci.* **41**:3200.

Marlhens, F., Bareil, C., Griffoin, J. M., Zrenner, E., Amalric, P., Eliaou, C., Liu, S. Y., Harris, E., Redmond, T. M., Arnaud, B., Claustres, M., and Hamel, C. P., 1997, Mutations in *RPE65* cause Leber's congenital amaurosis, *Nat Genet.* **17**:139.

Marmorstein, A. D., Marmorstein, L. Y., Rayborn, M., Wang, X., Hollyfield, J. G., and Petrukhin, K., 2000, Bestrophin, the product of the Best vitelliform macular dystrophy gene (VMD2), localizes to the basolateral plasma membrane of the retinal pigment epithelium, *Proc Natl Acad Sci U S A.* **97**:12758.

Morimura, H., Fishman, G. A., Grover, S. A., Fulton, A. B., Berson, E. L., and Dryja, T. P., 1998, Mutations in the RPE65 gene in patients with autosomal recessive retinitis pigmentosa or Leber congenital amaurosis, *Proc Natl Acad Sci U S A.* **95**:3088.

Mullen, R. J., and LaVail, M. M., 1976, Inherited retinal dystrophy: primary defect in pigment epithelium determined with experimental rat chimeras, *Science.* **192**:799.

Nagata, K., Ohashi, K., Nakano, T., Arita, H., Zong, C., Hanafusa, H., and Mizuno, K., 1996, Identification of the product of growth arrest-specific gene 6 as a common ligand for Axl, Sky, and Mer receptor tyrosine kinases, *J Biol Chem.* **271**:30022.

Nishikawa, S., Cao, W., Yasumura, D., Matthes, M. T., Lau-Villacorta, C., Steinberg, R. H., and LaVail, M. M., 1997, Comparing the ERG to retinal morphology in transgenic rats with inherited degenerations caused by mutant opsin genes [ARVO Abstract], *Invest Ophthalmol Vis Sci.* **38**:S33. Abstract nr 1415.

Petrukhin, K., Koisti, M. J., Bakall, B., Li, W., Xie, G., Marknell, T., Sandgren, O., Forsman, K., Holmgren, G., Andreasson, S., Vujic, M., Bergen, A. A., McGarty-Dugan, V., Figueroa, D., Austin, C. P., Metzker, M. L., Caskey, C. T., and Wadelius, C., 1998, Identification of the gene responsible for Best macular dystrophy, *Nat Genet.* **19**:241.

Saszik, S. M., Robson, J. G., and Frishman, L. J., 2002, The scotopic threshold response of the dark-adapted electroretinogram of the mouse, *J Physiol.* **543**:899.

Scott, R. S., McMahon, E. J., Pop, S. M., Reap, E. A., Caricchio, R., Cohen, P. L., Earp, H. S., and Matsushima, G. K., 2001, Phagocytosis and clearance of apoptotic cells is mediated by MER, *Nature.* **411**:207.

Sieving, P. A., Frishman, L. J., and Steinberg, R. H., 1986, Scotopic threshold response of proximal retina in cat, *J Neurophysiol.* **56**:1049.

Steinmetz, R. L., Polkinghorne, P. C., Fitzke, F. W., Kemp, C. M., and Bird, A. C., 1992, Abnormal dark adaptation and rhodopsin kinetics in Sorsby's fundus dystrophy, *Invest Ophthalmol Vis Sci.* **33**:1633.

Strauss, O., Stumpff, F., Mergler, S., Wienrich, M., and Wiederholt, M., 1998, The Royal College of Surgeons rat: an animal model for inherited retinal degeneration with a still unknown genetic defect, *Acta Anat.* **162**:101.

Sugawara, T., Sieving, P. A., and Bush, R. A., 2000, Quantitative relationship of the scotopic and photopic ERG to photoreceptor cell loss in light damaged rats, *Exp Eye Res.* **70**:693.

Thompson, D. A., Gyurus, P., Fleischer, L. L., Bingham, E. L., McHenry, C. L., Apfelstedt-Sylla, E., Zrenner, E., Lorenz, B., Richards, J. E., Jacobson, S. G., Sieving, P. A., and Gal, A., 2000, Genetics and phenotypes of *RPE65* mutations in inherited retinal degeneration, *Invest Ophthalmol Vis Sci.* **41**:4293.

Toda, K., Bush, R. A., Humphries, P., and Sieving, P. A., 1999, The electroretinogram of the rhodopsin knockout mouse, *Vis Neurosci.* **16**:391.

Varnum, B. C., Young, C., Elliott, G., Garcia, A., Bartley, T. D., Fridell, Y. W., Hunt, R. W., Trail, G., Clogston, C., Toso, R. J., and et al., 1995, Axl receptor tyrosine kinase stimulated by the vitamin K-dependent protein encoded by growth-arrest-specific gene 6, *Nature.* **373**:623.

Vollrath, D., Feng, W., Duncan, J. L., Yasumura, D., D'Cruz, P. M., Chappelow, A., Matthes, M. T., Kay, M. A., and LaVail, M. M., 2001, Correction of the retinal dystrophy phenotype of the RCS rat by viral gene transfer of *Mertk*, *Proc Natl Acad Sci U S A.* **98**:12584.

Weng, J., Mata, N. L., Azarian, S. M., Tzekov, R. T., Birch, D. G., and Travis, G. H., 1999, Insights into the function of Rim protein in photoreceptors and etiology of Stargardt's disease from the phenotype in *abcr* knockout mice, *Cell.* **98**:13.

MOUSE MODELS OF HUMAN RETINAL DISEASE CAUSED BY EXPRESSION OF MUTANT RHODOPSIN

A valuable tool for the assessment of novel gene therapies

May Nour[1] and Muna I. Naash [1,2]

1. RETINAL DISEASES AND GENE THERAPY

The genetic basis for a wide range of inherited retinal diseases has been identified in recent years. To date, over 118 retinal disease loci have been mapped, for which 55 genes have been isolated (Clarke et al. 2000). Of these diseases, inherited photoreceptor degenerations represent a significant proportion of diseases that cause serious visual impairments in individuals of all ages and ethnic backgrounds (Berson 1996). While most forms of photoreceptor degeneration are characterized by rod and/or cone dysfunction, individual subtypes are further classified based on the type of inheritance, pattern of visual loss, appearance of the retina, and the gene defect causing these symptoms (Dryja et al. 1990b; Berson 1996; Wang et al. 2001). One of the most prominent genes that has been shown to cause progressive photoreceptor degeneration when mutated, is the rhodopsin gene. Patients carrying a number of different mutations in rhodopsin have displayed clinical symptoms of retinitis pigmentosa, varying in the severity of visual impairment (Dryja et al. 1990a; Dryja et al. 1990b; Dryja et al. 1991).

Gene therapy may be the most promising treatment for a broad spectrum of inherited photoreceptor degenerations. Diseases amenable to treatment include retinitis pigmentosa and macular dystrophy. In terms of evaluating therapies for these diseases, the retina, and in particular the photoreceptors, present a unique challenge to achieving successful gene therapy. The careful assessment of gene therapy for patient application relies on the generation of animal models that closely mimic human disease presentation, the selection of an appropriate therapy, the cell-specific and efficient targeting, as well as the delivery of therapeutic agents with minimal damage to the eye. Fortunately, through genetic manipulation, and identification of spontaneous mutations, scientists have developed animal models for a variety of retinal diseases (Olsson et al. 1992; Goto et al. 1995; Goto

[1]Oklahoma Center for Neuroscience and [2]Department of Cell Biology, University of Oklahoma Health Sciences Center, 940 Stanton L. Young Blvd., Oklahoma City, OK 73104, USA. Corresponding author: M. I. Naash, E-mail: muna-naash@ouhsc.edu

Retinal Degenerations: Mechanisms and Experimental Therapy
Edited by LaVail *et al.*, Kluwer Academic/Plenum Publishers, 2003

et al. 1996; Penn et al. 2000; Hafezi et al. 2000; Rakoczy et al. 2002; Chang et al. 2002). Using these animal models, researchers have successfully delivered a number of potential treatments which have included a group of genes carried by modified viral vectors, liposomes, genetically modified cells, or by direct DNA transfer to the retina (Bennett et al. 1994; Bennett et al. 1996; Miyoshi et al. 1997; Bennett et al. 1997; Takahashi et al. 1999; Penn et al. 2000; Neuner-Jehle et al. 2000; Sarra et al. 2001; Bainbridge et al. 2001; Liang et al. 2001; Bok et al. 2002; Cashman et al. 2002). In an attempt to select the most promising therapy, researchers have also investigated the use of antisense or ribozyme technologies to reduce the expression of dominant negative acting gene products (Drenser et al. 1998; Hauswirth and Lewin 2000; Hauswirth et al. 2000; Garrett et al. 2001; Sullivan et al. 2002). Amongst others, studies have successfully applied ribozyme therapy to slow down the rate of photoreceptor degeneration by reducing the expression of the P23H mutant rhodopsin protein in the rat retina (Lewin et al. 1998; LaVail et al. 2000). Data from these promising pre-clinical experiments will lay the foundation for future clinical application.

2. TRANSGENIC MODELS FOR RHODOPSIN MUTATIONS

A genetic disease can result from the expression of a mutant protein or the failure to express its normal form. In recent years, the genetic basis for a wide range of inherited retinal diseases has been identified. Genomic analysis of patients with ADRP has provided convincing evidence that various subtypes of ADRP result from mutations in the gene encoding for rod opsin (Dryja et al. 1990a; Dryja et al. 1991; Inglehearn et al. 1991; Humphries et al. 1997; Berson et al. 2001). So far, over 100 different mutations in the rod opsin gene have been identified and associated with different forms of retinal disease (Figure 1).

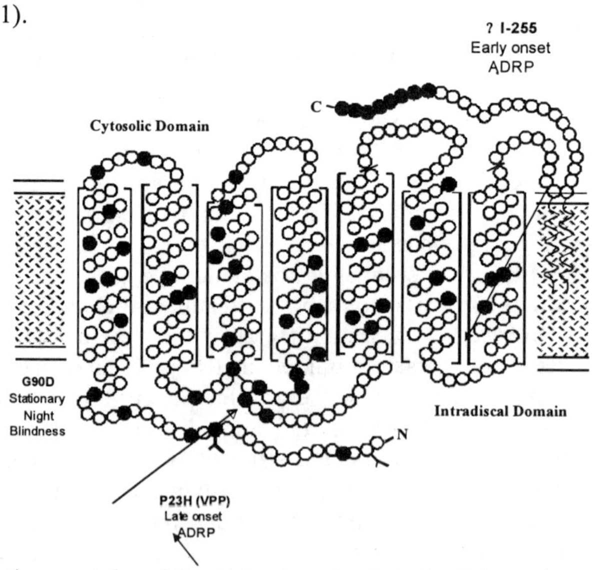

Figure 1. Schematic presentation of the rhodopsin molecule in the disk membrane. Amino acids are indicated by open circles. Closed circles are residues mutated in patients with retinal diseases. The lumen and cytoplasmic sides are indicated.

Parallel developments in molecular biology and transgenic mouse technology have enabled the generation of mouse models expressing these disease-causing mutations. The presence of such animal models, which closely mimic human visual dysfunction, allows for conducting extensive biochemical, histological, and physiological studies. These detailed studies will enable researchers to shed light on the pathophysiology of disease and directs them towards possibilities for therapeutic intervention.

We have generated several transgenic mouse models for a number of these retinal disease-causing mutations. The structural and functional defects observed in these models parallel the clinical presentation of patients carrying the same mutations. We describe three of these mutations here: the P23H mutation, which causes a late onset form of ADRP; the ΔI-255/256 deletion which causes an early onset form of ADRP; and the G90D mutation which causes congenital stationary night blindness (CSNB).

2.1. Transgenic Model for Late Onset ADRP

The P23H mutation in the opsin gene is one of the most prevalent mutations found in the United States, and accounts for over 30% of ADRP. To examine the link between the P23H genotype and its phenotypic expression, we have generated several lines of transgenic mice expressing a mouse opsin gene containing the P23H mutation (Naash et al. 1993). The structural and functional defects of P23H mice closely mimic those in patients carrying the same mutation (Olsson et al. 1992; Naash et al. 1993; Goto et al. 1995; Goto et al. 1996; Wu et al. 1998). Like ADRP patients with the P23H mutation, transgenic mice exhibit slowly progressive retinal degeneration that results in (a) an early-onset decline in rod-electroretinogram (ERG), (b) a greater loss of rods in the inferior retina, (c) an initial sparing of cones, which degenerate only at later stages of the disease, (d) a significantly slower rate of dark adaptation, and (e) sensitivity losses that can be ascribed solely to the lowered rhodopsin content in rods. Most importantly, in the rhodopsin hemizygous background, these mice express a normal level of total opsin, with equal amounts of mutant and endogenous protein. Functional similarities between the phenotype in the P23H mouse line and patients, qualifies this animal model as an excellent representative of human ADRP resulting from the P23H mutation in rhodopsin. One of our P23H mouse lines was generated by expressing three mutations: substitution of histidine for proline at position 23 (P23H); glycine for valine at position 20 (V20G); and leucine for proline at position 27 (P27L). These mice were expressed on the rhodopsin knockout background (Humphries et al. 1997) to determine whether or not the mutant protein is capable of supporting outer segment formation. Interestingly, in 1-month-old transgenic retina in the rhodopsin null background, the VPP protein localized to perinuclear regions of rod photoreceptors, without sign of outer segment formation. In contrast to these results in the rhodopsin null background, in the presence of wild-type rhodopsin, the VPP (P23H) protein was able to properly localize to the rod outer segment (Wu et al. 1998).

2.2. Transgenic Model for Early onset of ADRP

In Europe, a deletion of isoleucine at position 255 or 256 (ΔI-255/256) in the rhodopsin gene accounts for over 17% of familial ADRP cases (Inglehearn et al. 1991; Artlich et al. 1992). This highly conserved amino acid is located in the sixth transmembrane domain of the rhodopsin protein (Figure 1). Patients expressing this mutation suffer from (a) an early onset of night blindness that is usually detected during the first decade but can be present from birth, (b) cone dysfunction that is apparent by age 6 as evidenced by reduced visual acuity and color vision abnormalities, (c) a relatively mild pigment migration into the neural retina, and (d) the presence prominent choroidal vessels suggesting RPE thinning or depigmentation.

We have generated two transgenic lines expressing the ΔI-255/256 mutation with phenotypic changes similar to those observed in patients carrying the same mutation. More specifically, the transgenic animals exhibited fast photoreceptor degeneration and loss of ERG function affecting both rods and cones. At P5 to P10, the photoreceptor cells appeared to be developing normally. At P15, half of the photoreceptor cells were lost, and pyknotic photoreceptor nuclei, were frequently observed in the lower portion of the outer nuclear layer as an indication of ensuing cell death (Penn et al. 2000). By P20, only one row of photoreceptor nuclei remained in the outer nuclear layer, indicating that both rods and cones were affected by this mutation. Ultrastructural studies of the outer segment at the EM level showed a grossly abnormal orientation of the disc membranes in the transgenic retinas. We observed elongated disc membranes parallel to the long axis of the photoreceptor. Initially the retinal pigment epithelium (RPE) in transgenic eyes appeared to be normal, however as the degenerative course proceeded, the RPE also exhibited significant changes.

Since the loss of retinal capillaries is an inherited component in the late stages of ADRP, we have also examined the contribution of tissue hyperoxia to the vascular attenuation process found in ΔI-255/256 retinas, and have tested the ability of ambient hypoxia to reverse this attenuation. Capillary density was significantly different in the deep plexus of normoxic wildtype versus transgenic mice at P20 (Penn et al. 2000). There was an additional 65% reduction in capillary density within the deep plexus of normoxic transgenic mice between P20 and P27. Ambient hypoxia between days P20 and P27 reversed this trend, causing an increase in the deep capillary plexus density to nearly 100%. As highly metabolic photoreceptors die during progression of ADRP, the choroidal oxygen that they normally consume is free to diffuse into the inner retina. This presumably leads to higher than normal PO_2 in this region. In response to the increase in PO_2, the deep capillary plexus rapidly atrophies. Low ambient oxygen reduces the PO_2, which decreases the amount of oxygen diffusing into the inner retina from the choroid and reverses the atrophy of the deep capillary plexus. This treatment not only halted vessel atrophy, but also stimulates new vessel growth as shown by an increase in deep vessel plexus density from P20 to P27. The results of this study have important implications for strategies designed to treat cases of RP using retinal cell transplantation.

2.3. Transgenic Model for Stationary Night Blindness (SNB)

The G90D mutation in the opsin gene has been shown to co-segregate with an unusual trait of congenital night blindness (Rao et al. 1994). On average, patients carrying the G90D mutation show no signs of retinal degeneration up to age 33, normal rhodopsin density (Sieving et al. 1995), and an absolute threshold for visual perception elevated by 3 log units. Expression of this mutation in COS cells has been shown to allow the formation of an opsin molecule, able to constitutively activate transducin (Rao et al. 1994).

To define the basis of constitutive activation of G90D and its association with CSNB, we have generated eight transgenic lines expressing the G90D transgene at different levels. Three of these lines were fully characterized histologically, electrophysiologically, and photochemically. The ratio of transgene expression to endogenous was 1:1, 1.5:1, and 2.5:1. During early postnatal development, photoreceptors of transgenic retinas appeared normal. At P30, several lines retained 6 to 8 rows of nuclei. In these cases, outer and inner segments were apparent, and the RPE cell layer appeared normal. In one line, in which expression of the transgene was equivalent to endogenous (1:1), there were no signs of rod cell death until three months of age. However, the dark-adapted a-wave was significantly reduced. The common phenotype seen in these eight lines was a desensitization of the rod photoreceptor cells, while cone function appeared normal. This correlation between rod sensitivity and rhodopsin content suggests that the transgene product is constitutively active. We are continuing to examine this feature more closely in the rhodopsin knockout mouse, where endogenous rhodopsin is absent.

3. USING TRANSGENIC MICE TO ASSESS EFFECTIVENESS OF NOVEL GENE THERAPIES

While currently no successful therapies exist for the treatment of genetically-linked ADRP, gene therapy may serve as the best and most direct way for therapeutic intervention. The wide range of severity for disease symptoms displayed by ADRP patients with mutations in rhodopsin dictates a tailored approach to treatment. In this case, designed gene therapies that target a specific mutation could be administered prophylactically to prevent the onset of disease in cases where the phenotype appears later in age. For early onset diseases, these therapies maybe used to halt any further progression of existing retinal degeneration. Many important factors are to be taken into consideration while producing any type of gene therapy. These factors include the selection of an appropriate gene or number of genes for transfer, the cell-specific and efficient targeting of therapy and its delivery in physical proximity to the intended cells. However, one of the most vital factors in the assessment of such therapy for potential use in the treatment of human ADRP is the existence of a well-characterized animal model, representative of the clinical symptoms displayed by affected patients. This model allows for a more accurate *in vivo* evaluation of both the efficacy and safety of these novel therapies. We report here three mouse models carrying retinal disease-causing mutations in the rhodopsin gene. These models of rhodopsin mutation, namely P23H, ΔI-255/256 and G90D, display phenotypes that vary in severity and temporal onset of expression. Such features allow these mouse models to serve as an invaluable tool for testing the potential and window for therapeutic intervention with rhodopsin-directed gene therapy treatments.

4. ACKNOWLEDGMENTS

This research was funded by grants from the National Eye Institute (NEI) Grant EY10609 (MIN), Department of Ophthalmology NEI Core grant EY-01792, the Foundation Fighting Blindness (MIN, MRA). Dr. Naash is a recipient of the Research to Prevent Blindness James S. Adams Scholar. We thank the National Eye Institute and Foundation Fighting Blindness for the travel award provided to MN to attend this meeting.

5. REFERENCES

Artlich, A., Horn, M., Lorenz, B., Bhattacharga, S., and Gal, A., 1992, Recurrent 3-bp deletion at codon 255/256 of the rhodopsin gene in a German pedigree with autosomal dominant retinitis pigmentosa, *Am.J.Hum.Genet.* **50**:876-878.

Bainbridge, J. W., Stephens, C., Parsley, K., Demaison, C., Halfyard, A., Thrasher, A. J., and Ali, R. R., 2001, In vivo gene transfer to the mouse eye using an HIV-based lentiviral vector; efficient long-term transduction of corneal endothelium and retinal pigment epithelium, *Gene Ther.* **8**:1665-1668.

Bennett, J., Duan, D., Engelhardt, J. F., and Maguire, A. M., 1997, Real-time, noninvasive in vivo assessment of adeno-associated virus-mediated retinal transduction, *Invest Ophthalmol. Vis.Sci.* **38**:2857-2863.

Bennett, J., Tanabe, T., Sun, D., Zeng, Y., Kjeldbye, H., Gouras, P., and Maguire, A. M., 1996, Photoreceptor cell rescue in retinal degeneration (rd) mice by in vivo gene therapy, *Nat.Med.* **2**:649-654.

Bennett, J., Wilson, J., Sun, D., Forbes, B., and Maguire, A., 1994, Adenovirus vector-mediated in vivo gene transfer into adult murine retina, *Invest Ophthalmol. Vis.Sci.* **35**:2535-2542.

Berson, E. L., 1996, Retinitis pigmentosa: unfolding its mystery, *Proc.Natl.Acad.Sci. U.S.A* **93**:4526-4528.

Berson, E. L., Grimsby, J. L., Adams, S. M., McGee, T. L., Sweklo, E., Pierce, E. A., Sandberg, M. A., and Dryja, T. P., 2001, Clinical features and mutations in patients with dominant retinitis pigmentosa-1 (RP1), *Invest Ophthalmol. Vis.Sci.* **42**:2217-2224.

Bok, D., Yasumura, D., Matthes, M. T., Ruiz, A., Duncan, J. L., Chappelow, A. V., Zolutukhin, S., Hauswirth, W., and LaVail, M. M., 2002, Effects of Adeno-associated Virus-vectored Ciliary Neurotrophic Factor on Retinal Structure and Function in Mice with a P216L rds/peripherin Mutation, *Exp.Eye Res.* **74**:719-735.

Cashman, S. M., Sadowski, S. L., Morris, D. J., Frederick, J., and Kumar-Singh, R., 2002, Intercellular Trafficking of Adenovirus-Delivered HSV VP22 from the Retinal Pigment Epithelium to the Photoreceptors-Implications for Gene Therapy, *Mol.Ther.* **6**:813-823.

Chang, B., Hawes, N. L., Hurd, R. E., Davisson, M. T., Nusinowitz, S., and Heckenlively, J. R., 2002, Retinal degeneration mutants in the mouse, *Vision Res.* **42**:517-525.

Clarke, G., Heon, E., and McInnes, R. R., 2000, Recent advances in the molecular basis of inherited photoreceptor degeneration, *Clin.Genet.* **57**:313-329.

Drenser, K. A., Timmers, A. M., Hauswirth, W. W., and Lewin, A. S., 1998, Ribozyme-targeted destruction of RNA associated with autosomal-dominant retinitis pigmentosa, *Invest Ophthalmol. Vis.Sci.* **39**:681-689.

Dryja, T. P., Hahn, L. B., Cowley, G. S., McGee, T. L., and Berson, E. L., 1991, Mutation spectrum of the rhodopsin gene among patients with autosomal dominant retinitis pigmentosa, *Proc.Natl.Acad.Sci. U.S.A* **88**:9370-9374.

Dryja, T. P., McGee, T. L., Hahn, L. B., Cowley, G. S., Olsson, J. E., Reichel, E., Sandberg, M. A., and Berson, E. L., 1990a, Mutations within the rhodopsin gene in patients with autosomal dominant retinitis pigmentosa, *N.Engl.J.Med.* **323**:1302-1307.

Dryja, T. P., McGee, T. L., Reichel, E., Hahn, L. B., Cowley, G. S., Yandell, D. W., Sandberg, M. A., and Berson, E. L., 1990b, A point mutation of the rhodopsin gene in one form of retinitis pigmentosa, *Nature* **343**:364-366.

Garrett, K. L., Shen, W. Y., and Rakoczy, P. E., 2001, In vivo use of oligonucleotides to inhibit choroidal neovascularisation in the eye, *J.Gene Med.* **3**:373-383.

Goto, Y., Peachey, N. S., Ripps, H., and Naash, M. I., 1995, Functional abnormalities in transgenic mice expressing a mutant rhodopsin gene, *Invest Ophthalmol. Vis.Sci.* **36**:62-71.

Goto, Y., Peachey, N. S., Ziroli, N. E., Seiple, W. H., Gryczan, C., Pepperberg, D. R., and Naash, M. I., 1996, Rod phototransduction in transgenic mice expressing a mutant opsin gene, *J.Opt.Soc.Am.A* **13**:577-585.

Hafezi, F., Grimm, C., Simmen, B. C., Wenzel, A., and Reme, C. E., 2000, Molecular ophthalmology: an update on animal models for retinal degenerations and dystrophies, *Br.J.Ophthalmol.* **84**:922-927.

Hauswirth, W. W., LaVail, M. M., Flannery, J. G., and Lewin, A. S., 2000, Ribozyme gene therapy for autosomal dominant retinal disease, *Clin.Chem.Lab Med.* **38**:147-153.

Hauswirth, W. W. and Lewin, A. S., 2000, Ribozyme uses in retinal gene therapy, *Prog.Retin.Eye Res.* **19**:689-710.

Humphries, M. M., Rancourt, D., Farrar, G. J., Kenna, P., Hazel, M., Bush, R. A., Sieving, P. A., Sheils, D. M., McNally, N., Creighton, P., Erven, A., Boros, A., Gulya, K., Capecchi, M. R., and Humphries, P., 1997, Retinopathy induced in mice by targeted disruption of the rhodopsin gene, *Nat.Genet.* **15**:216-219.

Inglehearn, C. F., Bashir, R., Lester, D. H., Jay, M., Bird, A. C., and Bhattacharya, S. S., 1991, A 3-bp deletion in the rhodopsin gene in a family with autosomal dominant retinitis pigmentosa, *Am.J.Hum.Genet.* **48**:26-30.

LaVail, M. M., Yasumura, D., Matthes, M. T., Drenser, K. A., Flannery, J. G., Lewin, A. S., and Hauswirth, W. W., 2000, Ribozyme rescue of photoreceptor cells in P23H transgenic rats: long-term survival and late-stage therapy, *Proc.Natl.Acad.Sci.U.S.A* **97**:11488-11493.

Lewin, A. S., Drenser, K. A., Hauswirth, W. W., Nishikawa, S., Yasumura, D., Flannery, J. G., and LaVail, M. M., 1998, Ribozyme rescue of photoreceptor cells in a transgenic rat model of autosomal dominant retinitis pigmentosa, *Nat.Med.* **4**:967-971.

Liang, F. Q., Aleman, T. S., Dejneka, N. S., Dudus, L., Fisher, K. J., Maguire, A. M., Jacobson, S. G., and Bennett, J., 2001, Long-term protection of retinal structure but not function using RAAV.CNTF in animal models of retinitis pigmentosa, *Mol.Ther.* **4**:461-472.

Miyoshi, H., Takahashi, M., Gage, F. H., and Verma, I. M., 1997, Stable and efficient gene transfer into the retina using an HIV-based lentiviral vector, *Proc.Natl.Acad.Sci.U.S.A* **94**:10319-10323.

Naash, M. I., Hollyfield, J. G., al Ubaidi, M. R., and Baehr, W., 1993, Simulation of human autosomal dominant retinitis pigmentosa in transgenic mice expressing a mutated murine opsin gene, *Proc.Natl.Acad.Sci.U.S.A* **90**:5499-5503.

Neuner-Jehle, M., Berghe, L. V., Bonnel, S., Uteza, Y., Benmeziane, F., Rouillot, J. S., Marchant, D., Kobetz, A., Dufier, J. L., Menasche, M., and Abitbol, M., 2000, Ocular cell transfection with the human basic fibroblast growth factor gene delays photoreceptor cell degeneration in RCS rats, *Hum.Gene Ther.* **11**:1875-1890.

Olsson, J. E., Gordon, J. W., Pawlyk, B. S., Roof, D., Hayes, A., Molday, R. S., Mukai, S., Cowley, G. S., Berson, E. L., and Dryja, T. P., 1992, Transgenic mice with a rhodopsin mutation (Pro23His): a mouse model of autosomal dominant retinitis pigmentosa, *Neuron* **9**:815-830.

Penn, J. S., Li, S., and Naash, M. I., 2000, Ambient hypoxia reverses retinal vascular attenuation in a transgenic mouse model of autosomal dominant retinitis pigmentosa, *Invest Ophthalmol.Vis.Sci.* **41**:4007-4013.

Rakoczy, P. E., Zhang, D., Robertson, T., Barnett, N. L., Papadimitriou, J., Constable, I. J., and Lai, C. M., 2002, Progressive age-related changes similar to age-related macular degeneration in a transgenic mouse model, *Am.J.Pathol.* **161**:1515-1524.

Rao, V. R., Cohen, G. B., and Oprian, D. D., 1994, Rhodopsin mutation G90D and a molecular mechanism for congenital night blindness, *Nature* **367**:639-642.

Sarra, G. M., Stephens, C., De Alwis, M., Bainbridge, J. W., Smith, A. J., Thrasher, A. J., and Ali, R. R., 2001, Gene replacement therapy in the retinal degeneration slow (rds) mouse: the effect on retinal degeneration following partial transduction of the retina, *Hum.Mol.Genet.* **10**:2353-2361.

Sieving, P. A., Richards, J. E., Naarendorp, F., Bingham, E. L., Scott, K., and Alpern, M., 1995, Dark-light: model for nightblindness from the human rhodopsin Gly-90-->Asp mutation, *Proc.Natl.Acad.Sci.U.S.A* **92**:880-884.

Sullivan, J. M., Pietras, K. M., Shin, B. J., and Misasi, J. N., 2002, Hammerhead ribozymes designed to cleave all human rod opsin mRNAs which cause autosomal dominant retinitis pigmentosa, *Mol.Vis.* **8**:102-113.

Takahashi, M., Miyoshi, H., Verma, I. M., and Gage, F. H., 1999, Rescue from photoreceptor degeneration in the rd mouse by human immunodeficiency virus vector-mediated gene transfer, *J.Virol.* **73**:7812-7816.

Wang, Q., Chen, Q., Zhao, K., Wang, L., Wang, L., and Traboulsi, E. I., 2001, Update on the molecular genetics of retinitis pigmentosa, *Ophthalmic Genet.* **22**:133-154.

Wu, T. H., Ting, T. D., Okajima, T. I., Pepperberg, D. R., Ho, Y. K., Ripps, H., and Naash, M. I., 1998, Opsin localization and rhodopsin photochemistry in a transgenic mouse model of retinitis pigmentosa, *Neuroscience* **87**:709-717.

EVALUATION OF INNER RETINAL STRUCTURE IN THE AGED RCS RAT

Sherry Ball[1], Brett Hanzlicek[2], Melissa Blum[2], and Machelle Pardue[3]

1. INTRODUCTION

In retinal diseases such as retinitis pigmentosa (RP), photoreceptors degenerate while inner retinal layers are relatively spared. In these cases, it is thought that, if given a signal, the "healthy" inner retina would be capable of processing visual information in a somewhat normal fashion. Several laboratories are testing treatments designed to replace lost photoreceptors by transplanting healthy retinal tissue (Woch et al., 2001; Coffey et al., 2002) or implanting a light sensitive prosthetic device (Chow et al., 2001; Zrenner et al., 1999) into the subretinal space. These strategies will work only if the inner retina is intact and functional. Recent evidence in humans as well as rodent models of RP shows that, especially at later stages of disease progression, the inner retina begins to exhibit changes (Stone et al., 1992; Strettoi et al., 2002). Determining the nature of these changes is critical to understanding the disease process as well as developing treatments.

The Royal College of Surgeons (RCS) rat is widely used as a model for RP (LaVail, 2001). In the RCS retina a mutation in the merTK gene (D'Cruz, et al., 2000) results in an inability of the retinal pigment epithelium (RPE) to phagocytize shed rod outer segments (Mullen and LaVail, 1976). Photoreceptors in the RCS rat gradually degenerate beginning at around 3 weeks of age until a complete loss occurs at 6 months of age (Dowling and Sidman, 1962; LaVail and Battelle, 1975). Morphological studies have shown that, after photoreceptor degeneration, the inner retina remains normal in terms of thickness and cell and synapse number (Eisenfeld et al., 1984). Electroretinograms (ERGs) showed that some inner retinal activity was maintained throughout degeneration (Bush et al., 1995). However, subsequent studies have shown reactive changes in Müller cells (Hartig et al., 1995; Sheedlo et al., 1995), abnormal branching patterns in bipolar (Hanitzsch et al., 1998) and horizontal cells (Chu et al., 1993), and alterations in neurotransmitter levels (Okada et al., 2000; Gründer et al.,

[1]Research Service, Cleveland VA Medical Center, Department of Psychology, Case Western Reserve University, Cleveland, OH 44106, USA; [2] Cleveland VA Medical Center, Cleveland, OH, USA; [3]Atlanta VA Medical Center, Department of Ophthalmology, Emory University, Atlanta, GA 30033, USA. Corresponding author: S. Ball, E-mail: slb25@po.cwru.edu

2001). Here we use well-characterized antibodies to further assess inner retinal circuitry in the RCS rat.

2. METHODS

2.1. Animals

Pigmented RCS dystrophic rats and congenic controls were obtained from a breeding colony originating in the laboratory of M. LaVail at the University of California at San Francisco and maintained at the Cleveland VA Medical Center. Pigmented Long Evans rats purchased from Harlan Sprague Dawley were also used as normal controls. All animals were maintained on a 12/12 hr light/dark cycle with light levels at 4 cd/m^2. All procedures adhered to guidelines set by the Association for Research in Vision and Ophthalmology Statement for the Use of Animals in Ophthalmic and Vision Research.

2.2. Electroretinography

To follow loss of retinal function in the RCS rat, full field ERGs were measured using conventional methods (Ball et al., 2002). Briefly, after overnight dark adaptation, recordings were made to a single flash in anaesthetized (80 mg/ kg ketamine / 16 mg/kg xylazine) rats from 2 to 25 weeks of age, when a signal was no longer recordable.

2.3. Immunohistochemistry

Retinas were collected from rats at 1, 6, and 12 months of age after lethal injection (ketamine/xylazine) and immersion fixed in 4% paraformaldehyde for 30 minutes, embedded in OCT, cryoprotected, and frozen at –25° C. Vertical sections were cut on a Leica cryostat at 10 μm, mounted on slides, rinsed in PBS, and incubated in blocking solution. Following incubation in primary antibodies (listed in Table 1), sections were rinsed, incubated in Alexa Fluor 488 or 532 secondary antibodies, and coverslipped. For controls, sections were incubated without the primary antibody. All digital images were taken with an RT spot camera in central retina using identical exposure settings.

Table 1. Primary antibodies used in immunohistochemical experiments

Antibody	host	Source	Cntrn	Relevant references
PKC	mouse	Sigma	1:1,000	Greferath et al., 1990
Recoverin	Rabbit	Dizhoor	1:2,000	McGinnis et al., 1999
Bassoon	Mouse	Stressgen	1:1,000	Brandstätter et al., 1999
Synapsin	rabbit	Chemicon	1:1,000	Mandell et al., 1990
Parvalbumin	Mouse	Chemicon	1:1,000	Wässle et al., 1993
Calretinin	goat	Sigma	1:4,000	Gabriel & Witkovsky, 1998
Choline acetyltransferase	goat	Chemicon	1:1000	Voigt, 1986
Tyrosine hydroxylas	rabbit	Chemicon	1:1,000	Haverkamp & Wässle, 2000

3. RESULTS

3.1. Electroretinography

We tracked photoreceptor degeneration by recording ERGs in 5 RCS rats from 2 to 25 weeks of age. Representative waveforms for each age are presented in Figure 1. Over the period of photoreceptor degeneration the b-wave is reduced, followed by the appearance of an inner retinal response shown by others to be the scotopic threshold response (STR; Bush et al., 1995). Beyond 22 weeks of age the ERG was not distinguishable from noise.

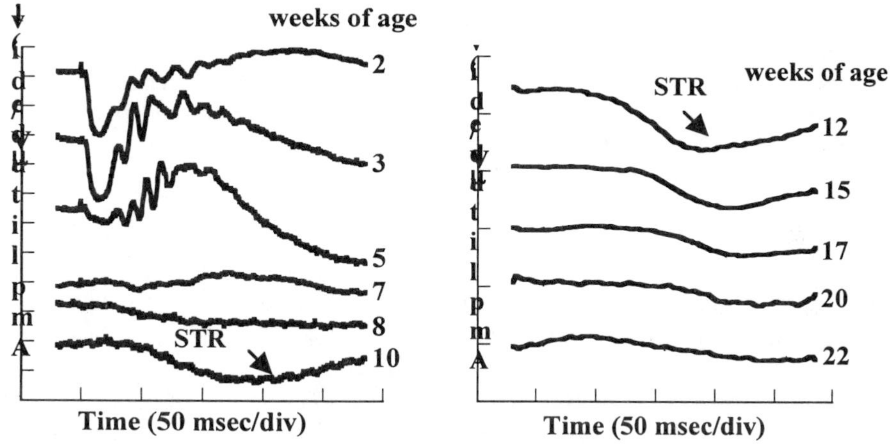

Figure 1. ERGs recorded from RCS rats to a bright flash (138 cd sec/m^2) under dark-adapted conditions. The b-wave gradually declines in amplitude followed by the appearance of a negative waveform or STR representing inner retinal activity. Note difference in magnitude between left and right plots.

3.2. Immunohistochemistry

3.2.1. *Bipolar Cells*

Recoverin labels two types of cone bipolar cells in normal rat retina (Dizhoor et al., 1991; Milam et al., 1993; Stepanik et al., 1993). In Figure 2A-D, these labeled cells are seen in the outer portion of the inner nuclear layer (INL) of RCS retinas before and after photoreceptor degeneration. The two cell types can be distinguished based on the location of synaptic terminals in inner and outer portions of the IPL corresponding to ON and OFF synaptic layers. As previously reported in normal retina, photoreceptor cell bodies label intensely while cells in the inner nuclear layer (INL) stain faintly (McGinnis et al., 1999). A comparable labeling pattern is seen in the RCS retina at 1 month of age (Fig. 2D). In contrast, by 5 months of age label is no longer seen in the outer retina as this layer has degenerated. However, a more striking difference is the presence of

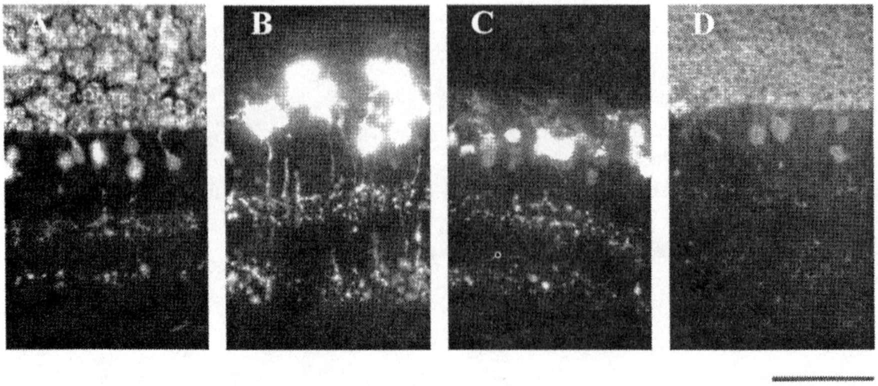

Figure 2. Cell bodies of on-and off-cone bipolar cells show intense expression of recoverin in the RCS retina at 1 (A), 5 (B) and 12 months (C) of age as compared to faint label in a normal rat at 12 months of age (D). All images were captured using identical exposure times and gain settings. Scale bar = 50 μm

prominent label in both populations of bipolar cells in RCS retina (Fig. 2A-C). Intensely labeled cell bodies can be seen in RCS retinas from 3 to 12 months of age while only faint label is observed in age-matched normal retinas at all ages examined (Fig 2D). These intensely labeled cells appear to maintain an overall normal organization pattern with cell bodies confined to outer portions of the INL and projections forming a bilaminar organization in the IPL.

We used a PKC antibody to examine rod bipolar cell morphology. As previously reported, after photoreceptor loss, rod bipolar cell dendrites labeled with PKC are no longer seen projecting into the OPL (Hanitzsch et al., 1998). As early as 3 months of age, we also observed atrophy in bipolar cell terminals as compared to normal retinas. However, by 12 months of age, no further changes were noted in cell body number or in axon projections and terminals in the RCS rat as compared to normal (data not shown).

3.2.2. *Amacrine cells*

To assess inner retinal circuitry, we investigated the pattern of calretinin, a calcium binding protein known to label a subpopulation of amacrine cells in the rat retina (Gabriel and Witkovsky, 1998). Figure 3A-D compares labeling patterns in degenerating RCS and normal retinas. Labeled cells are seen in the inner portions of the INL and a population of cells in the ganglion cell layer. Calretinin consistently revealed a tri-laminar organization within the IPL presumably corresponding to synaptic layers. In RCS rats aged 1-12 months (Fig. 3A-C), no changes were noted in the number, distribution or intensity of label in this subset of amacrine cells and these measures did not differ from normal retinas. Additionally, we labeled three different populations of amacrine cells in RCS and normal retinas at 1, 3, 6, and 12 months of age with antibodies against choline acetyltransferase, tyrosine hydroxylase, and parvalbumin.

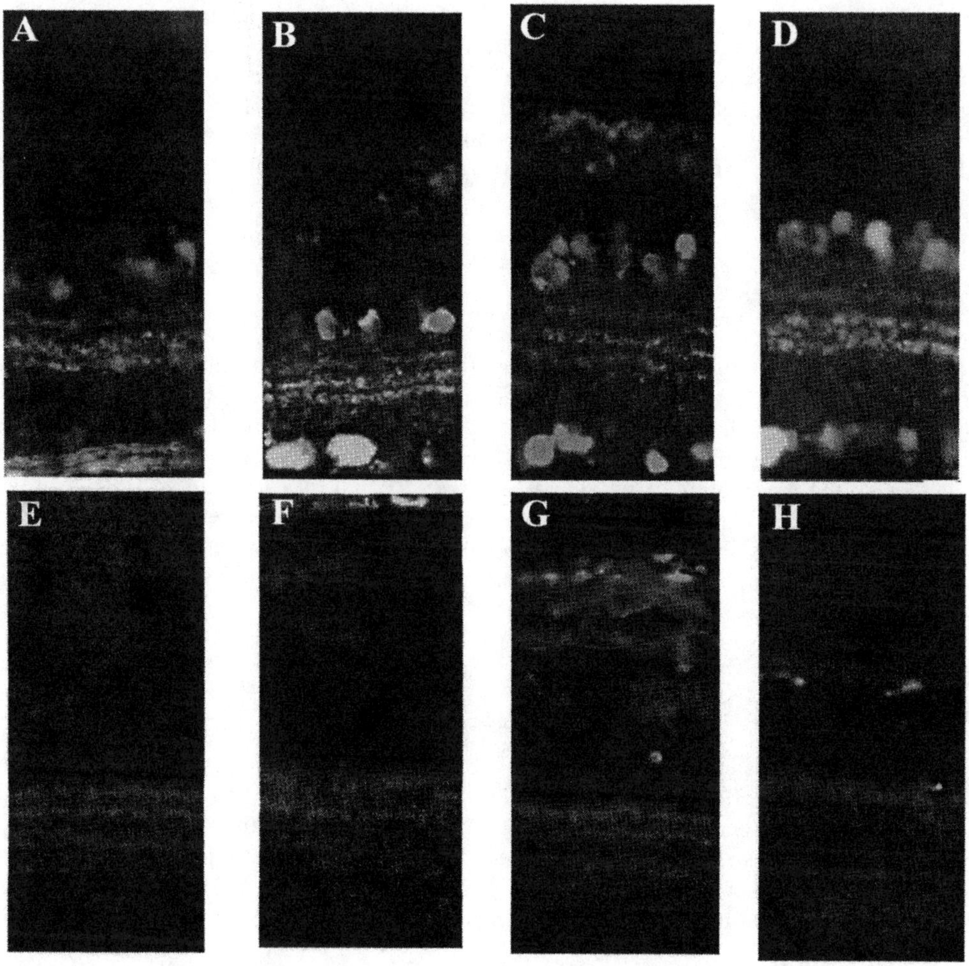

Figure 3. Distribution of amacrine cells (A-D) and synaptic proteins (E-H) in the inner retina of RCS (A-C & E-F) and normal (D & H) retinas. A comparable calretinin and bassoon labeling pattern is found in RCS retinas at 1(A & E), 5 (B & F), and 12 (C & G) month(s) of age, and in normal retina at 12 months (D & H). Scale bar = 50μm

All antibodies have been well defined in the rat retina and our results were comparable to those previously reported (Table 1). No differences in labeling patterns were noted between normal and RCS rat retinas (data not shown).

3.2.3. *Synaptic proteins in the IPL*

Bassoon is a component of the active zone in the IPL at conventional inhibitory GABAergic synapses made by amacrine cells (Brandstätter et al., 1999). In Figure 3E-H, bassoon labeled synaptic regions can be seen in the IPL of RCS rat retinas at 1 (E) , 5 (F)

and 12 (G) months of age as well as a 12-month-old normal (H) retina. As previously reported, label is seen throughout the IPL. The label forms three broad bands defined by a slightly higher-intensity label. This laminar pattern is noted in the RCS rats up to 12 months of age in a fashion comparable to normal retinas. We also used synapsin I to label conventional synapses in the IPL (Mandell et al., 1990) and noted no consistent differences in the amount of synapsin label in the IPL of RCS rats as compared to normal rats (data not shown).

4. DISCUSSION

Treatment strategies for RP such as retinal transplantation and implantation of a retinal prosthesis are currently being investigated. These strategies are based on the hypothesis that after the loss of the photoreceptor layer, inner retinal layers remain able to receive and process a visual signal. This has motivated further investigation into the integrity of the inner retina in human patients with RP and rodent models of this disorder. As evidence accumulates on alterations in the inner retina the need for a more detailed analysis becomes clear. We present an immunohistochemical analysis of the inner retina in the RCS rat throughout the process of degeneration and compare our results to normal age-matched retinas.

The most prominent change we observed was in the intensity of recoverin label in cone bipolar cells. These results caution that deficits in the cone pathway may be occurring at early stages of degeneration. As the function of this protein is unknown, a definitive explanation cannot be made. However, it is likely that the increase in label is due to changes in the cellular environment as a result of photoreceptor degeneration (McGinnis et al., 1999). We are continuing to investigate the nature of these changes in the cone bipolar cells and how to relate the loss of rod and cone function with changes in the respective bipolar cells. Our results on rod bipolar cell morphology were similar to those previously reported.. In both the RCS rat (Hanitzsch et al., 1998) and the *rd* mouse (Strettoi and Pignatelli, 2000) retinal rod bipolar dendrites and axon terminals gradually contract.

The integrity of amacrine cells and their processes in the IPL was assessed by labeling with well-established markers for amacrine cells and synaptic proteins. We found no alterations in the distribution of four different amacrine cell markers and two markers for synaptic proteins. These results are consistent with a similar study in the *rd* mouse (Strettoi et al., 2002). The maintained structural integrity in the inner retina may be explained by the time course in which amacrine cells develop. Cholinergic amacrine cells in the rat retina show a stratified projection pattern in the IPL at P0, well before bipolar cell processes have developed (Günham et al., 2002). As cells and their synaptic contacts (Haas et al., 1990; Sarthy and Bacon, 1985) are fully mature prior to the onset of degeneration, they may be the least susceptible to the loss of photoreceptors. One study has shown inner retinal changes prior to the loss of photoreceptors (Gründer et al., 2000). In that study the normal distribution pattern of the NR1 NMDA receptor was altered by P0. However, it is thought that these modifications may be directly attributable to the merTK mutation rather than an indirect result of altered RPE function and may thus represent an additional disease mechanism (Gründer et al., 2002).

We have presented evidence that inner retinal circuitry in the RCS rat remains anatomically organized to some degree at ages up to 12 months. The degree to which this circuitry is functional at this time point remains to be determined. As previously reported, there are clear reactive changes occurring in second order neurons following photoreceptor loss in the RCS rat (Fletcher and Kalloniatis, 1997; Hanitzsch et al., 1998; Villegas-Perez et al., 1998; Gründer et al., 2002). Although the consequences of these changes remain to be determined, the results presented here may prove useful for evaluating treatment strategies in which the RCS rat is used as an animal model.

5. ACKNOWLEDGEMENTS

We thank Dr. Neal Peachey for critically reviewing the manuscript, Dr. Dizhoor for providing the antibody for recoverin, and Michael Matthes and Doug Yasumura for supplying fixed retinal tissue.

6. REFERENCES

Ball, S.L., Powers, P.A., Shin, H-S., Morgans, C.W., Peachey, N.S., and Gregg, R.G., 2002, Role of the β_2 subunit of voltage-dependent calcium channels in the retinal outer plexiform layer, *Invest Ophthalmol Vis Sci.* **43**:1595-1603.

Brandstätter, J.H., Fletcher, E.L., Garner, C.C., Gundelfinger, E.D., and Wässle, H., 1999, Differential expression of the presynaptic cytomatrix protein bassoon among ribbon synapses in the mammalian retina, *Eur J Neurosci.* **11**:3683-3693.

Bush, R.A., Hawks, K.W., and Sieving, P.A., 1995, Preservation of inner retinal responses in the aged Royal College of Surgeons rat. Evidence against glutamate excitotoxicity in photoreceptor degeneration, *Invest Ophthalmol Vis Sci.* **36**:2054-2062.

Chow, A.Y., Pardue, M.T., Chow, V.Y., Peyman, G.A., Liang, C., Perlman, J.I., and Peachey, N.S., 2001, Implantation of silicon chip microphotodiode arrays into the cat subretinal space, *IEEE Trans Neural Sys Rehab Eng.* **9**:86-95.

Coffey P.J., Girman S., Wang S.M., Hetherington L., Keegan D.J., Adamson P., Greenwood J., Lund R.D., 2002, Long-term preservation of cortically dependent visual function in RCS rats by transplantation, *Nat Neurosci.* **5**:53-6.

Chu, Y., Humphrey, M.F., and Constable, I.J., 1993, Horizontal cells of the normal and dystrophic rat retina: A wholemount study using immunolabelling for the 28-kD calcium-binding protein, *Exp Eye Res.* **57**:141-148.

Dizhoor, A.M., Ray, S., Kumar, S., Niemi, G., Spencer, M., Brolley, D., Walsh, K.A., Philipov, P.P., Hurley, J.B., and Stryer, L., 1991, Recoverin: a calcium sensitive activator of retinal rod guanylate cyclase, *Science.* **251**:915-918.

D'Cruz, P.M., Yasumura, D., Weir, J., Matthes, M.T., Aberrahim, H., LaVail, M.M., and Vollrath, D., 2000, Mutation of the receptor tyrosine kinase gene Mertk in the retinal dystrophic RCS rat, *Hum Mol Gen.* **9**: 645-651.

Dowling, J.E., and Sidman, R.L., 1962, Inherited retinal dystrophy in the rat, *J Cell Biol.* **14**:73-109.

Eisenfeld, A.J., LaVail, M.M., and LaVail, J.H., 1984, Assessment of possible transneuronal changes in the retina of rats with inherited retinal dystrophy, *J Comp Neurol.* **223**:22-34.

Fletcher, E.L. and Kalloniatis, M., 1997, Neurochemical development of the degenerating rat retina, *J Comp Neurol.* **420**:98-112.

Gabriel, R. and Witkovsky, P., 1998, Cholinergic, but not the rod pathway-related glycinergic (AII), amacrine cells contain calretinin in the rat retina, *Neurosci Lett.* **247**:179-182.

Greferath, U., Grünert, U., Wässle, H., 1990, Rod bipolar cells in the mammalian retina show protein kinase C-like immunoreactivity, *J Comp Neurol* **301**:433-442.

Gründer, T., Kohler K., and Guenther, E., 2001, Alterations in NMDA receptor expression during retinal degeneration in the RCS rat, *Vis Neurosci.* **18**:781-787.

Gründer, T., Kohler, K., Kaletta, A., and Guenther, E., 2000, Distribution and developmental regulation of NMDA receptor subunit proteins in the outer and inner retina of the rat, *J Neurobiol.* **44**:333-342.

Günhan, E., Choudary, P.V., Landerholm, T.E., and Chalupa, L.M., 2002, Depletion of cholinergic amacrine cells by a novel immunotoxin does not perturb the formation of segregated on and off cone bipolar cell projections, *J Neurosci.* **22**:2265-2273.

Haas, C.A., DeGennaro, L.J., Müller, M., and Hollander, H., 1990, Synapsin I expression in the rat retina during postnatal development, *Exp Brain Res.* **82**:25-32.

Hanitzsch, R., Zeumer, C., Lichtenberger, T., and Wurziger, K., 1998, Impaired function of bipolar cells in the Royal College of Surgeons rat, *Acta Anatom.* **162**:119-126.

Hartig, W., Grosche, J., Distler, C., Grimm, D., El-Hifnawi, E., and Reichenbach, A., 1995, Alterations of Müller (glial) cells in dystrophic retinae of RCS rats, *J Neurocytol.* **24**:507-517.

Haverkamp, S., and Wässle, H., 2000, Immunocytochemical analysis of the mouse retina, *J Comp Neurol.* **424**:1-23.

LaVail, M.M., 2001, Legacy of the RCS rat: impact of a seminal study on retinal cell biology and retinal degenerative diseases, *Prog Brain Res.* **131**:617-27.

LaVail, M.M., and Battelle, B.A., 1975, Influence of eye pigmentation and light deprivation on inherited retinal dystrophy in the rat, *Exp Eye Res.* **21**:167-92.

Mandell, J.W., Townes-Anderson, E., Czernik, A.J., Cameron, R., Greengard, P., and De Camilli, P., 1990, Synapsins in the vertebrate retina: absence from ribbon synapses and heterogeneous distribution among conventional synapses, *Neuron.* **5**:19-33.

McGinnis, J.F., Stepanik, P.L., Chen, W., Elias, R., Cao, W., and Lerious, V., 1999, Unique retina cell phenotypes revealed by immunological analysis of recoverin expression in rat retina cells, J Neurosci. **55**:252-260.

Milam, A.H., Dacey, D.M., and Dizhoor, A.M., 1993, Recoverin immunoreactivity in mammalian cone bipolar cells, *Vis Neurosci.* **10**:1-12.

Mullen, R.J., and LaVail, M.M., 1976, Inherited retinal dystrophy: primary defect in pigment epithelium determined with experimental rat chimeras, *Science.* **192**:799-801.

Okada, M., Okuma, Y., Osumi, Y., Nishihara, M., Yokotani, K., and Ueno, H., 2000, Neurotransmitter contents in the retina of the RCS rat, *Graefe's Arch Clin Exp Ophthalmol.* **238**:998-1001.

Sarthy, P.V. and Bacon, W., 1985, Developmental expression of a synaptic vesicle-specific protein in the rat retina, *Dev Biol.* **112**:284-91.

Sheedlo, H.J., Jaynes, D., Bolan, A.L., and Turner, J.E., 1995, Müllerian glia in dystrophic rodent retinas: an immunocytochemical analysis, *Dev Brain Res.* **85**:171-180.

Stepanik, P.L., Lerious, V., and McGinnis, J.F., 1993, Developmental appearance, species and tissue specificity of mouse 23-kDa, a retinal calcium-binding protein (recoverin), *Exp Eye Res.* **57**:189-197.

Strettoi, E. and Pignatelli, V., 2000, Modificaitons of retinal neurons in a mouse model of retinitis pigmentosa, *Proceed Natl Acad Sci.* **97**:11020-11025.

Strettoi, E., Porciatti, V., Falsini, B., Pinatelli, V., and Rossi, C., 2002, Morphological and functional abnormalities in the inner retina of the *rd/rd* mouse, *J Neurosci.* **22**:5492-5504.

Stone, J.L., Barlow, W.E., Humayun, M.S., de Juan, E. Jr, and, Milam, A.H., 1992, Morphometric analysis of macular photoreceptors and ganglion cells in retinas with retinitis pigmentosa, *Arch Opthalmol.* **110**: 1634-1640.

Villegas-Perez, M.P., Lawrence, J.M., Vidal-Sanz, M., LaVail, M.M., and Lund, R.D., 1998, Ganglion cell loss in RCS rat retina: a result of compression of axons by contracting intraretinal vessels linked to the pigment epithelium, *J Comp Neurol.* **392**:58-77.

Voigt T., 1986, Cholinergic amacrine cells in the rat retina, *J Comp Neurol.* **248**:19-35.

Wässle, H., Grünert, U., and Rohrenbeck, J., 1993, Immunocytochemical staining of AII-amacrine cells in the rat retina with antibodies against parvalbumin, *J Comp Neurol.* **332**:407-420.

Woch, G., Aramant, R.B., Seiler, M.J., Sagdullaev, B.T., and McCall, M.A., 2001, Retinal transplants restore visually evoked responses in rats with photoreceptor degeneration, *Invest Ophthalmol Vis Sci.* **42**:1669-1676.

Zrenner, E., Stett, A., Weiss, S., Aramant, R.B., Guenther, E., Kohler, K., Miliczek, K.D, Seiler, M.J., and Haemmerle, H., 1999, Can subretinal microphotodiodes successfully replace degenerated photoreceptors? *Vision Res.* **39**:2555-67.

THE INTACT *XENOPUS LAEVIS* EYE RUDIMENT: A QUASI-IN VIVO SYSTEM FOR THE STUDY OF RETINAL DEVELOPMENT AND DEGENERATIONS

Monica M. Jablonski[1]

1. INTRODUCTION

The purpose of this chapter is to demonstrate the validity and value of the "quasi-in vivo" intact eye culture system from *Xenopus laevis* tadpoles as a model for the study of retinal development and degenerations.

2. METHODS

2.1 Culture paradigm

This methodology has been previously published by our laboratory (Jablonski, et al., 2001). *Xenopus laevis* embryos were obtained through induced breeding of adult frogs by injection of human chorionic gonadotropin. Embryos were staged by external morphologic criteria (Nieuwkoop and Faber, 1956). Eyes were removed from stage 33/34 or stage 35/36 *Xenopus* tadpoles. At these developmental stages, the eye rudiments have not yet been enveloped by the sclera-choroid layers; therefore, the retinal pigment epithelium (RPE) is the outermost cell layer. In some experimental paradigms, the RPE was left intact and supporting the neural retina, while in other paradigms, the RPE was removed. Other than removal of the RPE, eyes were cultured intact with minimal perturbation of the 3D structure of the eye. RPE-deprived, but otherwise intact eye rudiments were cultured under the following conditions: (1) in Niu-Twitty (NT) medium alone; (2) in NT containing lactose (5mM), an example of a permissive sugar (Jablonski and Ervin, 2000); (3) in NT containing pigment epithelium-derived factor (PEDF) (50ng/ml), an example of a neurotrophic factor (Jablonski, et al., 2000); and (4) in NT containing 20-mer antisense olignucleotides complementary to *rds*/peripherin using a paradigm previously described (Jablonski, 2001). Eyes were placed cornea-side down in

[1]Department of Ophthalmology, University of Tennessee Health Science Center, 956 Court Avenue, Suite D228, Memphis, TN 38163, USA. E-mail: mjablonski@utmem.edu

containing 20-mer antisense olignucleotides complementary to *rds*/peripherin using a paradigm previously described (Jablonski, 2001). Eyes were placed cornea-side down in 35 mm culture dishes containing 1.5-2 ml culture medium and were maintained in vitro for 3-4 days at 23 ^{0}C. RPE-supported eyes were used as control.

2.2. Outer segment grading

We have recently developed and published (Jablonski et al., 2001) a six-step grading classification of photoreceptor outer segment morphology based on the assessment of the percentage of photoreceptor outer segment membrane that was properly assembled. This grading system was applied to RPE-deprived eyes exposed to various neurosupportive factors as listed above. All structural analyses were performed on tissue sections taken exclusively from the posterior pole region of the retina. The grader was blinded to the experimental condition. The key criterion of this grading system is the level of outer segment membrane organization, with a grade of 4 indicating 100% organization of the outer segment and a grade of 0 indicating 0% organization.

2.3. Western blotting

Our protocol for this procedure has been previously published (Jablonski et al., 2000). To quantify proteins of interest, eyes from each experimental condition were solubilized in 1% sodium cholate detergent. Retinal extracts of equal total protein concentration were separated on 1-D SDS PAGE gels using standard protocols. The proteins was electroblotted to HyBondP membranes. Quantification of proteins was performed using the ECF Western blotting kit (Amersham Biosciences) according to the manufacturer's specifications. Quantification was determined on blots scanned on a Typhoon 9400 Imaging Device at 560nm. ImageQuant software version 3 was used to quantify the images. For each of the three repetitions, data were normalized to values obtained for eyes maintained with an intact RPE.

2.4. Immunohistochemistry

Our protocol for this procedure has been previously published (Jablonski et al., 2000). Eyes were fixed in Davidson fixative (32% ethanol, 2% formalin, 11% acetic acid), dehydrated, embedded in Unicryl (Electron Microscopy Sciences, Fort Washington, PA), cut at 1 μm thickness through the posterior pole of the retina and collected on microscope slides. The sections will be incubated in 5% serum in phosphate buffered saline (PBS), rinsed in PBS and incubated overnight in primary antibody. Gold-conjugated secondary antibodies followed by silver enhancement were applied to the tissue. Retinal sections were stained lightly with Toluidine blue O and viewed on a Nikon Eclipse E800 microscope equipped with Sensys Color Camera (Photometrics, Tucson, AZ) and images were collected using MetaMorph Imaging System software (Universal Imaging Corporation, West Chester, PA). Two images were collected from each retinal section: a brightfield image which shows the morphology of the tissue; and another image taken with epipolarized light which shows only the immunolabeling pattern. The epipolarized image was color enhanced and merged with the brightfield image.

3. RESULTS

The “quasi-in vivo” intact eye culture system from *Xenopus laevis* tadpoles is amenable to studies of retinal development and the evaluation of putative supportive factors. Retinal dysmorphogenesis can be induced by removal of the RPE cell layer, leaving the photoreceptors exposed the culture medium (Stiemke et al., 1994). The medium can then be supplemented with various factors, including permissive sugars (Jablonski and Ervin, 2000) and neurotrophic factors (Jablonski et al., 2000), and the effect upon retinal morphology, protein expression patterns and ability to support outer segment membrane assembly can be evaluated.

In RPE-supported retinas, development proceeded normally; photoreceptor outer segment membranes were properly assembled (Fig. 1A and 5). However, in RPE-deprived retinas, photoreceptor outer segments were markedly disorganized (Fig. 2A and 5). In parallel to these structural changes, immunolabeling patterns were altered compared

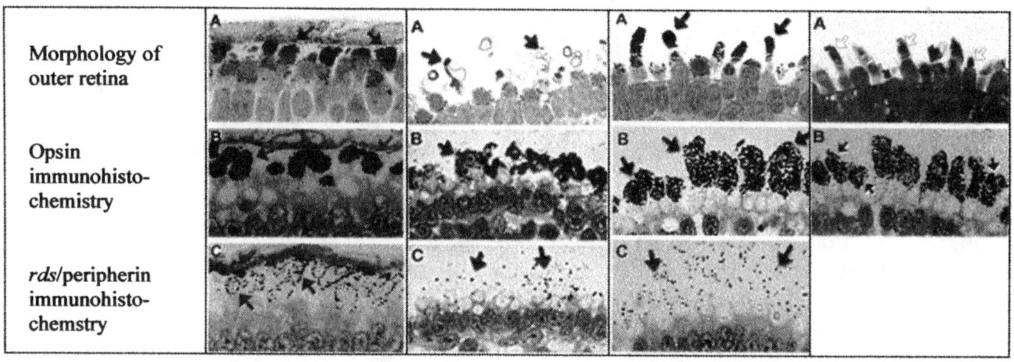

Figure 1. RPE-supported retina **Figure 2.** RPE-deprived retina **Figure 3.** Lactose-exposed retina **Figure 4.** PEDF-exposed retina.

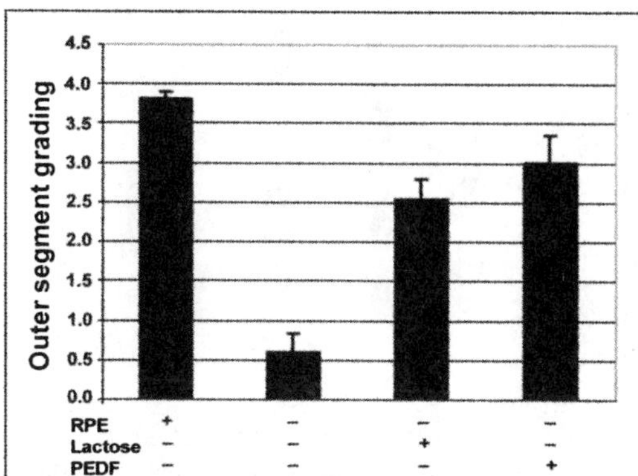

Figure 5. Support of photoreceptor outer segment assembly: grading.

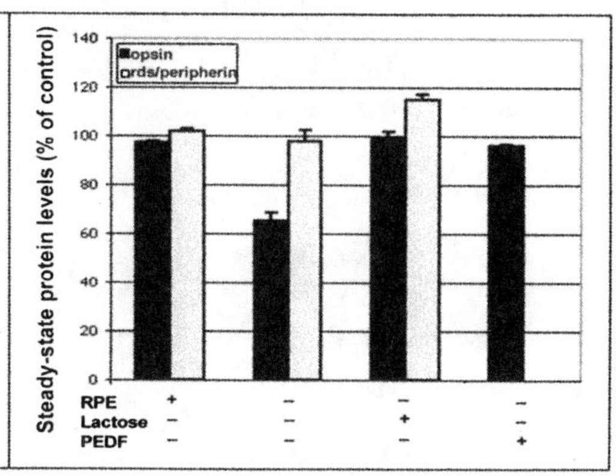

Figure 6. Steady state protein expression levels

to RPE-supported retinas. Heavy patches of opsin-positive label alternate with areas of irregularly shaped profiles of immunoreactivity (Fig. 2B). Steady-state opsin levels were reduced (Fig. 6). Also reflecting the altered outer segment conformation, *rds*/peripherin labeling was present in a random pattern (Fig. 2C). However, the steady-state level of *rds*/peripherin was unchanged (Fig. 6).

In RPE-deprived retinas, lactose promoted the formation of well-organized outer segment membranes (Fig. 3A and 5), and allowed for normal expression of both photoreceptor-specific proteins (Figs. 3B, 3C and 6). PEDF also promoted both the assembly of organized outer segments (Fig. 4A and 5) and the expression of normal opsin levels (Figs. 4B and 6).

When coupled with antisense oligonucleotides, our "quasi-in vivo" eye model can also be used to determine the function of any retina-expressed gene product (Jablonski, 2001). The oligonucleotide is delivered to the intact eye via the culture medium and inhibits the translation of the target mRNA, thus generating a loss-of-function model. A thorough understanding of the function of individual genes will provide insights on the role of specific gene products in retinal health and pathophysiology. This experimental approach will also allow for specific therapeutic interventions to be evaluated so that targeted treatments can be designed for individuals with specific genetic mutations.

Utilizing antisense oligonucleotides coupled with our intact *Xenopus* eye rudiment model, we have effectively demonstrated that we are able to downregulate the expression of a photoreceptor-specific protein, *rds*/peripherin, and generate a loss-of-function model upon which to further study the function of the *rds*/peripherin gene. The ultrastructure (Fig. 7) and protein expression patterns (Fig. 8) very closely resemble those previously documented in both the *rds* mouse and in human autosomal dominant retinitis pigmentosa due to peripherin/RDS mutations. An identical strategy can be applied to any gene correlated with a degenerative retinal phenotype.

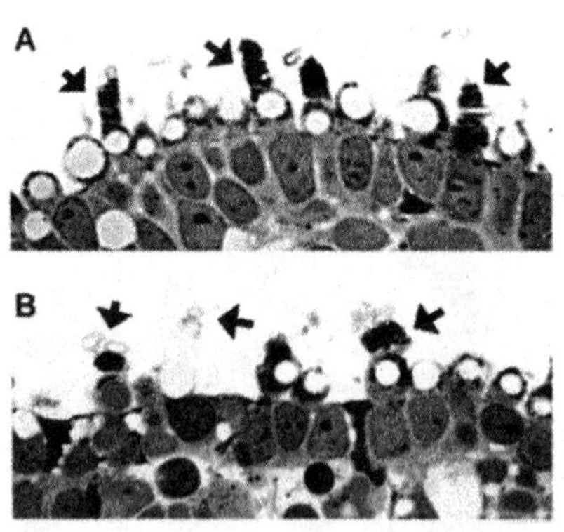

Figure 7. Morphology of retinas to sense control oligonucleotide (A) and xrds (62-81) (B).

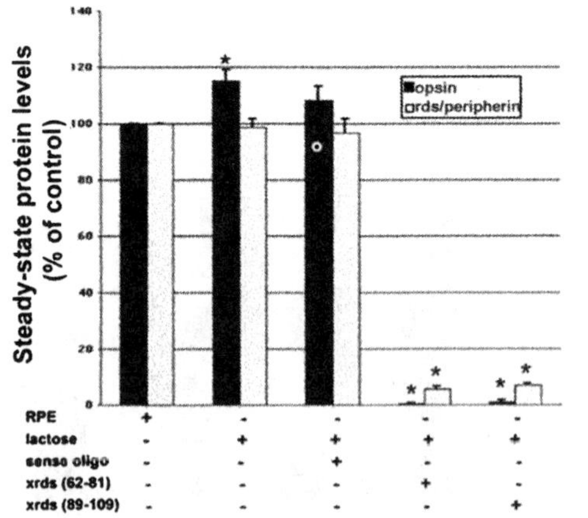

Figure 8. Protein expression profiles of eyes exposed to sense and antisense oligonucleotides.

4. DISCUSSION

The intact *Xenopus laevis* tadpole eye in vitro culture system in an excellent model to determine the molecular mechanisms that regulate retinal development and degenerations. The five main reasons that make this an excellent and highly desirable model can be summarized as follows:

1) *The Xenopus laevis model is very well established and characterized.* There are nearly 50 years worth of studies on the *Xenopus* model in the retinal cell biology and development literature (Nieuwkoop and Faber, 1956; Witkovsky et al., 1976; Kinney and Fisher, 1978; Hollyfield and Rayborn, 1979; Besharse and Pfenninger, 1980; Defoe and Besharse, 1985; Fliesler et al., 1985; Stiemke et al., 1994; Jablonski et al., 2000). In more recent years, the contributions of the *Xenopus* have grown to include its use as a transgenic (Bronchain et al., 1999; Green et al., 1999; Mani et al., 1999; Tam et al., 2000; Wheeler et al., 2000; Moritz et al., 2001) or gene knockdown model (Jablonski, 2001; Rungger-Brandle et al., 2002) to aid in the understanding retinal disease, thus indicating a continued appreciation for the value of *Xenopus* in the study of developmental retinal cell biology, as well as the dystrophic process.

2) *Amphibian photoreceptors are the only ones capable of elaborating significant outer segment material even in the absence of the RPE.* In the vast majority of vertebrate models, when photoreceptors are cultured either as isolated cells or as reaggregates following retinal cell dissociation, the assembly of outer segment membranes does not normally proceed past the elaboration of a few rudimentary membranous leaflets (LaVail and Hild, 1971; Adler, 1986; Adler, 1987; Lahav, 1987; Politi et al., 1988; Adler and Politi, 1989; Watanabe and Raff, 1990; Roof et al., 1991). Although a developing mouse eye organ culture system which supports significant outer segment elaboration in the presence of the RPE has been described (Caffé et al., 1989), even the mouse eye organ system does not allow for studing the effects of RPE removal on photoreceptor outer segments, because these do not develop in this system in the absence of the RPE. The removal and culture of intact *Xenopus laevis* eye rudiments is the only in vitro preparation in which photoreceptors elaborate expanses of outer segment membrane also in RPE-deprived retinas. Of course, the membranes produced under these conditions are disorganized (Stiemke et al., 1994), but abundant enough to make it possible to understand which proteins are indispensable to the production of *organized* outer segments.

3) *Our intact eye rudiment in vitro system mimics very closely the physiologic state.* The *Xenopus* intact eye culture system allows for comparison of retinal developmental parameters (e.g., photoreceptor outer segment membrane assembly, retinal cell gene expression, and protein expression profiles) between in vivo and in vitro retinal development and also between retinas allowed to mature in vitro in the presence or absence of a surrounding RPE. Another essential feature of our culture paradigm is that, at the developmental stages of our interest, the sclera has not yet formed around the eye rudiments. Therefore, it is extremely easy to access the RPE layer without disrupting the physiology or the anatomic integrity of the eye rudiments. Hence, the outer retina can be easily freed of the RPE, leaving the neuroepithelium intact with no requirement for dissociation of the retinal rudiments into single cells. Thus, photoreceptors continue to maintain their association with other inner retinal elements, i.e., developing Müller cells and second order neurons. Because the retina remains intact, photoreceptors do not

develop and differentiate in contact with an unusual substratum, such as plastic or coated surfaces. Accordingly, comparisons of photoreceptor development with and without the RPE can be assessed with minimal disturbance of photoreceptor interactions with other retinal elements. This is particularly important in light of evidence that indicates that the expression of specific differentiation markers of photoreceptors (Harris and Messersmith, 1992; Wohabrebbi et al., 2002) and Müller cells are dependent upon the appropriate cell-cell contact (Linser et al., 1982; Vardimon et al., 1991; Prada et al., 1998; Jablonski and Iannaccone, 2001).

4) *Our simple culture conditions make our system easy to interpret and to manipulate in a tightly controlled fashion.* Our *Xenopus* in vitro system allows for probing differentiating retinas with various factors in the absence of complex media. The NT culture medium in which the intact embryonic eyes are maintained is very simple, consisting of salts and water with no confounding serum or other additives, yet in this medium the intact eye is able to form an in vivo-like, well-differentiated retina. This is possible because *Xenopus* embryos, like those of all amphibians, contain nutrient stores in all cells of the embryo (Jacobson, 1967). Therefore, neuroretinal cells contain sufficient metabolites to allow for development, differentiation, and histogenesis in the absence of external nutrients. Specific components can be added to the culture medium and the resultant effects on photoreceptor outer segment assembly and retinal protein expression profiles can be monitored under tightly controlled conditions.

5) *Xenopus laevis photoreceptors are extremely similar to human ones.* Although *Xenopus* and humans or other mammals are phylogenetically distant, it has been clearly shown that the subcellular retinal architecture of *Xenopus laevis* and humans is very comparable, especially with respect to photoreceptors. In fact, the ultrastructure of the frog photoreceptor cell is nearly identical to that of human and primates, with approximately 15-20 calycal processes circumscribing the base of the outer segment and with periodic incisures that are approximately as numerous in the *Xenopus* as in human photoreceptors (Pedler and Tilly, 1967; Steinberg et al., 1980; Fetter and Corless, 1987). The striking similarity of *Xenopus* and human photoreceptor cells has been also reviewed and further characterized recently by Eckmiller (2000). This similarity extends to the proteins expressed within the retinas. For example, there is significant evolutionary conservation of pigment epithelium-derived factor, an RPE-secreted factor with neuronal development, differentiation and survival properties, across the vertebrate realm (Tombran-Tink et al., 1996). Moreover, within the *Xenopus* retina, we have documented the expression of many photoreceptor and Müller cell proteins that have been previously described in humans and mice (Jablonski et al., 1999; Jablonski et al., 2000; Jablonski and Iannaccone, 2001; Jablonski et al., 2001). These similarities are potentially of great importance when evaluating subcellular architectural and protein expression pattern differences that may have relevance to human retinal degenerative disease.

Because of the extensive literature on normal structure and function of the *Xenopus* retina both in vitro and in vivo, comparisons of the developmental and/or dysmorphic parameters under investigation can be made with certainty. In addition, over the past 10 years we have always normalized our in vitro results to the equivalent in vivo developmental stage. To date, all variables that we have investigated in vitro have shown no appreciable difference from the in vivo state (Jablonski et al., 1999; Jablonski et al., 2000; Jablonski and Iannaccone, 2000; Jablonski and Iannaccone, 2001; Jablonski et al., 2001). In summary, these lines of evidence, along with the fact that *Xenopus laevis* frogs are readily available, have low procurement and per diem caring costs, have a rapid breeding time frame, and provide large amounts of embryos at every breeding cycle (by the hundreds), make the *Xenopus* model highly desirable and extremely well suited for studies of retinal – and particularly photoreceptor – physiology and pathology.

5. ACKNOWLEDGMENTS

This investigation was generously supported by grants to MMJ by: The National Eye Institute, Bethesda, MD; Fight for Sight, Schaumburg, IL; the Mid-South Lions Sight and Hearing Service, Memphis, TN; and by an unrestricted grant to the University of Tennessee HSC Department of Ophthalmology from Research to Prevent Blindness, Inc., New York, NY. MMJ is a Research to Prevent Blindness Greve Scholar.

6. REFERENCES

Adler, R., 1986, Developmental predetermination of the structural and molecular polarization of photoreceptor cells, *Dev Biol.* **117**:520-527.

Adler, R., 1987, Nature and nurture in the differentiation of retinal photoreceptors and neurons, *Cell Differ.* 20:183-188.

Adler, R. and Politi, L., 1989, *Expression of a "survival crisis" by normal and rd/rd mouse photoreceptor cells in vitro*, in: *Inherited and environmentally induced retinal degenerations*, M. M. LaVail, R. E. Anderson, J. G. Hollyfield, eds. , Alan R. Liss, Inc., New York, pp. 169-181.

Besharse, J. C. and Pfenninger, K. H., 1980, Membrane assembly in retinal photoreceptors I. Freeze-fracture analysis of cytoplasmic vesicles in relationship to disc assembly, *J Cell Biol.* **87**:451-463.

Bronchain, O. J., Hartley, K. O. and Amaya, E., 1999, A gene trap approach in Xenopus, *Curr Biol.* **9**:1195-1198.

Caffé, A. R., Visser, H., Jansen, H. G. and Sanyal, S., 1989, Histopathic differentiation of neonatal mouse retina in organ culture, *Curr Eye Res.* **8**:1083-1092.

Defoe, D. M. and Besharse, J. C., 1985, Membrane assembly in retinal photoreceptors. II. Immunocytochemical analysis of freeze-fractured rod photoreceptor membranes using anti-opsin antibodies, *J Neurosci.* **5**:1023-1034.

Eckmiller, M. S., 2000, Microtules in a rod-specific cytoskeleton associated with outer segment incisures, *Vis Neurosci.* **17**:711-722.

Fetter, R. D. and Corless, J. M., 1987, Morphological components associated with frog cone outer segment disc margins, *Invest Ophthalmol Vis Sci.* **28**:646-657.

Fliesler, S. J., Rayborn, M. E. and Hollyfield, J. G., 1985, Membrane morphogenesis in retinal rod outer segments: inhibition by tunicamycin, *J Cell Biol.* **100**:574-587.

Green, C. B., Liang, M. Y., Steenhard, B. M. and Besharse, J. C., 1999, Ontogeny of circadian and light regulation of melatonin release in Xenopus laevis embryos., *Brain Res Dev Brain Res.* 117:109-116.

Harris, W. A. and Messersmith, S. L., 1992, Two cellular inductions involved in photoreceptor determination in the *Xenopus* retina, *Neuron.* **9**:357-372.

Hollyfield, J. G. and Rayborn, M. E., 1979, Photoreceptor outer segment development: Light and dark regulate the rate of membrane addition and loss, *Invest Ophthalmol Vis Sci.* **18**:117-133.

Jablonski, M. M., Wohabrebbi, A. and Ervin, C. S., 1999, Lactose promotes organized photoreceptor outer segment assembly and preserves expression of photoreceptor proteins in retinal degeneration, *Mol Vis.* **5**:16.

Jablonski, M. M. and Ervin, C. S., 2000, A closer look at lactose-mediated support of retinal morphogenesis, *Anat Rec.* **259**:205-214.

Jablonski, M. M. and Iannaccone, A., 2000, Targeted disruption of Müller cell metabolism induces photoreceptor dysmorphogenesis, *Glia.* **32**:192-204.

Jablonski, M. M., Tombran-Tink, J., Mrazek, D. A. and Iannaccone, A., 2000, Pigment epithelium-derived factor supports normal development of photoreceptor neurons and opsin expression after retinal pigment epithelium removal, *J Neurosci.* **20**:7149-7157.

Jablonski, M. M., 2001, Investigating the mechanisms of retinal degenerations with antisense oligonucleotides, *Doc Ophthalmol.* **102**:179-196.

Jablonski, M. M., Graney, M. J., Kritchevsky, S. B. and Iannaccone, A., 2001, Reliability assessment of a rod photoreceptor outer segment grading system, *Exp Eye Res.* **72**:573-579.

Jablonski, M. M. and Iannaccone, A., 2001, Lactose supports Müller cell protein expression patterns in the absence of the retinal pigment epithelium, *Mol Vis.* 7:27-35.

Jablonski, M. M., Tombran-Tink, J., Mrazek, D. A. and Iannaccone, A., 2001, Pigment epithelium-derived factor supports normal Müller cell development and glutamine synthetase expression after removal of the RPE, *Glia.* **35**:14-25.

Jacobson, A. G., 1967, *Amphibian cell culture, organ culture, and tissue dissociation*, in: *Methods in Developmental Biology*, F. H. Wilt, N. K. Wessells, eds., Thomas Y. Crowell, New York, pp. 531-542.

Kinney, M. S. and Fisher, S. K., 1978, The photoreceptors and pigment epithelium of the larval Xenopus retina: morphogenesis and outer segment renewal, *Proc R Soc Lond B.* **201**:149-167.

Lahav, M., 1987, In vitro model of retinal photoreceptor differentiation, *Trans Am Ophthalmol Soc.* **85**:600-638.

LaVail, M. M. and Hild, W., 1971, Histotypic organization of the rat retina in vitro, *Z Zellforsch.* 114:557-579.

Linser, P., Saad, A. D., Soh, B. M. and Moscona, A. A., 1982, Cell contact-dependent regulation of hormonal induction of glutamine synthetase in embryonic neural retina, *Prog Clin Biol Res.* **85** Pt B:445-458.

Mani, S. S., Besharse, J. C. and Knox, B. E., 1999, Immediate upstream sequence of arrestin directs rod-specific expression in Xenopus., *J Biol Chem.* **274**:15590-15597.

Moritz, O. L., Tam, B. M., Papermaster, D. S. and Nakayama, T., 2001, A functional rhodopsin-green fluorescent protein fusion protein localizes correctly in transgenic *Xenopus laevis* retinal rods and Is expressed in a time-dependent pattern, *J Biol Chem.* **276**:28242-28251.

Nieuwkoop, P. D. and Faber, J., 1956, *Normal Table of Xenopus laevis (Daudin)*, North Holland Publishing Co., Amsterdam.

Pedler, C. M. H. and Tilly, R., 1967, The fine structure of photoreceptor discs, *Vision Res.* **7**:829-836.

Politi, L. E., Lahav, M. and Adler, R., 1988, Development of neonatal mouse retinal neurons and photoreceptors in low density cell culture, *Invest Ophthalmol Vis Sci.* **29**:534-543.

Prada, F. A., Quesada, A., Dorado, M. E., Chmielewski, C. and Prada, C., 1998, Glutamine synthetase (GS) activity and spatial and temporal patterns of GS expression in the developing chick retina: relationship with synaptogenesis in the outer plexiform layer, *Glia.* **22**:221-236.

Roof, D., Hayes, A., Adamian, M., Marescalchi, P. and Heth, C., 1991, Photoreceptor development in rat neural retina-retinal pigment epithelium co-cultures, *Invest Ophthalmol Vis Sci.* **32**:1148.

Rungger-Brandle, E., Conti, A. and Rungger, D., 2002, Retinal deficiencies in Xenopus tadpoles carrying antisense morpholinos directed against Pax6, *Invest Ophthalmol Vis Sci*: E-Abstract #2444.

Steinberg, R. H., Fisher, S. K. and Anderson, D. H., 1980, Disc morphogenesis in vertebrate photoreceptors, *J Comp Neurol.* **190**:501-518.

Stiemke, M. M., Landers, R. A., Al-Ubaidi, M. R. and Hollyfield, J. G., 1994, Photoreceptor outer segment development in *Xenopus laevis*: Influence of the pigment epithelium, *Dev Biol.* **162**:169-180.

Tam, B. M., Moritz, O. L., Hurd L.B. and Papermaster, D. S., 2000, Identification of an outer segment targeting signal in the COOH terminus of rhodopsin using transgenic Xenopus laevis, *J Cell Biol.* **151**:1369-1380.

Tombran-Tink, J., Rodrigues, I. R., Chung, D., Linker, T., Englander, E. and Chader, G. J., 1996, Organization, evolutionary conservation, expression and unusual Alu density of the human gene for pigment epithelium-derived factor, a unique neurotrophic serpin, *Mol Vis.* **2**:11.

Vardimon, L., Fox, L. E., Cohen-Kupiec, R., Degenstein, L. and Moscona, A. A., 1991, Expression of v-src in embryonic neural retina alters cell adhesion, inhibits histogenesis, and prevents induction of glutamine synthetase, *Mol Cell Biol.* **11**:5275-5284.

Watanabe, T. and Raff, M. C., 1990, Rod photoreceptor development in vitro: intrinsic properties of proliferating neuroepithelial cells change as development proceeds in the rat retina, *Neuron.* **2**:461-467.

Wheeler, G. N., Hamilton, F. S. and Hoppler, S., 2000, Inducible gene expression in transgenic Xenopus embryos, *Curr Biol.* **10**:849-852.

Witkovsky, P., Gallin, E., Hollyfield, J. G., Ripps, H. and Bridges, C. D. B., 1976, Photoreceptor thresholds and visual pigment levels in normal and vitamin A-deprived Xenopus tadpoles, *J Neurophys.* **39**:1272-1287.

Wohabrebbi, A., Umstot, E. S., Iannaccone, A., Desiderio, D. M. and Jablonski, M. M., 2002, Downregulation of a novel photoreceptor protein correlates with improper outer segment assembly, *J Neurosci Res.* **67**:298-308.

STREPTOZOTOCIN-INDUCED DIABETES - A RAT MODEL TO STUDY INVOLVEMENT OF RETINAL CELL TYPES IN THE ONSET OF DIABETIC RETINOPATHY

Elisabeth Rungger-Brändle and André A. Dosso[1]

1. INTRODUCTION

Diabetic retinopathy (DR) is a potentially blinding complication of diabetes mellitus. After the initial diagnosis of diabetes, the deficiencies typical for DR, namely abnormal autoregulation of retinal arteries, increased blood flow, breakdown of the blood-retinal barrier, and disturbances in retinal function develop silently and remain clinically undetected for quite some time. A characteristic microangiopathy in the fundus is one of the first visible signs of DR. Macular edema, neovascularization, contraction of the vitreous and of proliferated fibrovascular tissue are late consequences in the diabetic eye. The severity of these lesions is closely related to the duration of diabetes, to blood glucose control, and to risk factors such as hypertension.

Dysfunctions of the neural parenchyma such as decrease in contrast sensitivity (Dosso et al., 1996) and color vision (Hardy et al., 1992) manifest themselves long before any signs of a retinal vasculopathy become visible. Given the difficulty to analyze the underlying cellular and molecular features in diabetic patients, most of our knowledge on these deficiencies relies on diabetic animal models. The widely used streptozotocin (STZ)-induced diabetic rat, in which pancreatic βcells are destroyed, causing insulin deficiency and systemic hyperglycemia, is ideally suited for studying the sequence of changes caused by a controlled duration of diabetes. Still, there is a substantial lack of knowledge regarding the initial alterations in the hyperglycemic retina. Here, we describe early structural changes in different retinal cell types that may help to identify early molecular changes leading to DR.

[1]Cell Biology Laboratory, University Eye Clinic, CH-1211 Geneva 14, Switzerland. Corresponding author: E. Rungger-Brandle, E-mail: Elisabeth.Rungger@hcuge.ch

Retinal Degenerations: Mechanisms and Experimental Therapy
Edited by LaVail *et al.*, Kluwer Academic/Plenum Publishers, 2003

2. MATERIALS AND METHODS

Diabetes was induced in pigmented Long Evans rats weighing between 300 and 350g, by an intravenous injection of STZ (60mg/kg). Housing of the rats was standard with food and water supplied ad libitum. The fluorescent light source, cycled at 12 hours on and 12 hours off, was dimmed by a curtain to an intensity of maximal 90 lux. Hyperglycemia (>20mM) was detected at day 3 postinjection. Rats were sacrificed at one and 2 weeks postinjection. The ARVO Statement for the Use of Animal Protection in Ophthalmic and Vision Research and the Swiss Law on Animal Protection were followed in all respects. Vascular hyperpermeability was monitored by intravenous injection of Evans blue as previously described (Rungger-Brändle et al., 2000). Immunofluorescence localization of the neurotransmitters glutamate and gamma amino butyric acid (GABA) was performed on deplasticized semithin sections, using specific rabbit antisera (dilution 1:20; Signature Immunologics, Salt Lake City). Secondary antibodies coupled to Texas red (Jackson ImmunoResearch, Milan Analytica, La Roche, Switzerland) were diluted 1:200. Sections were viewed in a Zeiss confocal microscope (LSM 410). The ultrastructural histochemical demonstration of glycogen particles (glycosomes; Rybicka, 1996) in retina and choroid was done as described (Rungger-Brändle et al., 1996). Thin sections were viewed in a Philips CM10 electron microscope.

3. RESULTS

Vascular hyperpermeability: Evans blue injection revealed minute leaky foci in the retinal vasculature to occur as early as one week after STZ injection. Figure 1a depicts such focal leakage from a capillary. Extent and frequency of leaky regions were markedly increased at 2 weeks postinjection (b).

Glycogen accumulation: At one week of diabetes, Müller cells contained clusters of glycosomes in their endfeet (Fig. 2a), whereas age-matched controls contained only few, dispersed particles (b). A few glycogen particles were also found in smooth muscle cells but not in the endothelium of retinal vessels of both control and diabetic animals (not

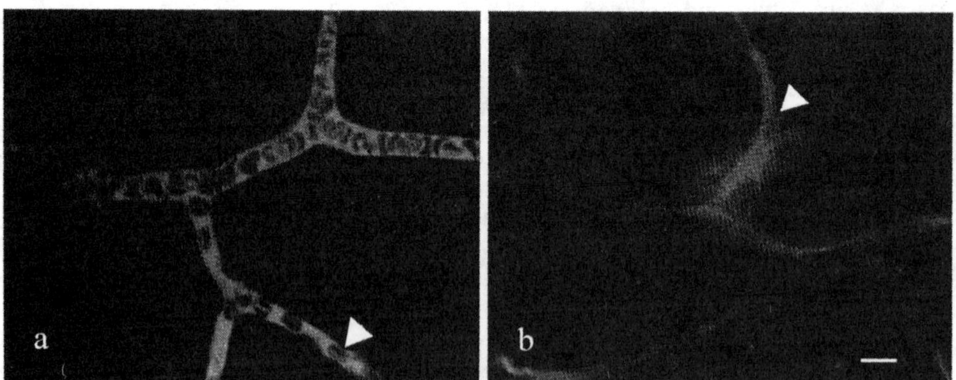

Figure 1. Hyperpermeability of retinal capillaries at one (a) and two (b) weeks after STZ injection. Retinal wholemounts after intravascular perfusion with Evans blue. Arrowheads denote leaky sites. Bar, 20 μm.

shown). Astrocytes were devoid of particles, even in the diabetic retina (not shown). In the choriocapillaris and choroid of diabetic individuals, glycosomes were abundant in smooth muscle cells and fibroblasts (c). They formed aggregates within the cytoplasm or were included in large vacuoles. In the control tissue, few isolated particles were detected in smooth muscle cells close to Bruch's membrane (d). Surprisingly, in both diabetic (c) and control (d) tissue the retinal pigment epithelium (RPE) was virtually devoid of glycosomes as were choroidal melanocytes.

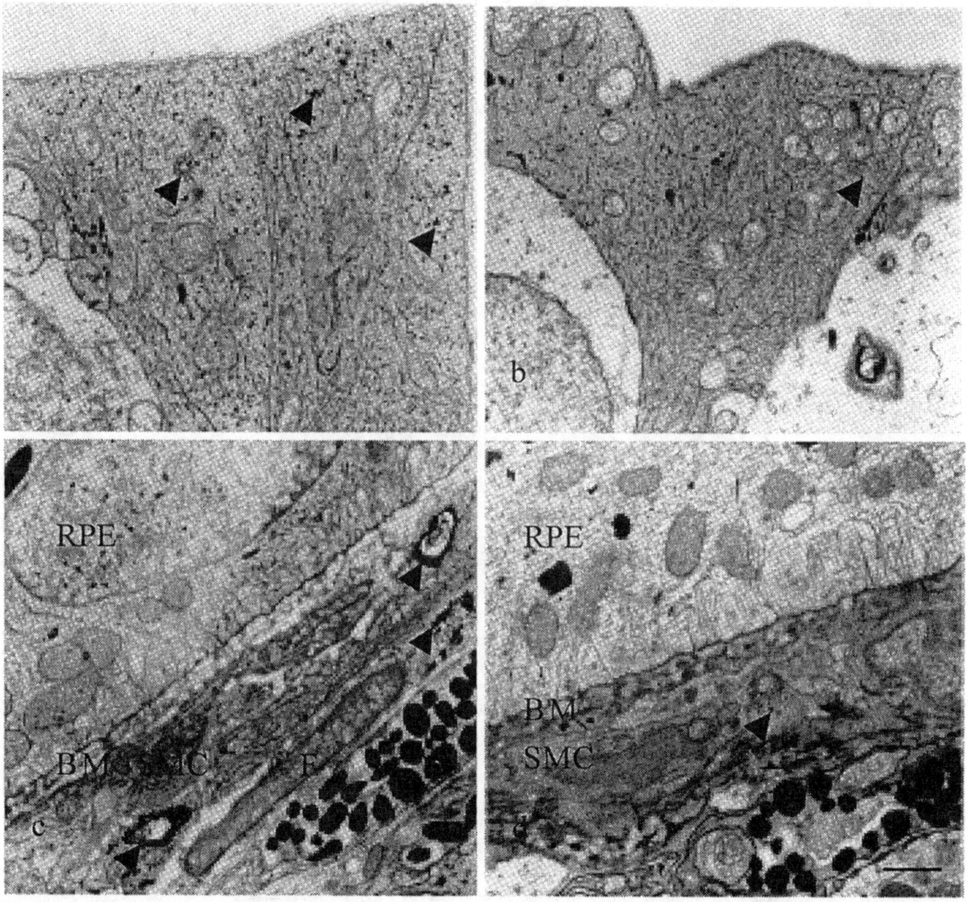

Figure 2. Histochemical demonstration of glycosomes in diabetic (a, c) and normal (b, d) retina, one weak after STZ injection. Electron micrographs. (a, b) Müller cell endfeet. (c, d) Basal aspect of RPE, Bruch's membrane (BM) and choroidal smooth muscle cells (SMC) and fibroblasts (F). Note clustered and dispersed glycosomes in diabetic and normal tissue, respectively. Bar, 1 μm.

Accumulation of neurotransmitters: Staining for GABA (Fig. 3a,b) and glutamate (c,d) revealed conspicuous accumulation of immunoreactive product in Müller cells at one week after STZ injection (a,c). GABA immunoreactivity was especially intense in the outer retinal layers up to the outer limiting membrane, whereas glutamate staining

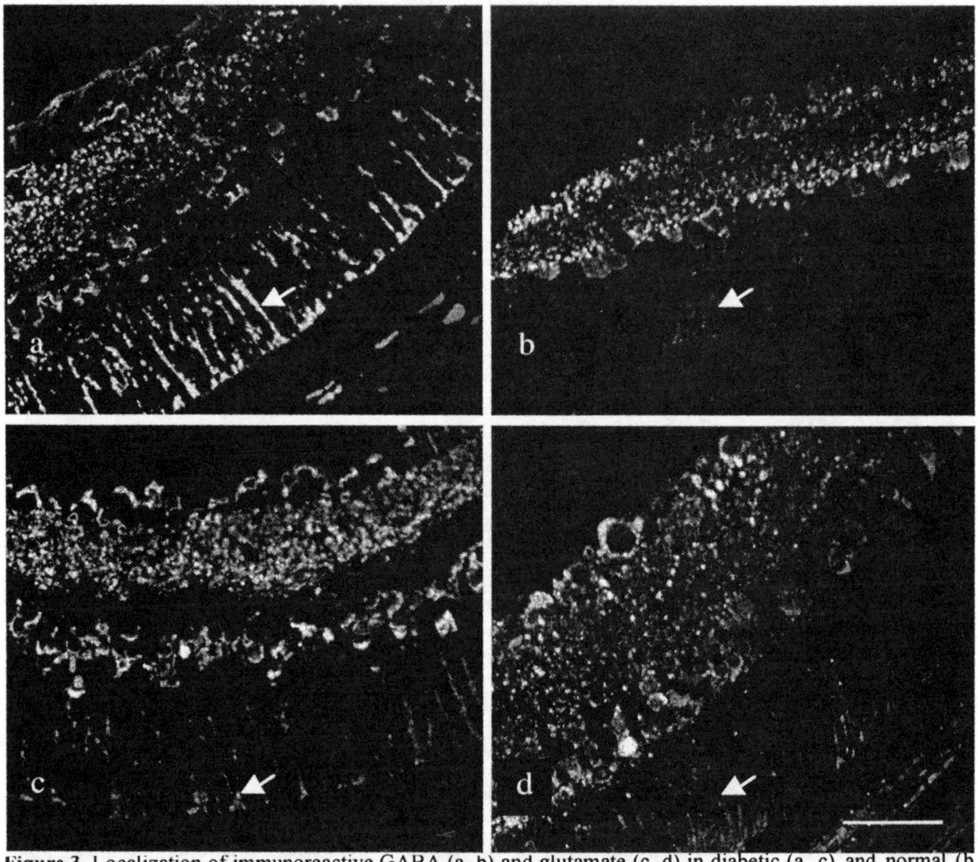

Figure 3. Localization of immunoreactive GABA (a, b) and glutamate (c, d) in diabetic (a, c) and normal (b, d) retina, one week after STZ injection. Fluorescence micrographs of semithin sections. Note marked staining of Müller cells in the outer nuclear layer (arrows) in the diabetic retina (a, c). In the normal outer nuclear layer, GABA (b) is inconspicuous and glutamate (d) confined to the intercellular space. Bar, 50 µm.

was less consistent in Müller cells and sometimes also localized to the extracellular space, similar to the situation in the control tissue (d). Immunoreactivity for GABA in the control retina was marked in amacrine cells and inner plexiform layer but virtually absent from the outer retinal layers (b).

4. DISCUSSION

In this study we describe some important early changes in the diabetic retina, namely increased permeability of the inner blood-retinal barrier, accumulation of particulate glycogen, and modulation of the localization of the neurotransmitters GABA and glutamate.

Hyperpermeability to Evans blue in retinal microvessels was local and, though confined to discrete sites, scattered over the whole vascular bed. With progression of diabetes, these leaky foci markedly increased in extent and number. This observation corroborates recent quantitative data on an early overall increase in vascular permeability in the diabetic retina (Do Carmo et al., 1998; Qaum et al., 2001). It is likely that hyperpermeability is caused by VEGF, the expression of which rapidly increases upon STZ injection (Qaum et al., 2001), leading to stimulation of the transcellular endothelial pathway (for review, see Bates et al., 2002). Moreover, VEGF is capable of activating the vascular endothelium by increasing I-CAM levels. Such an increase in ICAM-1 expression has indeed been noted after induction of diabetes (Lu et al., 1999; Miyamoto et al., 1999). Together with increased adhesion of neutrophils (Barouch et al., 2000), enhanced VEGF and ICAM-1 expression led to the recently emerged concept of DR as a low grade inflammatory disease (Adamis, 2002). It is likely that such a proinflammatory situation favors a rapid passage of blood glucose into the subendothelial space, entailing early metabolic alterations in the retinal parenchyma. Immediate targets may be the Müller glial cells forming the glia limitans of retinal vessels. Indeed, glycogen particles in Müller cell endfeet accumulate as early as one week after injection of STZ. Thus, the observed glycogen load by far precedes hyperplasia and GFAP expression in Müller cells (Rungger-Brändle et al., 2000). Glycosomes also accumulate in smooth muscle cells and fibroblasts of choriocapillaris and choroid, whereas RPE cells contain no glycosomes. This is surprising, since these cells have been shown to synthesize glycogen (Bok et al., 2002). However, as our data stipulate, RPE seems not to function as storage site. Apparently, RPE glycogen is labile and undergoes rapid turnover.

We also observed an important load of Müller cells with immunoreactive GABA and glutamate as early as one week after induction of diabetes. Accumulation of GABA was noted before but at a much later time point in the progression of DR (Ishikawa et al., 1996) and high levels of glutamate were found in both human vitreous (Ambati et al., 1997) and rat retina (Lieth et al., 1998; Kowluru et al., 2000). Intercellular accumulation of glutamate was observed in the diabetic retina of goto-kakizaki rats (Takeo-Goto et al., 2002). In our study, glutamate immunoreactivity in the normal retina was marked in all neuronal layers like in human (Crooks and Kolb, 1992), including the extracellular space between photoreceptors and Müller cells. By contrast, in the diabetic retina, it was most prominent within Müller glial cells. At first sight, this seems to contrast with a recent report on an early decrease in glutamate import in diabetic Müller cells (Li and Pouro, 2002) that would predict high glutamate levels in the extracellular space rather than within the glial cells. Yet, given the fundamental difference in the experimental approach used, the findings may be compatible. However, once within the Müller cells, glutamate is a substrate of various reactions (Poitry et al., 2000; Lieth et al., 2000; Qu et al., 2001; Kalloniatis and Napper, 2002). In the diabetic retina, several of these processes may be slowed down under hypoxic conditions and intracellular accumulation of unprocessed glutamate may occur in spite of a decreased transporter activity.

In conclusion, metabolic dysregulation in Müller glial cells may be an immediate consequence of vascular leakage and glucose entry into the retinal parenchyma. This metabolic deficiency may largely contribute to an early onset of diabetic neuropathy. Establishing a cell-specific temporal pattern of structural alterations in the retina of the STZ rat may provide a sound basis for targeting molecular markers during the time period of acute hyperglycemia that, in human patients, escapes detection by clinical testing.

5. ACKNOWLEDGMENTS

The authors would like to thank N. Gilodi and A. Conti for technical assistance and the Foundation Provisu, Switzerland, for financial support.

6. REFERENCES

Adamis, A.P., 2002, Is diabetic retinopathy an inflammatory disease? *Br. J. Ophthalmol.* **86**: 363.

Ambati, J., Chalam, K., Chawla, D., D'Angio, C., Guillet, E., Rose, S., Vanderlinde, R., and Ambati, B., 1997, Elevated gamma-aminobutyric acid, glutamate, and vascular endothelial growth factor levels in the vitreous of patients with proliferative diabetic retinopathy. *Arch. Ophthalmol.* **115**: 1161.

Barouch, F.C., Miyamoto, K., Allprot, J.R., Fujita, K., Bursell, S.-E., Aiello, L.P., Luscinskas, F.W., and Adamis, A.P., 2000, Integrin-mediated neutrophil adhesion and retinal leukostasis in diabetes. *Invest. Ophthalmol. Vis. Sci.* **41**: 1153.

Bates, D.O., Hillman, N.J., Williams, B., Neal, C.R., and Pocock, T.M., 2002, Regulation of microvascular permeability by vascular endothelial growth factors. *J. Anat.* **200**: 581.

Bok, D., Senanayake, P., Calabro, A., Hu, J.G., and Hollyfield, J.G., 2002, Glycogen synthesis in human RPE. *Invest. Ophthalmol. Vis. Sci. ARVO no. 4569*, **43:** 182.

Crooks, J., and Kolb, H., 1992, Localization of GABA, glycine, glutamate and tyrosine hydroxylase in the human retina. *J. comp. Neurol.* **315**: 287.

Do Carmo, A., Ramos, P., Reis, A., Proença, R., and Cunha-Vaz, J.G., 1998, Breakdown of the inner and outer blood retinal barrier in streptozotocin-induced diabetes. *Exp. Eye Res.* **67**: 569.

Dosso, A.A., Bonvin, E.R., Morel, Y., Golay, A., Assal, J.P., and Leuenberger, P.M., 1996, Risk factors with contrast sensitivity loss in diabetic patients. *Graefes Arch. Clin. Exp. Ophthalmol.* **234**: 300.

Hardy, K.J., Lipton, J., Foster, D.H., and Scarpello, J.H.B., 1992, Detection of colour vision abnormalities in uncomplicated type I diabetic patients with angiographically normal retinas. *Br. J. Ophthalmol.* **76**: 461.

Ishikawa, A., Ishiguro, S., and Tamai, M., 1996, Changes in GABA metabolism in streptozotocin-induced diabetic rat retinas. *Curr. Eye Res.* **15**: 63.

Kalloniatis, M., and Napper, G.A., 2002, Retinal neurochemical changes following application of glutamate as a metabolic substrate. *Clin. Exp. Optom.* **85**: 27.

Kowluru, R.A., Engerman, D.L., Case, G.L., and Kern, T.S., 2001, Retinal glutamate in diabetes and effect of antioxidants. *Neurochem. Int.* **38**: 385.

Li, Q., and Pouro, D.G., 2002, Diabetes-induced dysfunction of the glutamate transporter in retinal Müller cells. *Invest. Ophthalmol. Vis. Sci.* **43**: 3109.

Lieth, E., Barber, A., Xu, B., Dice, C., Ratz, M., Tanase, D., and Strother, J., 1998, Glial reactivity and impaired glutamate metabolism in short-term experimental diabetic retinopathy. *Diabetes* **42**: 815.

Lieth, E., Lanoue, K.F., Antonetti, D.A., Ratz, M., and The Penn State Retina Research Group, 2000, Diabetes reduces glutamate oxidation and glutamine synthesis in the retina. *Exp. Eye Res.* **70**: 723.

Lu, M., Perez, V.L., Naili, M., Miyamoto, K., Peng, H.-B., Liao, J.K., and Adamis, A.P., 1999, VEGF increases retinal vascular ICAM-1 expression in vivo. *Invest. Ophthalmol. Vis. Sci.* **40**: 1808.

Miyamoto, K., Khosrof, S., Bursell, S.-E., Moromizato, Y., Aiello, L.P., Ogura, Y., and Adamis, A.P., 2000, Vascular endothelial growth factor (VEGF)-induced retinal vascular permeability is mediated by intercellular adhesion molecule-1 (ICAM-1). *Am. J. Pathol.* **156**: 1733.

Miyamoto, K., Khosrof, S., Bursell, S.-E., Rohan, R., Murata, T., Clermont, A.C., Aiello, L.P., Ogura, Y., and Adamis, A.O., 1999, Prevention of leukostasis and vascular leakage in streptozotocin-induced diabetic retinopathy via intercellular adhesion molecule-1 inhibition. *Proc. Natl. Acad. Sci.* **96**: 10836.

Poitry, S., Poitry-Yamate, C., Ueberfeld, J., MacLeish, P.R., and Tsacopoulos, M., 2000, Mechanisms of glutamate metabolic signaling in retinal glial (Müller) cells. *J. Neurosci.* **20**: 1809.

Qaum, T., Xu, Q., Joussen, A.M., Clemens, M.W., Qin, W., Miyamoto, K., Hassessian, H., Wiegand, S.J., Rudge, J., Yancopoulos, G.D., and Adamis, A.P., 2001, VEGF-initiated blood-retinal barrier breakdown in early diabetes. *Invest. Ophthalmol. Vis. Sci.* **42**: 2408.

Qu, H., Eloqayli, H., Unsgard, G., and Sonnewald, U., 2001, Glutamate decreases pyruvate carboxylase activity and spares glucose as energy substrate in cultured cerebellar astrocytes. *J. Neurosci. Res.* **66**: 1127.

Rungger-Brändle, E., Kolb, H., and Niemeyer, G., 1996, Histochemical demonstration of glycogen in neurons of the cat retina. *Invest. Ophthalmol. Vis. Sci.* **37**: 702.

Rungger-Brändle, E., Dosso, A.A., and Leuenberger, P.M., 2000, Glial reactivity, an early feature of diabetic retinopathy. *Invest. Ophthalmol. Vis. Sci.* **41**: 1971.

Rybicka, K.K., 1996, Glycosomes - the organelles of glycogen metabolism. *Tissue and Cell* **28**: 253.

Takeo-Goto, S., Doi, M., Ma, N., Goto, R., Semba, R., and Uji, Y., 2002, Immunohistochemical localization of amino acids in the diabetic retina of goto-kakizaki rats. *Ophthalmic Res.* **34**: 139.

A2E, A FLUOROPHORE OF RPE LIPOFUSCIN: CAN IT CAUSE RPE DEGENERATION?

Janet R. Sparrow, Bolin Cai, Nate Fishkin, Young Pyo Jang, Sonja Krane, Heidi R.Vollmer, Jilin Zhou, and Koji Nakanishi[1]

1. INTRODUCTION

In atrophic age-related macular degeneration (AMD) and Stargardt disease, the death of retinal pigment epithelial (RPE) cell death leads to photoreceptor cell degeneration and visual impairment. Nevertheless, the cause of RPE atrophy is poorly understood. One factor that may place RPE cells at risk is the accumulation of critical levels of lipofuscin. Indeed, several lines of evidence indicate that the excessive accumulation of lipofuscin by RPE cells is significant in terms of the etiology of AMD. Firstly, histological analyses of human donor eyes (Wing et al., 1978; Weiter et al., 1986), in addition to fundus spectrophotometry (Delori et al., 1995a; Delori et al., 2001), and confocal ophthalmoscopy (von Rückmann et al., 1997), have shown that RPE cells overlying the macula, with the exception of RPE in the cone-rich fovea, exhibit the most pronounced age-related accumulation of fluorescent material. Lipofuscin levels in RPE cells are also topographically correlated with histopathological indicators of AMD (Feeney-Burns et al., 1984; Dorey et al., 1989) and with the loss of photoreceptor cells in aged eyes (Dorey et al., 1989). Interestingly, increased fundus autofluorescence at the borders of geographic atrophy is considered to represent an enhanced accumulation of RPE lipofuscin and to implicate the latter in the disease process (Holz et al., 1999; Holz et al., 2001). While the amassing of lipofuscin by RPE is a feature of aging, excessive accretion also occurs in Stargardt disease, some forms of retinitis pigmentosa and cone-rod dystrophy (Weingeist et al., 1982; Rabb et al., 1986; Lopez et al., 1990; Delori et al., 1995b; Kennedy et al., 1995).

2. A2E: A PRODUCT OF LIGHT RELATED RELEASE OF ALL-*TRANS*-RETINAL

The concept that a substantial portion of RPE lipofuscin forms as a consequence of

[1]Departments of Ophthalmology and Chemistry, Columbia University, New York, NY 10032, USA.
Corresponding author: J.R. Sparrow, E-mail: jrs88@columbia.edu

light related vitamin A recycling, is supported by reports that RPE lipofuscin is profoundly reduced in animals lacking the 11-cis and all-*trans*-retinal chromophores either due to dietary deficiency (Katz et al., 1986; Katz et al., 1992) or gene knockout (Katz and Redmond, 2001). Indeed, chromatographic analyses of chloroform/methanol extracts of human RPE by our group and others have revealed that the major hydrophobic constituent is a di-retinal conjugate (Eldred and Katz, 1988; Eldred and Lasky, 1993; Parish et al., 1998), given the name A2E (Sakai et al., 1996) (Fig. 1). A2E presents with an unusual pyridinium bis-retinoid structure: a rigid amphiphilic wedge with a hydrophilic head group and two hydrophobic retinoid-derived side arms (Sakai et al., 1996; Ren et al., 1997). The precursor of A2E is the phosphatidylpyridinium bisretinoid

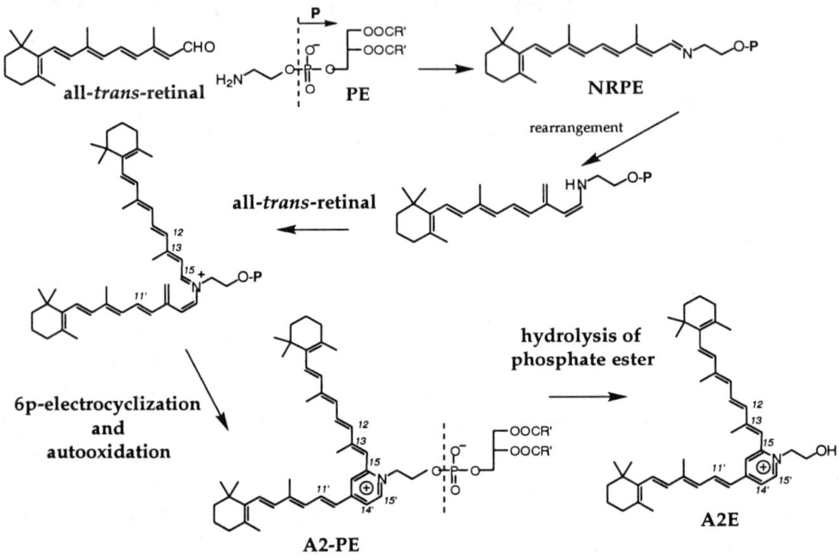

Figure 1. Biosynthetic pathway of A2-PE and A2E.

A2-PE (Parish et al., 1998; Liu et al., 2000), whose biosynthesis in photoreceptor outer segments (Liu et al., 2000; Mata et al., 2000) involves reactions between phosphatidylethanolamine (PE) and all-*trans*-retinal (Fig. 1). The Schiff base adduct formed between PE and a single molecule of all-*trans*-retinal likely serves as a physiological substrate of the photoreceptor-specific ABCR/RIM transporter (Sun et al., 1999; Weng et al., 1999; Ahn et al., 2000), the latter facilitating delivery of all-*trans*-retinal to all-*trans*-retinol dehydrogenase, the cytosolic enzyme charged with its reduction. However, reaction between the Schiff base conjugate and a second molecule of all-*trans*-retinal to generate A2-PE probably occurs randomly and under conditions when all-*trans*-retinal is allowed to accumulate. A2E is generated from A2-PE after phosphate hydrolysis, it is deposited in RPE cells secondary to the phagocytosis of shed outer segment membrane and it accumulates as lipofuscin because it cannot be degraded by RPE lysosomal enzymes (Eldred, 1998). Within the eye, A2E undergoes photoisomerization to generate iso-A2E, a C13-14 double bond isomer (Parish et al., 1998), and other minor cis-isomers (Ben-Shabat et al., 2002b) that are present as additional components of RPE lipofuscin.

Fluorophores of RPE such as A2E /isoA2E are amassed by the cells over the lifetime of an individual, although the rate of accumulation exhibits considerable variability amongst individuals and does not necessarily occur at a constant rate (Kitagawa et al., 1989; Delori, 1995; von Rückmann et al., 1997; Okubo et al., 1999; Sparrow et al., 1999; Delori et al., 2001). Under certain conditions, for instance, diminished activity in ABCR/RIM, the deposition A2E in RPE cells is exacerbated. Thus mice with a null mutation in ABCR exhibit accumulations of all-*trans*-retinal and A2-PE in outer segments and accentuated A2E levels in the RPE (Weng et al., 1999; Mata et al., 2000; Mata et al., 2001). Mutations in the ABCR gene account for the abundant deposits of A2E in RPE cells of patients with Stargardt disease (Allikmets et al., 1997; Mata et al., 2000), and may increase the susceptibility to atrophic age related macular degeneration (AMD) in a subset of patients (Allikmets et al., 1997; Allikmets, 2000). In addition to dysfunction of the ABCR transporter, light exposure is also a factor modulating the formation of A2-PE since dark rearing intervenes in the deposition of A2E in RPE of abcr -/- mice (Mata et al., 2000) and the formation of A2-PE in photoreceptor outer segments is augmented by exposing rats to bright light (Ben-Shabat et al., 2002b).

3. BLUE LIGHT MEDIATED DEGENERATION OF A2E-LADEN RPE

Light and the related release of all-*trans*-retinal is not just a determinant of the rate of accumulation of A2E; light is also involved in mechanisms that lead to the atrophy of A2E-laden RPE. Thus, accumulation of A2E within the lysosomal compartment of cultured RPE (Sparrow et al., 1999) has been shown to confer a susceptibility for blue light-induced apoptosis (Schutt et al., 2000; Sparrow et al., 2000; Sparrow and Cai, 2001), which is not manifest by cells devoid of A2E. Not only is the extent of cell death proportional to the A2E content of the cultures, the wavelength-dependency of this effect is consistent with the excitation spectra of A2E (Sparrow et al., 2000) and with the known susceptibility of RPE cells to blue light mediated damage in animal models (Ham et al., 1978; Ham et al., 1984; Ham et al., 1986; Putting et al., 1994; Busch et al., 1999a; Busch et al., 1999b). The cell death program which is initiated by blue light illumination of A2E involves the activation of caspase-3 (Sparrow and Cai, 2001), a downstream cysteine-dependent protease and is regulated by Bcl-2 (Sparrow and Cai, 2001), an anti-apoptotic protein situated in the outer mitochondrial membrane.

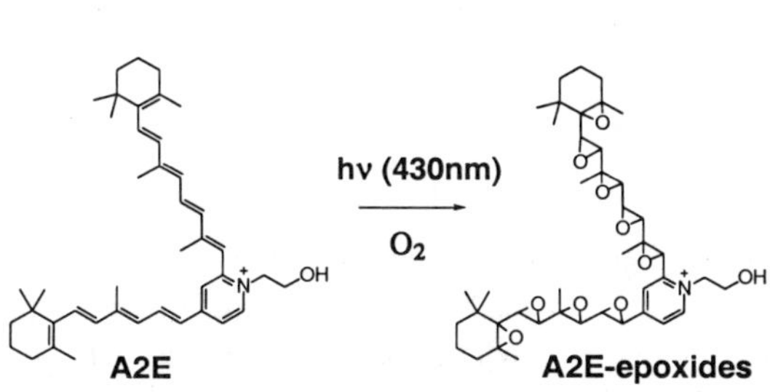

Figure 2. Illumination of A2E leads to epoxide ring formation (A2E-epoxides) within A2E side arms.

Evidence indicates that the generation of singlet oxygen upon photoexcitation of A2E is integral to the death of the cells. For instance, an enhancer (D_2O) and quenchers (histidine, DABCO, azide) of singlet oxygen modulate the incidence of nonviable A2E-laden RPE following blue light illumination (Sparrow et al., 2002). Significantly, the singlet oxygen which is generated by photosensitization of A2E becomes inserted into carbon-carbon double bonds of the retinoid-side arms of A2E to generate highly reactive epoxides (A2E-epoxides; Fig. 2) (Ben-Shabat et al., 2002a; Sparrow et al., 2002).

The formation of A2E-epoxides is significant since the considerable ring strain and electrophilicity of these three-membered oxygen and carbon containing rings makes them susceptible to reaction with nucleophilic macromolecules such as DNA and protein. Indeed, we have shown that at least one of the subcellular structures damaged by the interaction of blue light and A2E is DNA (Fig. 3) (Sparrow et al., 2003). The extent of injury, measured by single cell gel electrophoresis (comet assay) is proportional to exposure duration while the reduction in damage afforded by the singlet oxygen quencher sodium azide, revealed a role for oxidative mechanisms. Comet assay performed in conjunction with repair enzymes (Fpg and endonuclease III) to characterize DNA lesions and immunoperoxidase labeling with specific antibody indicated that guanine bases of DNA were oxidatively modified to generate 8-oxo-dG and perhaps other structurally related lesions. While singlet oxygen is an important cytotoxic product emerging from the photosensitization of A2E (Ben-Shabat et al., 2002a; Sparrow et al., 2002), A2E-epoxides generated from the photoxidation of A2E, also have the potential to attack intracellular targets.

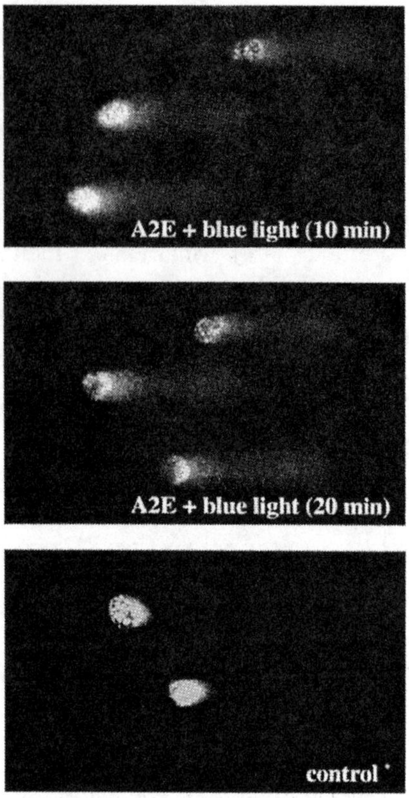

Figure. 3. DNA damage in blue light exposed A2E-laden RPE detected by comet assay. Strand breaks cause DNA to migrate out of the nucleus to form comet tails. Nuclei of control cells (unilluminated) remain intact.

4. CONCLUSIONS

Although AMD is a multifactorial disorder, our work suggests that the accumulation of A2E to certain intracellular levels, may be relevant to atrophic AMD, the form accounting for 12-20% of the blindness attributable to AMD (Bressler et al., 2000)). The mediation of light damage may not be the only mechanism by which A2E induces cell injury, since A2E can also perturb cell membranes (Sparrow et al., 1999; De and Sakmar, 2002). Nevertheless, the work discussed here underscores the unresolved issue of whether lifelong exposure to bright light contributes to AMD (West et al., 1989; Taylor et al., 1992). The disposition towards blue light damage conferred by A2E may be particularly significant in the elderly eye following cataract extraction when the latter procedure involves replacement of an aging yellow crystalline lens (Gaillard et al., 2000) with an intraocular lens (IOL) without similar blue light-absorbing capability.

Consequently, pseudophakic eyes may be at risk of progression to AMD (Liu et al., 1989; Pollack et al., 1996; Pollack et al., 1998). Whether or not unattenuated blue light is also injurious to the younger eyes which have accumulated substantial amounts of lipofuscin in the RPE (Feeney-Burns et al., 1984) before yellowing of the crystalline lens has occurred, remains to be determined.

5. ACKNOWLEDGEMENTS

The work was supported by National Institutes of Health Grants EY12951 and GM 34509, Macula Vision Research Foundation, American Health Assistance Foundation, the Charina Foundation and unrestricted funds from Research to Prevent Blindness.

6. REFERENCES

Ahn J, Wong JT, Molday RS, 2000, The effect of lipid environment and retinoids on the ATPase activity of ABCR, the photoreceptor ABC transporter responsible for Stargardt macular dystrophy. *J Biol Chem.* **275:**20399.

Allikmets R, 2000, Further evidence for an association of ABCR alleles with age-related macular degeneration. The international ABCR Screening Consortium. *Am J Hum Genet.* **67**:487.

Allikmets R, Shroyer NF, Singh N, Seddon JM, Lewis RA, Bernstein PS, Peiffer A, Zabriskie NA, Li Y, Hutchinson A, Dean M, Lupski JR, Leppert M, 1997, Mutation of the Stargardt disease gene (ABCR) in age-related macular degeneration. *Science* **277:**1805.

Ben-Shabat S, Itagaki Y, Jockusch S, Sparrow JR, Turro NJ, Nakanishi K, 2002a, Formation of a nona-oxirane from A2E, a lipofuscin fluorophore related to macular degeneration, and evidence of singlet oxygen involvement. *Angew Chem Int Ed.* ***41***: 814.

Ben-Shabat S, Parish CA, Vollmer HR, Itagaki Y, Fishkin N, Nakanishi K, Sparrow JR, 2002b, Biosynthetic studies of A2E, a major fluorophore of RPE lipofuscin. *J Biol Chem.* **277:**7183.

Bressler SB, Bressler NM, Gragoudas ES, 2000, Age-related macular degeneration: drusen and geographic atrophy. In: Albert DM, Jakobiec FA, Azar DT, Gragoudas ES (eds) *Principles and Practice of Ophthalmology*, vol 3. W.B. Saunders Co, Philadelphia, pp 1982.

Busch EM, Gorgels TGMF, Roberts JE, van Norren D, 1999a, The effects of two stereoisomers of N-acetylcysteine on photochemical damage by UVA and blue light in rat retina. *Photochem Photobiol.* **70:**353.

Busch EM, Gorgels TGMF, van Norren D, 1999b, Temporal sequences of changes in rat retina after UV-A and blue light exposure. *Vision Res.* **39:**1233.

De S, Sakmar TP, 2002, Interaction of A2E with model membranes. Implications to the pathogenesis of age-related macular degeneration. *J Gen Physio.l* **120:**147.

Delori FC, 1995, RPE lipofuscin in ageing and age-related macular degeneration. In: Coscas G, Piccolino FC (eds) *Retinal Pigment Epithelium and Macular Disease (Documenta Ophthalmologica)*, vol 62. Kluwer Academic Publishers, Dordrecht, The Netherlands, pp 37.

Delori FC, Dorey CK, Staurenghi G, Arend O, Goger DG, Weiter JJ, 1995a, In vivo fluorescence of the ocular fundus exhibits retinal pigment epithelium lipofuscin characteristics. *Invest Ophthalmol Vis Sci.* **36:**718.

Delori FC, Goger DG, Dorey CK, 2001, Age-related accumulation and spatial distribution of lipofuscin in RPE of normal subjects. *Invest Ophthalmol Vis Sci.* **42:**1855.

Delori FC, Staurenghi G, Arend O, Dorey CK, Goger DG, Weiter JJ, 1995b, In vivo measurement of lipofuscin in Stargardt's disease--Fundus flavimaculatus. *Invest Ophthalmol Vis Sci.* **36:**2327.

Dorey CK, Wu G, Ebenstein D, Garsd A, Weiter JJ, 1989, Cell loss in the aging retina. Relationship to lipofuscin accumulation and macular degeneration. *Invest Ophthalmol Vis Sci.* **30:**1691.

Eldred GE, 1998 Lipofuscin and other lysosomal storage deposits in the retinal pigment epithelium. In: Marmor MF, Wolfensberger TJ (eds) *The Retinal Pigment Epithelium: Function and Disease*. Oxford University Press, New York, pp 651.

Eldred GE, Katz ML, 1988, Fluorophores of the human retinal pigment epithelium: separation and spectral characterization. *Exp Eye Res.* **47:**71.

Eldred GE, Lasky MR, 1993, Retinal age pigments generated by self-assembling lysosomotropic detergents. *Nature* **361:**724.

Feeney-Burns L, Hilderbrand ES, Eldridge S, 1984, Aging human RPE: morphometric analysis of macular, equatorial, and peripheral cells. *Invest Ophthalmol Vis Sci.* **25:**195.

Gaillard ER, Zheng L, Merriam JC, Dillon J, 2000, Age-related changes in the absorption characteristics of the primate lens. *Invest Ophthalmol Vis Sci.* **41:**1454.

Ham WTJ, Allen RG, Feeney-Burns L, Marmor MF, Parver LM, Proctor PH, Sliney DH, Wolbarsht ML (eds), 1986, The involvement of the retinal pigment epithelium. CRC Press, Inc, Boca Raton.

Ham WTJ, Mueller HA, Ruffolo JJJ, Millen JE, S.F. C, Guerry RK, Guerry DI, 1984, Basic mechanisms underlying the production of photochemical lesions in the mammalian retina. *Curr Eye Res.* **3:**165.

Ham WTJ, Ruffolo JJJ, Mueller HA, Clarke AM, Moon ME, 1978, Histologic analysis of photochemical lesions produced in rhesus retina by short-wavelength light. *Invest Ophthalmol Vis Sci.* **17:**1029.

Holz FG, Bellman C, Staudt S, Schutt F, Volcker HE, 2001, Fundus autofluorescence and development of geographic atrophy in age-related macular degeneration. *Invest Ophthalmol Vis Sci.* **42:**1051.

Holz FG, Bellmann C, Margaritidis M, Schutt F, Otto TP, Volcker HE, 1999, Patterns of increased in vivo fundus autofluorescence in the junctional zone of geographic atrophy of the retinal pigment epithelium associated with age-related macular degeneration. *Graefe's Arch Clin Exp Ophthalmol.* **237:**145.

Katz ML, Drea CM, Robison WG, Jr., 1986, Relationship between dietary retinol and lipofuscin in the retinal pigment epithelium. *Mech Ageing Dev.* **35:**291.

Katz ML, Norberg M, Stientjes HJ, 1992, Reduced phagosomal content of the retinal pigment epithelium in response to retinoid deprivatioin. *Invest Ophthalmol Vis Sci.* **33:**2612.

Katz ML, Redmond TM, 2001, Effect of Rpe65 knockouton accumulation of lipofuscin fluorophores in the retinal pigment epithelium. *Invest Ophthalmol Vis Sci.* **42:**3023.

Kennedy CJ, Rakoczy PE, Constable IJ, 1995, Lipofuscin of the retinal pigment epithelium: a review. *Eye* **9:**763.

Kitagawa K, Nishida S, Ogura Y, 1989, In vivo quantitation of autofluorescence in human retinal pigment epithelium. *Ophthalmologica* **199**:116.

Liu IY, White L, LaCroix AZ, 1989, The association of age-related macular degeneration and lens opacities in the aged. *Am J Public Health* **79:**765.

Liu J, Itagaki Y, Ben-Shabat S, Nakanishi K, Sparrow JR, 2000, The biosynthesis of A2E, a fluorophore of aging retina, involves the formation of the precursor, A2-PE, in the photoreceptor outer segment membrane. *J Biol Chem.* **275:**29354.

Lopez PF, Maumenee IH, de la Cruz Z, Green WR, 1990, Autosomal-dominant fundus favimaculatus. Clinicopathologic correlation. *Ophthalmol.* **97** 798.

Mata NL, Tzekov RT, Liu X, Weng J, Birch DG, Travis GH, 2001, Delayed dark adaptation and lipofuscin accumulation in abcr+/- mice: implications for involvement of *ABCR* in age-related macular degeneration. *Invest Ophthalmol Vis Sci.* **42:**1685.

Mata NL, Weng J, Travis GH, 2000, Biosynthesis of a major lipofuscin fluorophore in mice and humans with ABCR-mediated retinal and macular degeneration. *Proc Natl Acad Sci U S A.* **97:**7154.

Okubo A, Rosa RHJ, Bunce CV, Alexander RA, Fan JT, Bird AC, Luthert PJ, 1999, The relationships of age changes in retinal pigment epithelium and Bruch's membrane. *Invest Ophthalmol Vis Sci.* **40:**443.

Parish CA, Hashimoto M, Nakanishi K, Dillon J, Sparrow JR, 1998, Isolation and one-step preparation of A2E and iso-A2E, fluorophores from human retinal pigment epithelium. *Proc Natl Acad Sci U S A.* **95:**14609.

Pollack A, Bukelman A, Zalish M, Leiba H, Oliver M, 1998, The course of age-related macular degeneration following bilateral cataract surgery. *Ophthalmic Surg Lasers* **29:**286.

Pollack A, Marcovich A, Bukelman A, Oliver M, 1996, Age-related macular degeneration after extracapsular cataract extraction with intraocular lens implantation. *Ophthalmology* **103:**1546.

Putting BJ, Van Best JA, Vrensen GFJM, Oosterhuis JA, 1994, Blue-light-induced dysfunction of the blood-retinal barrier at the pigment epithelium in albino versus pigmented rabbits. *Exp Eye Res* **58:**31.

Rabb MF, Tso MO, Fishman GA, 1986, Cone-rod dystrophy. A clinical and histopathologic report. *Ophthalmology* **93:**1443.

Ren RF, Sakai N, Nakanishi K, 1997, Total synthesis of the ocular age pigment A2E: a convergent pathway. *J. Am. Chem. Soc.* **119:**3619.

Sakai N, Decatur J, Nakanishi K, Eldred GE, 1996, Ocular age pigment "A2E": An unprecedented pyridinium bisretinoid. *J. Am. Chem. Soc.* **118:**1559.

Schutt F, Davies S, Kopitz J, Holz FG, Boulton ME, 2000, Photodamage to human RPE cells by A2-E, a retinoid component of lipofuscin. *Invest Ophthalmol Vis Sci.* **41:**2303.

Sparrow JR, Cai B, 2001, Blue light-induced apoptosis of A2E-containing RPE: involvement of caspase-3 and protection by Bcl-2. *Invest Ophthalmol Vis Sci* **42:**1356.

Sparrow JR, Nakanishi K, Parish CA, 2000, The lipofuscin fluorophore A2E mediates blue light-induced damage to retinal pigmented epithelial cells. *Invest Ophthalmol Vis Sci* **41:**1981.

Sparrow JR, Parish CA, Hashimoto M, Nakanishi K, 1999, A2E, a lipofuscin fluorophore, in human retinal

pigmented epithelial cells in culture. *Invest Ophthalmol Vis Sci* **40**:2988.
Sparrow JR, Zhou J, Ben-Shabat S, Vollmer H, Itagaki Y, Nakanishi K, 2002, Involvement of oxidative mechanisms in blue light induced damage to A2E-laden RPE. *Invest Ophthalmol Vis Sci.* **43:** 222.
Sparrow Jr, Zhou J, Cai B, 2003, DNA is a target of the photodynamic effects elicited in A2E-laden RPE by blue light illumination. *IOVS (in press)*
Sun H, Molday RS, Nathans J, 1999, Retinal stimulates ATP hydrolysis by purified and reconstituted ABCR, the photoreceptor-specific ATP-binding cassette transporter responsible for Stargardt disease. *J Biol Chem.* **274**:8269.
Taylor HR, West S, Munoz B, Rosenthal FS, Bressler SB, Bressler NM, 1992, The long-term effects of visible light on the eye [see comments]. *Arch Ophthalmol.* **110**:99.
von Rückmann A, Fitzke FW, Bird AC, 1997, Fundus autofluorescence in age-related macular disease imaged with a laser scanning ophthalmoscope. *Invest Ophthalmol Vis Sci.* **38**:478.
Weingeist TA, Kobrin JL, Watzke RC, 1982, Histopathology of Best's macular dystrophy. *Arch Ophthalmol.* **100**:1108.
Weiter JJ, Delori FC, Wing GL, Fitch KA, 1986, Retinal pigment epithelial lipofuscin and melanin and choroidal melanin in human eyes. *Invest Ophthalmol Vis Sci.* **27**:145.
Weng J, Mata NL, Azarian SM, Tzekov RT, Birch DG, Travis GH, 1999, Insights into the function of Rim protein in photoreceptors and etiology of Stargardt's disease from the phenotype in abcr knockout mice. *Cell* **98**:13.
West SK, Rosenthal FS, Bressler NM, Bressler SB, Munoz B, Fine SL, Taylor HR, 1989, Exposure to sunlight and other risk factors for age-related macular degeneration [see comments]. *Arch Ophthalmol* **107:** 875
Wing GL, Blanchard GC, Weiter JJ, 1978, The topography and age relationship of lipofuscin concentration in the retinal pigment epithelium. *Invest Ophthalmol Vis Sci.* **17**:601.

BRIGHT LIGHT INDUCES RETINAL DEGENERATION BY A TRANSDUCIN-INDEPENDENT MECHANISM

Elliott Brill[1], Sujatha Patnala[1], Janis Lem[1], and Martin Obin[2]

1. INTRODUCTION

Retinitis pigmentosa (RP) represents a collection of genetically diverse inherited retinal degenerations involving mutations in more than 140 different genes. Despite mutational differences, end stage photoreceptor cell death occurs by apoptotic mechanisms (Chang et al., 1993; Portera-Cailliau et al., 1994; Reme et al., 1998). Blocking apoptotic photoreceptor cell death might provide a therapeutically useful treatment regimen. However, relatively little is known about the apoptotic pathway or the pathological changes that trigger apoptosis in the course of the disease process.

The "equivalent light hypothesis" (Fain and Lisman, 1999; Fain and Lisman, 1993; Lisman and Fain, 1995) suggests that apoptotic cell death in a subset of retinal degenerations is mediated by excessive phototransduction signaling. In support of the hypothesis, mutations in the gene encoding rhodopsin have been identified that constitutively activate phototransduction signaling (Rao and Oprian, 1996; Rim and Oprian, 1995; Sieving et al., 2001). In addition, mutations in rhodopsin kinase (Cideciyan et al., 1998; Khani et al., 1998) or arrestin (Nakazawa et al., 1998; Xu et al., 1997) producing defects in the termination of the photoresponse have been associated with human retinal degenerations. These mutations might cause degeneration by abnormally prolonging phototransduction signaling.

The purpose of these experiments was to test the validity of the "equivalent light hypothesis" in light damage models of retinal degeneration. Exposure of rodents to continuous or bright cyclic light leads to retinal degeneration (Noell and Albrecht, 1971; Noell et al., 1966). We wanted to determine if rhodopsin photolysis alone, or photolysis coupled to transducin signaling caused degeneration (Figure 1). To answer this question, we exposed transgenic mice deficient for the rod photoreceptor G-protein α-transducin subunit ($Tr_{\alpha}^{-/-}$) (Calvert et al., 2000) to various light conditions. These animals showed

[1]Tufts-New England Medical Center, Boston, MA 02111, USA; [2]Jean Mayer United States Department of Agriculture-Human Nutrition Research Center on Aging at Tufts University, Boston, MA 02111, USA. Corresponding author: J. Lem, E-mail: jlem@lifespan.org

minimal changes in gross retinal morphology up to 18 months of age. However in the absence of the rod transducin α-subunit, rhodopsin signaling was uncoupled from the activation of the photoreceptor effector molecule, cyclic GMP phosphodiesterase and the mice recorded no detectable rod ERG responses.

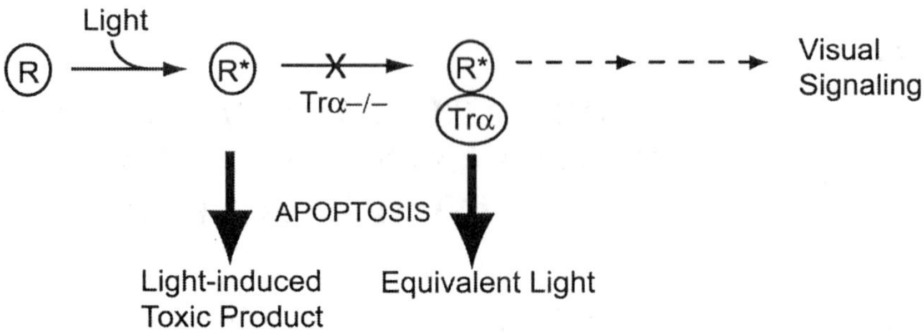

Figure 1. Apoptotic photoreceptor cell pathways. Light activation of rhodopsin may lead to apoptotic cell death by the production of a light-induced toxic product or by transducin-mediated cell signaling.

Cone function was unaffected (Calvert et al., 2000). If the equivalent light hypothesis is valid in the light damage paradigm, one would predict that Tr_{α}-/- mice would be protected from light-induced damage in comparison to wild type (WT) mice exposed to the same light conditions.

Alternatively, if comparable degenerative changes are observed in WT and Tr_{α}-/- mice, this would suggest that rhodopsin photolysis in the absence of visual signaling is sufficient to cause retinal degeneration. In these studies, we show that Tr_{α}-/- mice undergo degeneration comparable to that observed in WT mice with long-term exposure to bright cyclic light, suggesting that under these light damage conditions, photoreceptor degeneration does not require transducin signaling.

Previous studies in rats have demonstrated that susceptibility to light-induced photoreceptor degeneration can be reduced by "pre-conditioning" exposure to bright light (Li et al., 2001; Liu et al., 1998). Reductions in light-induced photoreceptor degeneration are likely to reflect the prolonged induction of neurotrophic factor expression that is observed in the retina in response to preconditioning (Liu et al., 1998). We wished to determine whether a similar adaptive response occurred in mice, and whether this protective adaptation required transducin signaling. Our studies showed that Tr_{α}-/- mice which were preconditioned in relatively bright, cyclic light were less susceptible to damage than mice that were not preconditioned. These results suggest that mice pre-exposed to bright cyclic light also undergo a protective adaptation response. This adaptive response occurs independently of transducin signaling.

2. RESULTS

To determine whether signal transduction is required for light-induced retinal degeneration, we compared light damage sensitivity of Tr_{α}-/- with WT mice by examining changes in retinal morphology under damaging light conditions. The lack of eye pigmentation in albino mice increases the amount of light entering the eye by approximately two orders of magnitude relative to that in pigmented animals (LaVail and Gorrin, 1987). Furthermore, albino mice reveal light damage on the order of hours, while pigmented mice can require weeks of light exposure before damage is observed. For these reasons, we examined retinal damage in albino Tr_{α}-/- mice derived from a 129SV (pigmented) x BALB/c (albino) founder.

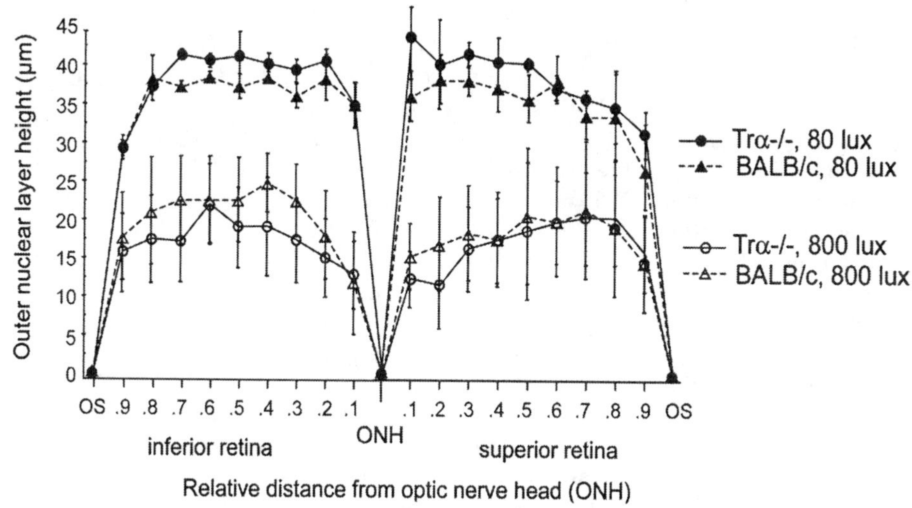

Figure 2. Outer nuclear layer height of Tr_{α}-/- and BALB/c mice exposed to 80 or 800 lux cyclic light for four months. Exposure to 800 lux caused comparable degeneration in both wild type BALB/c and Tr_{α}-/- mice.

2.1. 800 Lux Cyclic Light Exposure

We exposed WT and Tr_{α}-/- mice to 800 lux cyclic light (12hL:12hD) for four months, as we were interested in examining photoreceptor degeneration under physiologically relevant conditions[#]. Changes in outer nuclear layer (ONL) height, a method commonly used to quantify retinal degeneration, was evaluated. Light exposed WT mice showed a 40-50% reduction in ONL height, most significantly in the central and equatorial retina (Fig. 2). Light exposed Tr_{α}-/- mice exhibited slightly more damage in the inferior hemisphere than WT mice, although this difference was not statistically significant (P = 0.15). Thus, the presence or absence of transducin had no significant effect on the extent or pattern of light-induced ONL cell loss (P = 0.32 for effect of genotype). Reducing both exposure duration and intensity by half (450 lux cyclic light for eight weeks) still induced an overall 10-12% reduction in ONL height (mean ± sem) in bright light exposed Tr_{α}-/- mice (39.6 μm + 0.8μm) as compared with dim-light controls (44.6 μm + 0.5 μm; $P < 0.001$). Thus, the absence of transducin failed to protect mice against degeneration induced by long-term exposure to bright cyclic light.

[#] Conventional incandescent lighting is 300-500 lux

2.2. Light-damage Sensitivity of Light-adapted Tr_{α}-/- Mice

Light exposure history can influence susceptibility of the rat retina to light-induced photoreceptor degeneration. Both long-term rearing in cyclic bright light or brief (12h) pre-conditioning with intense light reduced the extent of retina damage induced by subsequent exposure to continuous bright light. This adaptive protective response was reversible by removing animals from the pre-conditioning light environment.

To determine whether light exposure history could increase resistance to light damage in Tr_{α}-/- mice, we compared sensitivity to light damage in dark-adapted and light preconditioned animals. Tr_{α}-/- mice were preconditioned by exposure to 300 lux cyclic light (12hL:12hD) for 3 days. Dark-adapted animals were exposed to no more than five lux cyclic red light (12hL:12hD) over a three day period. Both dark-adapted and light preconditioned animals were then exposed to 2,100 lux damaging light for 56 hours, then allowed to recover in either 300 (preconditioned animals) or five lux cyclic light (dark-adapted animals) for 40 hours. Preconditioned and dark-adapted animals were exposed to a second light damage insult consisting of 2,100 lux for 56 hours after which retinal morphology was assessed by light microscopy. Under these light conditions, both dark- and light-adapted mice were light damaged (Fig. 3). A normal mouse, unexposed to damaging light, possesses 10 to 12 rows of photoreceptor ONL. More than half the rows of ONL were lost in the central retina of light preconditioned or dark-adapted mice exposed to damaging light. However, dark-adapted mice showed a significantly more severe degeneration than did light-adapted mice, retaining only two to three rows of ONL in comparison to three to five rows of ONL in the central retina of light preconditioned animals ($P < 0.001$). These results suggest that a protective adaptive response was still functioning in Tr_{α}-/- mice. Thus, this protective adaptation mechanism does not require transducin signaling, but is likely to be modulated directly by light.

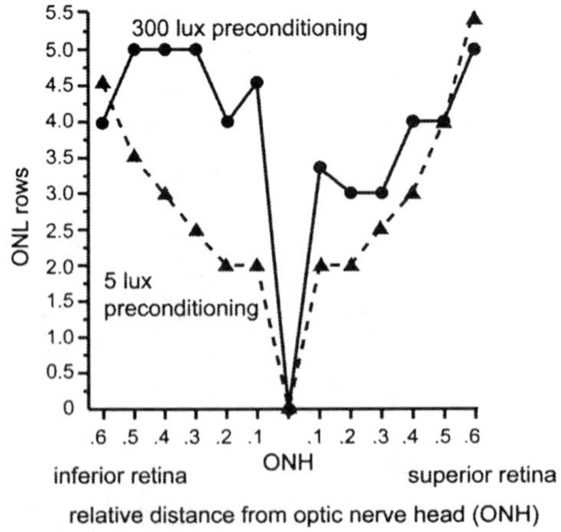

Figure 3. Resistance to light damage is observed in mice preconditioned with light prior to light damage insult.

3. DISCUSSION

The described studies were designed to rigorously test the validity of the equivalent light hypothesis. The results of our studies indicate that under the light exposure conditions used, photoreceptor degeneration does not occur by excessive phototransduction signaling. Degeneration was observed when animals were exposed to 800 lux cyclic light over a four-month period. Comparable degrees of degeneration were observed in Tr_{α}-/- and WT control animals (Fig. 2), demonstrating that degeneration does not require transducin-mediated signaling. Similarly, exposure of Tr_{α}-/- mice to eight weeks of light exposure at 450 lux caused photoreceptor cell degeneration. The results of these studies are consistent with results of Hao et al. (Hao et al., 2002), who showed that animals exposed to bright constant light undergo a transducin-independent degeneration involving induction of the AP-1 transcription factor. Hao et al. further described a transducin-dependent degenerative pathway that was utilized at low light intensities (Hao et al., 2002) in mice carrying genetic defects in rhodopsin shut-off. It is unlikely that we would detect this transducin-dependent pathway in our experiments, since the light intensities used are 40- to 100-fold higher than the low light intensities used by Hao and collaborators.

A number of laboratories have documented light-induced adaptive response in rats that confer protection from subsequent exposure to constant bright light (Li et al., 2001; Organisciak et al., 1998). Specifically, extended rearing or short-term preconditioning with bright light can reduce the extent of photoreceptor cell death induced by subsequent exposure to continuous bright light. Depending upon the duration of preconditioning, protection may reflect induction of antioxidant enzymes, neurotrophic factors and stress response pathways(Li et al., 2001; Liu et al., 1998; Organisciak et al., 1998; Penn et al., 1987) as well as reduced quantal capture resulting from photostasis (Penn and Williams, 1986). We tested whether protective adaptation was utilized by mice, and whether transducin-signaling played a role in regulating the adaptive response to light. The results of our studies indicate that Tr_{α}-/- mice preconditioned with 300 lux cyclic light over a three day period also show some degree of protection. This suggests that Tr_{α}-/- mice, like rats, undergo an adaptive protective response when reared under cyclic bright light. Irrespective of the biochemical features of this adaptation, our data suggest that this adaptive response occurs in the absence of transducin signaling.

4. MATERIALS AND METHODS

All procedures were carried out in accordance with the Association for Research in Vision and Ophthalmology statement concerning the use of animals in ophthalmic and vision research. We maintained albino lines of Tr_{α}-/- mice in a mixed background of 129SV and BALB/c. Albino Tr_{α}-/- and BALB/c mice were reared from birth in dim (5-80 lux) diffuse, white fluorescent light on a 12 h/12h light/dark cycle.

Experiments were initiated with mice aged 6 to 8 weeks. We exposed mice in reflective cages to diffuse white fluorescent light (ultraviolet impermeable diffuser). Average illuminance was measured on the cage floor, where food and water were provided. We exposed albino Tr_{α}-/- mice and BALB/c controls for 4 months to 800 lux of diffuse, cool, white fluorescent light, and maintained age-matched Tr_{α}-/- and BALB/c control mice in cyclic light at or below 80 lux. Temperatures were maintained at about 25 C during light exposure. We collected retinas immediately after light treatments (or after a recovery period in darkness) for histology and morphometry. Light exposures for preconditioned and dark-adapted control mice are as shown (Fig. 3).

We perfused Tr_{α}-/- and WT mice and enucleated and fixed eyes for 2 hours in formalin, glacial acetic acid and 70% ethanol (volume ratio 2:1:20). After removal of cornea and lens, eyecups were fixed overnight, processed and embedded in paraffin. We cut sections (4 um) along the vertical meridian and stained them with hematoxylin and eosin. The ONL thickness was measured by light microscopy at several equidistant loci in both the inferior and superior retinal hemispheres beginning at the optic nerve head. We assessed ONL thickness data by repeated-measures analysis of variance, using full interaction models and protected tests for multiple comparisons (Systat 9).

5. ACKNOWLEDGEMENTS

We thank F. Celestin of the SCOR in Ischemic Heart Disease Histology Core for assistance with histological sections. Support for this research was provided by a Fight For Sight Student Fellowship to E.B., the US National Institutes of Health (J.L. and M.S.O.), Massachusetts Lions Eye Research Fund (an institutional grant to the Department of Ophthalmology, New England Medical Center), and a Research to Prevent Blindness Special Scholar's Award (J.L.).

6. REFERENCES

Calvert P.D., Krasnoperova N.V., Lyubarsky A.L., Isayama T., Nicolo M., Kosaras B., Wong G., Gannon K.S., Margolskee R.F., Sidman R.L., Pugh E.N., Jr. Makino C.L., and Lem J., 2000, Phototransduction in transgenic mice after targeted deletion of the rod transducin alpha -subunit. *Proc Natl Acad Sci USA.* **97**:13913-8.

Chang G.-Q., Hao Y., and Wong F., 1993, Apoptosis: Final common pathway of photoreceptor death in rd, rds, and rhodopsin mutant mice. *Neuron* **11**:595-605.

Cideciyan A.V., Zhao X., Nielsen L., Khani S.C., Jacobson S.G., and Palczewski K., 1998, Null mutation in the rhodopsin kinase gene slows recovery kinetics of rod and cone phototransduction in man. *Proc Natl Acad Sci USA.* **95**:328-33.

Fain G.L., and Lisman J.E., 1999, Light, Ca2+, and photoreceptor death: new evidence for the equivalent-light hypothesis from arrestin knockout mice. *Invest Ophthalmol Vis Sci.* **40**:2770-2.

Fain G.L., and Lisman J.E., 1993, Photoreceptor degeneration in vitamin A deprivation and retinitis pigmentosa: the equivalent light hypothesis. *Exp Eye Res.* **57**:335-40.

Hao W., Wenzel A., Obin M.S., Chen C.-K., Brill E., Krasnoperova N.V., Eversole-Cire P., Kleyner Y., Taylor A., Simon M.I., Grimm C., Reme C.E., and Lem J., 2002, Evidence for two apoptotic pathways in light-induced retinal degeneration. *Nature Genetics* **32**:254-260.

Khani S.C., Nielsen L., and Vogt T.M., 1998, Biochemical evidence for pathogenicity of rhodopsin kinase mutations correlated with the Oguchi form of congenital stationary night blindness. *Proc Natl Acad Sci USA.* **95**:2824-7.

LaVail M.M., and Gorrin G.M., 1987, Protection from light damage by ocular pigmentation: Analysis using experimental chimeras and translocation mice. *Exp Eye Res.* **44**:877-889.

Li F., Cao W., and Anderson, R.E., 2001, Protection of photoreceptor cells in adult rats from light-induced degeneration by adaptation to bright cyclic light. *Exp Eye Research* **73**:569-577.

Lisman J., and Fain G., 1995, Support for the equivalent light hypothesis for RP *Nat Med.* **1**:1254-5.

Nakazawa M., Wada Y., and Tamai M., 1998, Arrestin gene mutations in autosomal recessive retinitis pigmentosa. *Arch Ophthalmol.* **116**:498-501.

Noell W.K., and Albrecht R., 1971, Irreversible effects of visible light on the retina: Role of vitamin A. *Science* **172**:76-80.

Noell W.K., Walker V.S., Kang B.S., and Berman S., 1966, Retinal damage by light in rats. *Invest Ophthalmol.* **5**:450-73.

Organisciak D.T., Darrow R.M., Barsalou L., Darrow R.A., Kutty R.K., Kutty G., and Wiggert B., 1998, Light history and age-related changes in retinal light damage. *Invest Ophthal Vis Sci.* **39**:1107-1116.

Penn J.S., Naash M.I., and Anderson R.E., 1987, Effects of light history on retinal antioxidants and light damage susceptibility in the rat. *Exp Eye Res.* **44**:205-15.

Portera-Cailliau C., Sung C.-H., Nathans J., and Adler R., 1994, Apoptotic photoreceptor cell death in mouse models of retinitis pigmentosa. *Proc Natl Acad Sci. USA.* **91**:974-978.

Rao V.R., and Oprian D.D., 1996, Activating mutations of rhodopsin and other G protein-coupled receptors. *Annu Rev Biophys Biomol Struct.* **25**:287-314.

Reme C.E., Grimm C., Hafezi F., Marti A., and Wenzel A., 1998, Apoptotic cell death in retinal degenerations. *Prog in Retin Eye Res.* **17**:443-464.

RetNet; The Univ. of Texas Health Sci. Ctr., 2002, http://www.sph.uth.tmc.edu/Retnet/disease.htm

Rim J., and Oprian D.D., 1995, Constitutive activation of opsin: interaction of mutants with rhodopsin kinase and arrestin. *Biochemistry* **34**:11938-45.

Sieving P.A., Fowler M.L., Bush R.A., Machida S., Calvert P.D., Green D.G., Makino C.L., and McHenry C.L., 2001, Constitutive "light" adaptation in rods from G90D rhodopsin: a mechanism for human congenital nightblindness without rod cell loss. *J Neurosci.* **21**:5449-60.

Xu J., Dodd R.L., Makino C.L., Simon M.I., Baylor D.A., and Chen J., 1997, Prolonged photoresponses in transgenic mouse rods lacking arrestin. *Nature* **389**:505-9.

DOES CONSTITUTIVE PHOSPHORYLATION PROTECT AGAINST PHOTORECEPTOR DEGENERATION IN *RPE65*$^{-/-}$ MICE?

Baerbel Rohrer[1,2], Zsolt Ablonczy[1], Sergei Znoiko[1], Michael Redmond[3], Jian-Xing Ma[1], and Rosalie Crouch[1].

1. INTRODUCTION

Photoreceptor development and survival is dependent on, in part, normal function of the light-induced signal transduction cascade. Over- or underproduction as well as mutations of components of the rod signaling transduction cascade have been reported to result in photoreceptor degeneration (e.g., Tan et al., (2001); Lem et al., (1999); Calvert et al., (2000); Redmond et al., (1998); Farber et al., (1988); Li et al., (1995)). One hypothesis that has been proposed by Fain and Lisman (1993) is that if these changes lead to continuous photoreceptor activation this pathogenic overstimulation of photoreceptors could cause photoreceptor degeneration. Free opsin (bleached opsin) is known to activate the photoreceptor signal transduction cascade at some small fraction of the activity of normal activated rhodopsin (Rh*) ($\sim 1x10^{-6}$), and it has therefore been suggested that this activity might lead to photoreceptor degeneration.

RPE65 is a major protein in the retinal pigment epithelium (RPE) (Hamel et al., 1993), where it is required for the regeneration of 11-*cis* retinal, the native ligand of cone and rod opsins (Redmond et al., 1998). Therefore, the photoreceptors of *Rpe65*$^{-/-}$ mice are almost completely depleted of 11-*cis* retinal, resulting in minimal levels of rhodopsin and the deterioration of their photosensitivity (Redmond et al., 1998). Despite a slow degeneration of the retina, photoreceptor function can be partially restored by supplying exogenous ligand (Van Hooser et al., 2000; Ablonczy et al., 2002), even as old as 18 months-of-age (Rohrer et al., 2002). Thus, the *Rpe65*$^{-/-}$ mouse is an excellent model in which to study three key factors that control the activity of rhodopsin, the supply of the ligand (e.g., Crouch and Ma, (2000); McBee et al., (2001)), the level of rhodopsin/opsin phosphorylation (e.g., Hurley et al., (1998); Mendez et al., (2000)) and the role of free opsin (e.g., Jones et al., (1996)).

[1]Departments of Ophthalmology and [2]Physiology and Neuroscience, Medical University of South Carolina, Charleston, SC; and [3]Laboratory of Retinal Cell and Molecular Biology, National Eye Institute, National Institutes of Health, Bethesda, MD, USA. Corresponding author: B. Rohrer, E-mail: rohrer@musc.edu

2. RESULTS

2.1. Retinal degeneration in RPE65$^{-/-}$ mice

Katz and Redmond (2001) reported previously that 1-year-old *Rpe65*$^{-/-}$ mice lose ~30% of their photoreceptor nuclei in the outer nuclear layer (ONL). We confirmed and extended this observation by analyzing 1.5-year-old *Rpe65*$^{-/-}$ mice, using both toluidine blue staining to analyze overall retinal morphology and immunohistochemistry to analyze single-cell morphology in 7 μm paraffin sections. As shown in Figure 1, the thickness of the ONL and the length of the inner (IS) and outer segments (OS) are drastically reduced in the aged *Rpe65*$^{-/-}$ mouse retina (Figure 1A, B). By 18 months-of-age, the ONL thickness and the IS/OS length are both reduced by ~50% in comparison to the 1-month-old *Rpe65*$^{-/-}$ retina (p = 0.0003 and 0.007, respectively) (see also Rohrer et al., (2002)). These anatomical changes correlated well with the changes in the measurable amount of regenerable rhodopsin levels (rho_{max}); defined as the total amount of rhodopsin regenerated from opsin using an excess of 11-*cis* retinal, measured using absorption difference spectrophotometry) (see also Rohrer et al., (2002)). On the other hand, it is generally assumed that the inner retinal cells do not suffer as a consequence of photoreceptor death. However, when transverse sections were stained with an antibody against protein kinase C (α and β subspecies) to label rod bipolar cells, (Figure 1C, D; PKC antibody 1:100; Amersham, Arlington Heights, IL; (Grunert et al., 1994)), the staining pattern revealed that by 18 months-of-age, which is late in the process of photoreceptor degeneration, rod bipolar cell dendrites sprouted into the outer nuclear layer (arrows in Figure 1D). Based on the ERG results from the *Rpe65*$^{-/-}$ mice, where the bipolar cell response is augmented (see Ablonczy et al., (2002) and Ekesten et al., (2001)) for further discussion), this sprouting response might be the morphological basis. These changes are different from results reported in the *rd* mouse, where the rapid photoreceptor degeneration leads to a reduction in rod bipolar cell dendrites (Strettoi et al., 2002) and concomitant changes in electroretinogram recordings (ERGs) that result

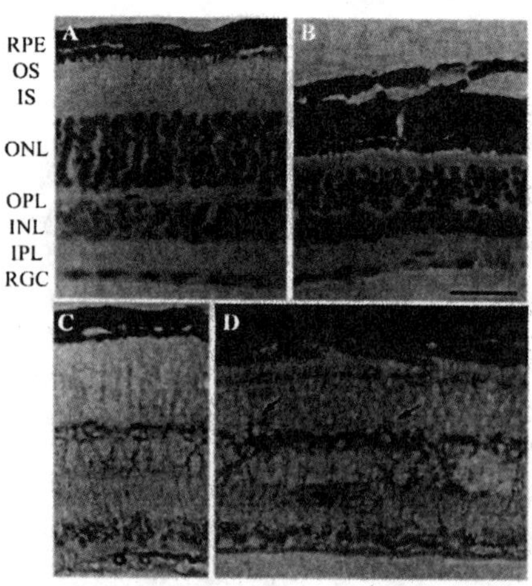

Figure 1. Retina structure in young and aged *Rpe65*$^{-/-}$ mice. Eyes from 1-month-old (**A, C**) and 18-month-old *Rpe65*$^{-/-}$ mice (**B, D**) were sectioned in transverse orientation. Thickness of retinal layers were compared in toluidine blue stained sections (**A, B**), and rod bipolar cell morphology was analyzed in sections stained for protein kinase C (**C, D**). Please note the dendritic sprouting in the bipolar cells of the aged animals (arrows in D). Abbreviations: INL, inner nuclear layer; IPL, inner plexiform layer; IS, inner segments; ONL, outer nuclear layer; OPL, outer plexiform layer; OS, outer segments; RGC, retinal ganglion cell layer; RPE, retinal pigment epithelium. The scale bar represents 25 μm. (Panels A and B, adapted from Rohrer et al., (2002); © Association for Research in Vision and Ophthalmology.)

from an impairment of bipolar cells. On the other hand, our results appear to be similar to data reported by Jansen and co-workers (1997), who showed in mouse models of slow retinal degeneration, that rod synaptic terminals get enlarged due to sprouting from the processes of the second order neurons. In addition, Aleman and co-workers (2001) have proposed that changes at the rod-bipolar cell synapse in early stages or retinal degeneration, augmenting the bipolar cell responses, are compensatory mechanisms for photoreceptor cell loss. These interesting observations in the *Rpe65*$^{-/-}$ mouse merit further investigation.

2.2. Electroretinogram recordings in *Rpe65*$^{-/-}$ mice

Photoreceptor function was assessed in single-flash ERGs after complete dark-adaptation, using recording conditions previously described in detail (Rohrer et al., 1999; Ablonczy et al., 2002). Although no endogenous levels of rhodopsin were measurable in the *Rpe65*$^{-/-}$ mice (detection limit: 0.225 pmol) (also see Ablonczy et al., (2002); Van

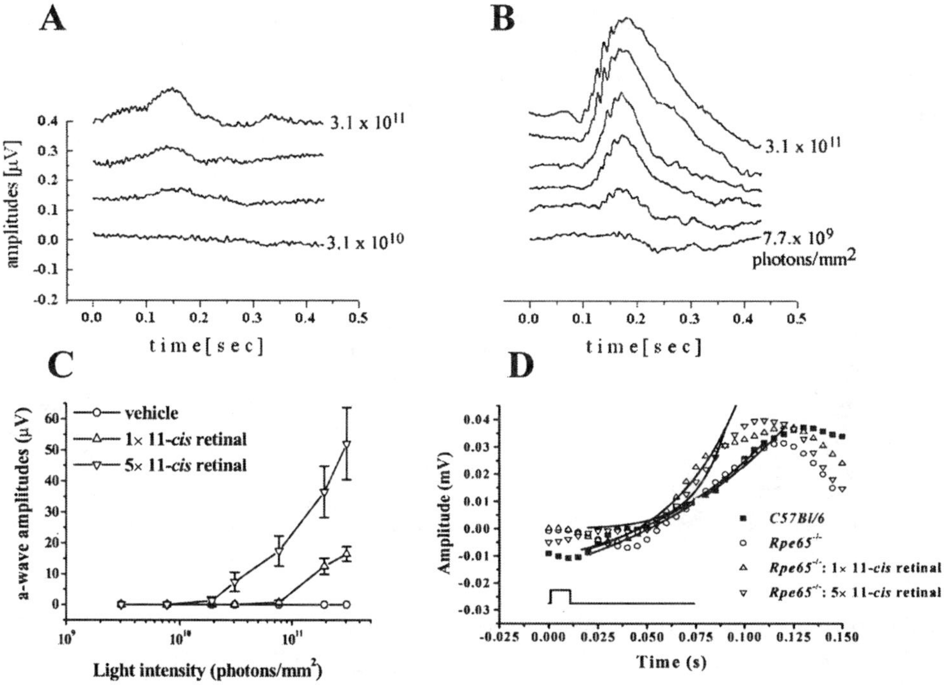

Figure 2. ERG analysis of untreated and 11-*cis* retinal injected *Rpe65*$^{-/-}$ mice. The *Rpe65*$^{-/-}$ photoreceptor responses only elicit a very small b-wave at maximum light intensities (**A**), whereas if the *Rpe65*$^{-/-}$ mice were treated with 11-*cis* retinal, both a- and b-waves could be recorded (**B**). Light intensities in Panels **A** and **B** ranged from 3.1×10^{10} to 3.1×10^{11} and 7.7×10^{9} to 3.1×10^{11} photons/mm^2, respectively. A-wave amplitudes (C) of *Rpe65*$^{-/-}$ mice treated with 1× (up-triangles) and 5× (down-triangles) retinal increased in a dose-dependent way with treatment. Time constant of a single exponential function fitted to the leading edge of the scotopic b-wave responses at threshold (~35 µV in amplitude) were obtained (**D**). Time constants did not differ between WT and *Rpe65*$^{-/-}$ mice ($p > 0.5$), but shortened significantly upon 11-*cis* retinal treatment ($p < 0.05$ for all comparisons). (Adapted from Ablonczy et al., (2002); © American Society for Biochemistry and Molecular Biology.)

Hooser et al., (2000)), small ERGS, consisting of b-waves only, could be recorded in *Rpe65*$^{-/-}$ mice (Figure 2A). Upon the addition of exogenous 11-*cis* retinal (animals were injected intraperitoneally (IP) with 0.05 (1×) or 0.25 μg/g bodyweight (5×) of 11-*cis* retinal (1 mg/mL in 10% ethanol, 10% bovine serum albumin, and 0.9% NaCl) or vehicle, and were kept in the dark), ERG response amplitudes increased significantly in a dose-dependent manner (Figure 2B, C). At the 5×_ dose, we determined that 8.8 ±0.81% of wild type levels of rhodopsin were regenerated (Figure 3), resulting in the formation of ~1.4 x 10^{13} molecules of functional rhodopsin per retina. These biochemical results correlate well with the ERG results, resulting in an a-wave recovery of 11.9 ±2.6% of the wild type levels (Figure 2C).

2.3. Opsin structure and post-translational modification in *Rpe65*$^{-/-}$ mouse retina

Opsins from both wild type and *Rpe65*$^{-/-}$ mice were mapped using mass-spectrometry (MS) (Ablonczy et al., 2000), and no differences between the two could be determined (Ablonczy et al, 2002), with the exception of an increased phosphorylation level in the dark-adapted *Rpe65*$^{-/-}$ mice (Figure 3; phosphorylation levels were analyzed by digestion of whole retinas with the Asp-N enzyme, which cleaves the opsin C-terminus, and the measurement of the soluble fraction by MS). 11-*cis* Retinal injections partially restored the endogenous rhodopsin levels (Figure 3, right-hand Y-axis) in a dose-dependent manner from undetectable to ~10%. Opsin phosphorylation was conversely partially reduced from 24 ±2% (vehicle injected) to 8 ±2% (5×) (Figure 3; left-hand Y-axis).

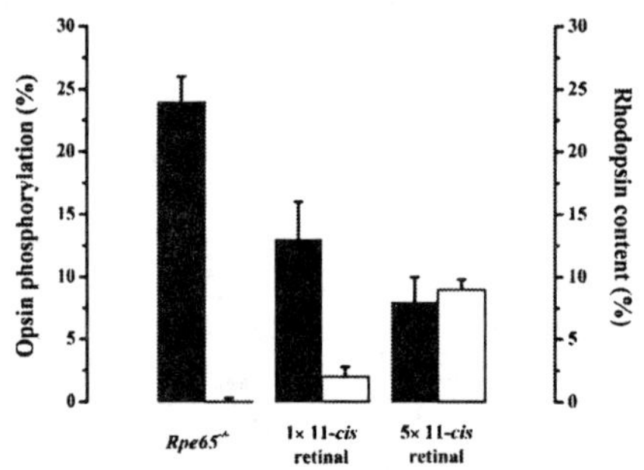

Figure 3. Levels of rhodopsin and opsin phosphorylation are dependent on 11-*cis* retinal. Rhodopsin levels (white bars) and opsin phosphorylation data (black bars) are expressed as % of the maximal amount. An inverse relation-ship exists between constitutive opsin phosphorylation and rhodopsin levels in relation to supplied 11-*cis* retinal. (Adapted from Ablonczy et al., (2002); © American Society for Biochemistry and Molecular Biology.

2.4. Opsin from *Rpe65*$^{-/-}$ mice is not constitutively active

Rpe65$^{-/-}$ mouse photoreceptors contain mostly opsin, which has never been exposed to its ligand ("virgin opsin") and was found to be constitutively phosphorylated, as well as minimal levels of rhodopsin. It was expected that low levels of rhodopsin should result in small ERGs; however, the influence of the virgin opsin on the photoreceptor response could not be predicted. From the literature it is know that bleached opsin can activate the photoreceptor signal transduction cascade with low efficiency (Cornwall and Fain, 1994), resulting in bleaching adaptation and increased response kinetics (Jones et al., 1996). Yet, when free opsin is phosphorylated is not constitutively active (Li et al., 1995). On the other hand, Ratto and co-workers (1991) presented data that suggest that free opsin that has never been exposed to its ligand, does not activate transducin. These researchers showed that in the rat, at postnatal day 13 (P13) when only 10% of their opsin is in the form of rhodopsin, whereas the remaining opsin is free opsin (virgin opsin) that has never been exposed to its ligand, the single photon responses were indistinguishable from those of adult animals. If the free opsin were contributing to bleaching adaptation, as in light-adapted rods, the single photon response would have been expected to be both reduced in amplitude and duration (Schneeweis and Schnapf, 2000).

The question we wished to investigate is the activity state of the free opsin in the *Rpe65*$^{-/-}$ rods. Using single flash ERGs, only very small b-wave responses could be recorded, which had rise times typical of dark-adapted ERGs (Figure 2D; b-wave kinetics were obtained from the time constant of a single exponential function fitted to the leading edge of the scotopic b-wave responses at threshold (~35 μV in amplitude), arguing that the underlying rod reponse kinetics are also those of dark-adapted rods. Van Hooser and coworkers (2002; see their Figure 4C) using single-cell recordings also reported that the kinetics of the *Rpe65*$^{-/-}$ rods resemble those of a dark-adapted rod. Taken together, if the free opsin were to act like "bleached" opsin, activating transducin (Jones et al., 1996), one should see a very fast response as reported by several laboratories for both cold-blooded and warm-blooded vertebrates (e.g., (Schneeweis and Schnapf, 2000; Baylor and Hodgkin, 1974; Cornwall and Fain, 1994)).

Upon 11-*cis* retinal treatment, we found that the b-wave responses had significantly shorter time constants (24.15 ±3.8 msec; Figure 2D), when compared to either uninjected *Rpe65*$^{-/-}$ (38.29 ±1.46 msec) or wild type mice (39.14 ±3.2 msec; p <0.05 for all comparisons). Shorter b-wave time constants in rod ERGs are an indication of light-adaptation ((Naarendorp and Akula, 2002); and Laura Frishman, personal communication). Similarly, 9-*cis* retinal injections were found to increase single rod photoreceptor kinetics (Van Hooser et al., 2002). The results from the opsin phosphorylation measurements, in which we showed that 11-*cis* retinal partially reduced opsin phosphorylation, suggest that the dephosphorylation releases opsin that then acts like bleached opsin creating background adaptation, thus speeding up the rod response.

3. SUMMARY

Despite the presence of this virgin opsin, *Rpe65*$^{-/-}$ rods are behaving like dark-adapted rods. These results argue that opsin which has not been exposed to 11-*cis* retinal and is constitutively phosphorylated, does not generate the activity generally associated with the bleached apoprotein. However, increased light-independent activation of transducin (due to bleached opsin) could be demonstrated after the addition of exogenous 11-*cis* retinal. We hypothesize that free opsin in the *Rpe65*$^{-/-}$ rods does not cause degeneration of rods by constitutive activation of the phototransduction cascade; but rather rods may die due to other causes such as the impairment of RPE function due to excess unprocessed retinyl-esters in the RPE.

4. ACKNOWLEDGEMENTS

Funding was provided by NIH grants EY-13520, EY-04939, EY08239, EY-12231; Foundation Fighting Blindness; and an unrestricted grant to MUSC from Research to

Prevent Blindness, Inc., New York, NY. The authors thank Luanna Bartholomew for editorial assistance.

5. REFERENCES

Ablonczy, Z., Knapp, D. R., Darrow, R., Organisciak, D. T., and Crouch, R. K., 2000, Mass spectrometric analysis of rhodopsin from light damaged rats, *Mol. Vis.* **6:**109-15.

Ablonczy, Z., Crouch, R. K., Goletz, P. W., Redmond, T. M., Knapp, D. R., Ma, J.-x., and Rohrer, B., 2002, 11-*cis* retinal reduces constitutive opsin phosphorylation and improves quantum catch in retinoid deficient mouse rod photoreceptors, *J. Biol. Chem.* **277:**40491-8.

Aleman, T. S., LaVail, M. M., Montemayor, R., Ying, G., Maguire, M. M., Laties, A. M., Jacobson, S. G., and Cideciyan, A. V., 2001, Augmented rod bipolar cell function in partial receptor loss: An ERG study in P23H rhodopsin transgenic and aging normal rats, *Vision Res.* **41**:2779-97.

Baylor, D. A., and Hodgkin, A. L., 1974, Changes in time scale and sensitivity in turtle photoreceptors, *J. Physiol.* **242:**729-58.

Calvert, P. D., Krasnoperova, N. V., Lyubarsky, A. L., Isayama, T., Nicolo, M., Kosaras, B., Wong, G., Gannon, K. S., Margolskee, R. F., Sidman, R. L., Pugh, E. N., Jr., Makino, C. L., and Lem, J., 2000, Phototransduction in transgenic mice after targeted deletion of the rod transducin alpha -subunit, *Proc Natl. Acad. Sci. USA.* **97:**13913-8.

Cornwall, M. C., and Fain, G. L., 1994, Bleached pigment activates transduction in isolated rods of the salamander retina, *J. Physiol.* **480:**261-79.

Crouch, R. K., and Ma, J., 2000, The role of vitamin A in visual transduction, in: *Vitamin A and Retinoids: An Update of Biological Aspects and Clinical Applications*, Birkhauser Verlag, Basel, Switzerland, pp. 59-72.

Ekesten, B., Gouras, P., and Salchow, D. J., 2001, Ultraviolet and middle wavelength sensitive cone responses in the electroretinogram (ERG) of normal and Rpe65$^{-/-}$ mice, *Vision Res.* **41**:2425-33.

Fain, G. L., and Lisman, J. E., 1993, Photoreceptor degeneration in vitamin A deprivation and retinitis pigmentosa: the equivalent light hypothesis, *Exp. Eye Res.* **57:**335-40.

Farber, D. B., Park, S., and Yamashita, C., 1988, Cyclic GMP-phosphodiesterase of rd retina: Biosynthesis and content, *Exp. Eye Res.* **46:**363-74.

Grunert, U., Martin, P. R., and Wassle, H., 1994, Immunocytochemical analysis of bipolar cells in the macaque monkey retina, *J. Comp. Neurol.* **348:**607-27.

Hamel, C. P., Tsilou, E., Harris, E., Pfeffer, B. A., Hooks, J. J., Detrick, B., and Redmond, T. M., 1993, A developmentally regulated microsomal protein specific for the pigment epithelium of the vertebrate retina, *J. Neurosci. Res.* **34:**414-25.

Hurley, J. B., Spencer, M., and Niemi, G. A., 1998, Rhodopsin phosphorylation and its role in photoreceptor function, *Vision Res.* **38:**1341-52.

Jansen, H. G., Hawkins, R. K., and Sanyal, S., 1997, Synaptic growth in the rod terminals of mice after partial photoreceptor cell loss: A three-dimensional ultrastructural study. *Microsc. Res. Tech.* **35**:96-105.

Jones, G. J., Cornwall, M. C., and Fain, G. L., 1996, Equivalence of background and bleaching desensitization in isolated rod photoreceptors of the larval tiger salamander, *J. Gen. Physiol.* **108:**333-40.

Katz, M. L., and Redmond, T. M., 2001, Effect of Rpe65 knockout on accumulation of lipofuscin fluorophores in the retinal pigment epithelium, *Invest. Ophthalmol. Vis. Sci.* **42:**3023-30.

Lem, J., Krasnoperova, N. V., Calvert, P. D., Kosaras, B., Cameron, D. A., Nicolo, M., Makino, C. L., and Sidman, R. L., 1999, Morphological, physiological, and biochemical changes in rhodopsin knockout mice, *Proc. Natl. Acad. Sci. USA.* **96:**736-41.

Li, T., Franson, W. K., Gordon, J. W., Berson, E. L., and Dryja, T. P., 1995, Constitutive activation of phototransduction by K296E opsin is not a cause of photoreceptor degeneration, *Proc. Natl. Acad. Sci. USA.* **92:**3551-5.

McBee, J. K., Palczewski, K., Baehr, W., and Pepperberg, D. R., 2001, Confronting complexity: The interlink of phototransduction and retinoid metabolism in the vertebrate retina, *Prog. Retin. Eye Res.* **20**:469-529.

Mendez, A., Burns, M. E., Roca, A., Lem, J., Wu, L. W., Simon, M. I., Baylor, D. A., and Chen, J., 2000, Rapid and reproducible deactivation of rhodopsin requires multiple phosphorylation sites, *Neuron* **28:**153-64.

Naarendorp, F., and Akula, J., 2002, Rod-driven b-waves of the rat electroretinogram isolated on dime and moderately intense backgrounds, *Invest. Ophthalmol. Vis. Sci.* **CD:**Abstr. No. 1818.

Ratto, G. M., Robinson, D. W., Yan, B., and McNaughton, P. A., 1991, Development of the light response in neonatal mammalian rods, *Nature* **351:**654-7.

Redmond, T. M., Yu, S., Lee, E., Bok, D., Hamasaki, D., Chen, N., Goletz, P., Ma, J. X., Crouch, R. K., and Pfeifer, K., 1998, Rpe65 is necessary for production of 11-*cis*-vitamin A in the retinal visual cycle, *Nat. Genet.* **20**:344-51.

Rohrer, B., Korenbrot, J. I., LaVail, M. M., Reichardt, L. F., and Xu, B., 1999, Role of neurotrophin receptor TrkB in the maturation of rod photoreceptors and establishment of synaptic transmission to the inner retina, *J. Neurosci.* **19**:8919-30.

Rohrer, B., Znoyko, S., Ablonczy, Z., Goletz, P., Ma, X.-J., and Crouch, R., 2002, Correlation of regenerable opsin with rod ERG signal in RPE65-/- mice during development and aging, *Invest. Ophthalmol. Vis. Sci.* **43**:in press.

Schneeweis, D. M., and Schnapf, J. L., 2000, Noise and light adaptation in rods of the macaque monkey, *Vis. Neurosci.* **17**:659-66.

Strettoi, E., Porciatti, V., Falsini, B., Pignatelli, V., and Rossi, C., 2002, Morphological and functional abnormalities in the inner retina of the rd/rd mouse, *J. Neurosci.* **22**:5492-504.

Tan, E., Wang, Q., Quiambao, A. B., Xu, X., Qtaishat, N. M., Peachey, N. S., Lem, J., Fliesler, S. J., Pepperberg, D. R., Naash, M. I., and Al-Ubaidi, M. R., 2001, The relationship between opsin overexpression and photoreceptor degeneration, *Invest. Ophthalmol. Vis. Sci.* **42**589-600.

Van Hooser, J. P., Aleman, T. S., He, Y. G., Cideciyan, A. V., Kuksa, V., Pittler, S. J., Stone, E. M., Jacobson, S. G., and Palczewski, K., 2000, Rapid restoration of visual pigment and function with oral retinoid in a mouse model of childhood blindness, *Proc. Natl. Acad. Sci. USA.* **97**:8623-8.

Van Hooser, J. P., Liang, Y., Maeda, T., Kuksa, V., Jang, G. F., He, Y. G., Rieke, F., Fong, H. K., Detweiler, P. B., and Palczewski, K., 2002, Recovery of visual functions in a mouse model of Leber congenital amaurosis, *J. Biol. Chem.* **277**:19173-82.

LIGHT-INDUCED PHOTORECEPTOR DAMAGE TRIGGERS DNA REPAIR: DIFFERENTIAL FATE OF RODS AND CONES

M. Soledad Cortina, William C. Gordon, Walter J. Lukiw, and Nicolas G. Bazan[1]

1. SUMMARY

Intense light triggers cell death in cones and rods by two distinct mechanisms: rapid, necrosis-like cell death occurs in cones; prolonged, apoptotic death is displayed by rods. However, the areas of cone and rod loss coincide, suggesting some interrelationship. DNA fragmentation in rods occurs in two waves. Our evidence suggests a repair mechanism that triggers up-regulation of the enzyme DNA polymerase ß, implying a rod-specific mechanism that initially attempts DNA repair, but which subsequently fails, resulting in a second wave of damage leading to apoptotic rod cell death. This study suggests two different but coincidental pathways lead to photoreceptor death. Definition of these two pathways could suggest useful pharmacologic sites toward which compounds could be targeted to promote cell survival in damaging retinal degenerative diseases. In addition, maintenance or enhancement of existing in-house repair mechanisms could provide additional approaches for the rescue of neurons with stress-induced nuclear damage.

2. INTRODUCTION

Exposure to intense light causes photoreceptor death. Light-induced photoreceptor death is mediated by rhodopsin (Grimm et al., 2000), and the extent of its bleaching, as well as its regeneration and visual transduction proteins (Organisciak et al., 1991; Keller et al., 2001), determine the degree of damage. Changes in cell morphology appear. Following initial swelling of mitochondria there is gradual DNA fragmentation and cell breakdown in photoreceptors (Szczesny et al., 1995). DNA fragmentation, one of the

[1]Neuroscience Center of Excellence and Department of Ophthalmology, Louisiana State University Health Sci. Center, New Orleans, LA 70112, USA. Corresponding author: N. G. Bazan, E-mail: nbazan@lsuhsc.edu

Retinal Degenerations: Mechanisms and Experimental Therapy
Edited by LaVail *et al.*, Kluwer Academic/Plenum Publishers, 2003

hallmarks of apoptosis, can be demonstrated by DNA ladders and TUNEL labeling of photoreceptor nuclei, but this is predominately a rod response (Gordon et al., 2002). Cones comprise only 2.7% of the photoreceptive elements in rat retina. Thus, cone photoreceptors have not been extensively studied in light-damage models. In slowly developing degenerative retinal diseases that result in extensive rod loss, cones persist, finally dropping out many months later (Jimenez et al., 1996). However, a detailed explanation of the process of cone loss has not been described. Bright light-induced photoreceptor degeneration results in complete loss of rods *and* cones in the superior central region of the rat retina within 48 h (Gordon et al., 2002). This is not a slow degenerative effect, which suggests the possibility that the retinal response here may be different from that shown in previous studies. Cones do not persist. Moreover, there are fewer photoreceptors lost after 10 days than the number of cells exhibiting prior DNA fragmentation (Gordon et al., 2002). Thus, some photoreceptors, those that have entered an apoptotic pathway, escape, recovering from the initial light treatment. Studies (e.g., Szczesny et al., 1995) have demonstrated retinal recovery, but specific mechanisms have not been described. In this study we monitored the ultrastructure of rod and cone cell death, DNA fragmentation, and the repair enzyme DNA polymerase ß (Pol ß). This study indicates that rod and cone death can occur by two different pathways, and that there are mechanisms involved in the repair of DNA fragmentation, arresting the apoptotic process. Therefore, light-induced retinal damage up-regulates enzymes that repair DNA.

3. MATERIALS AND METHODS

3.1. Light-Damage Model

Grimm et al. (2000) have shown in rodents that light-induced photoreceptor death is mediated through green-sensitive (500 nm λ_{max}) rhodopsin. With this in mind, our light-damage model was developed to satisfy the following conditions: presentation of wavelengths across the absorbance spectrum of rhodopsin, exclusive photoreceptor damage, 50% cell loss, short light treatment, and rapidly completed damage. White fluorescent light supplies more than adequate wavelength coverage, 34% of which is absorbed by the green rod (500 nm λ_{max}) rhodopsin (figure 1A). With this model, 5-h light treatment results in approximately 50% of photoreceptors dropping out by about 48 h with little or no subsequent loss (figure 1B).

3.2. Light-Damage Protocol

Albino (male, Sprague-Dawley) rats were dark adapted for 48 h before exposure to 5 h of light. After treatment they were returned to darkness and killed at 6-h intervals for 5 days, or after 10 days. The light source consisted of an 8-light array of 10-inch circular fluorescent bulbs (22 W; Cool White, FC8T9/CW; General Electric, Fairfield, CT; 18 klux; 270 $\mu E\ m^{-2}\ s^{-1}$). Control rats were kept in the dark for 2 days then killed.

This study was approved by the LSUHSC Institutional Animal Care and Use Committee and was in accordance with the ARVO and National Institutes of Health guidelines for the use of animals.

3.3. Histology and TUNEL

A TdT-mediated fluorescein-12-dUTP nick-end labeling (TUNEL) kit (Promega, Madison, WI) was used as directed; all nuclei were counterstained with propidium iodide. To determine the number of TUNEL-positive nuclei, sequential fields at 150-μm intervals along a vertical meridian through the optic nerve were digitized and the number of labeled nuclei counted.

The extent of final photoreceptor loss was determined from the 10-day survival animal group. The surviving cells within the outer nuclear layer (ONL) were counted as described above, and retinal profiles, representing the thickness of the ONL, were obtained and compared with those of dark-control retinas.

3.4. Immunohistochemistry

Frozen sections were stained with an anti-rat monoclonal antibody against DNA Pol ß (Alpha Diagnostics International, San Antonio, TX) and an anti-rat monoclonal proliferating cell nuclear antigen (PCNA) antibody (Santa Cruz Biotechnology, Santa Cruz, CA) to show DNA repair and cell proliferation, respectively. This was followed by incubation with a biotinylated secondary antibody and visualized by the DAB reaction. Slides were examined by bright-field microscopy.

3.5. Western Blot

Detection of DNA Pol ß and PCNA by Western-blot analysis was performed as previously described (Gordon et al., 2002). Gels were probed with mouse monoclonal anti-DNA Pol ß (Alpha Diagnostics International, San Antonio, TX) and anti-PCNA (Santa Cruz Biotechnology, Santa Cruz, CA) primary antibodies. Bands for DNA Pol ß were detected at 39 kDa and for PCNA at 36 kDa.

3.6. DNA Laddering and PCR

The apoptotic DNA laddering kit was used as suggested (Roche Molecular Biochemicals, Indianapolis, IN). DNA ladders were visualized and the 180- to 200-bp bands quantified with UV light. For PCR analysis, total retinal DNA was isolated and purified. Analysis of rat retinal low-molecular-weight DNA fragments with rat SINE-specific primers (highly repetitive short interspersed nuclear elements) was performed as previously described (Gordon et al., 2002).

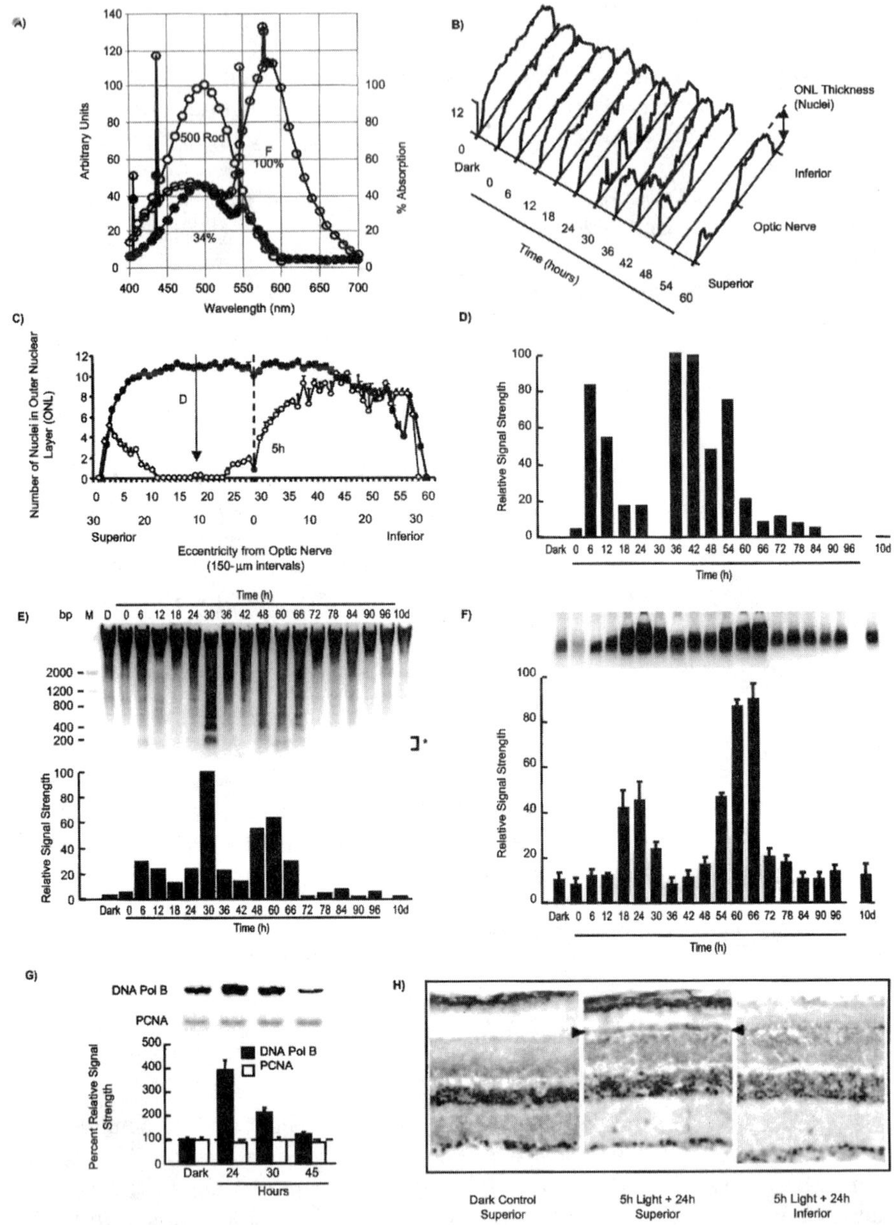

Figure 1. Characterization of the light-damage model (A-C); Photoreceptor DNA fragmentation time series (D-F); Light-induced DNA polymerase ß up-regulation and localization (G-H). A-H reprinted from Gordon et al., 2002 with permission. **A)** Emission spectrum of cool white fluorescent light (F, triangles); absorption curve for the rat 500 nm $_{max}$ rod rhodopsin (500 rod, open circles); absorption curve for this rhodopsin in fluorescent light (filled circles). The rat rod visual pigment absorbs 34% of the available fluorescent light (100%). **B)** Light damage retinal time course. Each curve is a single retinal profile of the thickness of the outer nuclear layer from

the superior margin, through the optic nerve, to the inferior margin. Profiles demonstrate cell loss in the superior central retina with dropout occurring at about 36-48 h after light treatment. No additional damage occurs. **C)** Retinal profile curves showing thickness of the outer nuclear layer (number of nuclei) in vertical meridian sections. Dark-control retinas (D, filled circles, n = 38) generally exhibit thicknesses of 12 nuclei. In 5-h light-treated retinas, after 10 days (5 h, open circles, n = 3), extensive cell loss (55.5%), centered within the superior central retina, occurred. Retinal profiles drop to zero at the optic nerve (dashed line). However, if sections are tangential to the optic nerve, some nuclei are present. Micrographs in figures 1H and 2A-D were obtained from the center of the light-damage region (arrow). **D)** Representative time course of TUNEL-labeled photoreceptor cell nuclei following 5 h of light treatment. Total numbers of labeled nuclei for each retina were counted and plotted. DNA fragmentation occurred in two waves that peaked approximately 24 h apart and were separated by a period of very low DNA damage. **E)** Representative time course of DNA fragmentation after 5 h of light. DNA gel ladders are shown at top with a low-DNA-mass ladder marker (M). Base pair (bp) fragment sizes are indicated. Density measurements of the 180- to 200-bp band of each lane (*) are plotted below. Two peaks of DNA fragmentation occur about 24 h apart. **F)** Time course of SINE analysis following 5 h of light. Rat highly repetitive SINE PCR of total retinal DNA was employed. SINE-specific primers were used to detect mononucleosome-sized and low-molecular-weight DNA fragments, which, when quantified, gave an estimate of the amount of DNA fragmentation. These values are plotted at the bottom (n = 3). One typical gel is shown above. Two peaks of DNA fragmentation were observed. **G)** Up-regulation of DNA polymerase ß following 5 h of light. Immunolabeled Western blot analysis of whole retinas, comparing 24, 30, and 45 h post-treatment with dark-control retinas (bottom), indicated a peak (four times dark values) at 24 h. Values were normalized so all dark controls equaled 100. A representative Western blot is shown at top. Proliferating cell nuclear antigen (PCNA, cyclin) was used as a standard and did not change. **H)** Immunohistochemical localization of DNA polymerase ß in 5-h light-treated retina. This repair enzyme was only observed in photoreceptor inner segments (indicated by arrowheads in middle image) in the superior central portion of the retina at 24 h post-light. No label was found at other times or in dark-control retinas. PCNA was not found.

4. RESULTS

4.1. Rod Response

4.1.1. Normal Distribution and Region of Cell Loss

Sections of dark-adapted control retinas were examined from the superior retinal margin, through the optic disk, to the inferior margin, and ONL thicknesses were plotted to produce retinal profiles. These consistently demonstrated thicknesses of 11-12 photoreceptor nuclei. Ten days after light treatment, photoreceptor nuclei had disappeared from the superior central "sensitive" region of the retina (figure 1C).

4.1.2. Ultrastructure

A time course of light-induced rod photoreceptor damage within the sensitive region was established by electron microscopy and compared to dark-control retinas (figure 2A). Immediately following light treatment, rod inner segment mitochondria were observed to have shortened and rounded up, but no other changes were observed in these cells. However, at 1 h post-light treatment, inner segment cytoplasm was darkened and the

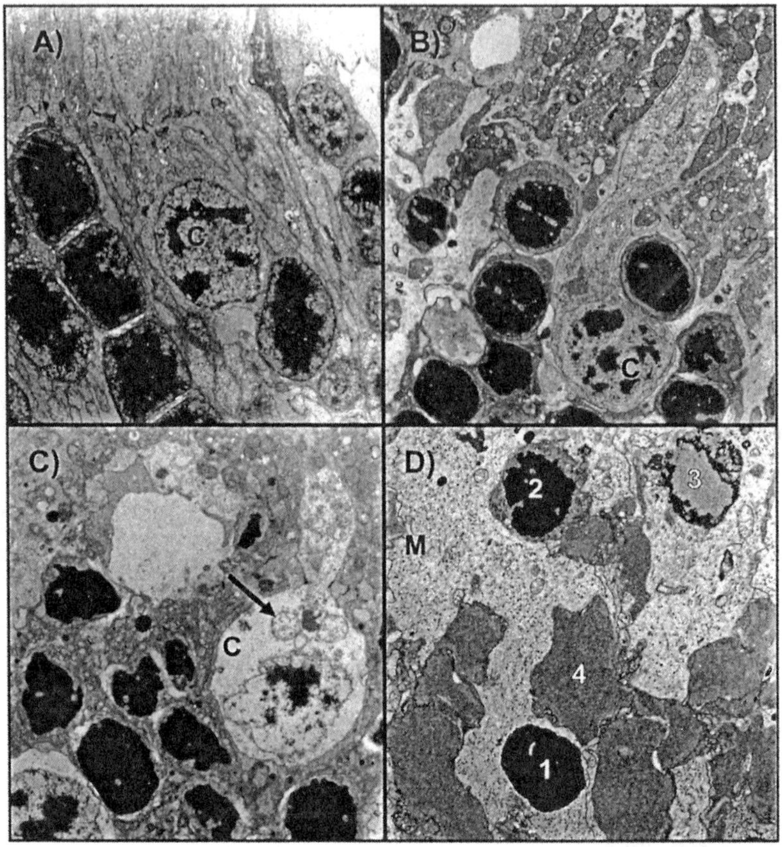

Figure 2. Electron micrographs of the outer nuclear layer (ONL) within the superior central sensitive region of the rat retina, demonstrating changes in cone and rod nuclei at different times following 5 h of light treatment. Micrographs are generally oriented with inner segments up and are shown at the same size (original magnification x3000). **A)** Dark-control retina showing a cone nucleus (C) with several typical dense regions within granular chromatin. Less than 3% of rat photoreceptor nuclei are cones and are localized within the upper third of the ONL; most occur just proximal to the outer limiting membrane. Rod nuclei are slightly smaller and very dense, surrounded by a thin border of lighter chromatin. These cells are organized into columns of about 12 nuclei. **B)** Photoreceptor nuclei shown 1 h after light treatment. A cone cell (C) is shown with a portion of the inner segment. Swelling has commenced. All mitochondria within the inner segments have rounded up, but those within the cone are more severely affected. Rod nuclei have rounded appearances with a slightly thicker surrounding region of lighter chromatin. **C)** At 6 h post-treatment cones (C) have become extremely swollen. Little ground cytoplasm remains and mitochondria (c.f. arrow) have no internal structure. The cone nucleus retains some of the dense chromatin but the less dense chromatin has dispersed. The nuclear membrane has begun to fragment. After 12 h no cones are present within this region of the retina. Rod nuclei, in general, appear irregular as the surrounding lighter chromatin thickens, but otherwise appear about the same. **D)** By 36 h post-light rod nuclei present several appearances, depending on the progression of apoptosis. Some, just beginning this process, appear as dense spherical packages (1), while others show an irregular profile as the lighter chromatin region thickens and begins to invade the denser area (2). Eventually, the central chromatin becomes lighter with remnants of the dense nuclear material forming a surrounding ring (3). Finally, only lighter material remains in very irregular nuclei that appear forced together (4). Plastic sections stained with toluidine blue will not reveal this last stage. During this process, the Müller cells (M) expand as photoreceptors disintegrate.

endoplasmic reticulum vesiculated, mitochondria were rounded and enlarged with distorted inner membranes, and nuclei appeared dark and rounded within dense perinuclear cytoplasm (figure 2B). Little change was noted at 2 h after light, but outer segments had begun to darken, and in some cases, to round up. By 3 h, inner segments were filled with swollen mitochondria, but nuclei appeared unchanged. At 6 h post-light, rod nuclei had small central light regions and an irregular outline (figure 2C), and at 18 h some nuclei had become light and granular, surrounded by a thin, irregular, electron-opaque ring. Many nuclei appeared as compressed polygons with thin, dark extensions between adjacent cells. Rod nuclei within the sensitive region became uniformly light, irregular, and occasionally pressed into each other by 36 h (figure 2D), and outer segments had become dense and irregular. Thereafter, most photoreceptor nuclei could not be found; only a few dense rod nuclei were observed.

4.1.3. DNA Fragmentation

Throughout these studies the fragmentation of DNA was observed. Only nuclei within the ONL became TUNEL-positive.

4.1.3a. TUNEL Response Time Course Fluorescing TUNEL-positive nuclei were counted (ONL thickness) every 150 μm along each retinal section for each time point after light treatment. Some photoreceptor nuclei first appeared TUNEL-positive 6 h post-treatment, but the height of labeling occurred in a biphasic manner, first peaking around 24 h, followed by another 24-30 h later (figure 1D). Fragmented or nicked photoreceptor DNA was concentrated around a region in the superior central retina, 1.2 mm from the optic nerve, and appeared only sparsely in the inferior retina. Following the second wave of nuclear damage, the number of TUNEL-positive cells declined rapidly.

4.1.3b. DNA Laddering Retinas from these time courses were also used to determine the extent of DNA fragmentation laddering. Comparisons of the 180- to 200-bp band densities, representative of nucleosomal fragments characteristic of apoptosis, with the same regions of dark-control retinas revealed one peak at 30 h and another about 24 h later (figure 1E).

4.1.3c. Low-Molecular-Weight DNA Fragmentation Low-molecular-weight retinal SINE DNA fragments from each time point were quantified and a time course for these nuclear fragments was generated (figure 1F). Increases in relative signal intensity of 5-fold and 10-fold over dark values were found at 24 and 60 h, respectively.

4.2. Cone Response

4.2.1. Normal Distribution and Region of Cell Loss

Cones represent only 2.7% of total rat photoreceptors (LaVail, 1976) and are dispersed throughout the retina with no regions of concentration. Cones are easily distinguished from rods, having nuclei with several isolated patches of densely staining chromatin within lighter, granular spheres, while rod nuclei contain a single, irregular, dense region surrounded by a thin band of lighter, granular material. Cone nuclei are distributed throughout the outer third of the ONL, tending to occupy the distal-most position just below the inner segments. Evaluation of retinal sections after light treatment demonstrates that cones near the superior margin and those within the inferior portion of the retina remain, while cones within the "sensitive" superior central region are lost after about 12 h; the retinal region of cone loss matches that of rods.

4.2.2. Ultrastructure

A time course of light-induced cone photoreceptor damage within the sensitive region was also determined by electron microscopy. Cone nuclei are larger than those of rods and contain only small, isolated regions of dense chromatin within an otherwise light granular material, while rod nuclei are smaller with very dense chromatin surrounded by a thin granular material. Cone nuclei are easily identified (figure 2A). Following the 5-h light treatment, inner segment mitochondria appeared shortened and rounded, and by 1 h inner segments appeared slightly swollen (figure 2B). By 2 h, the perinuclear cytoplasm and its mitochondria appeared slightly swollen, and at 3 h post-light, inner segments showed swelling and vesiculation of endoplasmic reticulum. Cone perinuclear cytoplasm continued to swell, becoming clear and empty, but nuclei remained as in dark controls. Mitochondria associated with cone nuclei were extremely swollen after 4 h, and the cytoplasm of inner segments and nuclear regions continued to swell. Finally, after 6 h, perinuclear regions appeared empty and the plasma membranes fragmented (figure 2C). By 12 h, electron-lucent, granular cone nuclei had begun to disperse. No cone photoreceptor nuclei could be identified 18 h after light treatment (figure 2D).

4.3. PCNA and DNA Polymerase ß localization

Retinas collected throughout the time course were examined for the presence of DNA Pol ß and PCNA. The nuclear polymerase increased briefly then declined, while PCNA remained constant. Western blots demonstrated that DNA Pol ß was elevated 4-fold above dark values at 24 h post-light treatment, declined to 2-fold by 30 h, and had almost returned to dark levels by 45 h (figure 1G). Immunohistochemistry of dark-control retinas and of 24, 30, and 45 h post-treatment retinas localized DNA Pol ß within photoreceptor inner segments, but only in the 24-h animals and only in the superior retina (figure 1H). Neither DNA Pol ß nor PCNA could be localized by immunohistochemistry at other time points.

5. DISCUSSION

It has been well established that photoreceptors can be damaged by bright light, with excessive duration or intensity triggering apoptosis. In rodents, damaged photoreceptors center around a superior central retinal region 1.2 mm from the optic nerve (Howell et al., 1982). The present study, as well as most others, have derived their data chiefly from this region. Retinal nuclear thickness profiles of the ONL along vertical sections through the light-sensitive region demonstrate the degree and extent of cell loss. By 48 h, after 5 h of light exposure, all photoreceptor nuclei within this area drop out; cell loss within the sensitive region was 100%, with only about 50% loss across the entire retinal section. This is similar to other studies in which these profiles are examined (e.g., Rapp and Smith, 1992).

Photoreceptor damage has been described at the ultrastructural level, but retinas are usually viewed only during the cell-death interval and after cell loss, and then compared with dark controls. However, several studies have utilized time courses to define morphologic changes (e.g., Szczesny et al., 1995) that demonstrate typical apoptotic figures of rods as endoplasmic reticulum vesiculates, mitochondria round up, DNA condenses and blebs, and cells break down. This process is prolonged and occurs within our light-treated retinas in a similar manner. However, very short times have not usually been examined.

Within the first 6 h following light damage, cone photoreceptors undergo a process very unlike that described for rods. Cones appear to become osmotically stressed; the perinuclear cytoplasm swells, and the plasma membranes rupture in a necrosis-like manner similar to that described for glutamate-induced hypothalamic neurons (Olney, 1971) and inner retinal neurons (Sisk and Kuwabara, 1985). Cone death occurs very rapidly while rod death is prolonged over several days.

The process and time course of this cone death subsequent to light damage are unlike the process of photoreceptor death described in models of retinal genetic defects, which occurs over an extended period of time. For example, the retinal degeneration mouse (*rd/rd*) contains a rod mutation in the ß subunit of cGMP-phosphodiesterase (Bowes et al., 1990), resulting in rod loss with some cone survival (LaVail et al., 1997) within the superior retina (Garcia-Fernandez et al., 1995). The pattern and rate of photoreceptor degeneration also vary among species and defects. Briefly, degeneration in the *rd/rd* mouse proceeds from central to peripheral retina (Carter-Dawson et al., 1978) but in the slow mouse (*rds*) it occurs in an opposite manner (Sanyal et al., 1980). The inferior retina degenerates fastest in the RCS rat (LaVail and Battelle, 1975), while the progressive rod-cone dystrophy (*prcd*) and cone degeneration (*cd*) dogs retain their peripheral retina (Aguirre and Acland, 1988). Conversely, light-damaged rats and mice can lose cells from the entire retina, but maximum loss occurs within the superior central region (Howell et al., 1982), suggesting that photoreceptor death as a result of a genetic mutation is different from death triggered by excessive light. Despite the occurrence of two very different degenerative pathways, the region of cone loss coincides with that of rod degeneration, implying in our model of light damage that rod and cone deaths are linked either through ocular optics or by cell signaling. Regardless, the pathways through which cone and rod cell death occur subsequent to light damage are different.

In our DNA-fragmentation studies we examined retinas at 6-h intervals following light treatment. Quantification of nuclear DNA SINE and DNA ladders, and analysis of sections by the TUNEL method throughout these time courses, revealed that nuclear DNA damage occurred in two waves, the peaks approximately 24 h apart. There are two possible explanations. First, there are two populations of photoreceptors that undergo nuclear fragmentation at different times. Because 97% of photoreceptors in rat retina are green-sensitive rods, it is unlikely that identical cell types would exhibit two different times for fragmentation. However, there is a large decrease in DNA damage following the first peak, and there is inadequate time for cells to have degenerated. If there is no cell loss and little DNA damage, fragmentation must have been repaired. Moreover, the total number of 24-and 48-h TUNEL-positive nuclei does not equal the number of cells lost by 10 days post-light treatment (Gordon et al., 2002). Thus, some TUNEL-positive nuclei must have been mended. The second possibility assumes one population of photoreceptors undergoing nuclear fragmentation twice. Again, the large decrease in label shortly following the first peak occurs prior to the onset of massive cell loss, suggesting this reduction occurs as a result of DNA repair. Both possibilities suggest the presence of a DNA-repair mechanism within photoreceptors, and the time course of damage and cell dropout demonstrates this to be a process of rods.

Photoreceptors can recover from damaging light if the stimulus is at or only slightly above the threshold for cell damage (Moriya et al., 1986). An early increase in single-strand DNA breaks was followed by delayed DNA fragmentation in mouse and gerbil ischemic brains, suggesting activation of a DNA-repair mechanism (Tobita et al., 1995). Finally, after bilateral common carotid artery occlusion in gerbils, by 3 days after 6 minutes of ischemia, CA3 hippocampal neurons were TUNEL-positive. However, by day 7, no TUNEL label remained and there was no decrease in cell numbers, implying that DNA had been repaired (Gordon et al., 1999).

There is also support for a biphasic light-induced response by retina. Single- and double-stranded DNA breaks were quantified in the presence and absence of dimethylthiourea, an antioxidant, following light. DNA quantification demonstrated a 10% cell loss after 24 h of light and a 40% dropout in the following 2 weeks in darkness. Some cells may recover from slight damage, and activated oxygen species may trigger the initial damage with apoptosis causing dropout during the second phase (Specht et al., 1999).

Our findings and those of others prompted our search for up-regulation of DNA polymerases. There are four major DNA-repair mechanisms that have been described, each utilizing a specific polymerase (Wood et al., 2001). However, only DNA Pol ß repairs short gaps (Wang and Korn, 1980) and is up-regulated during oxidative stress (Chen et al., 1998). Thus, the up-regulation and presence of Pol ß indicates that a repair process has been triggered. It is significant that it occurs mainly within the superior retina in photoreceptor inner segments only after 24 h post-light treatment. Moreover, there is a single peak of occurrence that coincides with the first wave of nuclear DNA fragmentation. Therefore, when photoreceptor DNA is first fragmented, a repair mechanism may be triggered and Pol ß up-regulated. This results in DNA nick repair and the retina becoming TUNEL-negative. However, this mechanism, apparently, does not persist; perhaps the repair genes are also damaged. Nuclear DNA continues to become damaged, but is not mended. We suggest that the second wave of fragmentation results in rod photoreceptor apoptotic cell death.

In summary, intense light triggers cell death in cones and rods by two very different mechanisms. Rapid, necrosis-like cell death occurs in cone photoreceptors within 6-12 h, while prolonged apoptotic death is displayed by rods. However, the retinal regions of cone and rod cell loss coincide, suggesting some interrelationship between the two processes. During rod cell death, DNA fragmentation occurs as a biphasic event. Our evidence, using our light-damage model, suggests the presence of a photoreceptor DNA-repair mechanism and is supported by the up-regulation of the repair enzyme DNA Pol ß, whose peak coincides only with the first TUNEL-positive fragmentation wave and which localizes in the region of subsequent photoreceptor cell loss at 24 h post-light treatment. This implies a rod-specific mechanism that initially attempts DNA repair, but subsequently fails, resulting in a second wave of fragmentation leading to apoptotic rod cell death. This study suggests two coincident retinal pathways that lead to photoreceptor death, and offers the opportunity to define separately the processes that cause cone and rod cell death. Maintenance or enhancement of in-house repair mechanisms can provide additional approaches for the rescue of neurons with nuclear stress-induced damage. In addition, definition of these two separate pathways could provide pharmacologic targets to promote cell survival in damaging retinal degenerative diseases.

6. ACKNOWLEDGMENT

The Authors' research was supported by United States Public Health Service grant R01EY05121 from the National Eye Institute, National Institutes of Health, Bethesda, MD; the Eye, Ear, Nose, and Throat Foundation of Louisiana; and the Ernest C. and Yvette C. Villere Chair for the Study of Retinal Degeneration.

7. REFERENCES

Aguirre, G. D. and Acland, G. M., 1988, Variation in retinal degeneration phenotype inherited at the *prcd* locus, *Exp. Eye Res.* **46**:663.

Bowes, C., Li, T., Danciger, M., Baxter. L. C., Applebury, M. L., and Farber, D. B., 1990, Retinal degeneration in the *rd* mouse is caused by a defect in the β subunit of rod cGMP-phosphodiesterase, *Nature.* **347**:677.

Carter-Dawson, L. D., LaVail, M. M., and Sidman, R. L., 1978, Differential effect of the *rd* mutation on rods and cones in the mouse retina, *Invest. Ophthalmol. Vis. Sci.* **17**:489.

Chen, K.-H., Yakes, F.M., Srivastava, D.K., Singhal, R.K., Sobol, R.W., Horton, J.K., Van Houten, B., and Wilson, S.H., 1998, Up-regulation of base excision repair correlates with enhanced protection against a DNA damaging agent in mouse cell lines, *Nucleic Acids Res.* **26**:2001.

García-Fernández, J. M., Jimenez, A. J., and Foster, R. G., 1995, The persistence of cone photoreceptors within the dorsal retina of aged retinally degenerate mice (*rd/rd*): Implications for circadian organization, *Neurosci. Lett.* **187**:33.

Gordon, W. C., Colangelo, V., Bazan, N. G., and Klatzo, I., 1999, *Aspects of the maturation phenomenon observed by the TUNEL method, in: Maturation Phenomenon in Cerebra Ischemia II,* Ito, U., Kirino, T., Kuroiwa, and T., Klatzo, I., eds., Springer-Verlag, Berlin, pp. 15-23.

Gordon, W. C., Casey, D. M., Lukiw, W. J., and Bazan, N. G., 2002, DNA damage and repair in light-induced photoreceptor degeneration, *Invest. Ophthalmol. Vis. Sci.* **43**:3511.

Grimm, C., Wenzel, A., Hafezi, F., Yu, S., Redmond, T. M., and Remé, C. E., 2000, Protection of Rpe65-deficient mice identifies rhodopsin as a mediator of light-induced retinal degeneration, *Nat. Genet.* **25**:63.

Howell, W. I., Rapp, L. M., and Williams, T. P., 1982, Distribution of melanosomes across the retinal pigment epithelium of a hooded rat: Implications for light damage, *Invest. Ophthalmol. Vis. Sci.* **22**:139.

Jimenez, A.-J., García-Fernández, J.-M., González, B., and Foster, R. G., 1996, The spatio-temporal pattern of photoreceptor degeneration in the aged *rd/rd* mouse retina, *Cell Tissue Res.* **284**:193.

Keller, C., Grimm, C., Wenzel, A., Hafezi, F., and Remé, C. E., 2001, Protective effect of halothane anesthesia on retinal light damage: Inhibition of metabolic rhodopsin regeneration, *Invest. Ophthalmol. Vis. Sci.* **42**:476.

LaVail, M. M., 1976, Survival of some photoreceptor cells in albino rats following long-term exposure to continuous light, *Invest. Ophthalmol.* **15**:64.

LaVail, M. M. and Battelle, B. A., 1975, Influence of eye pigmentation and light deprivation on inherited retinal distrophy in the rat, *Exp. Eye Res.* **21**:167.

LaVail, M. M., Matthes, M. T., Yasumura, D., and Steinberg, R. H., 1997, Variability in rate of cone degeneration in the retinal degeneration (*rd/rd*) mouse, *Exp. Eye Res.* **65**:45.

Moriya, M., Baker, B. N., and Williams, T. P., 1986, Progression and reversibility of early light-induced alterations in rat retinal rods, *Cell Tissue Res.* **246**:607.

Olney, J. W., 1971, Glutamate-induced neuronal necrosis in the infant mouse hypothalamus, *J. Neuropath. Exp. Neurol.* **30**:75.

Organisciak, D. T., Xie, A., Wang, H. M., Jiang, Y. L., Darrow, R. M., and Donoso, L. A., 1991, Adaptive changes in visual cell transduction protein levels: Effect of light, *Exp. Eye Res.* **53**:773.

Rapp, L. M. and Smith, S. C., 1992, Morphologic comparisons between rhodopsin-mediated and short-wavelength classes of retina light damage, *Invest. Ophthalmol. Vis. Sci.* **33**:3367.

Sanyal, S., DeRuiter, A., and Hawkins, R., 1980, Development and degeneration of retina in *rds* mutant mice: Light microscopy. *J. Comp. Neurol.* **194**:193.

Sisk, D. R. and Kuwabara, T., 1985, Histologic changes in the inner retina of albino rats following intravitreal injection of monosodium L-Glutamate, *Graefe's Arch. Clin. Exp. Ophthalmol.* **223**:250.

Specht, S., Leffak, M., Darrow, R. M., and Organisciak, D. T., 1999, Damage to rat retinal DNA induced *in vivo* by visible light, *Photochem. Photobiol.* **69**:91.

Szczesny, P. J., Munz, K., and Remé, C. E., 1995, Light damage in the rat retina patterns of acute lesions and recovery, in: *Cell and Tissue Protection in Ophthalmology,* Pleyer, Y., Schmidt, K., and Thiel, H-J., eds., series *New Strategies in Prevention and Therapy,* Schmidt, K., series ed., Hippokrates Verlag, Stuttgart, pp.163-175.

Tobita, M., Nagano, I., Nakamura, S., Itoyama, Y., and Kogure, K., 1995, DNA single-strand breaks in postischemic gerbil brain detected by *in situ* nick translation procedure, *Neurosci. Lett.* **200**:129.

Wang, T. S.-F., and Korn, D., 1980, Reactivity of KB cell deoxyribonucleic acid polymerases ✂ and ✍ with nicked and gapped deoxyribonucleic acid, *Biochemistry.* **19**:1782.

Wood, R. D., Mitchel, M., Sgouros, J., and Lindahl, T., 2001, Human DNA repair genes, *Science.* **291**;1284.

MITOCHONDRIAL DELETIONS IN NORMAL AND DEGENERATING RAT RETINA

Arturo Bravo-Nuevo[1], Neal Williams[1], Scott Geller[2], and Jonathan Stone[1]

1. SUMMARY

Photoreceptor death by apoptosis is the central pathology of most forms of retinal degeneration. Mitochondria play key roles in apoptosis, releasing both signals which induce apoptosis (cytochrome c, caspases) and signals which inhibit apoptosis (Bcl-2). Because mitochondria are the site of oxidative metabolism they are also a major site of formation of the toxic oxygen intermediates which form as oxygen is recruited into the oxidative phosphorylation pathway. Previous studies have shown that deletions in mtDNA accumulate in postmitotic tissues (central nervous, muscle) and that their accumulation is accelerated by oxidative stress (such as hypoxia) (Takeda et al. 1996;Lee et al. 1994; Merril et al. 1996; Englander et al. 1999). It seems possible therefore that mitochondria are a site at which oxidative stress induces the death of retinal neurones. This study investigates the accumulation of mtDNA deletions in the rat retina, in both normal (non-degenerative) and degenerative strains. Deletions were undetectable in Sprague-Dawley albino rats (24 months) but were detected at 15 months in the rapidly degenerating RCS strain. The appearance of deletions in the RCS strain, in which retinal oxygen tension is known to rise as the degeneration progresses, gives support to the ideas that oxidative stress is a factor in mtDNA deletions, and in the progress of the late stages of the degeneration.

2. INTRODUCTION

Mitochondria are prominent in photoreceptors, clustering in the inner segment and in the axon terminals. These locations place mitochondria as close as possible (Mervin 2001) to the photoreceptors' two sources of oxygen, the choriocapillaris and the deep

[1]Institute for Biomedical Research and the Department of Anatomy and Histology, University of Sydney, Sydney, NSW, Australia 2006; [2]School of Optometry, University of California, Berkeley, CA 94720, USA. Corresponding author: A. Bravo-Nuevo, E-mail: arturo@anatomy.usyd.edu

capillary plexus of the retinal circulation (Yu et al. 1999). The choriocapillaris is a particularly rich source of oxygen, since the fast flow of blood through this profuse capillary bed keeps pO_2 near arterial levels (Chan-Ling et al. 1993), and the corresponding concentration of mitochondria, in the inner segment, is the densest concentrations in the retina (Wong-Riley et al. 1998). The oxygen-orientation of mitochondria in photoreceptors is evident in their migration along retinal Müller cells, following an induced oxygen gradient (Germer et al. 1998). It is in mitochondria that oxygen is recruited for the cycle, by the enzyme cytochrome oxidase; mitochondria are as a consequence the principal site at which reactive oxygen intermediates form. Since mitochondria also release of signals induce apoptosis (Green et al. 1998; Daugas et al. 2000) it seems likely that they are major site at which oxidative stress induces cell death. We have postulated (the oxygen toxicity hypothesis) (Stone et al. 1999) that depletion of photoreceptors, such as occurs in the early stages of a retinal degeneration leads to an increase in oxygen tension in the outer retina and that increase has recently been demonstrated in the RCS rat (Yu et al. 2000).

To test whether retinal mitochondria are under oxidative stress sufficient to cause them damage we have examined normal and degenerating rat retinas for deletions in the mitochondrial genome (mtDNA). Previous work has established that deletions occur in mtDNA, and accumulate during adult life, particularly in post-mitotic cells and that their accumulation is accelerated by oxidative stress (Takeda et al. 1996; Lee et al. 1994; Merril et al. 1996; Englander et al. 1999).

3. MATERIALS AND METHODS

3.1. Animals

Rats were studied from 2 strains, albino Sprague-Dawley (SD) and the naturally degenerating Royal College of Surgeons (RCS) strain. Animals were raised in cyclic light (12h 5 lux/12h dark). Ages studied were P(postnatal day) 30, 150, 220, 430 (RCS rat) and P30, 200, 430, 550, 740 (SD). Eyes were obtained from 3 animals at each age, and the retinas were dissected free, without fixation. Each group of 6 retinas was pooled and the DNA extracted. We found that when performing DNA extraction by the phenol-chloroform method a sticky layer (presumably of glycosaccharides) interfered with the DNA extraction. To avoid this, we used Phase lock Gel Heavy 2ml tubes (Eppendorf AG, Hamburg, Germany) which allow the aqueous (nucleic acid) phase to be poured off. The concentration of the DNA was measured using a UV spectrophotometer and then normalised to 100ng/μl.

3.2. Designing and cloning the competitor

Two sets of primers were designed to perform nested PCR; they are illustrated in Fig 1. The first PCR used the primers L1(7687-7709) and H2(13641-13658) and the second PCR used primers L2 (7980-7999) and H4(13103-13124) (Fig 1). An extra primer that annealed to a site within the deletion was designed (WH)(8791-8809). PCR using L1/WH or L2/WH provided products which give a measure of undeleted mtDNA.

For quantitation of deleted mtDNA, we developed a competitive PCR protocol, which employed a construct designed to amplify with the L2/H4 primers, but to yield a slightly smaller product. To obtain the competitive construct, we designed two additional primers (LO2(8180-8202) and HO2(8084-8103) (Fig 1)) and added a BamHI restriction site to the 5′end of each. The two PCR products of L1/HO2 and LO2/H2 amplification were both cloned separately into the vector pGEM-T EASY (Promega) and each used to transform DH5_ competent cells (GibcoBRL). One of the cloned PCR products was then subcloned into the other product vector, ligating at the BamHI, site to yield a sequence which is 74bp shorter than the deletion amplicon of L2/H4 but still included the L1, L2, H1 and H4 primer annealing sites (see Fig 1). Specifically, the nested PCR procedure yielded a 313bp product from deleted mtDNA, and a 239bp product from the competitor construct (Fig 1). Serial dilutions of the competitor, ranging from 1ng/µl to 10^{-9}ng/µl, were mixed with the sample DNA. PCR products were then run on a 2% agarose gel, stained with ethidium bromide and the image analysed for quantification with ImageQuant software (Amersham Biosciences).

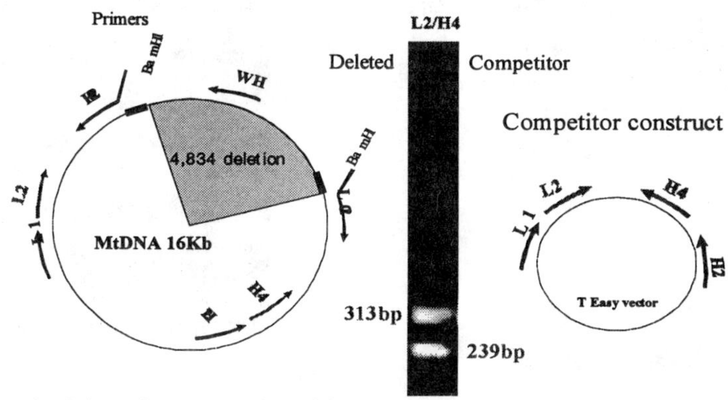

Figure 1. Schematic representation of the mtDNA and competitor construct used for nested, competitive PCR, showing the locations (not to scale) of the nested primers L1, L2, H2 and H4. The primers HO2 and LO2 were used for competitor construction. The insert photograph shows the secondary PCR amplification products (primer pair, L2/H4) from deleted mtDNA and the competitor construct, of 313 bp and 239 bp, respectively.

4. RESULTS

4.1. Products of competitive PCR protocol

The protocol used applied nested PCR (first L1 /H2 then L2/H4) to amplify products from a mixture of two DNA templates. One 'template' was total DNA extracted from the retina, which gave a 313pb product amplified from mtDNA carrying the 4,834bp deletion. The second template was the competitive construct, from which gave a 239bp product (gel in Fig 1). The relative amounts of these 2 products are determined by the amounts of each template in the mixture. Where only the shorter, competitor band was visible, it was assumed that the competitor was much more frequent that the deleted mtDNA, and vice versa. Where the two bands were approximately equal, it was assumed that the concentration of deleted mtDNA approximately equalled the concentration of the competitor.

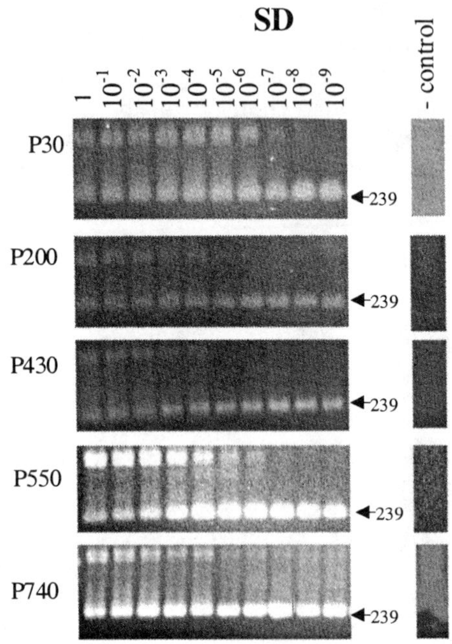

Figure 2A. Each panel shows a single gel, displaying PCR products from a competitive PCR protocol, using DNA from non-degenerative Sprague-Dawley albino rats. Successive ages are represented by each panel. The deletion product does not appear even in the oldest age

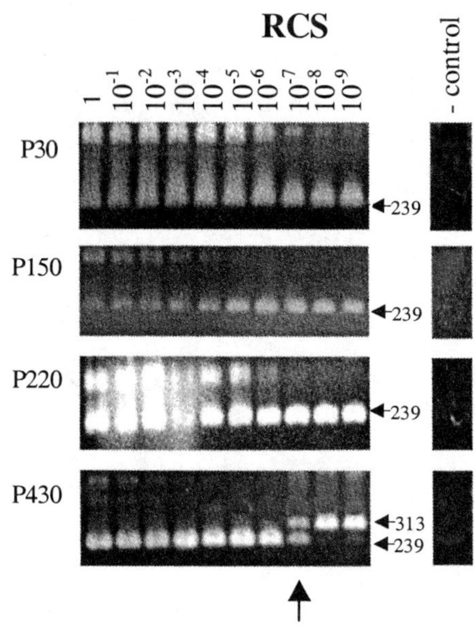

Figure 2B. A similar analysis to that in Fig 3A, for the degenerative RCS rat. The heaviest band represents the product L1/H4. The 239bp band is the competitor product and is present in all lanes. The 313bp band is the deletion product L2/H4. The deletion product can be detected at 10^{-7} ng/µl dilution of the competitor at P430 (arrow).

When this protocol was applied to DNA from non-degenerative rats aged P30 –740 and with dilutions of the competitor up to 10^{-9} ng/µl (Fig 2A), the 313bp band amplified from mtDNA was not seen even at the oldest age (P740). The 239bp 'competitor band is present in all lanes.

When the same analysis was applied to the DNA from retinas of RCS rats up to P430 (Fig 2B), At P430 (bottom panel in Fig 2B), the mtDNA band was clear at a competitor dilution of 10^{-7} ng/µl and prominent at higher dilutions.

These trends are shown in graphical form in Fig 3. For each strain and dilution, the concentration of mtDNA in the original extract was calculated and then plotted. Deleted DNA was detected only in the RCS rat, at the oldest age examined (P430).

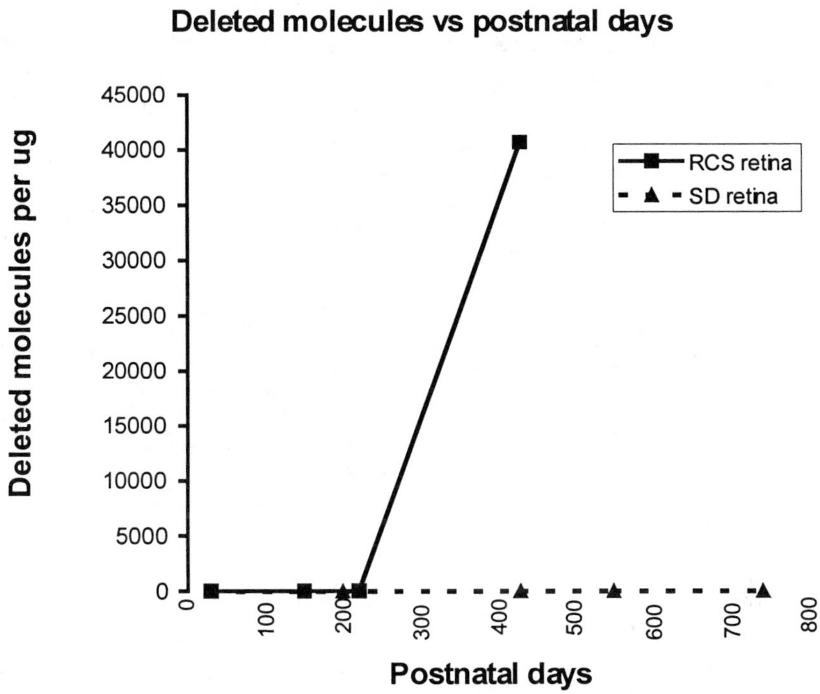

Figure 3. Estimates of the number of mtDNA molecules with the 4,834bp deletion/μg DNA for the two rat strains, as a function of age. Deleted mtDNA molecules were not detected in the normal non-degenerative strain (SD). In the RCS strain deletions were detected at P430 (up to 4.07×10^4 deleted mitochondrial molecules per _g of DNA).

5. DISCUSSION

5.1. What destabilises the mtDNA of the RCS rat?

The repetition of base-pair sequences at intervals along mtDNA creates the possibility of deletions, due to the annealing of the 2 direct repeated sequences leading to a recombination (Fig 1, 4) (Mita et al. 1990). In the rat, the repetition of a 16bp sequence (5′ cctgagccctaataat 3′) in rat mtDNA results in the production of mtDNA lacking the 4,834bp between the two sites (Moore et al. 1998). The rate at which this deletion occurs has been linked to oxidative stress (Wei et al. 1996; Kakimoto et al. 2002), to UV light (Pang et al. 1994; Berneburg et al. 2000, 1998; Birch-Machin et al. 1998) and to the level of production of reactive oxygen/oxidative species (ROS) (Berneburg, 1999) during the recruitment of oxygen into the oxidative phosphorylation pathway. In the central nervous system, as in other tissues, regions that are exposed to high levels of ROS accumulate high levels of mtDNA deletions (Lee et al. 1994; Cortopassi et al. 1992).

The present results confirm previous reports that mtDNA deletions may be undetectable in normal retina (Soong et al. 1996). Nevertheless deletions do accumulate with the time, in particular in the RPE (Barreau et al. 1996), being higher in peripheral

Retina. Barron et al. 2001, showed further that, in cases of macular degeneration, the frequency of mtDNA deletions rises in central retina. The present study provides evidence that unlike in the non-degenerative SD strain, in the RCS strain in which the retina degenerates spontaneously, mtDNA deletions occur, and reach high concentrations.

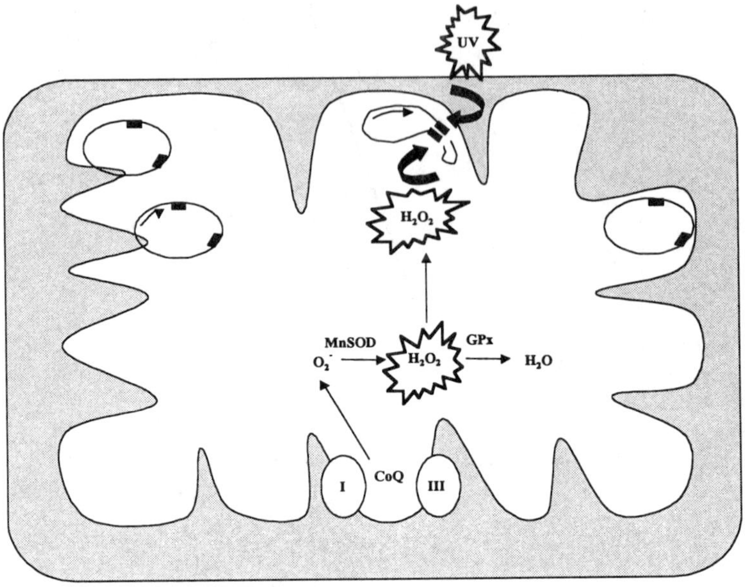

Figure 4. ROS such as H_2O_2 are formed during the oxidative phosphorylation at the inner membrane of the mitochondria. mtDNA is linked to the inner membrane of the mitochondrion and therefore at risk of being affected by the ROS or UV light. These two factors are known to enhance the number of mtDNA deleted molecules.

5.2. What factor causes this higher rate of deletion?

The present results do not provide direct evidence on this point. Nevertheless, analyses of oxygen consumption in the retina predict that, as the photoreceptor population is depleted oxygen levels in the outer retina will rise (reviewed in Stone et al. 1999) whatever the cause of the depletion. For the RCS rat, this rise has been documented experimentally (Yu et al. 1999). Further, hyperoxia is associated with an increase in ROS production (Gille 1992; Fehrenbach et al. 2001), and ROS are known to be associated with the accumulation of mtDNA deletions (Wei et al. 1996). In brief, depletion-induced hyperoxia of the outer retina may be a cause of the accelerated accumulation of mtDNA deletions in the RCS retina.

Since mitochondria are a major site of release of cell death signals, also associated with oxidative stress, it is possible that hyperoxia-induced mitochondrial damage is a key step in the progressive nature of photoreceptor degenerations.

5.3. Deletions must occur early if they are to accumulate

Deletions in mtDNA become prominent where several conditions are met. First, and self-evidently, it is necessary that the deletions should not interfere with mtDNA replication. In fact, most of the factors which regulate the replication of mtDNA are encoded in nuclear DNA, so that mtDNA deletions do not hinder mtDNA replication (Clayton 2000). Second, and also self–evidently, the deletions must occur at a rate which exceeds the repair rate of DNA repair mechanisms. The repair mechanisms for mtDNA are considered less efficient that those of nuclear DNA (Englander et al. 1999). Third, in many tissues, deletions remain at low levels in many tissues unless a factor such as UV light or oxidative stress acts to accelerate the frequency of deletions (Pang et al. 1994; Birch-Machin et al. 1998; Varanasi et al. 2001). Fourth, cell division appears to dilute deleted mtDNA (Collombet et al. 1996; van de Corput et al. 1997) so that deletions are more readily detected in post-mitotic cells, such as neurons of the central nervous system (including the retina). Fifth, the shorter length of deleted mtDNA is considered to reduce the time required for replication, giving deleted molecules a competitive advantage (Kowald 1999). Finally, multiple replications of mitochondria must occur, and therefore a period of time (Kowald et al. 1993; Elson et al. 2001) must pass, before deletions accumulate to detectable levels.

This last point is a prediction from mathematical models of mitochondrial replication. It implies that the high levels of deletions we have detected at late stages of RCS degenerations (at P430 and thus months after the degeneration has become severe) are the result of damage caused at earlier stages of the degeneration, and then reinforced during adult life. The point deserves testing with other markers of mitochondrial damage.

6. ACKNOWLEDGEMENTS

We thank the National Eye Institute for the travel award provided to ABN to attend this meeting.

7. REFERENCES

Barreau, E., J. Y. Brossas, et al., 1996, Accumulation of mitochondrial DNA deletions in human retina during aging, Invest. Ophthal. Vis. Sci. **37**:384-91.

Barron, M. J., M. A. Johnson, et al., 2001, Mitochondrial abnormalities in ageing macular photoreceptors, Invest. Ophthal. Vis. Sci. **42**:3016-22.

Berneburg, M. and J. Krutmann, 1998, Mitochondrial DNA deletions in human skin reflect photo- rather than chronologic aging. [letter; comment.]. J. Invest. Dermatol. **111**:709-10.

Berneburg, M. and J. Krutmann, 2000, Photoaging-associated large-scale deletions of mitochondrial DNA. Methods in Enzymol. **319**:366-76.

Berneburg, M., S. Grether-Beck, et al., 1999, Singlet oxygen mediates the UVA-induced generation of the photoaging-associated mitochondrial common deletion. J. Biol. Chem. **274**:15345-9.

Birch-Machin, M. A., M. Tindall, et al., 1998, Mitochondrial DNA deletions in human skin reflect photo- rather than chronologic aging. J. Invest. Dermatol. **110**:149-52.

Chan-Ling, T., 1993, Retinopathy of Prematurity: Origins in the architecture of the retina. Prog. Ret. Res. **12**:155-178.

Clayton, D. A., 2000, Transcription and replication of mitochondrial DNA. Hum. Reprod. **15** Suppl 2:11-7.

Collombet, J. M., G. Mandon, et al., 1996, Accumulation of mitochondrial DNA deletions in myotubes cultured from muscles of patients with mitochondrial myopathies. Mol. Gen. Genet. **253**:182-8.

Cortopassi, G. A., D. Shibata, et al., 1992, A pattern of accumulation of a somatic deletion of mitochondrial DNA in aging human tissues. Proc. Natl. Acad. Sci. USA **89**:7370-4

Daugas, E., S. A. Susin, et al., 2000, Mitochondrio-nuclear translocation of AIF in apoptosis and necrosis. FASEB J. **14**:729-39

Elson, J. L., D. C. Samuels, et al., 2001, Random intracellular drift explains the clonal expansion of mitochondrial DNA mutations with age. Am. J. Hum. Gen. **68:**802-6.

Englander, E. W., G. H. Greeley, Jr., et al., 1999, Hypoxia-induced mitochondrial and nuclear DNA damage in the rat brain. J. Neurosci. Res. **58**:262-9.

Fehrenbach, E. and H. Northoff, 2001, Free radicals, exercise, apoptosis, and heat shock proteins. Exerc. Immunol. Rev. **7**: 66-89.

Germer, A., B. Biedermann, et al., 1998, Distribution of mitochondria within Muller cells--I. Correlation with retinal vascularization in different mammalian species. J. Neurocytol. **27**:329-45.

Germer, A., J. Schuck, et al., 1998, Distribution of mitochondria within Muller cells--II. Post-natal development of the rabbit retinal periphery in vivo and in vitro: dependence on oxygen supply. J. Neurocytol. **27**:347-59.

Gille, J. J., 1992, Cell culture models for oxidative stress: superoxide and hydrogen peroxide versus normobaric hyperoxia. Mutat. Res. **275**:405-14.

Green, D. R. and G. P. Amarante-Mendes, 1998, The point of no return: mitochondria, caspases, and the commitment to cell death. Results Probl. Cell Differ. **24**: 45-61.

Kakimoto, M., T. Inoguchi, et al., 2002, Accumulation of 8-hydroxy-2'-deoxyguanosine and mitochondrial DNA deletion in kidney of diabetic rats. Diabetes **51**:1588-95.

Kowald, A., 1999, The mitochondrial theory of aging: do damaged mitochondria accumulate by delayed degradation? [see comments.]. Exp. Gerontol. **34**:605-12.

Kowald, A. and T. B. Kirkwood, 1993, Mitochondrial mutations, cellular instability and ageing: modelling the population dynamics of mitochondria. Mutat. Res. **295**:93-103.

Lee, H. C., C. Y. Pang, et al., 1994, Differential accumulations of 4,977 bp deletion in mitochondrial DNA of various tissues in human ageing. Biochim. Biophys. Acta. **1226**:37-43.

Mervin, K., 2001, Regulation of photoreceptor death in the mammalian retina. PhD thesis. University of Sydney.

Merril, C. R., S. Zullo, et al., 1996, Possible relationship between conditions associated with chronic hypoxia and brain mitochondrial DNA deletions. Arch. Biochem. Biophys. **326**:172-7.

Mita, S., R. Rizzuto, et al., 1990, Recombination via flanking direct repeats is a major cause of large-scale deletions of human mitochondrial DNA. Nucleic Acids Research **18**: 561-7.

Moore, C. A., J. Gudikote, et al., 1998, Mitochondrial DNA rearrangements, including partial duplications, occur in young and old rat tissues. Mutat. Res. **421**:205-17.

Pang, C. Y., H. C. Lee, et al., 1994, Human skin mitochondrial DNA deletions associated with light exposure. Arch. Biochem. Biophys. **312**:534-8.

Soong, N. W., M. H. Dang, et al., 1996, Mitochondrial DNA deletions are rare in the free radical-rich retinal environment. Neurobiol. of Aging **17**:827-31.

Stone, J., J. Maslim, et al., 1999, Mechanisms of photoreceptor death and survival in mammalian retina. Prog. Ret. Eye Res. **18**:689-735.

Takeda, N., Y. Ota, et al., 1996, Myocardial adaptive changes and damage in ischemic heart disease. Ann. N.Y. Acad. Sci. **793**:282-8.

van de Corput, M. P., J. M. van den Ouweland, et al., 1997, Detection of mitochondrial DNA deletions in human skin fibroblasts of patients with Pearson's syndrome by two-color fluorescence in situ hybridization. J. Histochem. Cytochem. **45**:55-61.

Varanasi, S. S. and H. K. Datta, 2001, Southern analysis of mitochondrial DNA in cortical bone of elderly patients undergoing knee and hip arthroplasty. J. Pathol. **193**:557-62.

Wei, Y. H., S. H. Kao, et al., 1996, Simultaneous increase of mitochondrial DNA deletions and lipid peroxidation in human aging. Ann. N.Y. Acad. Sci. **786**:24-43.

Wong-Riley, M. T., Z. Huang, et al., 1998, Neurochemical organization of the macaque retina: effect of TTX on levels and gene expression of cytochrome oxidase and nitric oxide synthase and on the immunoreactivity of Na+ K+ ATPase and NMDA receptor subunit I. Vis. Res. **38**:1455-77.

Yu, D. Y., S. J. Cringle, et al., 1999, Intraretinal oxygen distribution in the rat with graded systemic hyperoxia and hypercapnia. Invest. Ophthalmol. Vis. Sci. **40**:2082-7.

Yu, D. Y., S. J. Cringle, et al., 2000, Intraretinal oxygen levels before and after photoreceptor loss in the RCS rat. Invest. Ophthalmol. Vis. Sci. **41**:3999-4006.

QUANTITATIVE PCR ANALYSIS OF FosB mRNA EXPRESSION AFTER SHORT DURATION OXYGEN AND LIGHT STRESS

Scott F. Geller[1] and Jonathan Stone[2]

1. SUMMARY

Quantitative polymerase chain reaction (QPCR) was used to examine changes in FosB mRNA expression in models of oxygen and light stress to the retina. C57BL/6 mice or Sprague-Dawley (SD) albino rats were subjected to several experimental paradigms: short-term light or oxygen stress, extended hyperoxia (75% oxygen), or a model of oxygen-induced retinopathy (OIR). Control animals were subjected to room air and 5 lux cyclic light. FosB expression dramatically increases in response to light stress as well as in a model of OIR, but not in response to sustained 75% oxygen. These data suggest that both hypoxia and light stress induce expression of FosB in the retina.

2. INTRODUCTION

Transcription factors are among the earliest transcripts induced by cellular stress. The Fos and Jun proteins (c-Fos, FosB, Fra-1, Fra-2, c-Jun, JunB, and JunD) are members of the activator protein-1 (AP-1) family of basic leucine-zipper transcription factors. AP-1 proteins can differentially homo- or hetero-dimerize, bind DNA regulatory elements, and regulate the expression of numerous genes (Herdegen and Leah, 1998; Kovary and Bravo, 1991; Rauscher et al., 1988; Schuermann et al., 1991). These proteins are known to be global regulators of gene expression, influencing complex cellular processes like proliferation, maturation, and differentiation, and have been extensively studied in many areas of biology including the visual system (Herdegen and Leah, 1998).

[1]School of Optometry, University of California, Berkeley, CA 94720-2020, USA; [2]Institute for Biomedical Research and the Department of Anatomy and Histology, University of Sydney, Sydney, NSW, Australia 2006; Corresponding author: S. F. Geller, E-mail: sgeller@socrates.berkeley.edu

One of the most well documented attributes of AP-1 factors is that their rapid expression is routinely associated with both acute and chronic cellular stress (Abate et al., 1990; Miano et al., 1993). However, a detailed understanding of AP-1 gene expression is blurred by several common characteristics of these factors. Expression of individual family members is tissue specific (Herdegen and Leah, 1998; Jochum et al., 2001; Redemann-Fibi et al., 1991) and may serve different functional roles in such tissues (Jochum et al., 2001; Lucibello et al., 1990). Each member appears to be uniquely developmentally regulated and can be induced by a variety of different stimuli (Condorelli et al., 1993; Gillardon et al., 1994; Trejo et al., 1992). Finally, many AP-1 factors can both positively and/or negatively regulate the expression of most AP-1 family members (Lazo et al., 1992), depending on the dimer's constituents.

c-Fos and c-Jun are expressed in the retina and are induced by both retinal detachment (Geller et al., 2001) and intense, short-term light damage (Grimm et al., 2000; Reme et al., 1998). More specifically, FosB is expressed in the retina after optic nerve injury (Redemann-Fibi et al., 1991; Robinson, 1994) and in cultured cells following hypoxia (Bergman et al., 2002). Increased expression of FosB has been associated with both ischemic injury (Kamme et al., 1995) and apoptosis (Yamamura et al., 2000), a phenomenon known to occur in the retina following hypoxia (Maslim et al., 1997; Stone et al., 1999) and ischemia (Joo et al., 1999). Lastly, c-Fos is expressed in the retina in a circadian rhythm (Sagar and Sharp, 1990), and FosB circadian expression is apparent in the suprachiasmatic nucleus (Guido et al., 1996; Smith et al., 2001), the retina-recipient nucleus through which retinal output entrains the circadian rhythm. In the current study we have challenged animals with hypoxia, hyperoxia, or bright light for (up to 24 hours), and used QPCR to quantify changes in FosB expression in the retina.

3. MATERIALS AND METHODS

3.1 Animals

All procedures used were in accordance with the ARVO Statement for the Use of Animals in Ophthalmic and Vision Research and were approved by the University of Sydney Animal Ethics Committee. C57BL/6 (c57) mice and Sprague-Dawley (SD) rats were used in all the studies. For oxygen studies, animals were placed, in normal rodent cages, into plexiglass chambers connected to an OXYCYCLER (BioSpherix Ltd., Redfield, NY), and were subjected to altered oxygen environments for various periods. All animals were maintained on a 12-hour on/off light cycle (5 lux), and were allowed food and water *ad libitum*. For light-damage studies, rats were subjected to 1500 lux light for 6 hours and mice to 2000 lux for 1, 6, or 24 hours.

3.2. Tissue Isolation

Animals were deeply anesthetized with halothane and cervically dislocated. The eyes were removed and the retinas were dissected immediately, pooled (see figure legends), and homogenized in Tri Reagent (Sigma, Australia). Total RNA was precipitated, quantified, and reversed transcribed with M-MLV reverse transcriptase according to the manufacturer's protocol (Promega, Australia).

3.3. Primers

Primers were designed to 28S ribosomal RNA (Forward: 5'-gcggtacacctgtcaaac-3', Reverse: 5'-agaggcgttcagtcataatc-3') and FosB (Forward: 5'-tcatcaccctccgccgagtctcagt-3', Reverse: 5'-gcttgccccgccagtgctgtag-3'). The 28S and FosB primer pairs gave products of 502 and 305 base pairs (bp), respectively, and amplified single bands from total RNA as determined by ethidium bromide staining and QPCR Dissociation Curve analysis (see Fig. 1; Applied Biosystems). PCR primers were designed using Prime GCG (Wisconsin Package), and all bioinformatics analyses were conducted on BioManager.com, provided by ANGIS (Australian National Genomics Information Service).

3.4. PCR

PCR was performed according to standard thermal cycling parameters using 50ng (1μL) of reverse transcribed total RNA as template in 25μL reactions. Cycling conditions were: 2 minutes at 94° (initial denaturation), 30 seconds at 60° (annealing), 1 minute at 72° (extension), and 15 seconds at 94° (melting). All reactions included 2.5mM Mg^{++} and 0.5U of Taq polymerase (Abgenics, Australia).

3.5. Quantitative PCR (QPCR)

QPCR was performed on an Applied Biosystems 7700 thermal cycler using the Applied Biosystems 2X SYBR green master mix. Primer annealing and extension was done at 60°C (1 minute), followed by denaturation for 15 seconds at 94°. The thermal cycler repeated the PCR for 40 cycles. Analysis of FosB expression was normalized to 28S ribosomal RNA (rRNA) expression because of its stability under hypoxic conditions (Zhong and Simons, 1999). In brief, a threshold value of fluorescence was chosen such that all samples, controls and experimentals, were exponentially amplifying. The intensity of fluorescence is then proportional to the number of original transcripts in the sample cDNA. All reactions were made from "master mixes" to reduce pipetting inaccuracy between samples, and all samples were also normalized to an internal fluorescent control present in the 2X SYBR green master mix (Applied Biosystems). Finally, all calculations were normalized to expression in control animals (assigned a value of 1).

4. RESULTS

4.1. PCR

The primers used (Section 2.3) amplified a single amplicon for both FosB and the internal control, 28S rRNA. The FosB primers yielded a single 305bp amplicon from both mouse and rat retinal cDNA's after 40 cycles of amplification. When denatured by increasing the reaction temperature, both the 305bp FosB amplicon (Fig. 1B) and the 502bp 28S amplicon showed single peaks of denaturation, indicating that the RT-PCR was specific for the desired mRNA transcript. During this process the temperature was ramped between the annealing temperature (60°C) and 95°C over 5 minutes, collecting fluorescence data as frequently as possible. Both FosB (Fig. 1B) and 28S (Fig. 1C)

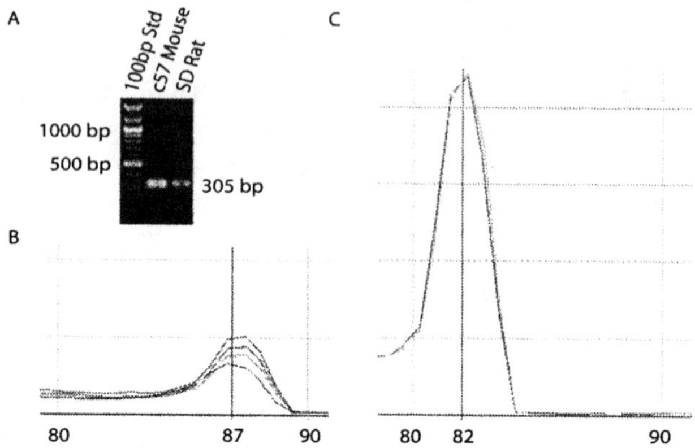

Figure 1. FosB PCR amplicon visualized by ethidium bromide after 40 cycles of amplification (A) and analysed by QPCR Dissociation Curve (Applied Biosystems) analysis (B). Figure 1C shows the Dissociation Curve analysis for 28S rRNA, which serves as an RNA template loading control.

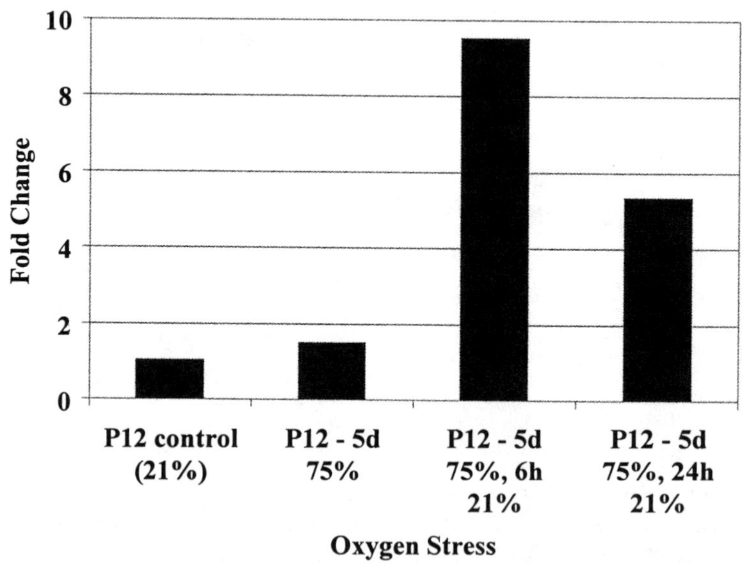

Figure 2. Retinal FosB expression in a model of oxygen-induced retinopathy. FosB is induced rapidly after returning the animal to room air. Four retinas from four animals were pooled in each sample, except for the control animals (8 retinas from 8 animals).

products showed a single peak of denaturation at a temperature consistent with their G+C nucleotide content; 64.1% for FosB and 50.1% for 28S.

4.2. QPCR Analysis of FosB Expression in Oxygen and Light Stress

FosB is rapidly induced in mouse retinas in a model of OIR (Fig. 2) (Smith et al., 1994). Mice were maintained in 21% oxygen from birth until postnatal day (P) 7, placed into 75% oxygen for 5 days, and then returned to room air (21% oxygen). The period in hyperoxia arrests the development of retinal vessels, with the result that when the animals are returned to room air the retina becomes severely hypoxic. Hyperoxia (75%) alone did not change FosB expression significantly (Fig. 2), but when the animal was returned to 21% oxygen for 6 or 24 hours, FosB expression increased 9.5 and 5.25 fold, respectively.

Figure 3 shows QPCR analysis of FosB expression in adult mice subjected to 0, 1, 2, or 3 weeks of continuous 75% oxygen. Similar to data from the developing mouse retina (Fig. 2), hyperoxia alone did not stimulate FosB expression.

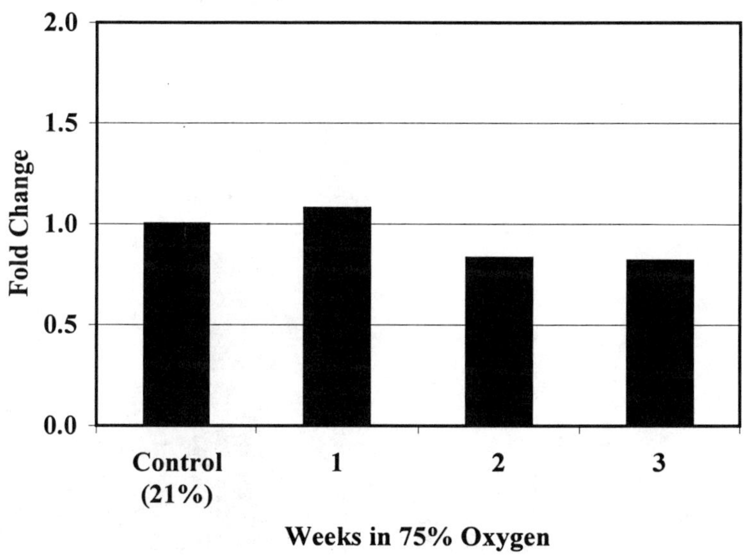

Figure 3. Retinal FosB expression after sustained 75% oxygen treatment in the adult mouse retina. Four retinas from four animals were pooled in each sample.

The effect of light stress on FosB expression in mouse retina is shown in Figure 4. Bright light (2,000 lux) induced a 10-fold increase in FosB expression within 1 hour. After 6 or 24 hours of continuous bright light exposure, FosB expression remained 3 times higher than control levels.

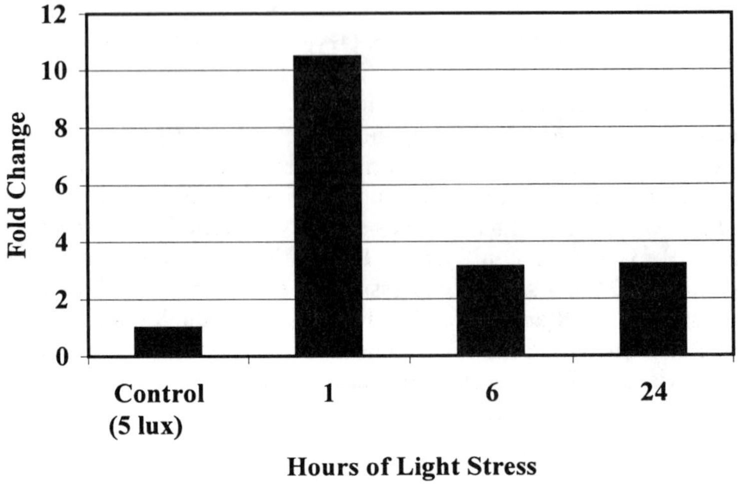

Figure 4. Retinal FosB expression in adult pigmented mice exposed to 2000 lux constant light. Four retinas from four animals were pooled in each sample.

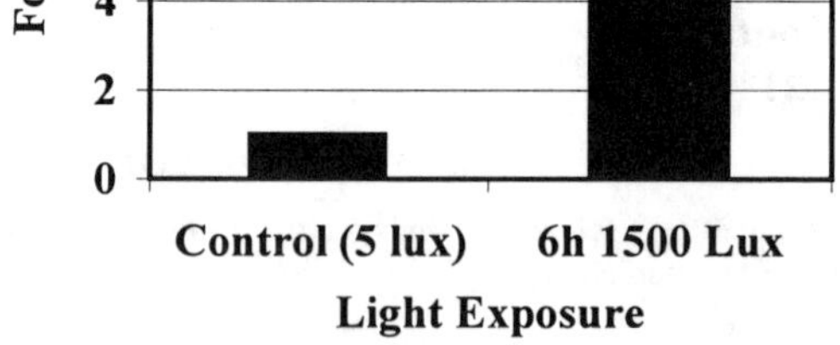

Figure 5. FosB expression in the albino rat retina subjected to light stress. Four retinas from four animals were pooled in each sample.

In the rat retina (Fig. 5), 6 hours of light exposure induced an 11-fold increase in FosB expression. This change in FosB expression is similar to that observed at 1 hour of light stress in the mouse (Fig. 4).

5. DISCUSSION

A persistent supply of oxygen is critical for normal cellular function in the vast majority of eukaryotic cells. Multiple mechanisms are likely to be involved in coordinating a tightly regulated response to an occasional [acute] lack of oxygen if the animal is to remain physiologically "normal." Hypoxic stress affects the expression of transcripts involved with cellular metabolism, gene transcription, erythropoiesis, and angiogenesis, among others (Dachs and Tozer, 2000; Ozaki et al., 1999). Identifying how oxygen stress affects the expression of transcription factors is a key step towards understanding secondary cellular responses

The power of oxygen to cause retinal disease has been known since the role of hyperoxia in the pathogenesis of retinopathy of prematurity was elucidated in the 1950's. More recently, hypoxia has been shown to be a factor in the pathogenesis of retinopathy of detachment (Lewis et al., 1999; Mervin et al., 1999) and in diabetic retinopathy (Arden et al., 1998; Linsenmeier et al., 1998). Conversely hyperoxia may be a factor in photoreceptor degenerations (Penn et al., 2000; Stone et al., 1999; Yu et al., 1999), and oxidative stress may play a role in age-related macular degeneration (AMD) (Beatty et al., 2000; Hartnett et al., 1996). The potential of oxygen to perturb the vasculature stems from the involvement of oxygen-regulated factors (particularly vascular endothelial growth factor; VEGF) in the normal formation of retinal vasculature (Stone et al., 1995), and the over- or under-regulation of these mechanisms when oxygen levels are abnormally low or high. The present data are a step towards identifying the detail of the mechanisms induced by changes in oxygen levels.

Transcription factor expression can be critical step in regulating cellular and molecular responses to both chronic and acute stimuli. Knockout animals indicate that single transcription factors can modulate "global" cellular events such as apoptosis. Such findings are true for the retina where the lack of c-Fos prevents photoreceptor apoptosis following light damage (Wenzel et al., 2000).

FosB has not been studied in detail in the retina, though it is known to be present in hypoxic retinal cells in culture (Xu et al., 1999), and is expressed during stress response in other tissues (Gillardon et al., 1994; Kamme et al., 1995; Miano et al., 1993). Recent data suggests that an increase in AP-1 family expression may not be pro-apoptotic, but rather, may correlate with a reduced capacity to protect [photoreceptor] cells from light damage (Wenzel et al., 2002). Our data provide no evidence as to whether FosB expression is protective or detrimental to the retina. Nevertheless, these data do suggest that both oxygen-induced hypoxia and light stress result in an increase in FosB mRNA expression. The induction may be the result of each stimulus (oxygen and light stresses) converging on a common second messenger signalling pathway(s) leading to increased expression, or it may be the result of different transcriptional elements becoming activated, resulting in FosB promoter activation. It seems reasonable to propose that FosB expression modulates AP-1 function and AP-1-regulated gene expression, and may be involved with the induction of an endogenous protective mechanism and/or the progressive loss of photoreceptor cells after oxygen or light stress.

6. ACKNOWLEDGEMENTS

The authors wish to thank Professor Nick Hunt for access to the QPCR machine. We would also like to thank the Foundation Fighting Blindness (FFB; Australia), the National Health and Medical Research Council (NH&MRC; Australia), and the Medical Foundation of the University of Sydney for grant support. Further, we thank the National Eye Institute for their travel support of SFG to attend this meeting.

7. REFERENCES

Abate C, Patel L, Rauscher FJ, 3rd, Curran T., 1990, Redox regulation of fos and jun DNA-binding activity in vitro. *Science* **249**:1157-1161.

Arden GB, Wolf JE, Tsang Y., 1998, Does dark adaptation exacerbate diabetic retinopathy? Evidence and a linking hypothesis. *Vision Res.* **38**:1723-1729.

Beatty S, Koh H, Phil M, Henson D, Boulton M., 2000, The role of oxidative stress in the pathogenesis of age-related macular degeneration. *Surv Ophthalmol.* **45**:115-134.

Bergman MR, Cheng S, Honbo N, Piacentini L, Karliner JS, Lovett DH., 2003, A functional AP-1 site regulates MMP-2 transcription by cardiac cells through interaction with JunB/Fra1 and JunB/FosB heterodimers. *Biochem J.* **369**:485-96

Condorelli DF, Dell'Albani P, Amico C, Kaczmarek L, Nicoletti F, Lukasiuk K, et al., 1993, Induction of primary response genes by excitatory amino acid receptor agonists in primary astroglial cultures. *J Neurochem.* **60**:877-885.

Dachs GU, Tozer GM., 2000, Hypoxia modulated gene expression: angiogenesis, metastasis and therapeutic exploitation. *Eur J Cancer.* **36**:1649-1660.

Geller SF, Lewis GP, Fisher SK., 2001, FGFR1, signaling, and AP-1 expression after retinal detachment: reactive Muller and RPE cells. *Invest Ophthalmol Vis Sci.* **42**:1363-1369.

Gillardon F, Eschenfelder C, Uhlmann E, Hartschuh W, Zimmermann M., 1994, Differential regulation of c-fos, fosB, c-jun, junB, bcl-2 and bax expression in rat skin following single or chronic ultraviolet irradiation and in vivo modulation by antisense oligodeoxynucleotide superfusion. *Oncogene.* **9**:3219-3225.

Grimm C, Wenzel A, Hafezi F, Reme CE., 2000, Gene expression in the mouse retina: the effect of damaging light. *Mol Vis.* **6**:252-260.

Guido ME, Rusak B, Robertson HA., 1996, Expression of fosB mRNA in the hamster suprachiasmatic nucleus is induced at only selected circadian phases. *Brain Res.* **739**:132-138.

Hartnett ME, Weiter JJ, Staurenghi G, Elsner AE., 1996, Deep retinal vascular anomalous complexes in advanced age-related macular degeneration. *Ophthalmology* **103**:2042-2053.

Herdegen T, Leah JD., 1998, Inducible and constitutive transcription factors in the mammalian nervous system: control of gene expression by Jun, Fos and Krox, and CREB/ATF proteins. *Brain Res - Brain Res Reviews* **28**:370-490.

Jochum W, Passegue E, Wagner EF., 2001, AP-1 in mouse development and tumorigenesis. *Oncogene* **20**:2401-2412.

Joo CK, Choi JS, Ko HW, Park KY, Sohn S, Chun MH, et al., 1999, Necrosis and apoptosis after retinal ischemia: involvement of NMDA-mediated excitotoxicity and p53. *Invest Ophthalmol Vis Sci.* **40**:713-720.

Kamme F, Campbell K, Wieloch T., 1995, Biphasic expression of the fos and jun families of transcription factors following transient forebrain ischaemia in the rat. Effect of hypothermia. *Eur J Neurosci.* **7**:2007-2016.

Kovary K, Bravo R., 1991, Expression of different Jun and Fos proteins during the G0-to-G1 transition in mouse fibroblasts: in vitro and in vivo associations. *Mol Cell Biol.* **11**:2451-2459.

Lazo PS, Dorfman K, Noguchi T, Mattei MG, Bravo R., 1992, Structure and mapping of the fosB gene. FosB downregulates the activity of the fosB promoter. *Nucleic Acids Res.* **20**:343-350.

Lewis G, Mervin K, Valter K, Maslim J, Kappel PJ, Stone J, et al., 1999, Limiting the proliferation and reactivity of retinal Muller cells during experimental retinal detachment: the value of oxygen supplementation. *Am J Ophthalmol.* **128**:165-172.

Linsenmeier RA, Braun RD, McRipley MA, Padnick LB, Ahmed J, Hatchell DL, et al., 1998, Retinal hypoxia in long-term diabetic cats. *Invest Ophthalmol Vis Sci.* **39**:1647-1657.

Lucibello FC, Slater EP, Jooss KU, Beato M, Muller R., 1990, Mutual transrepression of Fos and the glucocorticoid receptor: involvement of a functional domain in Fos which is absent in FosB. *Embo J.* **9**:2827-2834.

Maslim J, Valter K, Egensperger R, Hollander H, Stone J., 1997, Tissue oxygen during a critical developmental period controls the death and survival of photoreceptors. *Invest Ophthalmol Vis Sci.* **38**:1667-1677.

Mervin K, Valter K, Maslim J, Lewis G, Fisher S, Stone J., 1999, Limiting photoreceptor death and deconstruction during experimental retinal detachment: the value of oxygen supplementation. *Am J Ophthalmol.* **128**:155-164.

Miano JM, Vlasic N, Tota RR, Stemerman MB., 1993, Localization of Fos and Jun proteins in rat aortic smooth muscle cells after vascular injury. *Am J Pathol.* **142**:715-724.

Ozaki H, Yu AY, Della N, Ozaki K, Luna JD, Yamada H, et al., 1999, Hypoxia inducible factor-1alpha is increased in ischemic retina: temporal and spatial correlation with VEGF expression. *Invest Ophthalmol Vis Sci.* **40**:182-189.

Penn JS, Li S, Naash MI., 2000, Ambient hypoxia reverses retinal vascular attenuation in a transgenic mouse model of autosomal dominant retinitis pigmentosa. *Invest Ophthalmol Vis Sci.* **41**:4007-4013.

Rauscher FJ, 3rd, Voulalas PJ, Franza BR, Jr., Curran T., 1988, Fos and Jun bind cooperatively to the AP-1 site: reconstitution in vitro. *Genes Dev.* **2**:1687-1699.

Redemann-Fibi B, Schuermann M, Muller R., 1991, Stage and tissue-specific expression of fosB during mouse development. *Differentiation* **46**:43-49.

Reme CE, Grimm C, Hafezi F, Marti A, Wenzel A., 1998, Apoptotic cell death in retinal degenerations. *Prog Retin Eye Res.* **17**:443-464.

Robinson GA., 1994, Immediate early gene expression in axotomized and regenerating retinal ganglion cells of the adult rat. *Brain Res Mol Brain Res.* **24**:43-54.

Sagar SM, Sharp FR., 1990, Light induces a Fos-like nuclear antigen in retinal neurons. *Brain Res Mol Brain Res.* 7:17-21.

Schuermann M, Jooss K, Muller R., 1991, fosB is a transforming gene encoding a transcriptional activator. *Oncogene.* **6**:567-576.

Smith LE, Wesolowski E, McLellan A, Kostyk SK, D'Amato R, Sullivan R, et al., 1994, Oxygen-induced retinopathy in the mouse. *Invest Ophthalmol Vis Sci.* **35**:101-111.

Smith M, Burke Z, Humphries A, Wells T, Klein D, Carter D, et al., 2001, Tissue-specific transgenic knockdown of Fos-related antigen 2 (Fra-2) expression mediated by dominant negative Fra-2. *Mol Cell Biol.* **21**:3704-3713.

Stone J, Itin A, Alon T, Pe'er J, Gnessin H, Chan-Ling T, et al., 1995, Development of retinal vasculature is mediated by hypoxia-induced vascular endothelial growth factor (VEGF) expression by neuroglia. *J Neurosci.* **15**:4738-4747.

Stone J, Maslim J, Valter-Kocsi K, Mervin K, Bowers F, Chu Y, et al., 1999, Mechanisms of photoreceptor death and survival in mammalian retina. *Prog Retin Eye Res.* **18**:689-735.

Trejo J, Chambard JC, Karin M, Brown JH., 1992, Biphasic increase in c-jun mRNA is required for induction of AP-1-mediated gene transcription: differential effects of muscarinic and thrombin receptor activation. *Mol Cell Biol.* **12**:4742-4750.

Wenzel A, Grimm C, Marti A, Kueng-Hitz N, Hafezi F, Niemeyer G, et al., 2000, c-fos controls the "private pathway" of light-induced apoptosis of retinal photoreceptors. *J Neurosci.* **20**:81-88.

Wenzel A, Iseli HP, Fleischmann A, Hafezi F, Grimm C, Wagner EF, et al., 2002, Fra-1 substitutes for c-Fos in AP-1-mediated signal transduction in retinal apoptosis. *J Neurochem.* **80**:1089-1094.

Xu L, Ma J, Seigel GM, Ma JX., 1999, l-Deprenyl, blocking apoptosis and regulating gene expression in cultured retinal neurons. *Biochem Pharmacol.* **58**:1183-1190.

Yamamura Y, Hua X, Bergelson S, Lodish HF., 2000, Critical role of Smads and AP-1 complex in transforming growth factor-beta -dependent apoptosis. *J Biol Chem.* **275**:36295-362302.

Yu AY, Shimoda LA, Iyer NV, Huso DL, Sun X, McWilliams R, et al., 1999, Impaired physiological responses to chronic hypoxia in mice partially deficient for hypoxia-inducible factor 1alpha. *J Clin Invest.* **103**:691-696.

Zhong H, Simons JW., 1999, Direct comparison of GAPDH, beta-actin, cyclophilin, and 28S rRNA as internal standards for quantifying RNA levels under hypoxia. *Biochel Biophys Res Com.* **259**:523-526.

METABOLIC MODULATION OF VISUAL SENSITIVITY

Robert B. Barlow[1], Bart Farell[2], and Mukhtar Khan[3]

1. INTRODUCTION

Vision is an energy demanding process. Changes in metabolic levels can affect how we see. The earliest evidence for the metabolic modulation of human vision dates back to the pioneering days of aviation when pilots reported darkened visual fields while flying at high altitudes without oxygen masks (McFarland and Evans,1939). Studies at sea level showed that lowering the partial pressure of oxygen decreased visual sensitivity (McFarland and Forbes, 1940) and ingesting glucose reversed the effects of hypoxia (McFarland et al, 1945). Lowering blood glucose under normal oxygen conditions can decrease sensitivity (McCrimmon et al, 1998).

Human visual sensitivity also changes with time of day with lower sensitivity at night (Bassi and Powers, 1986; Barlow et al., 1997). Interestingly, the hormonal control of glucose homeostasis exhibits a circadian rhythm (Aschoff, 1979) with lower glucose utilization at night (Clore, et al., 1989; Van Cauter et al., 1991 & 1997). Ingesting glucose at night can return visual sensitivity to high daytime levels (Barlow, et al., 1997) indicating that both oxygen and glucose can modulate visual sensitivity.

The effects of glucose and oxygen on vision are not surprising. Photoreceptors are high-energy consumers. With dense aggregates of mitochondria, rods and cones produce prodigious amounts of ATP to power Na^+/K^+ pumps that generate gradients for photoreceptor dark currents. High blood flow in the choriocapillaris appears to provide the highest possible partial pressure of oxygen to the photoreceptors (Steinberg, 1987). But the flow seems just sufficient (Linsenmeier and Braun, 1992) because glycolysis is required even under normal oxygen conditions (Ames and Gurian, 1963; Winkler, 1981). Glycolysis is a relatively inefficient source of energy, but it is essential and can compensate for low oxygen (McRipley et al., 1997). Is the high metabolic demand of the retina related to the modulatory effects of oxygen and glucose on visual sensitivity?

We present here preliminary measurements of changes in human visual sensitivity as a function of the time of day and blood glucose levels and consider where they may originate in the visual system. We suggest that an important locus is the macular region

[1]Center for Vision Research, Department of Ophthalmology, Upstate Medical University, 750 East Adams Street, Syracuse, NY 13210, USA; [2]Institute for Sensory Research, Syracuse University, Syracuse, NY, USA; [3]Department of Medicine, Upstate Medical University, Syracuse, NY 13210, USA. Corresponding author: R. B. Barlow, E-mail: barlowr@upstate.edu

of the retina. Further studies of its metabolic and physical properties may yield insights into the causes of some forms of retinal disease such as age-related macular degeneration.

2. CONTRAST SENSITIVITY CHANGES WITH TIME OF DAY

Figure 1 shows that the threshold for detecting contrast can be three times (0.5 log units) higher at night. These changes may reflect an endogenous circadian rhythm in visual sensitivity. However we did not measure visual thresholds under constant environmental conditions, either constant light or constant darkness. Thus it is not clear whether these changes reflect the action of an endogenous circadian oscillator or the effects of environmental factors that would generate a diurnal rhythm in threshold. A study carried out in constant darkness reported a circadian rhythm in absolute threshold mediated by the rod system (Bassi and Powers, 1986). Our studies focused on vision mediated by cones.

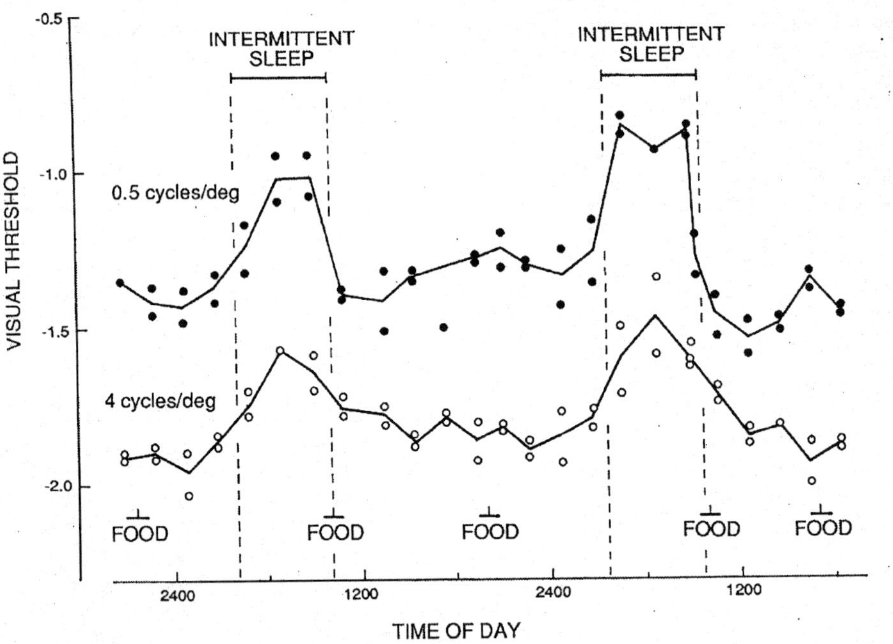

Figure 1. Visual threshold for detecting contrast increases at night. Secured on a bite bar the subject viewed sine-wave gratings displayed on a computer monitor (0.5 and 4 cycles/deg; 2mm artificial pupil; 10^0 visual angle; $100cd/m^2$). Thresholds were measured every two hours using a method of successive estimation with the QUEST algorithm using two-alternative force choice (Watson and Pelli, 1983). The tested eye was patched to minimize changes in sensitivity caused by changes in environmental lighting (Barlow, et al., 1997).

Decreases in threshold in Figure 1 appear to follow food ingestion as well as termination of sleep. To determine whether these changes correlate with changes in body metabolism we measured blood glucose as well as thresholds for detecting contrast. Figure 2 shows that blood glucose began to decrease near the onset of sleep (after 2400h) when visual thresholds began to increase. The opposite occurs near the termination of sleep (1000h). The nocturnal decline in blood glucose level is indicative of major metabolic events in the body controlled by circadian oscillators (Glore et al, 1989). The corresponding nocturnal increases in visual threshold in Figures1 and 2 roughly correlate with decreases in blood glucose and perhaps are caused by them.

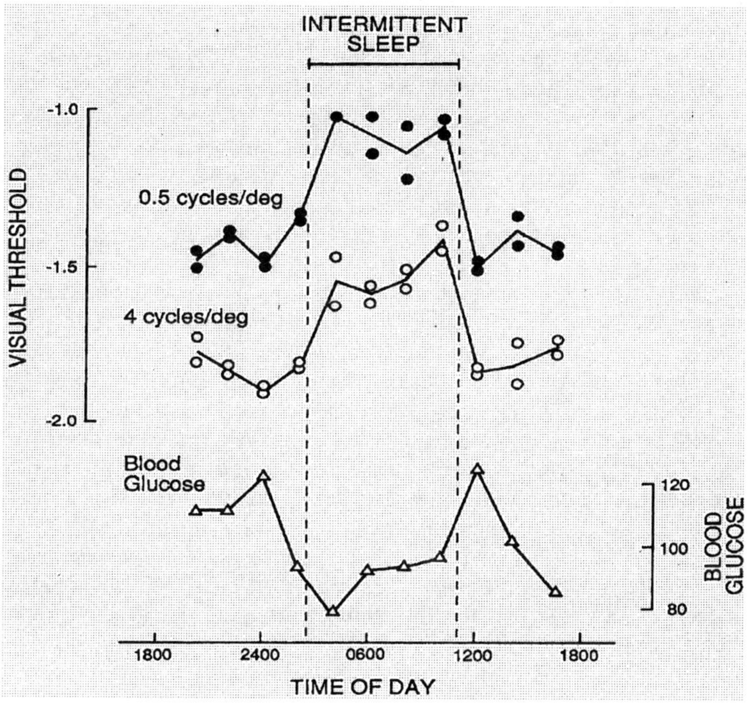

Figure 2. Blood glucose correlates inversely with visual threshold. A decrease in blood glucose from ~120mg/dl at 2400h to 80mg/dl at 0400h correlates with about a 3-fold increase in threshold for detecting sine-wave gratings of different spatial frequencies (Barlow et al., 1997). Visual thresholds were measured every 2h following the procedures described in the legend of Figure 1.

3. CONTRAST SENSITIVITY CHANGES WITH BLOOD GLUCOSE LEVEL

To test whether visual threshold is inversely related to blood glucose we woke subjects during the middle of the night, tested their contrast sensitivity, and gave them oral glucose (75g, Trutol Solution). Typically, within 40 min of ingestion blood glucose levels doubled (80 to 160mg/dl) and were associated with about a 2-fold reduction in threshold (data not shown). Interestingly, the induced hyperglycemia evoked strong insulin responses that occasionally drove blood glucose below euglycemic levels without associated changes in visual threshold. In fact, after glucose ingestion visual thresholds often continued to decline toward daytime levels as blood glucose returned to preingestion levels. The uncoupling of visual threshold and blood glucose under these conditions is not understood. However, the increases in threshold at night and the action of glucose to reverse them are consistent findings.

To further examine the relationship between contrast sensitivity and blood glucose we injected insulin to lower blood glucose levels. Using the same psychophysical techniques as in Figures 1and 2, we measured contrast sensitivity before, during and after insulin injection while assessing blood glucose with finger sticks. Although variability was considerable, insulin-induced hypoglycemia increased threshold in each test with a maximum of a 10-fold increase. Euglycemic levels were on average ~100mg/dl before insulin injection and maximal hypoglycemic levels were between 45 and 50mg/dl after injection. We found considerable subject to subject variability that may be age related; however, our sample is too small to draw any firm conclusions.

As noted above the important roles of glucose and oxygen in human vision have been known for more than 60 years. McFarland and Forbes (1940) showed then that hypoglycemia (~55m/dl) could elevate dark-adapted threshold 2.5 fold (0.4 log units). More recently McCrimmon et al. (1996) reported that about the same level of hypoglycemia reduced the ability of subjects to detect differences in contrast under photopic conditions, and Ewing et al. (1997) found that similar hypoglycemia impaired contrast sensitivity in adults with type 1 diabetes. Both studies assessed contrast sensitivity by presenting subjects with a series of gratings having different levels of contrast, and both detected a decrease in sensitivity as we did. However the different methodologies used in these studies and ours preclude a direct comparison of the changes in visual sensitivity.

What is the influence of oxygen on visual sensitivity? In a preliminary study we adjusted the partial pressure of inhaled oxygen with nitrogen replacement while assessing blood oxygen saturation with a pulse oximeter. Using the same psychophysical techniques described above, we found that changes in oxygen saturation in the range of 53% to 98% evoked nearly proportional changes in contrast threshold. A decrease in contrast sensitivity of about 0.1 log unit corresponded to a 10% decrease in blood oxygen saturation. We found as did McFarland et al (1945) that ingestion of glucose reversed the effects of low oxygen, that is, glucose restored visual sensitivity to levels measured under conditions of normal oxygen.

4. THE ORIGIN OF METABOLIC EFFECTS

Changes in metabolism could affect contrast sensitivity by acting at any or all stages along the visual pathway. We carried out preliminary tests of two loci: the retina (Barlow et al, 2003) and the occipital lobes of the brain (Barlow et al, 1997).

4.1. Brain

Does neuronal activity in central visual pathways exhibit diurnal changes corresponding to those in metabolism? To examine this possibility we used functional magnetic resonance imaging (fMRI, gradient echo planar sequence) to measure the BOLD response of the occipital lobes to a visual stimulus of an alternating (5Hz) grid of vertical and horizontal bars. In the imager subjects viewed the alternating grid in 15sec on/off cycles while multislice scans were made of their occipital poles. Typically, subjects were scanned at 2300h, allowed to fall asleep in the scanner, wakened at 0230h, rescanned, given glucose (75gm, Trutol), and rescanned 40min later. With this procedure the BOLD signal recorded from the occipital lobes declined after 3h of sleep when blood glucose had decreased about 30% to 61±13mg/dl and contrast sensitivity decreased 3-fold. After ingesting glucose, the BOLD signal increased to daytime levels as blood glucose levels increased about 80% to 135±14mg/dl (n=6). We conclude that cortical activation by visual stimuli decreases at night and then increases after ingestion of glucose (Barlow et al, 1997). The diurnal changes in metabolism affecting contrast sensitivity (Figs. 1 and 2) appear to correlate with changes in the activity of visual cortex. Changes in metabolic state may act directly on the visual cortex, on the retinal input to the cortex, or both.

4.2. Retina

Do the metabolic effects on visual sensitivity begin at the retina? We explored this question by recording the multifocal electroretinogram (mERG) from subjects under various metabolic conditions (Barlow et al, 2003). The mERG yields a topographical representation of retinal responsiveness (Sutter and Tran, 1992). We measured the mERG at various blood glucose levels set with a glucose clamp (Simonson et al, 1985) and determined contrast sensitivity as described above. Moderate hypoglycemia (50-60mg/dl) decreased by up to 50% the amplitude of the "b-wave" (first order analysis of mERG) generated by central retina (0^0to3^0) but not that generated by peripheral retina (beyond 3^0).

Effects of hypoglycemia on the ERG have also been detected in anesthestized cats but the results are mixed. Macaluso et al. (1992) reported that moderate hypoglycemia (54-74mg.dl) caused a 30% decrease in b-wave amplitude; whereas, McRipley et al. (1997) found mimimal effects (<5%). On the other hand, Sannita et al. (1993) reported in human studies that the ERG b-wave was affected by small, physiological changes in blood glucose suggesting that the human retina is more sensitive to metabolic changes than the cat retina. In these studies the ERGs are generated by all regions of the retina acting together. They are spatially insensitive and difficult to compare with mERGs that produce 2-D maps of retinal responses. Nevertheless, a consistent finding among the various studies, cat and human alike, is that metabolic state can affect the sensitivity of the retina as measured by the ERG or mERG. Our recordings of the mERG show that hypoglycemia affected the sensitivity of central more than peripheral retina.

4.3. Central Retina

Does hypoglycemia affect the contrast sensitivity of central more than peripheral vision? We do not yet have a definitive answer, but initial psychophysical tests point in this direction. Anecdotal reports also point to a greater sensitivity of central vision to hypoglycemia. For example, diabetics commonly report transient blurred vision when their blood glucose levels are low. Type 1 diabetics using insulin pumps (n=4) reported that when moderate hypoglycemic their central vision becomes "fuzzy" making reading or driving a car difficult. When severely hypoglycemic, each of the four reported a transient central "blind spot" or scotoma, roughly 10^0 to 15^0 in size, that vanishes after ingesting food and/or glucose. They estimate that central scotomas occur when their blood glucose is in the range of 40mg/dl. In our preliminary study of the effects of hypoglycemia on contrast sensitivity one subject reported a transient central scotoma when his blood glucose was ~45mg/dl. Interestingly, two nondiabetics who are marathon runners reported blurred central vision and occasionally a central blind spot when their energy stores were heavily depleted. Others have detected effects of hypoglycemia on the vision of type 1 diabetics (Ewing et al.,1997), but none reported scotomas or preferential effects on central vision. Figure 3 attempts to simulate the combined visual perceptions of four type 1 diabetics when they are moderately and severely hypoglycemic.

Taken together the results of our psychophysical and electrophysiological tests lead to several tentative conclusions: (1) metabolic state can affect responses of the human retina, (2) metabolic effects on occipital lobe activity may be caused in part by effects on retinal responses. and (3) metabolic stress (hypoglycemia) affects central more than peripheral retina in humans.

Figure 3. Simulation of the visual effects caused by AMD and moderate and severe hypoglycemia. AMD (upper right) can lead to a loss of vision in the central field. Moderate hypoglycemia can blur the central field (lower left), and severe hypoglycemia can lead to a central scotoma (lower right) similar in size to that caused by AMD. The hypoglycemic effects are transient and vanish upon food ingestion. We thank the National Eye Institute for permission to reproduce the upper panels from their website (www.nei.nih.gov) and modify them to simulate the effects of hypoglycemia in the lower panels.

4.4. Supply Oxygen and Glucose to the Central Retina

Why should central retina be more sensitive than peripheral retina to hypoglycemia? Could the high density of photoreceptors and paucity of blood supply in the central retina be contributing factors? Nourishment of photoreceptor outer segments depends primarily on blood flow in the choriocapillaris, but that of inner segments and the energy-demanding plexiform layers relies more on the retinal vasculature. Because retinal vessels do not penetrate the central retina, oxygen and glucose presumably reach the macula by diffusion, long-range diffusion in the case of cells in the central fovea. The uptake of oxygen may be enhanced by neuroglobin, an intracellular respiratory protein, localized in the ellipsoid region of photoreceptors and plexiform layers of the retina (Schmidt et al., 2003). Glucose has no known assistance and must find its way from blood capillaries through tissue to cells of the central retina. From these considerations we hypothesize that the supply of glucose to the central retina is diffusion limited. If this is the case, then a decline in blood glucose levels would preferentially impact the nourishment of neurons in the central retina.

5. POSSIBLE IMPLICATIONS FOR RETINAL DISEASE: AN HYPOTHESIS

Can the metabolic modulation of visual sensitivity shed any light on the causes of retinal disease? Could the sensitivity of the central retina to metabolic stress provide any clues as to why the macula ages more rapidly than the rest of the retina?

Age-related macular degeneration (AMD), the progressive loss of function of the central retina, is the leading cause of blindness among the elderly in the United States and other developed countries. It is a complex, multifactorial disease that damages both rods and cones in the macula with associated abnormalities in the retinal pigment epithelium (RPE) and choriocapillaris (Curcio et al., 2000). It is also characterized by the accumulation beneath the macula of proteinaceous debris, termed drusen. The "dry" form of AMD (~90% of cases) causes slow deterioration of vision, and the "wet" form, characterized RPE atrophy and choriodal neovascularization, leads to severe loss of central vision. Little is known about what causes the prevalent form of AMD, but several hypotheses have been suggested. One is that AMD results from oxidative injury to the RPE and choriocapillaris (Winkler et al., 1999). Oxidative stress may be caused by a combination of long-term exposure to light (Young, 1988), high rate of respiration mediated by densely packed mitochondria in rods and cones, and/or high concentrations of polyunsaturated fatty acids in the outer retina. Antioxidants in the macula, lutein and zeaxanthin, restrict the degree of oxidative damage by reactive oxygen species, but it is believed that throughout life oxidative damage can accumulate, "aging" the retina, and, if excessive, contribute to the onset of AMD (Winkler, 1999). This hypothesis is supported by the recent finding that proteins extracted from drusen in the eyes of AMD donors exhibit oxidative damage (Crabb et al., 2002). The role of oxidative stress in AMD is also consistent with the beneficial effects of antioxidants (Snodderly, 1995; AREDS, 2001). Another hypothesis for the loss of photoreceptors in AMD is a deficiency of a retinoid derivative caused by impaired transfer of nutrients across Bruch's membrane (Moore et al., 1995). Several studies have suggested an association of AMD with various vascular disorders, including hypertension (Sperduto and Hiller, 1986), atherosclerosis (Vingerling et al., 1995), and reduced choroidal blood flow (Friedman et al. 1995). One or more of these proposed mechanisms may shed light on the development of AMD, but none explain why the disease is localized to the central retina.

We offer another hypothesis: AMD is in part a metabolic disorder. We suggest that the diffusion properties of retinal tissue limit the rate of supply of nutrients to the avascular macula, metabolically stressing centrally located rods and cones and contributing to their eventual loss. Distant from capillaries of the retinal vasculature, the centrally located rods and cones may be at a metabolic disadvantage compared to photoreceptors in the peripheral retina. How long-term metabolic stress could trigger apoptosis in an aging retina is not understood. Several observations suggest the involvement of mitochondria. For example, Barron et al. (2001) found that mutated mitochondrial DNA (mtDNA) are more abundant in the center than in the periphery of the aging human retina. Also interesting is the involvement of the macula in more than half of patients with mitochondrial disease (Sue et al., 1997; Isashiki et al., 1998). Mutations in mtDNA that cause Leber's hereditary optic neuropathy impair central vision, but the causes are not known (Brown and Wallace, 1994). The molecular/cellular mechanisms underlying photoreceptor loss in the aging retina are also not known, however, the role of mtDNA mutations in inducing apoptosis (Susin et al., 1998) may provide a clue.

Our working hypothesis that AMD is a metabolic disorder evolved in part from observations of the effects of hypoglycemia on the central vision of diabetics. We do not mean to imply that diabetics are at risk for AMD; indeed no consistent relationship has been found between diabetes and AMD (Mitchell and Wang, 1999). Rather we reasoned that the transient impairment of central vision experienced by diabetics when hypoglycemia may be indicative of the sensitivity of the macula to metabolic stress. If this is the case, then metabolically stressed photoreceptors may be primary foci that trigger the multiple complications of AMD or they may act in concert with RPE and choroidal abnormalities to produce this multifactorial retinal disease.

6. ACKNOWLEDGMENTS

Supported the National Eye Institute (EY00667 and EY12286), Research to Prevent Blindness, Joslin Center (Syracuse, NY), and the Lions of Central NY.

7. REFERENCES

Ames, A., III, and Gurian, B. S., 1963, Effects of glucose and oxygen deprivation on function of isolated mammalian retina. *J. Neurophysiology*, **26**, 617-634.

Aschoff, J. , 1979, Circadian rhythms: general features and endocrinologic aspects. In Endocrine Rhythms (Krieger, D. T., ed.) NewYork: Raven Press, pp. 1-6.

Age-related Eye Disease Study Research Group. A randomized, placebo-controlled, clinical; trial of high-dose supplementation with vitamens C and E, beta carotene, and zinc for age-related macular degeneration and vision loss. AREDS Report No.8, *Arch. Ophthalmol.* **119**: 1417-1436.

Barlow, R.B., Boudreau, E.A., Moore, D.C., Huckins, S.C., Lindstrom, A.B., Farell, B., 1997,. Glucose and time of day modulate human contrast sensitivity and fMRI signals from visual cortex. *Invest. Ophthal. Vis. Sci.* **38**: S735.

Barlow, R.B., Khan, M., and Farell, B., 2003, Metabolic Modulation of Human Visual Sensitivity. *Invest. Ophthal. Vis. Sci.* **44**: E-Abstract.

Barron, M.J., Johnson, M.A., Andrews, R.M., Clarke, M.P, Griffiths, P.G, Bristow, E., Lang-Ping,P., Durham, S., Turnbull, D.M., 2001, Mitochondrial abnormalities in ageing macular photoreceptors. *Invest Ophthal. Vis Sci.* **42**(12):3016-22.

Bassi, C. J., and Powers, M. K., 1986, Daily fluctuations in the detectability of dim lights by humans. *Physiology and Behavior*, **38** , pp. 871-877.

Brown, M.D., and Wallace, D.G. , 1994, Spectrum of mitochrondrial DNA mutations in Leber hereditary optic neuropathy. *Clin Neurosci.* **2**: 138-45.

Clore, J., Nestler, J. and Blackard, W., 1989 Sleep-associate sleep fall in glucose disposal and hepatic glucose output in normal humans. *Diabetes.* **38**, 285-290.

Crabb, J.W., Miyagi, M., Gu, X., Shadrach, K., West, K.A., Sakaguchi, H., Kamei, M., Hasan, A., Yan, L., Rayborn, M., Salomon, R.G., and Hollyfield, J.G., 2002, Drusen proteome analysis: An approach to the etiology of age-related macualar degeneration, *Proc. Nat. Aca. Sci.,* **99**: 14682-14687.

Curcio, C.A., Owsley, C., Jackson, G.R., 2000, Spare the rods, save the cones in aging and age-related maculopathy. *Invest Ophthalmol Vis Sci.* **41**(8):2015-8.

Friedman, E., Krupsky, S., Lane, A., 1995, Ocular blood flow velocity in age-related macular degeneration. Ophthalmology, **102**:640-646.

Ewing F.M., Deary, I.J., McGrimmon , R.J., Strachen, M.W., Frier B.M., 1998, Effect of acute hypoglycemia on visual information processing in adults with type 1 diabetes mellitus. *Physiol. Behav.* 64:653-60.

Isashiki, Y., Nakagawa, M., Ohba, N., Kamimura, K., Sakoda, Y, et al., 1998, Retinal manifestations in mitochondrial diseases associated with mitochondrial DNA mutation. *Acta Ophthalmol Scand.* **76**(1):6-13.

Linsenmeier, R.A. and Braun, R.D., 1992, Oxygen distribution and consumption in the cat retina during normoxia and hypoxemia. *J Gen Physiol.* **99**(2):177-97.

Macaluso, C., Onoe, S., Niemeyer, G., 1992, Changes in glucose level affect rod function more than cone function in the isolated, perfused cat eye. *Invest. Opthal. & Vis. Sci.*, **33**, 2798-2807.

McGrimmon, R.J., Deary, I.J., Huntly, B.J.P., MacLoed, K.J., Frier, B.M., 1996, Visual information processing during controlled hypoglycaemia in humans. *Brain* **119**:1277-87.

McFarland, R. A., and Evans, J. N., 1939, Alterations in dark adaptation under reduced oxygen tensions. *Am. J. Physiol.*, **127**, 37.

McFarland, R. A., and Forbes, W. H., 1940, The effects of variations in the concentration of oxygen and of glucose on dark adaptation. *J. Gen. Physiol.*, **24**, 69.

McFarland, R.A., Halperin, M.H., and Niven, J.I., 1945, Visual thresholds as an index of the modification of the effects of anoxia by glucose *Amer. J. Physiol* **144**:378.

McRipley, M.A., Ahmed, J., Chen, E.P., Linsenmeier, R.A., 1997, Effects of adaptation level and hypoglycemia on function of the cat retina during hypoxemia. *Vis Neurosci* **14**(2):339-50.

Mitchell, P., and Wang, J.J., 1999, Diabetes, fasting blood glucose and age-related maculopathy: The Blue Mountains Eye Study. *Aust N Z J Ophthalmol.* **27**(3-4):197-9.

Moore, D.J., Hussain, A.A., and Marshall, J., 1995, Age-related variation in the hydraulic conductivity of Bruch's membrane. *Invest Ophthalmol Vis Sci.* **36**(7):1290-7.

Sannita, W.G., Balestra, V., DiBon, G., Gambaro, M., Malfatto, L., and Rosadini, G., 1993, Spontaneous variations of flash-electroretinogram and retinal oscillatory potentials in healthy volunteers are correlated to serum glucose. *Clin. Vision Sci.*, 8; 147-158.

Schmidt, M., Giessl, A., Laufs, T., Hankeln, T., Wolfrum, U., Burmester, T., 2003, How does the eye breathe? A neuroglobin-mediated oxygen supply in the mammalian retina. *J Biol Chem.***278**(3):1932-5.

Sperduto, R.D., Hiller, R., 1986, Systemic hypertension and age-related maculopathy in the Framingham Study. Arch. Ophthalmol. **104**(2):216-219.

Simonson, D.C., Tamborlane, W.V., DeFronzo, R.A., and Sherwin, R.S., 1985, Intensive Insulin therapy Reduces Counterregulatory Hormone responses to Hypoglycemia in Patients with Type I diabetes *Annals of Internal Medicine* **130**: 184-190.

Snodderly, D.M., 1995, Evidence for protection against age-related macular degeneration by carotenoids and antioxidant vitamins. *Am J Clin Nutr.* **62**(6 Suppl):1448S-1461S.

Steinberg, R.H. 1987, Monitoring Communications Between Photoreceptors and Pigment Epithelial Cells: Effects of "Mild" Systemic Hypoxia. *Invest. Opthal. & Vis. Sci.*, **28**:1888-1904.

Sue, C.M., Mitchell, P., Crimmins, D.S., Moshegov, C., Byrne, E., Morris, J.G., 1997, Pigmentary retinopathy associated with the mitochondrial DNA 3243 point mutation. *Neurology.* **49**(4):1013-7.

Susin, S.A., Zamzami, N., Kroemer, G., 1998, Mitochondria as regulators of apoptosis: doubt no more *Biochim Biophys Acta* **1366**: 151-165.

Sutter, E.E. and Tran, D., 1992, The field topography of ERG components in man--I. The photopic luminance response. *Vision Res.* **32**(3):433-46.

Vingerling J.R., Dielemans I, Bots ML, Hofman A, Grobbee DE, de Jong PT., 1995. Age-related macular degeneration is associated with atherosclerosis. The Rotterdam Study. *Am J Epidemiol.* **142**(4):404-9.

Van Cauter, E., Desir, D., Decoster, C., Fery, F., and Balasse, E. O. (1988) Nocturnal decrease in glucose tolerance during constant glucose infusion, *J. Clin. Endocrinol. Metab.*, **69**, No. 3, 604-610.

Van Cauter, E., Blackman, J. D., Roland, D., Spire, J. P., Refetoff, S., Polonsky, K. S. (1991) Modulation of glucose regulation and insulin secretion by circadian rhythmicity and sleep, *J. Clin. Invest.*, **88**, 934-942.

Watson, A.B., and Pelli, D.G.,1983, QUEST: a Bayesian adaptive psychometric method. *Percept.Psychophys.* **33**(2):113-20.

Winkler, B. S., 1981, Glycolytic and oxidative metabolism in relation to retinal function. *J. Gen. Physiol.*, **77**, 667-692.

Winkler, B. S., Boulton,M.E., Gottsch, J.D. Sternberg, P., 1999, Oxidative damage and age-related macular degeneration. *Molecular Vision.* **5**:32.

Young, R., 1988, Solar radiation and age-related macular degeneration. *Surv. Ophthalmol.* **32**: 252-269.

MITOCHONDRIAL UNCOUPLING PROTEINS: REGULATORS OF RETINAL CELL DEATH

Colin J. Barnstable[1], Man Li[1], Rajini Reddy[1], and Tamas L. Horvath[2]

1. INTRODUCTION

Many forms of neuronal degeneration are caused by events leading to increased oxidative stress, elevated levels of reactive oxygen radicals, and particularly the superoxide radical. Such elevated levels can initiate the irreversible cascade of reactions causing apoptotic cell death (Halliwell, 1992). In the neural components of the eye, oxidative stress can affect a number of cell types. In the inner retina there is extensive evidence that such stress occurs in a number of forms of glaucoma and optic nerve disorders that result in ganglion cell death. As energetically demanding cells, photoreceptors are also thought to be susceptible to oxidative stress induced by metabolic conditions, light or changes in intracellular metabolism induced by specific mutations (Sandbach et al., 2001).

The levels of reactive oxygen radicals are the sum of the contributions of two sources. The first source is a wide array of exogenous insults. The second source is the endogenous production in mitochondria during oxidative phosphorylation. Under normal aerobic conditions between 0.4% and 4% of molecular oxygen is converted into superoxide (Melov et al., 1999). Reduction of either exogenous or endogenous sources of free radical oxygen production will improve neuronal survival.

2. MITOCHONDRIAL UNCOUPLING PROTEINS

Uncoupling proteins encoded by nuclear DNA are located in the inner membrane of the mitochondria and their primary mechanism of action is thought to be to leak hydrogen protons from the intermembrane space to the matrix of the mitochondria (Nicholls and Locke, 1994). This decreases the proton gradient that serves as the driving force for ATP synthesis by ATP synthase (see Fig. 1). The energy released by this leak pathway is

[1]Department of Ophthalmology and Visual Science, and [2]Department of Obstetrics and Gynecology, Yale University School of Medicine, New Haven, CT 06520, USA. Corresponding author: C. J. Barnstable, E-mail: colin.barnstable@yale.edu

Retinal Degenerations: Mechanisms and Experimental Therapy
Edited by LaVail *et al.*, Kluwer Academic/Plenum Publishers, 2003

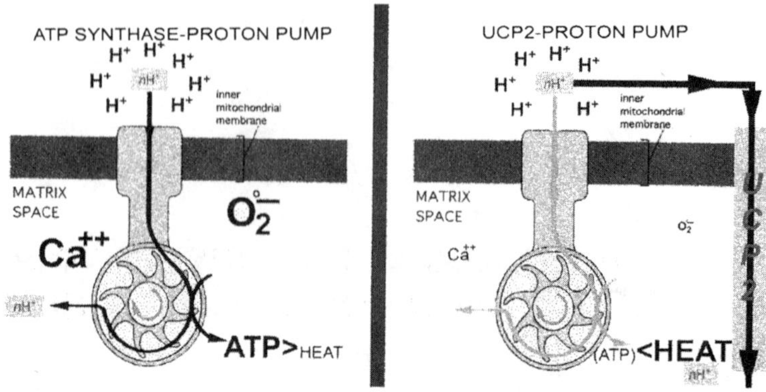

Figure 1. Schematic illustration of functioning ATP synthase on the left and the potential effects of UCP2 on the right. The proton gradient drives ATP synthase and the production of ATP. If that proton gradient is decreased due to the activity of UCP2, ATP production is decreased, energy is dissipated in the form of heat (thermogenesis), superoxide (O_2^{o-}) production is decreased and calcium efflux is increased. However, despite the decreased ATP production by individual mitochondria, because uncoupling supports mitochondrial proliferation, neurons will overall have more ATP available. Adapted from Horvath et al., 2003.

Mouse Uncoupling proteins

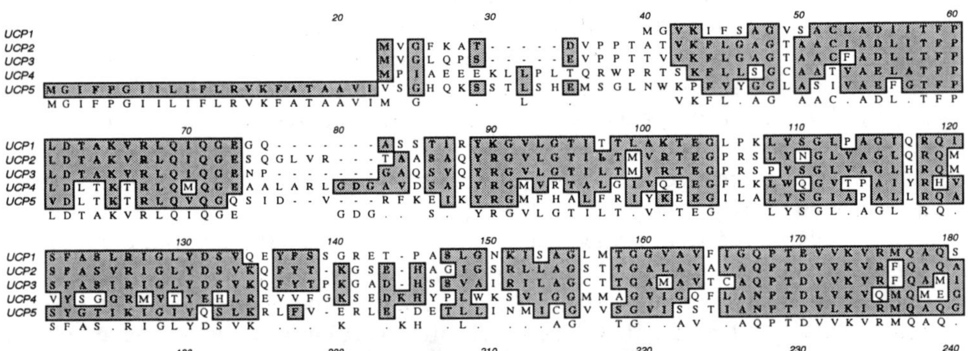

Figure 2. Alignment of mouse UCP protein sequences. The GenBank Accession numbers used were AK002759 (UCP1), NM011671 (UCP2), NM009464 (UCP3), AK014394 (UCP4) and AF155812 (UCP5). The consensus sequence is shown below and identical residues are shaded in gray.

dissipated in the form of heat. Recent findings suggest that uncoupling protein action can also diminish production of superoxides and block calcium entry to the mitochondrial matrix (Echtay et al., 2002; Castilho et al., 1999).

The most well characterized uncoupling protein, UCP1, is expressed solely in the brown adipose tissue and is responsible for thermogenesis in small rodents (Nicholls and Locke, 1994. Because brown adipose tissue is virtually insignificant for normal primate, including human, physiology this mechanism was thought to have little widespread importance until recently. In the last few years, however, several other members of the uncoupling protein family have been discovered and found to promote partial uncoupling of oxidation from phosphorylation *in vitro*. In addition to UCP1, this gene family now includes UCP2, UCP3, UCP4 and brain mitochondrial carrier protein 1 (BMCP1 or UCP5) (Bouillaud et al., 1985; Fleury et al., 1997; Boss et al., 1997; Vidal-Puig et al., 1997; Mao et al., 1999; Sanchis et al., 1998).

Comparison of the protein sequences of mouse UCPs clearly shows multiple regions of homology (Figure 2). A more comprehensive phylogenetic analysis has suggested that UCP4 is most representative of the original uncoupling protein (Hanak and Jezek, 2001). Each uncoupling protein has a characteristic mitochondrial localization sequence at the N-terminus. Each has three domains of approximately 100 amino acids and each domain has two potential transmembrane spanning regions. Both the N- and C-termini face the intermembranous space of the mitochondria. The mechanisms regulating uncoupling protein activity are not fully understood but both fatty acids and Coenzyme Q are essential cofactors.

3. UNCOUPLING PROTEINS ARE NEUROPROTECTIVE IN THE CNS

Only three of the five uncoupling proteins have been shown to have physiological uncoupling activity (UCP1, UCP2 and UCP3). Of these, UCP2 is the only one expressed in many regions of the CNS (Horvath et al., 1999; Diano et al., 2000). UCP2 is upregulated following acute brain injury and this has been interpreted as part of a protective response (Bechmann et al., 2002). To provide further evidence that UCP2 is protective we have tested the cell death in CA1 of hippocampus following drug-induced epileptic seizures in wild type mice and mice expressing a UCP2 transgene. The wild type mice showed approximately 50% dying cells whereas the UCP2 transgenic animals showed less than 10% dying cells. A similar dramatic protection has been found in an MPTP-induced primate model of Parkinson's disease in which increasing levels of the co-factor Coenzyme Q increased UCP2 activity. In the MPTP-treated animals the loss of substantia nigral dopamine neurons was approximately 70% but in the Coenzyme Q pretreated animals the same MPTP treatment did not cause any statistically significant loss of cells over controls. Together, these types of experiments indicate that UCP2 expression and activation can provide substantial protection to neurons subjected to a variety of injuries.

4. EXPRESSION OF UNCOUPLING PROTEINS IN RETINA

Given that many of the cell types in retina can be damaged by oxidative stress, we determined whether any of the uncoupling protein genes were expressed, and if so in which cell types. First, we carried out RT-PCR with RNA samples from adult wild type and rd1 retinas using primer pairs specific for each UCP isoform. For all positive results, bands were sequenced to verify their identity. The results show that UCP 2, 3, 4 and 5 were all detected in wild type mouse retina (Figure 3, *left*). All four genes were also detected in the *rd1* retinal cDNA (not shown), suggesting that their expression does not depend on an intact rod photoreceptor layer.

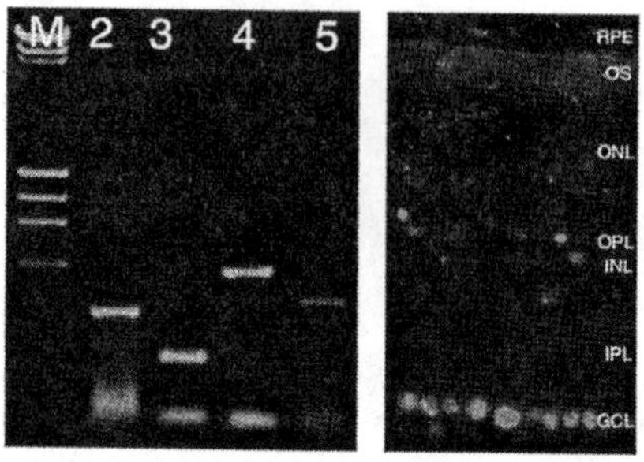

Figure 3. *Left.* RT-PCR amplification of UCP2, UCP3, UCP4 and UCP5 from wild type retinal cDNA. Each set of primers gave a specific band of the expected size whose identity was confirmed by DNA sequencing. *Right.* Immunocytochemical labeling with an antibody raised against UCP2 (C-20, Santa Cruz Biotechnology, Santa Cruz, CA). Apart from some non-specific labeling of blood vessels and some indistinct labeling over the outer retina, the only clear labeling was a punctate labeling over cell bodies in the ganglion cell layer.

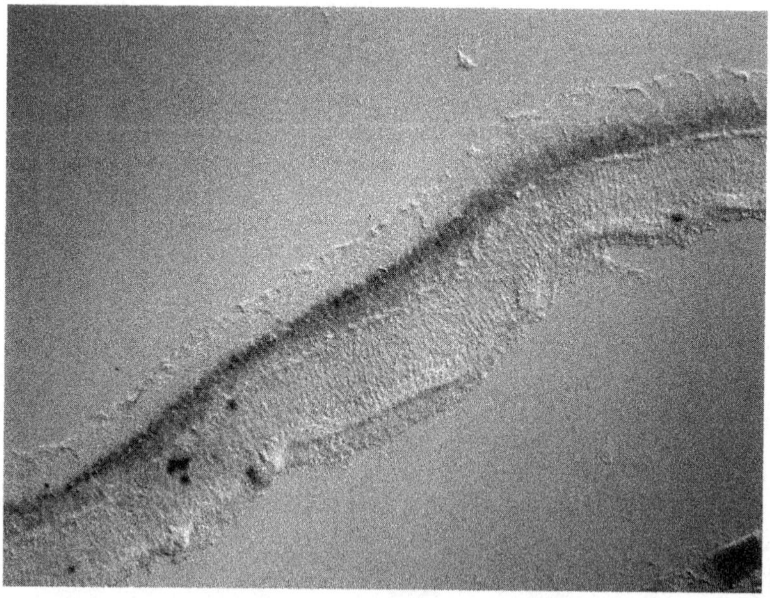

Figure 4. Digoxigenin-labeled antisense probe recognizing mouse UCP5 hybridized to adult mouse retina section. Label appears over the inner portion of the INL, suggesting that the gene is expressed primarily in amacrine cells.

Using antibodies against UCP2, we detected specific labeling over the ganglion cell layer (Figure 3, *right*). At least some of these were ganglion cells as judged by double labeling with antibodies against neurofilaments (not shown).

To further analyze expression of the uncoupling proteins we carried out in situ hybridization using digoxigenin labeled riboprobes. An example of expression is shown in Figure 4. UCP5 RNA was clearly expressed in the inner retina with labeling of amacrine and ganglion cells. To date we have not observed strong expression of any UCP gene in the photoreceptor layer.

5. UNCOUPLING PROTEINS MODULATE PROGRAMMED CELL DEATH IN GANGLION CELLS.

During the first postnatal week, mouse retinal ganglion cells undergo a substantial amount of cell death that is associated with the refinement of their projections to target areas. When we examined the cell density in the ganglion cell layers of wild type and UCP2 transgenic mice, we found that the transgenic animals had up to 50% more cells. While these counts did not distinguish between ganglion and amacrine cells, the two possible explanations for this result are that the transgenic animals made more cells or that they underwent diminished cell death. To test this we examined retinas from wild type and UCP2 transgenic mice at different ages. Using cresyl violet staining of pyknotic nuclei, TUNEL staining or staining for activated caspase 3 we observed a decrease in apoptotic cells during the first postnatal week in the transgenic animals as compared with wild type.

6. CONCLUSIONS

There is now strong evidence that mitochondrial uncoupling proteins play an important neuroprotective role in many parts of the mammalian CNS. UCP2 can be regulated at the transcriptional and post-transcriptional levels and it appears to be part of a dynamic response mechanism to a wide variety of brain injuries and diseases. Because many cell types in the retina are known to be susceptible to the deleterious effects of oxidative stress, we began this study expecting to find robust expression of uncoupling proteins in many retinal cells, particularly photoreceptors. It was, therefore, somewhat surprising that we could find no evidence for expression of any of these proteins in photoreceptors. Although our results should probably be confirmed by more sensitive methods, such as PCR analysis of single rods, it would appear that photoreceptors require tightly coupled mitochondria. Perhaps the inability of these cells to reduce endogenous superoxide production accounts for the observed increase in cell death when metabolic activity or other mild insults occur in the retina (Maslim et al., 1997). If photoreceptors cannot effectively regulate oxidative stress through intrinsic mitochondrial uncoupling proteins, it should, perhaps, not be surprising that they have a dependence upon the neuroprotective activity of factors secreted by the retinal pigment epithelium (Cao et al., 1999; Cao et al., 2001).

7. ACKNOWLEDGEMENTS

Original work mentioned in this article has been supported by grants from the NIH, from the Allene Reuss Memorial Trust, the Kemper and the Foresight Foundations and from Research to Prevent Blindness, Inc.

8. REFERENCES

Bechmann, I., Diano, S., Warden, C.H., Bartfai, T., Nitsch, R., and Horvath, T.L. Brain mitochondrial uncoupling protein 2 (UCP2): a protective stress signal in neuronal injury. Biochem. Pharmacol. **64**, 363-367 (2002).

Boss, O., Samec, S., Paolini-Giacobino, A., Rossier, C., Dulloo, A., Seydoux, J., Muzzin, P., Giacobino, J.-P. Uncoupling protein-3: a new member of the mitochondrial carrier family with tissue-specific expression. FEBS Lett. **408**, 39-42 (1997).

Bouillaud, F., Ricquier, D., Thibault, J., and Weissenbach, J.. Molecular approach to thermogenesis in brown adipose tissue: cDNA cloning of the mitochondrial uncoupling protein. Proc. Natl. Acad. Sci. USA **82**, 445-448 (1985).

Cao, W., Tombran-Tink, J., Chen, W., Mrazek, D., Elias, R., and McGinnis, J.F. Pigment epithelium-derived factor protects cultured retinal neurons against hydrogen peroxide-induced cell death. J. Neurosci. Res. **57**, 789-800 (1999).

Cao, W., Tombran-Tink, J., Elias, R., Sezate, S., Mrazek, D., and McGinnis, J.F. In vivo protection of photoreceptors from light damage by pigment epithelium-derived factor. Invest. Ophthalmol. Vis. Sci. **42**, 1646-1652 (2002).

Castilho, R.F., Hansson, O., Ward, M.W., Budd, S.L., and Nicholls, D.G. Mitochondrial control of acute glutamate excitotoxicity in cultured cerebellar granule cells. J. Neurosci. **18**, 10277-10286 (1998).

Diano, S., Urbanski, H.F., Horvath, B., Bechmann, I., Kagiya, A., Nemeth, G., Naftolin, F., Warden, C.H. and Horvath, T.L. Mitochondrial Uncoupling Protein 2 (UCP2) in the Nonhuman Primate Brain and Pituitary. Endocrinology **141**, 4226-4238 (2000).

Echtay, K.S., Roussel, D., St-Pierre, J., Jekabsons, M.B., Cadenas, S., Stuart, J.A., Harper, J.A., Roebuck, S.J., Morrison, A., Pickering, S., Clapham, J.C. and Brand, M.D. Superoxide activates mitochondrial uncoupling proteins. Nature **415**, 96-99 (2002).

Fleury, C., Neverova, M., Collins, S., Raimbault, S., Champigny O., Levi-Meyrueis, C., Bouillaud, F., Seldin, M.F., Surwit, R.S., Ricqier, D., and Warden, C.H. Uncoupling protein-2: a novel gene linked to obesity and hyperinsulinemia. Nat. Genet. **15**, 269-272 (1997).

Halliwell, B. Reactive oxygen species and the central nervous system. J Neurochem. **59**,1609 –1623 (1992).

Hanak, P. and Jezek, P. Mitochondrial uncoupling proteins and phylogenesis--UCP4 as the ancestral uncoupling protein. FEBS Letters. **495**, 137-141 (2001).

Horvath, T.L, Diano, S., and Barnstable, C.J. Mitochondrial uncoupling protein 2 in the central nervous system. Biochem. Pharmacol. (In Press).

Horvath, T.L., Warden, C.H., Hajos, M., Lombardi, A., Goglia, F., Diano, S. Brain uncoupling protein2: uncoupled neuronal mitochondria predict thermal synapses in homeostatic centers. J Neurosci. **19**, 10417–10427 (1999).

Mao, W., Yu, X.X., Zhong, A., Li, W., Brush, J., Sherwood, S.W., Adams, S.H., Pan, G. UCP4, a novel brain-specific mitochondrial protein that reduces membrane potential in mammalian cells. FEBS Lett. **443**, 326-330 (1999).

Maslin, J., Valter, K., Egensperger R., Hollander, H., Stone, J. Tissue oxygen during a critical developmental period controls the death and survival of photoreceptors. Invest. Ophthalmol. Vis. Sci. **38**, 1667-1677 (1997).

Melov, S., Coscun, P., Patel, M., Tuinstra, R., Cottrell, B., Jun, A.S., Zastawny, T.H., Dizdaroglu, M., Goodman, S.I., Huang, T.T., Miziorko, H., Epstein, C.J., and Wallace, D.C. Mitochondrial disease in superoxide dismutase 2 mutant mice. Proc. Natl. Acad. Sci. (USA). **96**, 846-851, (1999).

Miroux, B., Collins, S., Seldin, M., Richard, D., Warden, C., Bouillaud, F., and Ricquier, D. BMCP1, a novel mitochondrial carrier with high expression in the central nervous system of humans and rodents, and respiration uncoupling activity in recombinant yeast. J. Biol. Chem. **273**, 34611-34615 (1998).

Nicholls, D.G. and Locke, R.M. Thermogenic mechanism in brown fat. Physiol. Rev. **64**: 1-64 (1984).

Sanchis, D., ·Fleury, C., Chomiki, N., Goubern, M., Huang, Q., Neverova, M., Grégoire, F., Easlick, J., Raimbault, S., Lévi-Meyrueis, C., Miroux, B., Collins, S., Seldin, M., Richard, D., Warden, C. Bouillaud, F., and Ricquier, D. BMCP1, a novel mitochondrial carrier with high expression in the central nervous system of humans and rodents, and respiration uncoupling activity in recombinant yeast. J. Biol. Chem. **273**;34611-34615 (1998).

Sandbach, J.M., Coscun, P.E., Grossniklaus, H.E., Kokoszka, J.E., Newman, N.J., and Wallace, D.C. Ocular Pathology in mitochondrial superoxide dismutase (Sod2)-deficient mice. Invest. Ophthalmol. Vis. Sci. **42**, 2173-2178 (2001).

Vidal-Puig, A., Solanes, G., Grujic, D., Flier, J.S., and Lowell, B.B.. UCP3: an uncoupling protein homologue expressed preferentially and abundantly in skeletal muscle and brown adipose tissue. Biochem. Biophys. Res. Commun. **235**, 79-82 (1997).

ENERGY DEPLETION HYPOTHESIS FOR RETINITIS PIGMENTOSA

Marion S. Eckmiller[1]

1. INTRODUCTION

Enormous advances have been made in our understanding of human retinal degenerations, eg., numerous gene defects that cause retinitis pigmentosa (RP) have been identified and shown in many cases to encode photoreceptor cell proteins that participate in phototransduction (Clarke et al., 2000; Phelan and Bok, 2000; Pugh and Lamb, 2000). Although RP is genetically heterogeneous, the similar phenotype of typical RP (disturbed dark adaptation that progresses to night blindness, then complete blindness) has led researchers to speculate about a common underlying pathological mechanism. The equivalent light hypothesis for RP, formulated because constant or excessive light can cause photoreceptor degeneration in animals, has received some experimental support (Fain and Lisman, 1999). The pathological defect in RP has also been suggested to result from disturbance of the cytoskeletal systems in photoreceptor outer segments (OS) (Eckmiller, 1997), more specifically from disturbance of the special microtubule-containing cytoskeletal system at the multiple incisures in many rod OS (Eckmiller, 2000). For unknown reasons, rods are more susceptible than cones to degeneration by both light and gene mutation. It is particularly puzzling that gene defects in myosin VIIa cause Usher syndrome (RP with hearing and vestibular defects) in humans but do not cause visual disturbances or blindness in mice or zebrafish (Gibson et al., 1995; Weil et al., 1995; Ernest et al., 2000). This puzzle has not been clarified by studies showing that myosin VIIa is localized in rodent rods and cones at the base of the ciliary axoneme, where it participates in transporting proteins (eg., opsin) into the OS (Liu et al., 1999).

This chapter presents immunofluorescent evidence in amphibian photoreceptors that myosin VIIa is localized at cytoskeletal systems that are somewhat different in rods and cones, i.e., myosin VIIa is associated with OS axonemes and also with the microtubules in the cytoskeletal system at the multiple incisures of rod OS. The cytoskeletal systems in photoreceptor OS in human retinas closely resemble those in amphibian retinas.

[1]Vogt Brain Research Institute, University Clinic, Heinrich Heine University of Düsseldorf, Postfach 101007, D-40001 Düsseldorf, Germany. E-mail: eckmill@hirn.uni-duesseldorf.de

Accordingly, the distribution of myosin VIIa is consistent with the suggestion that RP results from a pathological disturbance of the cytoskeletal system at the multiple incisures of human rod OS (Eckmiller, 2000), which can clarify the different susceptibility to degeneration of human rods and cones, and the diverse retinal phenotypes resulting from myosin VIIa defects in different species. Synthesizing these findings with other available evidence, I speculate about the function performed by this special cytoskeletal system at the multiple incisures of rod OS in many retinas. This leads me to formulate the novel hypothesis that the progressive disturbance in rod function in RP can be due to local energy depletion, which suggests therapeutic treatment for some RP by local energy supplementation (part of this work has been presented in abstract form: Eckmiller, 2001).

2. MATERIALS AND METHODS

Retinas from African clawed toads (*Xenopus laevis*) were mechanically dissociated, fixed, applied to glass slides, and processed for indirect immunofluorescent staining of myosin VIIa using previously described methods (Eckmiller 2000). Experimental slides were incubated with polyclonal antibody to myosin VIIa (Affinity Bioreagents, CO, USA) from rabbits immunized with a synthetic peptide to a 15 amino acid sequence (identical in mouse and human myosin VIIa) from the myosin VIIa N-terminal domain; control slides were incubated with antibody competitively pre-inhibited with the neutralizing synthetic peptide. All slides were incubated with fluorochrome-conjugated goat anti-rabbit immunoglobin and examined as previously described (Eckmiller 2000).

3. RESULTS AND DISCUSSION

As shown in Figure 1, the OS of rods (A and B) on experimental slides displayed small, regularly spaced puncta of anti-myosin VIIa fluorescence, which occurred along parallel lines that coincided with their multiple incisures. Rod OS also had a bright, continous streak of fluorescence (arrow in B) that extended part way to the distal end of the OS and resembled the axoneme. The only fluorescence in the OS of cones (B and C) on experimental slides was a bright, continuous fluorescent streak that extended to the OS tip and resembled the axoneme. There was no significant fluorescence in OS on control slides, showing that the labelling on experimental slides was specific for myosin VIIa.

In amphibian retinas, the OS of rods and cones have microtubules in their ciliary axonemes and rod OS also have individual microtubules at the mouth of their multiple incisures (Eckmiller, 2000). Thus, myosin VIIa occurs in amphibian photoreceptor OS at precisely those locations where microtubules are present in the cytoplasm; its distribution along the microtubules is continuous at axonemes, punctate at rod multiple incisures.

These findings are consistent with previous reports of a spot of myosin VIIA at the axoneme base in rodent photoreceptor OS (Liu et al., 1999), and significantly extend those findings by revealing myosin VIIa along the entire length of the axoneme and also at multiple incisures in rod OS. The additional localization of myosin VIIa in the present study was made possible by the use of methods that stabilize microtubules and the retinas of amphibians, whose rod OS have multiple incisures. These findings provide convincing evidence that myosin VIIa is one molecular component of complex microtubule-based cytoskeletal systems in the OS of amphibian photoreceptors.

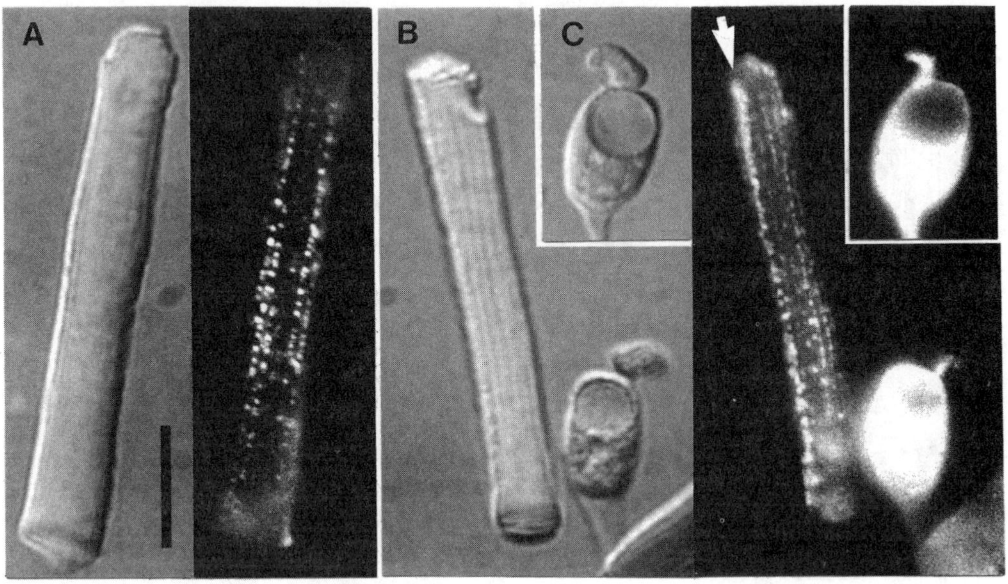

Figure 1. Light micrographs of anti-myosin VIIa fluorescence in photoreceptors and OS on experimental slides, each photographed with Nomarski optics (left) and epifluorescence (right). Magnification bar = 10 μm.

The cytoskeletal systems organized around microtubules in photoreceptor OS have numerous components; so far the same proteins have been identified at ciliary axonemes and multiple incisures (Eckmiller, 2000, 2002; Eckmiller et al., 2002), implying that these cytoskeletons perform functions that are similar. The pattern of myosin VIIa labelling in amphibian photoreceptor OS closely resembles that of several other identified protein components of these cytoskeletal systems (Eckmiller, 2000). Most features of the cytoskeletal system at the axoneme are the same in rods and cones (though the relative length of their axonemes and OS differ: Eckmiller, 1997). However, the cytoskeletal systems in amphibian rods and cones differ dramatically in that an additional cytoskeletal system occurs only in rods. The cytoskeleton at rod OS multiple incisures is unique in several respects (Eckmiller, 2000), eg., it has no direct connection to the IS, contains separate individual microtubules, and is composed of some proteins distributed along the microtubules in a punctate pattern. The cytoskeleton at rod OS multiple incisures is quite similar to, but more complex than, that at OS axonemes. Accordingly, the cytoskeleton at rod OS multiple incisures can be expected to perform a function that is quite similar to, but more complex than, that at OS axonemes.

The diverse retinal phenotypes resulting from gene defects in myosin VIIa in humans versus mice and zebrafish imply that their photoreceptors differ in some important unidentified manner. The present identification of myosin VIIa as a component of all the microtubule-based cytoskeletal systems in the OS of amphibian photoreceptors provides a new explanation for these differences between species. As illustrated in Figure 2, the cytoskeletal systems in the OS of cones in these species are not markedly different, but the number of incisures in the OS of their rods differs (Eckmiller, 2000): human rods have multiple incisures, mouse and fish rods have at most one incisure (at the cilium).

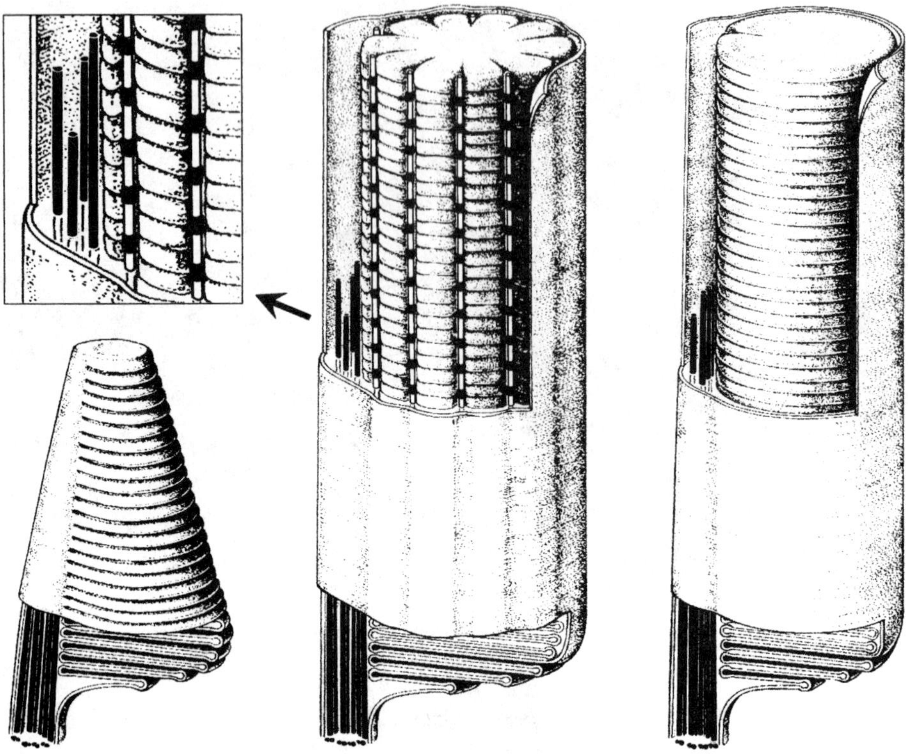

Figure 2. Schematic drawings showing the distribution of myosin VIIa (in black) at the microtubules in the OS of a cone (left), a rod having multiple incisures (middle), and that expected in a rod without multiple incisures (right). In all cases, myosin VIIa occurs at all heights of the microtubules in the ciliary axoneme. In the OS of rods having multiple incisures (as in amphibian, primate, and human retinas), myosin VIIa also occurs as puncta at regularly-spaced heights along the microtubules at multiple incisures (see inset). This additional myosin VIIa is not present in cone OS and is not expected in rod OS lacking multiple incisures (as in fish and rodent retinas).

With regard to their microtubule-based cytoskeletal systems (presumed to likewise contain myosin VIIa), the OS of rods in fish and rodent retinas thus resemble cone OS.

The present findings may have clinical relevance, because the similar features of photoreceptor OS in amphibian and human retinas make it likely that myosin VIIa is distributed in human photoreceptor OS as shown here for amphibian photoreceptor OS. Eckmiller (2000) has suggested that the function of the rod-specific cytoskeletal system at the multiple incisures of human rods is disturbed in typical RP. Assuming that this suggestion is correct, it implies that Usher syndrome can be caused by a defect in an important structure and/or function associated with puncta of myosin VIIa distributed along the microtubules at multiple incisures in human rod OS. This cytoskeletal system and this punctate distribution of myosin VIIa are not present in the OS of cones, the OS of rods in mouse retinas, or the OS of rods in zebrafish retinas. This special cytoskeletal system (having a punctate distribution of myosin VIIa) is uniquely present at the multiple incisures in the OS of rods in amphibian, primate, and human retinas. These findings provide compelling, albeit indirect, support for the suggestion that a pathological defect in the cytoskeletal system at the multiple incisures of human rod OS can mediate RP.

What is the function of the cytoskeletal system at the multiple incisures of rod OS? There are many clues. This function is apparently not needed in cones but only in rods, and only in rods of certain vertebrate species. It presumably represents a modification of a somewhat similar function performed by the more fundamental cytoskeletal system at the ciliary axoneme in all rods and cones. Disturbance of this function can cause the phenotype of typical RP. Many proteins participate in this function, eg., those localized at multiple rod OS incisures, those that undergo biochemical reactions with these proteins, and many of those encoded by RP-causing genes (Eckmiller, 1999; Clarke et al., 2000; Phelan and Bok, 2000; Pugh and Lamb, 2000; Eckmiller, 2001; Eckmiller et al., 2002).

Rods are more susceptible to degeneration than cones. Light has been implicated in causing photoreceptor degeneration, but darkness has not. What might be special about rods and light? Cones do not saturate even in bright light, but rods are very sensitive at scotopic light levels, become less sensitive at mesopic light levels, and become saturated at photopic light levels (Rodieck, 1998; Fain et al., 2001). Thus, a unique situation is expected to occur in rods in the retinas of humans and diurnal animals during their daily exposure to daylight; this situation does not occur in cones and may not occur in rods in nocturnal animals. It may be fruitful to focus on the processes taking place within the OS of rods at photopic light levels, when light can apparently bleach rhodopsin and activate phototransduction but this serves no useful purpose because the rods are fully saturated.

4. THE ENERGY DEPLETION HYPOTHESIS

Phototransduction is known to proceed in the OS of normal rods as a sequence of coupled biochemical reactions (Figure 3). The absorption of light by rhodopsin (Rh) in the OS disk membranes catalyzes the activation of transducin (T), which activates phosphodiesterase (PDE), which decreases the concentration of cyclic GMP (cGMP) in the cytoplasm surrounding the microtubules, which closes the cyclic nucleotide gated channels (CNGC) in the plasma membrane and hyperpolarizes the OS (Rodieck, 1998; Pugh and Lamb, 2000; Burns and Baylor, 2001). In the dark each of the activated proteins becomes deactivated, effectively halting the reaction sequence, and guanylate cyclase (GC) resynthesizes cGMP, which opens the CNGC and depolarizes the OS.

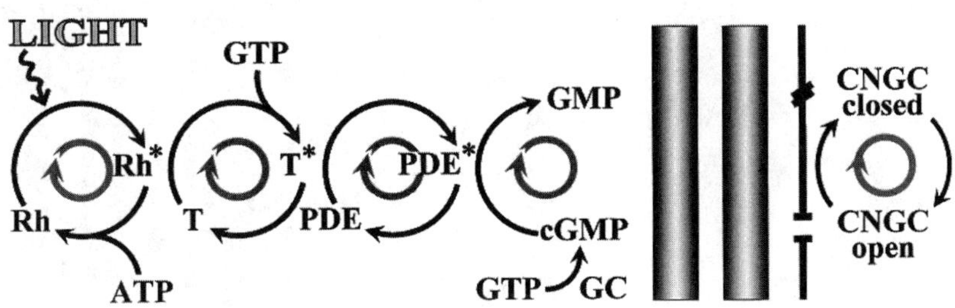

Figure 3. Scheme of the sequence of coupled biochemical reactions of rod phototransduction, with longitudinal microtubules (two vertical tubules) in the cytoplasmic space between the reactions occuring on disk membranes (eg., Rh activation) and on the plasma membrane (CNGC closure); see text for abbreviations. Light absorption is highly amplified (circular arrow in each reaction), uses energy as ATP/GTP, and produces a large response.

Photoreceptors also adapt or adjust their response to the ambient light level (Rodieck, 1998; Pugh and Lamb, 2000; Burns and Baylor, 2001; Fain et al., 2001). Dark adapted rods are highly sensitive because at low light levels the phototransduction cascade greatly amplifies their response to individual photons, i.e., they have a high amplification factor (effect per absorbed photon). When light level increases, this factor decreases and rods become less sensitive. At photopic light levels, all the CNGC in rods are closed, i.e., rods are fully saturated and in effect unresponsive to light; their responsiveness can return at lower light levels. Cones are responsible for photopic vision because they do not saturate but remain sensitive in steady bright light.

By synthesizing these disparate of lines of evidence, I have formulated the following novel hypothesis regarding the pathological mechanism that can underly photoreceptor degeneration in some forms of human RP: 1) In many normal vertebrate retinas, rods have a special mechanism (Figure 4) for massively downregulating and essentially uncoupling the individual steps of phototransduction at photopic light levels. 2) In these retinas the OS of rods have multiple incisures with microtubule-based cytoskeletal systems that mediate this downregulation/uncoupling process. 3) This downregulation/uncoupling process is disturbed as a result of gene defects that cause typical RP. 4) The pathological disturbance of this downregulation/uncoupling process in RP causes phototransduction in these rods to remain coupled or upregulated at photopic light levels, which (particularly because of the high light levels) necessarily results in an abnormally high expenditure of energy within rods and local energy depletion. 5) This abnormal local energy depletion in rods triggers apoptosis and leads to rod cell death, as in typical RP.

The logical neccessity for rods to have a mechanism as described above has not been previously recognized, but this hypothesis is consistent with evidence from a variety of sources. Intermediates of the rhodopsin cycle other than activated rhodopsin can activate the cascade at low efficiency, causing bleaching adaptation (Fain et al., 2001). In rodent rods prolonged bright light induces translocation of T from the OS to the IS that reduces amplification considerably (Sokolov et al., 2002). Defective rod dark adaptation is a common early symptom in typical RP, many of the same proteins used in phototransduction are also involved in dark adaptation, and many of these are encoded by genes whose defect can cause RP: Rh, subunits of T, subunits of PDE, RP GTPase regulator or RPGR, and subunits of CNGC (the three proteins at rod OS disk rims are not

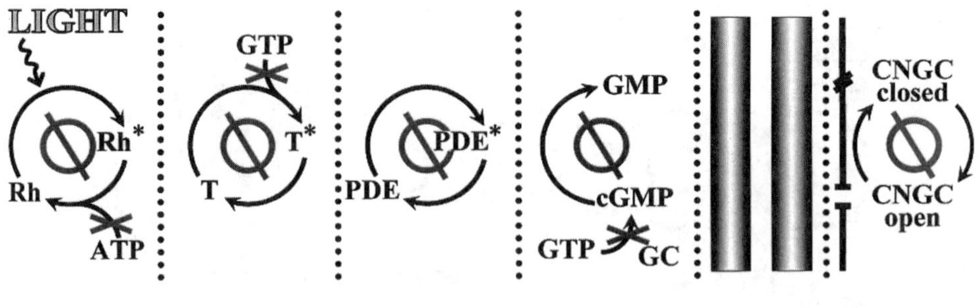

Figure 4. Scheme of proposed mechanism for downregulation and uncoupling of phototransduction in rods that are fully saturated by bright light. The reaction steps are uncoupled from each other (indicated by vertical dotted lines separating the reactions) and each individual reaction is massively downregulated (blocking of each reaction), so that light absorption is not amplified, produces essentially no response, and uses little or no energy.

discussed here because their functions are not well understood) (Rodieck, 1998; Eckmiller, 1999; Clarke et al., 2000; Phelan and Bok, 2000; Pugh and Lamb, 2000; Burns et al., 2001; Eckmiller, 2001; Fain et al., 2001; Eckmiller et al., 2002). Biochemical studies have identified myosin VIIa as a protein kinase A anchoring protein that can serve as a scaffold for assembling multiprotein signal transduction complexes at the cytoskeleton (Küssel-Andermann et al., 2000; Diviani and Scott, 2001). The function of myosin VIIa in normal photoreceptors is not well understood, but its known features are appropriate for it to contribute to adaptation by participating in protein transport and/or translocation along the microtubule-based cytoskeletal systems within photoreceptor OS.

Furthermore, it is known that the effectiveness of each step of rod phototransduction can be modulated during dark adaptation and that some proteins from each step can become associated with the microtubule-based cytoskeletal systems in OS. According to the above hypothesis these phenomena are causally related, i.e., phototransduction is modulated because the appropriate proteins become associated with these OS cytoskeletal systems. Specifically, the effectiveness of the first step of phototransduction (Rh activation/deactivation) is modulated by transfer of arrestin between the IS and the OS, which in rodent rods occurs via translocation along the axoneme (McGinnis et al., 2002). The effectiveness of the second step of phototransduction (T activation/deactivation) is modulated in rodent rods by translocation of α- and β-transducin between the IS and the OS, presumably via the ciliary axoneme (Sokolov et al., 2002). The effectiveness of the third step of phototransduction (PDE activation/deactivation) is modulated by the binding of δ-PDE, which binds to RPGR (Cook et al., 2001), which is localized along the microtubules in both kinds of cytoskeletal systems in rod OS (Eckmiller et al., 2002). The effectiveness of the final step of phototransduction (CNGC opening/closing) is modulated by the binding to the CNGC (which are likely tethered to the microtubules of OS cytoskeletons: Eckmiller, 1999) of calmodulin (which is localized at the microtubules in OS cytoskeletal systems: Eckmiller, 2002).

Rods are thus known to exhibit numerous individual biochemical reactions whose combined effect would be appropriate to regulate adaptation, eg., to downregulate and/or uncouple phototransduction as light level increases. Because most studies have investigated rod phototransduction in retinas of non-primate mammals (whose rod OS lack multiple incisures), the rod OS cytoskeletal system involved in adaptation in these studies is the axoneme. The rod OS axoneme apparently contributes to adaptation by modulating phototransduction in a partial, graded manner that varies continuously with light level. The cytoskeleton at the multiple incisures of rod OS is expected to perform a function that is quite similar, but different in several respects. The adaptation mechanism of the special cytoskeletal system at rod OS multiple incisures is expected to modulate phototransduction more abruptly and extensively, eg., by a mechanism that is expected to become quickly effective after prolonged exposure to bright light and to completely uncouple/downregulate phototransduction, reducing the amplification factor to zero.

Assuming that degeneration of rods in typical RP results from gradual depletion of energy, this could be ameliorated by local energy supplementation, eg., by delivery to the subretinal space of metabolites such as sugars, pyruvate or other metabolic intermediates, or compounds with high energy phosphate bonds (eg., ATP or GTP). The primary physiological source of energy used by photoreceptors in the intact eye is glucose, which is normally supplied via the choroidal and retinal blood vessels but can also reach the subretinal space following intravitreal delivery or retrobulbar application; rod OS have transporters for glucose uptake from the subretinal space (Hsu and Molday, 1994; Niemeyer, 1997; Stone et al., 1999). Accordingly, at early stages RP might be delayed or ameliorated by oral intake of a diet that is high in carbohydrates and sugars. At later stages of disease progression, an effective practical therapy for RP could be local energy supplementation to the retina, eg., by intraocular or retrobulbar delivery of glucose.

The energy depletion hypothesis for RP is consistent with results from many studies of energy and glucose metabolism in retinas and photoreceptors (Hsu and Molday, 1994; Niemeyer, 1997; Rodieck, 1998; Stone et al., 1999; Pepe, 2001). Metabolic studies reporting that the total use of energy by the retina and photoreceptors is normally larger in the dark than light (Sickel, 1972) might be construed as inconsistent with the energy depletion hypothesis. This is not the case because the hypothesis predicts that energy depletion occurs in retinas that are genetically, and therefore also functionally, abnormal.

In summary, this study has shown that myosin VIIa is associated with all the microtubule-based cytoskeletal systems in the OS of amphibian photoreceptors, which closely resemble the OS of human photoreceptors. These findings provide strong support for the suggestion that a disturbance in the special cytoskeletal system at the multiple incisure of rod OS can mediate pathological photoreceptor degeneration in RP. Based on these and other reserarch findings, I have formulated a novel energy depletion hypothesis for typical RP. According to this hypothesis the function of the cytoskeletal system at multiple incisures in rod OS is to massively downregulate/uncouple phototransduction in prolonged bright light, gene defects that cause RP disturb the function of this cytoskeletal system in human rods, this disturbance leads to local energy depletion within rods, and the pathological effects of RP may be ameliorated by local energy supplementation.

5. ACKNOWLEDGEMENTS

The author thanks B. Rohbeck and C. Opfermann for valuable technical assistance and the FAUN-Stiftung/Nürnberg for financial support.

6. REFERENCES

Burns, M. E., and Baylor, D. A., 2001, Activation, deactivation, and adaptation in vertebrate photoreceptor cells, *Ann. Rev. Neurosci.* **24**:779-805.

Clarke, G., Héon, E., and McInnes, R. R., 2000, Recent advances in the molecular basis of inherited photoreceptor degeneration, *Clin. Genet.* **57**:313-329.

Cook, T. A., Ghomashchi, F., Gelb, M. H., Florio, S. K., and Beavo, J. A., 2001, The δ subunit of type 6 phosphodiesterase reduces light-induced cGMP hydrolysis in rod outer segments, *J. Biol. Chem.* **276**:5248-5255.

Diviani, D. and Scott, J. N., 2001, AKAP signaling complexes at the cytoskeleton, *J. Cell Sci.* **114**:1431-1437.

Eckmiller, M. S., 1997, Morphogenesis and renewal of cone outer segments, *Prog. Ret. Eye Res.* **16**:401-441.

Eckmiller, M. S., 1999, Diverse localization of cyclic nucleotide gated channels in the outer segments of rods and cones, in: *Retinal Degenerative Diseases and Experimental Therapy*, J. G. Hollyfield, R. E. Anderson, and M. M. Lavail, eds., Plenum Press, New York, pp. 449-460.

Eckmiller, M. S., 2000, Microtubules in a rod-specific cytoskeleton associated with outer segment incisures, *Vis. Neurosci.* **17**:711-722.

Eckmiller, M. S., 2001, Myosin VIIa is associated with different cytoskeletal systems in the outer segments of rods and cones, [ARVO Abstract], *Invest. Ophthalmol. Vis. Sci.* **42**(4):S296, Abstract nr 1599.

Eckmiller, M. S., 2002, Calmodulin immunolocalization in outer segments of *Xenopus laevis* photoreceptors, *Cell Tissue Res.* **308**:439-442.

Eckmiller, M. S., Wright, A. F., and Manson, F., 2002, Colocalization of RPGR, RPGRIP, and microtubules in different cytoskeletal systems in the outer segments of rods and cones, [abstract], Annual Meeting Abstract and Program Planner assessed at www.arvo.org. *Association for Research in Vision and Ophthalmology*, Abstract 3739.

Ernest, S., Rauch, G.-J., Haffter, P., Geisler, R., Petit, C., and Nicolson, T., 2000, Mariner is defective in myosin VIIA: a zebrafish model for human hereditary deafness, *Hum. Molec. Gen.* **9**:2189-2196.

Fain, G. L., and Lisman, J. E., 1999, Light, Ca++, and photoreceptor death: new evidence for the equivalent light hypothesis from arrestin knockout mice, *Invest. Ophthalmol. Vis. Sci.* **40**:2770-2772.

Fain, G. L., Metthews, H. R, Cornwall, M. C., and Koutalos, Y., 2001, Adaptation in vertebrate photoreceptors, *Physiol. Rev.* **81**:117-151.

Gibson, F., Walsh, J., Mburu, P., Varela, A., Brown, K. A., Antonio, M., Beisel, K. W., Steel, K. P., and Brown, S. D. M., 1995, A type VII myosin encoded by the mouse deafness gene Shaker-1, *Nature* **374**: 62-64.

Hsu, S.-C., and Molday, R. S., 1994, Glucose metabolism in photoreceptor outer segments, *J. Biol. Chem.* **269**:17954-17959.

Küssel-Andermann, P., El-Amraoui, A., Safieddine, S., Hardelin, J. P., Nouaille, S., Camonis, J., and Petit, C., 2000, Unconventional myosin VIIA is a novel A-kinase-anchoring protein, *J. Biol. Chem.* **275**:29654-9.

Liu, X., Udovichenko, I. P., Brown, S. D. M, Steel, K. P., and Williams, D. S., 1999, Myosin VIIa participates in opsin transport through the photoreceptor cilium, *J. Neurosci.* **19**:6267-6274.

McGinnis, J. F., Matsumoto, B., Whelan, J. P., and Cao, W., 2002, Cytoskeleton participation in subcellular trafficking of signal transduction proteins in rod photoreceptor cells, *J. Neurosci. Res.* **67**:290-297.

Niemeyer, G., 1997, Glucose concentration and retinal function, *Clinical Neurosci.* **4**:327-335.

Pepe, I. M., 2001, Recent advances in our understanding of rhodopsin and phototransduction, *Prog. Ret. Eye Res.* **20**:733-759.

Phelan, J. K., and and Bok, D.,2000, A brief review of retinitis pigmentosa and the identified retinitis pigmentosa genes, *Mol. Vis.* **6**:116-124.

Pugh, E. N., and Lamb, T. D., 2000, Phototransduction in vertebrate rods and cones: molecular mechanisms of amplification, recovery and light adaptation, in: *Handbook of Biological Physics* Vol. 3, D. G. Stavenga, W. J. DeGrip, and E. N. Pugh, eds., Elsevier, Amsterdam, pp. 183-255.

Rodieck, R. W., 1998, *The First Steps in Seeing* , Sinauer Associates, Massachusetts.

Sickel, W., 1972, Retinal metabolism in dark and light, in: *Physiology of Photoreceptor Organs*, M. G. F. Fuortes, ed., Springer-Verlag, Berlin, pp. 667-727.

Sokolov, M., Lyubarsky, A. L., Strissel, K. J., Savchenko, A. B., Govardovskii, V. I., Pugh, E. N., and Arshavsky, V. Y., 2002, Massive light-driven translocation of transducin between the two major compartments of rod cells: a novel mechanism of light adaptation, *Neuron* **33**:95-106.

Stone, J., Maslim, J., Valter-Kosci, K., Mervin, K., Bowers, F., Chu, Y., Barnett, N., Provis, J., Lewis, G., Fisher, S. K., Bisti, S., Gargini, C., Cervetto, L., Merin, S., and Pe'er, J., 1999, Mechanisms of photoreceptor death and survival in mammalian retina, *Prog. Ret. Eye Res.* **18**:689-735.

Weil, D., Blanchard, S., Kaplan, J., Guilford, P., Gibson, F., Walsh, J., Mburu, P., Varela, A., Levilliers, J., Weston, M. D., Kelley, P. M., Kimberling, W. J., Wagenaar, M., Levi-Acobas, F., Larget-Piet, D., Munnich, A., Steel, K. P., Brown, S. D. M., and Petit, C., 1995, Defective myosin VIIA gene responsible for Usher syndrome type 1B, *Nature* **374**:60-61.

FUNCTIONAL STUDIES OF AIPL1

Potential role of AIPL1 in cell cycle exit and/or differentiation of photoreceptors

Dayna T. Akey[1], Xuemei Zhu[2], Michael Dyer[3], Amin Li[2], Adam Sorensen[4], Taeko Fukada-Kamitani[5], Stephen P. Daiger[5], Cheryl Craft[2], Tetsu Kamitani[5], and Melanie M. Sohocki[4]

1. INTRODUCTION

Leber congenital amaurosis (LCA) is a genetically heterogeneous, autosomal recessive retinal degenerative disease responsible for approximately 5% of all inherited retinopathies (Kaplan et al., 1990). LCA is often considered the most severe form of childhood retinopathy, and infants with this disease are usually blind at birth. To date, mutations in six genes have been found to cause LCA and two additional loci have been mapped (www.sph.uth.tmc.edu/RetNet/). Recently, we identified the fourth LCA associated gene, the aryl hydrocarbon receptor interacting protein-like 1 (*AIPL1*) (Sohocki et al., 2000a). Mutations in AIPL1 are estimated to account for approximately 7-9% of LCA cases worldwide and have been found to cause autosomal dominant cone-rod dystrophy (Sohocki et al., 2000b). Mutations in the aryl hydrocarbon receptor-interacting protein-like 1 (*AIPL1*) gene have been found to cause autosomal recessive Leber congenital amaurosis (LCA), an early onset form of retinal degeneration. As *AIPL1* is a novel gene, it is necessary to determine its normal function to better understand how mutations in this gene lead to disease.

AIPL1 encodes a 384 amino acid protein that contains three tetratricopeptide (TPR) motifs (Sohocki et al., 2000a). These degenerate 34 amino acid motifs have been found in proteins that participate in a variety of multi-protein complexes important for diverse biological processes (Blatch et al., 1999). The human and primate forms of the AIPL1 protein also contain a carboxy terminal proline-rich region, which may serve to regulate

[1]University of Cincinnati, Cincinnati, OH 45267, USA; [2]Keck School of Medicine of the University of Southern California, Los Angeles, CA 90033, USA; [3]St. Jude Children's Research Hospital, Memphis, TN 38105, USA; [4]Columbia University, New York, NY 10032, USA; [5]University of Texas, Houston, Texas, 77030, USA.
Corresponding author: M. M. Sohocki, E-mail: ms2241@columbia.edu

the recruitment or exchange of proteins in multi-protein complexes (Sohocki et al., 2001; Kay et al., 2000). Furthermore, sequence analysis has shown that the AIPL1 protein shares considerable homology (49% identity, 69% similarity) to the aryl hydrocarbon receptor-interacting protein (AIP) (Sohocki et al., 2000a). AIP possesses a chaperone-like function that helps to mediate the stabilization and nuclear transport of the aryl hydrocarbon receptor (AhR) (Ma et al., 1997; Kazlauskas et al., 2000). Binding of AIP to the AhR within the cytoplasm stabilizes and protects the AhR from ubiquitination (Kazlauskas et al., 2000). Collectively, these data suggest that AIPL1 may be important in protein trafficking and/or protein folding and stabilization.

2. METHODS

2.1. Identification of AIPL1-interacting proteins

The full-length bovine Aipl1 (984 bp, 328 amino acids) was used as bait to screen a bovine retinal cDNA library in the pGAD10 vector (Clontech). The bovine Aipl1 protein sequence is highly conserved when compared to human AIPL1 (88% identity and 94% similarity), however unlike human AIPL1, it does not contain the C-terminal 56 amino acid proline-rich region (Sohocki et al., 2001).

The yeast two-hybrid screen was performed with a sequential transformation procedure using the YEASTMAKER lithium acetate method (Clontech) as described previously (Zhu et al., 1998). Plates contained 5mM of 3-aminotriazole, an inhibitor of constitutive expression of the HIS protein within AH109 cells that suppresses background growth. Yeast two-hybrid re-tests were performed by transforming the isolated pGAD10 bovine retinal cDNA plasmids into AH109 with either the pGBKT7-Aipl1 vector or the empty pGBKT7 vector (negative control).

2.2. Analysis of interactions between wild-type and mutant human AIPL1 and NUB1 in yeast.

Mutations were introduced using the QuickChange site directed mutagenesis kit (Stratagene), and deletion constructs including the pre-TPR motif AIPL1 sequence (PRE, encoding amino acids 1-117), the TPR motif region only (TPR, amino acids 118-297) or the post-TPR sequence (POST, amino acids 298-384) were developed. Interactions were quantitatively analyzed using liquid *o*-nitrophenyl-ß-D-galactopyranoside (OPNG), according to published methods (Ausubel et al., 1994-2001). For growth assays, the transformants were first plated on –L/-W agar, then resulting colonies were pooled together by scraping into –L/-W/-H medium. These cells were used to seed liquid cultures at 1000 cells/ ml and cultures were grown for 48 hours. Data represent the mean ± standard deviation (SD) of a representative experiment done in triplicate and are represented as a percentage of the OD_{600} of yeast grown in selective medium supplemented with histidine.

2.3. Co-immunoprecipitation

For co-immunoprecipitation of endogenous NUB1, COS-M6 cells were transfected with empty pcDNA3 vector (Invitrogen) or plasmid encoding FLAG-tagged AIPL1, p53,

HHR23, or RanGAP1. Cells were harvested 20 hours after transfection by scraping and the total cell lysate prepared in TBS buffer (20 mM Tris-HCl (pH 7.5), 137 mM NaCl, 0.5% NP-40) containing Complete^{TM} protease inhibitor cocktail (Roche). For immunoprecipitation, the supernatant was incubated with anti-FLAG M2 agarose beads (Sigma), and washed 3 times with TBS buffer. Precipitated proteins were solubilized and analyzed by immunoblotting using rabbit anti-NUB1 antibody to detect co-immunoprecipitated endogenous NUB1.

For coimmunoprecipitation from retinoblastoma Y79, cells were harvested, washed twice with PBS and frozen at -80°C. Immunoprecipitation was performed with the Protein A Agarose IP Kit (KPL). The immunoprecipitates were collected by centrifugation, washed three times, solubilized and subjected to immunoblot analysis with anti-NUB1 or anti-AIPL1. Proteins were resolved by SDS-PAGE and electrophoretically transferred to Immobilon-P membranes (Millipore) using previously published protocols (Craft et al., 1998). Two identical blots were made, incubated first with anti-NUB1 and anti-AIPL1 rabbit polyclonal antibodies, respectively, then with horseradish peroxidase conjugated goat anti-rabbit IgG (Bio-Rad Laboratories) and detected using an Enhanced Chemiluminescence Kit (Amersham).

2.4. Immunohistochemistry.

Rabbit antisera against the peptide of AIPL1 (aa 371-384, ATEPPPSPGHSLQH) was produced by Alpha Diagnostic International, Inc. The peptide conjugate was injected into rabbits, and as the best ELISA results were obtained from the bleed at 9 weeks, this serum was affinity purified against the peptide using a Sepharose 4-B column.

Immunohistochemistry of adult retinal sections was performed as previously described with a few modifications (Zhu et al., 2000). After deparaffination and hydration through graded alcohols, the sections were heated in 0.01% sodium citrate buffer, pH 6.0 for antigen retrieval. After cooling to room temperature and rinsing with PBS, they were blocked with 3% H_2O_2 and 5% normal goat serum before incubation with anti-AIPL1 peptide antibody (1:50-200). Following the washing steps, the sections were reacted with biotinylated anti-rabbit IgG and with VECTASTAIN *Elite* ABC Reagent (Vector Laboratories). After washing, the sections were reacted with freshly prepared peroxidase substrate (diaminobenzidine tetrahydrochloride), rinsed with water, counterstained in hematoxylin QS, dehydrated through graded alcohols, and cleared with xylene. Examination and photography were performed using a Nikon light microscope and a Spot digital camera.

3. RESULTS

3.1. Identification of AIPL1-interacting proteins

Approximately 2 X 10^6 clones were screened for Aipl1-interacting proteins using a yeast-two hybrid system. A total of 57 clones grew on selection plates containing synthetic dropout medium lacking Leucine (-L), Tryptophan (-W), and Histidine (-H). All 57 colonies were then re-streaked onto selection plates and 17 demonstrated activation of all three reporters (*HIS3*, *MEL1*, and *ADE2*), which indicates the presence of a strong protein-protein interaction. Bovine cDNA plasmids were isolated from these colonies and re-tests were performed independently with each of the plasmids along with the Aipl1-pGBKT7 bait plasmid or an empty pGBKT7 plasmid. Clones that were capable of activating the reporter genes without the presence of Aipl1 were eliminated from further studies, leaving nine potential Aipl1-interacting proteins. Two of these clones corresponded to bovine Nub1 (GenBank accession no. AF514279), and three of the clones corresponded to bovine Aven. The bovine Nub1 and Aven clones are only capable

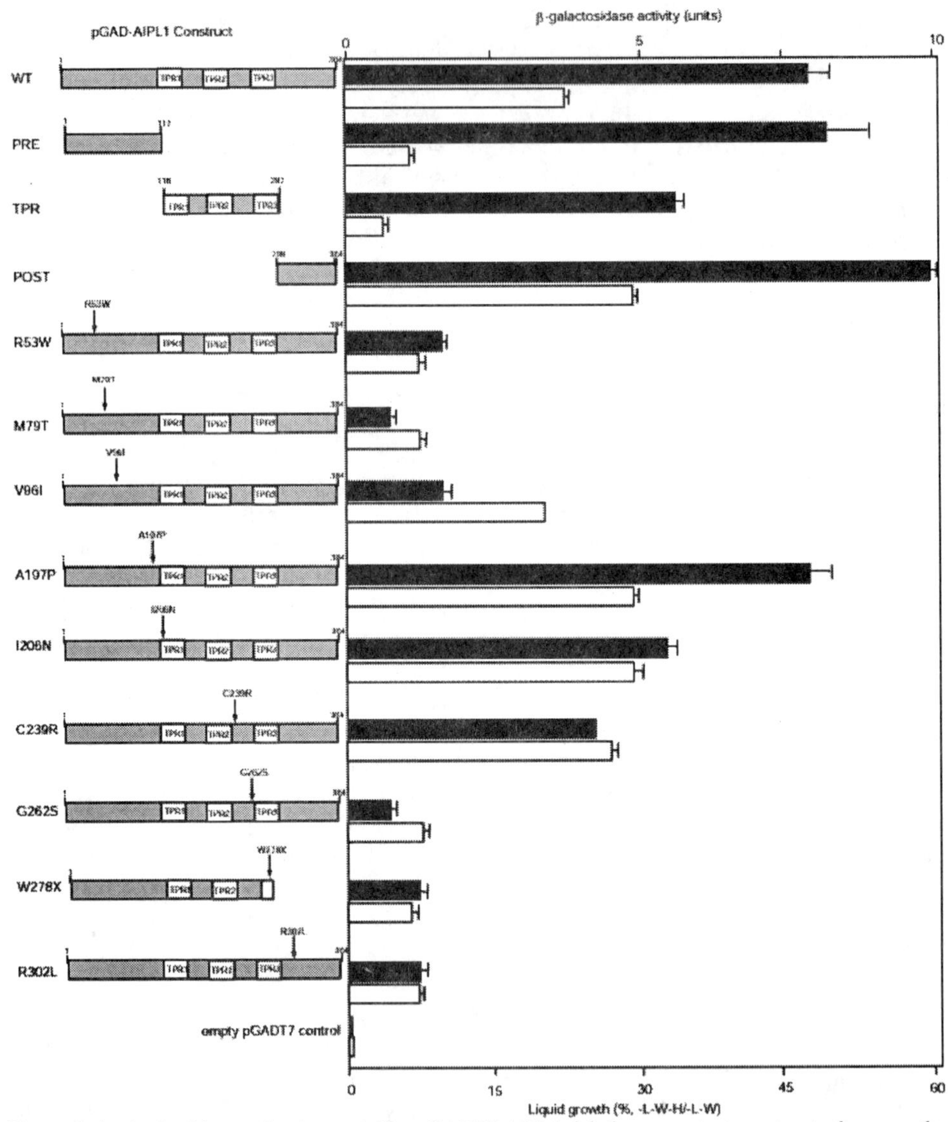

Figure 1. Analysis of interaction in yeast. The pGADT7-AIPL1 deletion or mutant constructs shown on the left were co-transformed with construct pGBKT7-NUB1 and the activation of the *HIS3* and *LacZ* reporter genes were determined (right) by a liquid growth assay (white bars) and a _-galactosidase assay (black bars), respectively.

of activating expression of the reporter genes in the presence of the bovine Aipl1-pGBKT7 plasmid, thereby demonstrating that these interactions are specific for Aipl1.

3.2. Analysis of human protein interactions in yeast

We previously identified a number of AIPL1 mutations in patients with recessive Leber congenital amaurosis (Sohocki et al., 2000a; Sohocki et al, 2000b). To investigate the affect that these mutations may have on the interaction between AIPL1 and NUB1, nine LCA associated mutations were introduced into the pGADT7-AIPL1 construct. The ability of the mutant AIPL1 proteins to interact with the protein encoded by pGBK-NUB1 was determined through quantitative _-galactosidase assays and growth in liquid dropout media (-L, -W, -H), with the empty pGADT7 vector co-transformed with pGBKT7-NUB1 as a control (Fig. 1). The majority of the mutations significantly reduced the interaction between the proteins. However, there was not a significant reduction of the interaction with the A197P mutation. Although the mutations I206N and C239R appeared to cause reduced interaction with NUB1 by the quantitative _-galactosidase assay, the growth assay did not indicate a significant deviation from wild-type control. In addition, deletion constructs were used to determine regions of the AIPL1 protein involved in the interaction with NUB1. While each construct was capable of interacting with NUB1, the construct encoding the C-terminal region of AIPL1 appeared to have the highest binding activity.

3.3. Co-immunoprecipitation of AIPL1 and NUB1

Cell lysates probed with anti-FLAG antibody indicated a band corresponding to FLAG-tagged AIPL1 at approximately 46 kDa (Fig. 2, top panel, Lane 2), confirming translation of the AIPL1 protein. Endogenous NUB1 was detected in all transfected COS cells at an expected molecular weight of approximately 66 kDa (Fig. 2, middle panel, Lanes 1-5). When AIPL1 was immunoprecipitated using the FLAG antibody, endogenous NUB1 was found to co- immunoprecipitate (Fig. 2, bottom panel). Co-immunoprecipitation of endogenous NUB1 and transfected mutant AIPL1 was used to confirm the results of the assays and ensure that reduced binding was caused due to the mutation and not reduced protein stability. The mutants R53W, M79T, V96I, G262S, W278X and R302L failed to immunoprecipitate NUB1 (data not shown).

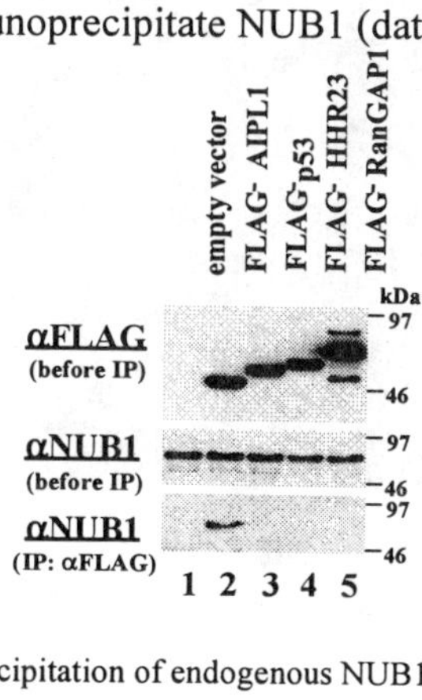

Figure 2. Co-immunoprecipitation of endogenous NUB1 with FLAG-AIPL1.

The AIPL1-NUB1 interaction was also assayed within cells of retinal origin using the retinoblastoma Y79 cell line, which expresses both AIPL1 and NUB1. Immunoprecipiation was performed with either preimmune serum, anti-AIPL1 serum, or affinity-purified anti-AIPL1 antibody, followed by immunoblot analysis with either the NUB1 (Fig. 3A) or AIPL1 antibody (Fig. 3B). The affinity purified AIPL1 antibody co-precipitated a band of ~46 kDa and ~50 kDa corresponding to the AIPL1 (Fig. 3B, lane 3) and NUB1 proteins (Fig. 3A, lane 3), respectively. The size of the NUB1 protein in retina is approximately 16 kDa smaller than that in other tissues. This result is consistent with identification of alternatively sliced NUB1 clones amplified from retinal cDNA libraries (ClonTech 7449-1, ATCC 77428, ATCC 87256 libraries; unpublished data).

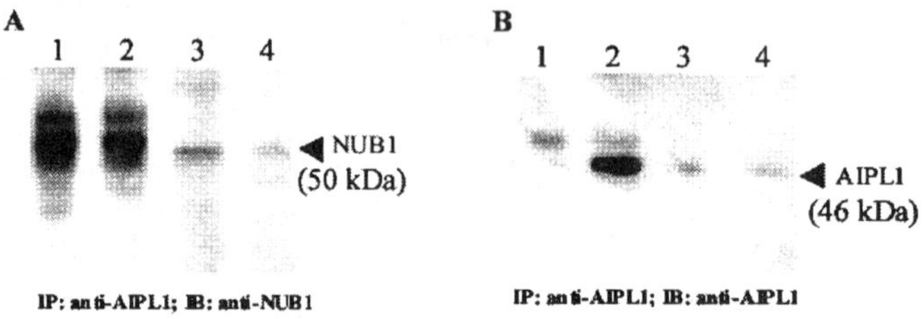

Figure 3. Co-immunoprecipitation in Y79 cell line. Immunoprecipitates with preimmune serum (lane 1), anti-AIPL1 serum (lane 2) and affinity-purified anti-AIPL1 antibody (lane 3) were separated by SDS-PAGE and transferred for immunoblot analysis (IB) with either anti-NUB1 (A) or unpurified anti-AIPL1 (B). Total Y79 cell lysate (lane 4) was also loaded on the gel.

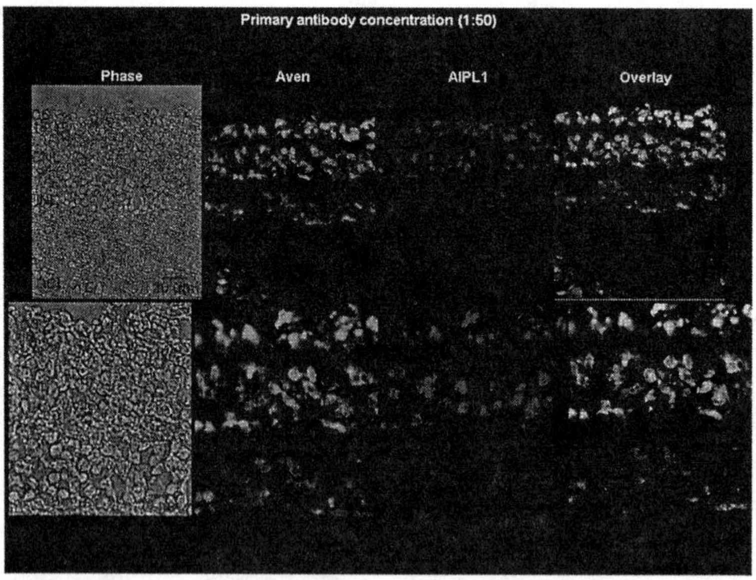

Figure 4. Coimmunolocalization of AIPL1 and Aven in differentiated retina.

3.4. Immunohistochemistry

Rabbit anti-human antibodies against an AIPL1 peptide sequence including the last 13 amino acids of the protein were developed. Immunolocalization using the affinity-purified antibody indicates that the sub-cellular location of the AIPL1 protein overlaps with published NUB1 localization, within the nucleus and cytoplasm of the cell. In addition, we have shown co-localization of AIPL1 and Aven in the adult retina.

4. DISCUSSION

Using a yeast-two hybrid approach, we determined that AIPL1 interacts with NUB1 (NEDD8 Ultimate Buster-1), a protein involved in regulating cell-cycle progression and proliferation. Furthermore, we have shown with co-immunoprecipitation experiments in cells of retinal origin that AIPL1 specifically interacts with a 50 kDa NUB1 protein, which is 16 kDa smaller than that present in other tissues.

NUB1 functions in cell cycle progression through down-regulating NEDD8 expression post-translationally by targeting NEDD8 and its conjugates for proteasomal degradation (Kito et al., 2001). NEDD8 is involved in the degradation of many proteins including Cyclin D1 and p27kip1, two important proteins in the regulation of cell cycle progression in the developing retina (Dyer et al., 2000; Dyer et al., 2001). Furthermore, previous studies have demonstrated that overexpression of NUB1 can inhibit cell growth by up to 83%, providing further support of the importance of NUB1 in cell growth and proliferation (Kito et al., 2001). Given what is known about NUB1, it is also possible that NUB1 facilitates proteolytic degradation of AIPL1. Further studies will be needed to fully understand what role the AIPL1-NUB1 interaction has in normal vision and/or retinal development and how mutations in AIPL1 affect the function of the AIPL1-NUB1 interaction. Experiments such as coexpression of AIPL1 and NUB1 within mammalian cells will be useful to determine if NUB1 targets AIPL1 to the proteosome, and to determine if AIPL1 expression leads to an increase in cytoplasmic stability of NUB1 or a change in the localization of NUB1 within the cell. Furthermore, it will be important to determine how LCA-causing mutations affect the normal localization patterns and stability of the individual AIPL1 and NUB1 proteins, as well as the AIPL1-NUB1 complex.

A second AIPL1-interacting protein, the anti-apoptotic protein, Aven, was identified independently in the yeast two hybrid screen three times. Aven impairs Apaf-1 mediated activation of caspases by interfering with the ability of Apaf-1 to self-associate. We have shown that AIPL1 and Aven colocalize in the adult retina, and we are pursuing confirmation of the interaction using co-immunoprecipitation (Chau et al., 2000).

5. CONCLUSIONS

Based on the data presented here, we hypothesize that AIPL1 plays a role in cytosolic stability and/or nuclear transport of NUB1 during regulation of cell cycle progression during photoreceptor development. This function would be consistent with the severe, early-onset blindness observed in patients with Leber congenital amaurosis caused by mutations of AIPL1. Additional studies will be necessary to confirm the interaction of AIPL1 with Aven; however, should this interaction be confirmed, a potential role of AIPL1 in regulation of apoptosis will be suggested.

6. ACKNOWLEDGEMENTS

We are especially grateful to Dr. Wolfgang Baehr for providing the bovine yeast-two hybrid-retinal cDNA library. The authors thank Chelsea G. Castellano, Taaha Shaikh, Bruce Brown, and Jodi Watson for technical assistance. Supported by the Kirchgessner Foundation, the Knights Templar Eye Foundation, the Foundation Fighting Blindness, Fight for Sight, Research Division of Prevent Blindness America, the RPB, and NIH grants EY00395 (CMC) and EY03040 and the generous continued support of Mary D. Allen and William R. Acquavella. CMC is the Mary D. Allen Professor for Vision Research, Doheny Eye Institute, and MMS is the William R. Acquavella Scholar of Ophthalmic Research, Columbia University. We thank the National Eye Institute and the Foundation Fighting Blindness for the travel award provided to MMS to attend this meeting.

7. REFERENCES

Ausubel, F.M., Brent, R., Kingston, R.E., Moore, D.D., Seidman, J.G., Smith, J.A., and Struhl, K., eds., 1994-2001, *Current protocols in molecular biology*. John Wiley & Sons, Inc., New York, NY. Vol. 2, pp. 13.6.2-13.6.5.

Blatch, G.L. and Lassle, M., 1999, The tetratricopeptide repeat: a structural motif mediating protein-protein interactions. *BioEssays*. **21**:932-939.

Chau B.N., Cheng E.H., Kerr D.A., and Hardwick J.M., 2000, Aven, a novel inhibitor of caspase activation, binds Bcl-xL and Apaf-1. *Mol Cell*. **6**:31-40.

Craft, C.M., Xu, J., Slepak, V.Z., Zhan-Poe, X., Zhu, X., Brown, B., and Lolley, R.N., 1998, PhLPs and PhLOPs in the phosducin family of G beta gamma binding proteins. *Biochemistry*. **37**:15758-15772.

Dyer, M.A. and Cepko, C.L., 2000, p57(Kip2) regulates progenitor cell proliferation and amacrine interneuron development in the mouse retina. *Development*. **127**:3593-605.

Dyer, M.A. and Cepko, C.L., 2001, p27Kip1 and p57Kip2 regulate proliferation in distinct retinal progenitor cell populations. *J Neurosci*. **21**:4259-71.

Kaplan, J., Bonneau, D., Frezal, J., Munnich, A., and Dufier, J.L., 1990, Clinical and genetic heterogeneity in retinitis pigmentosa. *Hum. Genet*. **85**:635-642.

Kay, B.K., Williamson, M.P., and Sudol, M., 2000, The importance of being proline: the interaction of proline-rich motifs in signaling proteins with their cognate domains. *FASEBJ*. **14**:231-231.

Kazlauskas, A., Poellinger, L., and Pongratz, I., 2000, The immunophilin-like protein XAP2 regulates ubiquitination and subcellular localization of the dioxin receptor. *J. Biol.Chem*. **275**:41317-41324.

Kito, K., Yeh E.T.H., and Kamitani, T., 2001, NUB1, a NEDD8-interacting protein, is induced by interferon and down-regulates the NEDD8 expression. *J. Biol. Chem*. **276**:20603-20609.

Ma, Q. and Whitlock J.P., 1997, A novel cytoplasmic protein that interacts with the Ah receptor, contains tetratricopeptide repeat motifs, and augments the transcriptional response to 2, 3,7,8-tetrachlorodibenzo-p-dioxin. *J. Biol. Chem*. **272**:8878-8884.

Sohocki, M.M., Bowne, S.J., Sullivan, L.S., Blackshaw, S., Cepko, C.L., Payne, A.M., Bhattacharya, S.S., Khaliq, S., Mehdi, S.Q., Birch D.G., *et al*., 2000a, Mutations in the novel photoreceptor-pineal gene on 17p cause Leber congenital amaurosis. *Nat. Genet*. **24**:79-83.

Sohocki, M.M., Perrault, I., Leroy, B.P., Payne, A.M., Dharmaraj, S., Bhattacharya, S.S., Kaplan, J., Maumenee, I.H., Koenekoop, R., Meire, F.M., *et al*., 2000b, Prevalence of AIPL1mutations in inherited retinal degenerative disease. *Molec. Genet. Metab*. **70**:142-150.

Sohocki, M.M., Sullivan, L.S., Tirpak, D.L., and Daiger, S.P., 2001, Comparative analysis of aryl-hydrocarbon receptor interacting protein-like 1 (Aipl1), a gene associated with inherited retinal disease in humans. *Mamm Genome*. **12**:566-8.

Zhu, X. and Craft, C.M., 1998, Interaction of phosducin and phosducin isoforms with a 26S proteasomal subunit, SUG1. *Mol. Vis.* **4**:13-20.

Zhu, X. and Craft, C.M., 2000, Modulation of CRX transactivation activity by phosducin isoforms. *Mol. Cell Biol.* **20**:5216-26.

PHOTORECEPTOR DEGENERATION IN PRO23HIS AND S334TER TRANSGENIC RATS

Donald Lee[1], Scott Geller[1], Natalie Walsh[1], Krisztina Valter[1], Doug Yasumura[2], Michael Matthes[2], Matthew LaVail[2], and Jonathan Stone[1]

1. INTRODUCTION

Transgenic animals, in which the transgene has been engineered to mimic mutations known to cause photoreceptor degenerations in humans, provide powerful models of the diseases known as retinitis pigmentosa. Here we describe the time course of photoreceptor death in rats in which degeneration is caused by a rhodopsin transgene containing the P23H or S334ter mutations.

The transversion of cytoseine to adenine in codon 23 of the rhodopsin gene manifests as a substitution (the P23H mutation) of a histidine residue for a proline in the 23rd amino acid of opsin from the N terminus of the protein, within the lumen of the outer segment disc. It was the first disease-relevant human rhodopsin gene mutation to be identified (Dryja et al. 1990; Gal et al. 1997). The proline in position 23 is known to be highly conserved in opsin and related G protein binding proteins receptors, and it is now established that this mutation is responsible for one form of ADRP (Dryja et al. 1991). The S334ter mutation of opsin is a truncation that removes the last 15 amino acids from the carboxy terminus. It is representative of a class of ADRP mutations affecting the end of the rhodopsin that faces the cytoplasm (Mendez et al. 2000).

Present results show that, in all 3 lines studied, photoreceptor death began in the second postnatal week, reached a maximum about 3w after birth and then declined sharply. The most intense period of photoreceptor death was thus in early neonatal life. It was striking that the time course of this wave of photoreceptor death matches a much more limited wave of photoreceptor death which occurs in normal retinal development. It is possible that the mechanisms which determine the temporal pattern of the degeneration in the normally developing retina also influence the photoreceptor degenerations caused by rhodopsin mutations.

[1]Department of Anatomy and Histology and Institute of Biomedical Research, University of Sydney, Sydney NSW 2006 Australia; [2]Department of Anatomy and Ophthalmology, Beckman Vision Center, University of California School of Medicine, San Francisco, CA 94143, USA. Corresponding author: J. Stone, E-mail: jonstone@anatomy.usyd.edu.au

2. METHODS

2.1. Material

Eyes from animals in the Pro23His line 1 (P23H-1) and S334ter line 5 (S334ter-5) were obtained from colonies at the University of California School of Medicine, San Francisco. Eyes of the Pro23His line 3 (P23H-3) were obtained from a colony at the University of Sydney, established with stock suppled from UCSF. Eyes were fixed by immersion in 4% paraformaldehyde in phosphate buffered saline (PBS) at pH 7.4 for 1.5 hours and cryoprotected in 15% sucrose for 12 hours. The eyes were then cryosectioned at 20μm. Labelling for the fragmentation of nuclear DNA characteristic of apoptosis was applied using the TUNEL technique first described by Gavrieli, et al (1992) following previously published protocols (Maslim et al. 1997).

2.2. Quantitative Analysis

Sections were surveyed systematically for the frequency of TUNEL+ profiles under a fluorescence microscope. We used a 20 x objective which gave a field of view with a diameter of 1mm. For each animal studied, counts were made in 3 to 5 sections. Each section was surveyed in 1 mm steps from one edge to the other. Counts were recorded separately for the outer nuclear layer (ONL), inner nuclear layer (INL) and inner retina (layers internal to the INL). The counts from the several sections were then tallied and averaged as TUNEL+ profiles/mm of retina. For each age of each transgenic line, 4-7 animals were studied. Means and standard deviations for each age group were then calculated.

3. RESULTS

The division of the neuroblast layer into the inner nuclear layer (INL) and the outer nuclear layer (ONL) was almost complete by postnatal day (P) 8, the division remaining incomplete only in the most peripheral 1mm of retina. Counts of TUNEL+ profiles were commenced at this age.

In the INL, TUNEL+ profiles were numerous at P8 in all strains (3 transgenic, SD control), and then declined to low adult levels by P12 (Figures 1 and 2A-C, E). In the ONL, the frequency of TUNEL+ profiles was low at P8 in all strains, then rose to a maximum at P18 to P22 and then decreased (Figures 1, 2A to C). Quantitatively, there appeared to be no significant difference between strains in the INL death rats (Figure 2E).

In all three transgenic strains, TUNEL+ frequencies in the ONL were low at P8 and then increased to a maximum at P17-20, before decreasing (Figures 2A-C) In the P23H-1 and P23H-3 strains, for which adult eyes were available (Figures 2A, B) the decrease was almost as prominent as the increase, and TUNEL+ frequencies in the adult (older than P40) were < 10/mm. When the data for the 3 transgenic lines and for the Sprague-Dawley control were superimposed (Figure 2D), it was apparent that in all 4 strains photoreceptor death was most intense in the period P10 to P40. Peak death rates were 6 to 12 times higher in the transgenics than in the control.

The S334ter-5 and P23H-1 lines showed the most severe pattern of photoreceptor degeneration, ONL TUNEL+ profiles appearing (at P12) and peaking (at P17) earlier than in the P23H-3 or SD control strains (Figure 2D). Overall however, there was a considerable temporal coincidence of photoreceptor death among the four strains considered.

4. DISCUSSION

A feature of the present observations is that, in all 3 transgenic lines studied, photoreceptor degeneration was most intense in neonatal life, in the period P10 to P25. Thus photoreceptor death begins at most a few days before eye opening (P14). The period of intense photoreceptors death in the transgenic strains matches closely a period of 'naturally occurring' photoreceptor death which has been observed during normal retinal development in the rat (Maslim et al. 1997). Maslim et al. (1997) and Valter et al. (1998) noted that during this 'critical period' photoreceptors are particularly vulnerable to hypoxia. The present data suggest that, during the same period, photoreceptors are vulnerable to stress induced by rhodopsin mutations.

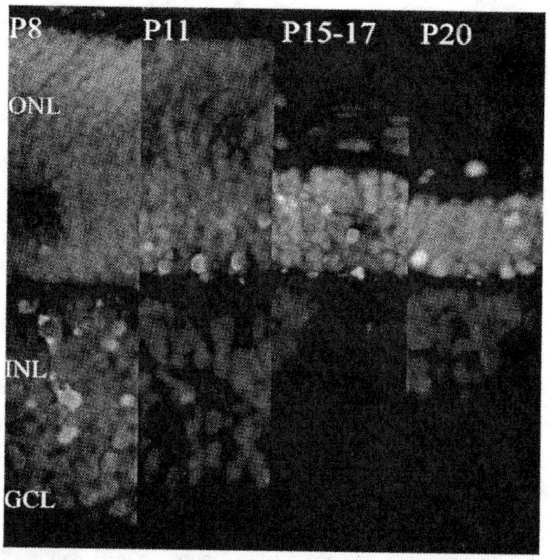

Figure 1: Cell death in the developing retina of rhodopsin-mutant strain (S334-ter line 5). Sample sections of retina are shown for 4 ages. At P8, TUNEL+ (dying) cells were relatively numerous in the INL (left panel). At subsequent ages, the frequency of TUNEL+ profiles declined in the INL, but rose in the ONL, peaking in this strain at P15-17. The blue label is a DNA-specific dye which shows cellular structure of the retina. Note that the ONL thins rapidly with age in this strain.

The present data raise the possibility that human photoreceptor degenerations resulting from rhodopsin mutations also begin during development. If, as (Dreher and Robinson 1988) have argued, eye-opening can be taken as a reference point by which developmental programs in different mammalian species can be aligned, the

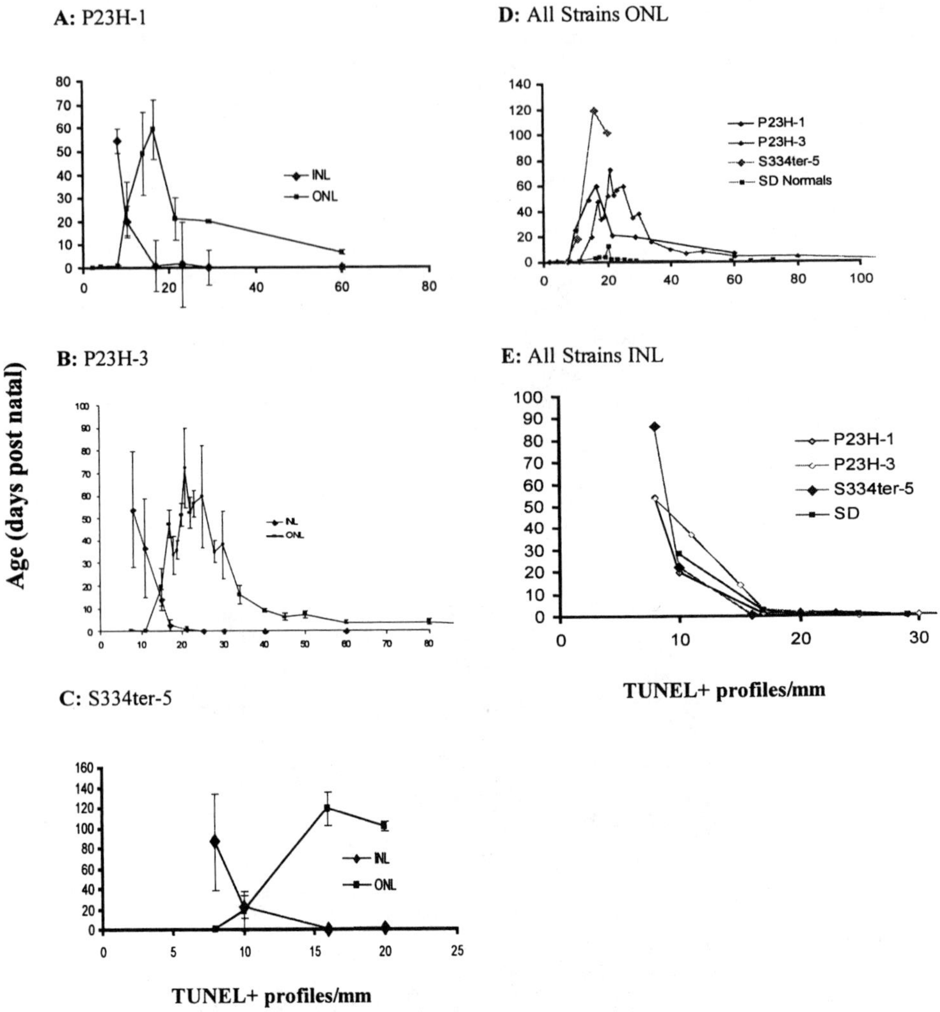

Figure 2: Patterns of cell death in the 4 strains studied, with INL and ONL data shown separately. In all strains, the frequency of TUNEL profiles in the INL was high at the earliest age examined (P8) and declined rapidly to low adult levels by P16. In all strains cell death in the ONL was low at P8, rose to a maximum at P18 –P20 and then declined. Data for the 3 transgenic strains are shown separately in Figures 2A-C. The error bars represent ± 1 standard deviation. Data for the ONL are combined in Figure 2D, with data for the control SD strain added. Data for the INL are combined in Figure 2E with data for the SD added.

corresponding period in human is in the fifth to sixth month of gestation (Robinson 1966; Dreher and Robinson 1988; Bron et al. 1997).

This suggestion is of course speculative. In the transgenic rats used, the mutated rhodopsin gene is supernumerary, and may be present in more than one copy. Further, some evidence is available that over-production of rhodopsin, either normal or mutant, can lead to photoreceptor degeneration (Olsson et al. 1992; Chen et al. 1995). Nevertheless, the time of onset of human photoreceptor degenerations is not known, and the possibility that those caused by rhodopsin mutations commence in late gestation deserves testing.

The present results raise the question why photoreceptor death begins as early as it does. Several authors have considered how mutant rhodopsin might cause stress to the photoreceptor, sufficient to induce death. The Pro23His mutation may cause an abnormality in the tertiary structure of rhodopsin which results in a blockade of its transport from its site of synthesis (the inner segment) to the outer segment (Dryja et al. 1991). In support of this hypothesis, the inner segment of the photoreceptor in a mouse model of the Pro23His mutation has been reported to be dilated (Olsson et al. 1992). Other postulates of the mechanism of photoreceptor death caused by the Pro23His mutation include faulty trafficking within the signal transduction cascade and defective outer segment disc morphogenesis (Machida et al. 2000). Rhodopsin mutations that truncate the carboxy terminus of rhodopsin remove phosphorylation sites from the carboxy terminus, resulting in a prolonged activation of rhodopsin following light exposure (Mendez et al. 2000). These various mechanisms are not age-specific and a simple explanation for the onset of photoreceptor death at P10-P20 is that this is period over which rat photoreceptors commence their expression of rhodopsin and its incorporation into the outer segment, and over which photoreceptor function commences. It is also the period over which retinal metabolism accelerates (Graymore 1959).

The present data also raise the question why photoreceptor degeneration slows markedly in the 4th postnatal week, and is further reduced at 3 months (P100). Answers to this question require understanding of how many photoreceptors survive into adulthood in the different strains (described by LaVail et al., unpublished). In the P23H-1 line the ONL is 1-2 cells thick by P100 (thus 30% of control levels), while in the P23H-3 line the ONL is 2-4 cells thick (thus 55% of control levels) and still functional according to Machida et al, (2000). The slowing of photoreceptor death in these strains may then result from the upregulation of factors which protect surviving photoreceptors.

TUNEL+ profile frequencies and ONL thinning are now widely used techniques of assessing photoreceptor death. The data sets obtained with the 2 techniques are complementary, one indicating the surviving population, the other the proportion of that population undergoing apoptosis. ONL thickness provides a better index of the structural state of the retina, while the TUNEL technique is more sensitive in detecting the onset of degeneration and in demonstrating the rate of degeneration at the time the retina is fixed. In the present data, the TUNEL technique has revealed an early burst of photoreceptor degeneration in transgenic strains, and poses novel questions about the determinants of the onset and slowing of this intense period of photoreceptor loss.

5. ACKNOWLEDGEMENTS

Ms Tania Novikova provided expert technical assistance. The work was supported by grants from Retina Australia, the Medical Foundation of the University of Sydney and the National Health and Medical Research Council of Australia.

6. REFERENCES

Bron, A. J., R. C. Tripathi, et al., 1997, *Wolff's Anatomy of the Eye and Orbit*, London, Chapman and Hall Medical.

Chen, J., C. L. Makino, et al., 1995, Mechanisms of rhodopsin inactivation in vivo as revealed by a COOH-terminal truncation mutant, *Science* **267**:374-7.

Dreher, B. and S. R. Robinson, 1988, Development of the retinofugal pathway in birds and mammals: evidence for a common 'timetable', *Brain Behav Evol* **31**:369-90.

Dryja, T. P., L. B. Hahn, et al., 1991, Mutation spectrum of the rhodopsin gene among patients with autosomal dominant retinitis pigmentosa, *Proc Natl Acad Sci U S A* **88**:9370-4.

Dryja, T. P., T. L. McGee, et al., 1990, A point mutation of the rhodopsin gene in one form of retinitis pigmentosa, *Nature* **343**:364-6.

Gal, A., E. Apfelstedt-Sylla, et al., 1997, Rhodopsin Mutations in Inherited Retinal Dystrophies and Dysfunctions, *Prog Ret Eye Res* **16**:51-79.

Graymore, C., 1959, Metabolism of the developing retina, *Br J Ophthalmol* **43**:34-39.

Machida, S., M. Kondo, et al., 2000, P23H rhodopsin transgenic rat: correlation of retinal function with histopathology, *Invest Ophthalmol Vis Sci* **41**:3200-9.

Maslim, J., K. Valter, et al., 1997, Tissue oxygen during a critical developmental period controls the death and survival of photoreceptors, *Invest Ophthalmol Vis Sci* **38**:1667-77.

Mendez, A., M. E. Burns, et al., 2000, Rapid and reproducible deactivation of rhodopsin requires multiple phosphorylation sites, *Neuron* **28**:153-64.

Olsson, J. E., J. W. Gordon, et al., 1992, Transgenic mice with a rhodopsin mutation (Pro23His): a mouse model of autosomal dominant retinitis pigmentosa, *Neuron* **9**:815-30.

Robinson, R. J., 1966, Assessment of gestational age by neurological examination, *Arch Dis Child* **41**:437-447.

Valter, K., J. Maslim, et al., 1998, Photoreceptor dystrophy in the RCS rat: roles of oxygen, debris, and bFGF, *Invest Ophthalmol Vis Sci* **39**:2427-42.

RETINAL DEGENERATION CAUSED BY MUTATIONS IN TULP1

Quansheng Xi, Gayle J.T. Pauer, Karen A.West, John W. Crabb, and Stephanie A. Hagstrom[1]

1. INTRODUCTION

TULP1 is a member of a family of four proteins named TULPs for tubby-like proteins, defined by the highly conserved C-terminal half of their primary sequences (60 – 90% amino acid identity). This protein family includes TUB, TULP1, TULP2, and TULP3, all of which have different N-termini and different cellular expression patterns (North et al., 1997). Other than members of the TULP family, database searches do not reveal any significant homology with known proteins or functional motifs. Their physiological functions are unknown but two have been linked to photoreceptor degeneration. Mutations in the human *TULP1* gene were found to be a cause of retinitis pigmentosa (Hagstrom et al., 1998; Banerjee et al., 1998). Loss of Tulp1 function in mice, introduced by gene knockout, also causes photoreceptor degeneration (Hagstrom et al., 1999; Ikeda et al., 2000). A naturally occurring recessive null mutation in the *tubby* gene in mice causes photoreceptor and cochlear degeneration and adult-onset obesity (Ohlemiller et al., 1995).

In a previous study, we explored the role of Tulp1 in the retina by analyzing the tissue distribution of the protein in normal mice and the photoreceptor disease phenotype in *tulp1* knockout mice (Hagstrom et al., 1999). We determined that Tulp1 is found exclusively in photoreceptors, localizing primarily in the inner segments and connecting cilium. In addition, *tulp1 -/-* mice develop an early-onset, progressive photoreceptor degeneration with involvement of both rods and cones, which recapitulates the retinal disease described in patients having defects in the same gene. In 17-day-old mice, the rod and cone opsins, normally targeted to the outer segment, were aberrantly localized to the plasma membranes of inner segments, perinuclear cytoplasm and synaptic regions. At the same age, an abnormal accumulation of rhodopsin-bearing extracellular vesicles was found surrounding the ellipsoid region of the inner segments.

[1]Cole Eye Institute, Cleveland Clinic Foundation, Cleveland, OH 44195, USA. Corresponding author: S. A. Hagstrom, E-mail: hagstrs@ccf.org

It is evident that TULP1 has an important role in the health and physiology of photoreceptor cells. Determining the cellular function of TULP1, and other family members, is important not only for understanding TULP proteins normal function but also for understanding the mechanism that causes retinal degeneration. There have been several recent studies aimed at identifying the function of TULP proteins. Postulated functions include mediation of insulin signaling, gene transcription and ribosomal RNA synthesis. Based upon the localization of Tulp1and the retinal phenotype observed in *tulp1 -/-* mice, we hypothesize that TULP1 is involved in the transport of proteins from their site of synthesis in the inner segment to their final destination in the outer segment of photoreceptor cells.

As a step toward examining this hypothesis, we are pursuing immunoprecipitation studies of retinal homogenates to identify proteins that interact with TULP1. Immunoprecipitated proteins are identified by LC MS/MS using a quadrupole-time of flight (QTOF) mass spectrometer. Several of the identified proteins are involved in various aspects of vesicle transport or protein movement in cells. These results suggest that TULP1 may function in intracellular trafficking of proteins and provide a first step toward defining the mechanism underlying photoreceptor degeneration caused by mutations in *TULP1*.

2. METHODS

2.1. Antibodies and Western Blots

Western blots were performed as described previously using mouse polyclonal antibody, m-Tulp1N, peroxidase-conjugated secondary antibodies, and detection by chemiluminescence (Hagstrom et al., 2001).

2.2. Immunoprecipitation

Two bovine retinas were homogenized and lysed in 5 ml of NP-40 lysis buffer supplemented with proteinase inhibitors. The suspension was centrifuged and the supernatant removed from the pellet. The bovine retinal lysates were used for immunoprecipitation analysis using Tulp1 polyclonal antibodies (Hagstrom et al., 2001) (Figure 1). Approximately 100 ug of Tulp1 antibody was coupled to 60 ul of protein A beads by incubating for 1 hour. The beads were washed and resuspended with 600 ul of 0.2 M sodium borate (pH 9.0) followed by the addition of 4 mg of dimethylpimelimidate. The reaction was stopped by adding 0.2M ethanolamine (pH 8.0). The coupled beads were then washed three times with 100 mM glycine (pH 3.0) followed by washing with PBS three times and finally resuspended in 60 ul of PBS.

For immunoprecipitation, 1 mg of total protein from retinal lysate, determined by Bradford method, was added to ~50 ul of Tulp1 antibody coupled to the protein A beads. The binding reaction occurred overnight at 4°C. As a negative control, protein A beads alone were allowed to bind to retinal lysate. The reactions were centrifuged, supernatants removed, and the beads washed four times with lysis buffer and resuspended in 30 ul lysis buffer. Immunoprecipitated antigens were analyzed by SDS-PAGE analysis on a

10% gel (Figure 2). The gel was stained with colloidal coomassie blue (Gel Code Blue, Pierce), and the bands were excised for protein identification.

2.3. Protein Identification by Mass Spectrometry

Identification of proteins by liquid chromatography tandem mass spectrometry (LC MS/MS) was as described in detail elsewhere (Crabb et al., 2002; West et al., 2003). Briefly, gel bands were excised, stain washed away, proteins digested in the gel with trypsin and peptides were extracted for mass spectrometric analysis. LC MS/MS was performed using a Cap LC system (Micromass) and a quadrupole time-of-flight mass spectrometer (QTOF2, Micromass, Beverly, MA). Protein identifications from MS/MS data utilized Micromass software ProteinLynx™ Global Server, MassLynx™, Ver 3.5, and the Swiss-Protein and NCBI protein sequence databases.

3. RESULTS

To explore the physiologic function of TULP1 in photoreceptor cells, we immunoprecipitated a Tulp1-containing complex from retinal homogenates and identified possible interacting proteins. Bovine retinas were used for these experiments since we can easily obtain large quantities of fresh tissue (Lawson Co.). To confirm reactivity of m-Tulp1N antibody with bovine Tulp1, western blot analysis was performed. Figure 1 illustrates that the antibody detects bovine, human, murine, and porcine Tulp1, and is suitable for immunoprecipitation studies on bovine retinal homogenates.

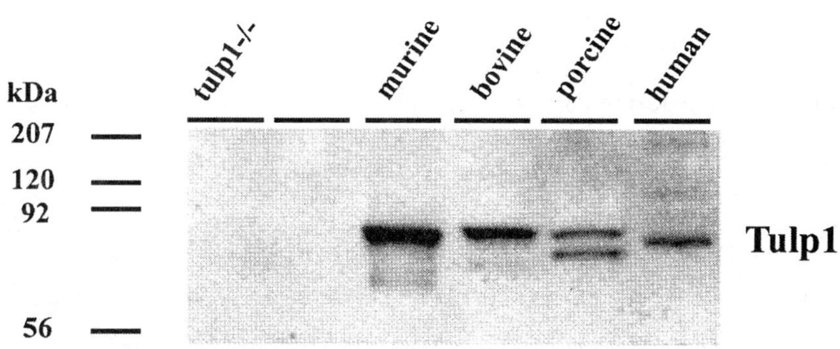

Figure 1. Western blot analysis of Tulp1 expression using Tulp1 antibodies. Each lane represents retinal homogenates from a different species. Tulp1 is detected in all wild-type retinas tested, including bovine, but not in the *tulp1-/-* mouse retina.

Immunoprecipitation experiments were performed on approximately 1 mg of total protein from bovine retinal lysate using approximately 100 ug of m-Tulp1N antibody. Data reported are the results from two independent experiments.

Ten bands, in addition to Tulp1, were detected by SDS-PAGE analysis of the anti-Tulp1 immunoprecipitation products (Figure 2). The gel bands were excised, digested

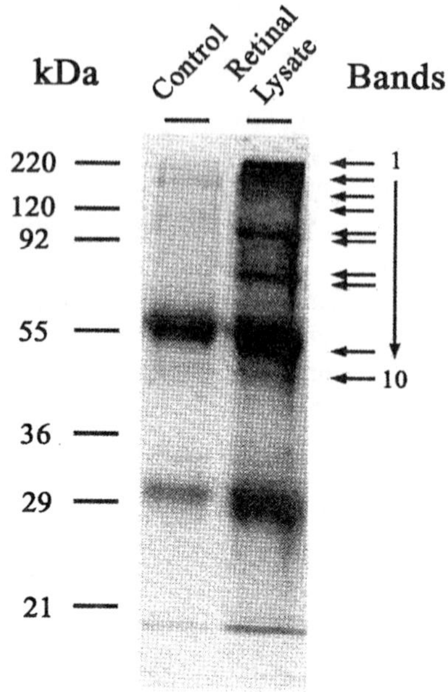

Figure 2. SDS-PAGE analysis of anti-Tulp1 immunoprecipitation products. The molecular weight markers are labeled on the left hand side of the gel. The control lane contains retinal lysate after binding with protein A beads alone. The retinal lysate lane contains anti-Tulp1 immunoprecipitation samples incubated with protein A beads. Bands absent in the control lane were excised, digested with trypsin, and analyzed by LC MS/MS. Gel band detection was performed with colloidal coomassie blue stain.

Table 1. Proteins Identified by LC MS/MS in Retinal Lysate Immunoprecipitates

Protein No.	M_{calc} kDa	M_{obs} kDa	Acces. No.	Identification
1	270.6	>220	P46821	Microtubule Associated Protein 1B
2	191.5	160	Q00610	Clathrin Heavy Chain
3	139.7	130	P12661	Interphotoreceptor Retinoid Binding Protein
4	113.4	100	Q02218	2-Oxoglutarate Dehydrogenase
5	97.4	95	Q05193	Dynamin-1
6	72.4	90	P37727	Rab Gerynl Gerynl Transferase
7	72.9	75	Q63100	Dynein Intermediate Chain
8	57.7	60	P14618	Pyruvate Kinase
9	51.1	55	P05209	Tubulin
10	41.7	50	P02571	Actin

with trypsin and proteins identified by LC MS/MS sequence analysis (Table 1). Identified proteins included: Microtubule Associated Protein 1B, Clathrin Heavy Chain, Interphotoreceptor Retinoid Binding Protein, 2-Oxoglutarate Dehydrogenase, Dynamin-1, Rab Gerynl Gerynl Transferase, Dynein Intermediate Chain, Pyruvate Kinase, Tubulin and Actin.

The interaction between Tulp1 and the candidate proteins are currently being confirmed by *in vitro* pull down experiments and *in vivo* co-immunoprecipitation and co-localization studies.

4. DISCUSSION

The amino acid sequence of TULP1 does not provide clues as to the function of the protein. We have identified several proteins that may interact with TULP1 and provide clues to its physiological role in photoreceptor cells. Several of the identified proteins are involved in various aspects of vesicle transport or protein movement.

MAP1B is a neuronal microtubule associated protein known to bind and stabilize microtubules and is also involved in the trafficking of vesicles and organelles. MAP1B is localized to the inner segments of photoreceptors, similar to TULP1 (Meixner et al., 2000). In addition, mutations in this gene cause retinal defects in mice. Interestingly MAP1A was recently reported to be a genetic modifier of the hearing loss seen in *tubby* mice suggesting that the cochlear degeneration may be caused by disrupted protein interactions with MAP1A (Ikeda et al., 2002). It is possible that disrupted protein interactions between TULP1 and MAP1B may also be involved in the retinal degeneration seen in both *tubby* and *tulp1-/-* mice.

Dynamin-1 is a neuronal specific GTPase that interacts with microtubules and is required for endocytosis of vesicles from the plasma membrane. This protein contains a pleckstrin homology domain that binds phosphatidylinositol 4,5 – biphosphate to mediate protein localization to the membrane. The same mechanism of binding the plasma membrane through phosphatidylinositols has been shown for another member of the TULP family, Tubby (Santagata et al., 2001). Further experiments are required to decipher whether TULP1 is a part of this signal transduction mechanism.

Dynein is a molecular motor that moves vesicles along the microtubules and is involved in intracellular motility and protein transport. Sung et al. has shown that the C-terminus of rhodopsin binds Dynein and is a part of the mechanism that transports rhodopsin to the outer segment (Tai et al., 1999). Confirmation of TULP1 binding to Dynein would support involvement of TULP1 in the transport of proteins from the inner segment to the outer segment of photoreceptor cells.

Our preliminary immunoprecipitation results have identified several candidate binding partners for TULP1. Growing evidence suggests that TULP1 may function in intracellular trafficking or movement of proteins synthesized in the inner segment to the outer segment of photoreceptor cells. Further studies are required to confirm this possible role and to determine at which step TULP1 is a required participant. These current results provide a first step toward defining the mechanism underlying photoreceptor degeneration caused by mutations in *TULP1*.

5. ACKNOWLEDGMENTS

The authors thank the Foundation Fighting Blindness and the National Eye Institute for the Travel Award to Stephanie Hagstrom to attend this meeting.

6. REFERENCES

Banerjee, P., Kleyn, P. W., Knowles, J. A., Lewis, C. A., Ross, B. M., Parano, E., Kovats, S. G., Lee, J. J., Penchaszadeh, G.,K., Ott, J., Jacobson, S. G., and Gilliam, T. C., 1998, TULP1 mutation in two extended Dominican kindreds with autosomal recessive retinitis pigmentosa, *Nat. Genet.* **18**:177-179.

Crabb, J. W., Miyagi, M., Gu, X., Shadrach, K., West, K. A., Sakaguchi, H., Kamei, M., Hasan, A., Yan, L., Rayborn, M. E., Salomon, R. G., and Hollyfield, J. G., 2002, Drusen proteome analysis: an approach to the etiology of age-related macular degeneration. *Proc. Natl. Acad.Sci. U.S.A.* **99**:14682-14687.

Hagstrom, S. A., North, M. A., Nishina, P. M., Berson, E. L., and Dryja, T. P., 1998, Recessive mutations in the gene encoding the tubby-like protein TULP1 in patients with retinitis pigmentosa, *Nat. Genet.* **18**:174-176.

Hagstrom, S. A., Duyao, M., North, M. A., and Li, T., 1999, Retinal degeneration in tulp1-/- mice: vesicular accumulation in the interphotoreceptor matrix, *Invest. Ophthalmol. Vis. Sci.* **40**:2795-2802.

Hagstrom, S. A., Adamian, M., Scimeca, M., Pawlyk, B. S., Yue, G., and Li, T., 2001, A role for the tubby-like protein 1 in rhodopsin transport, *Invest. Ophthalmol. Vis. Sci.* **42**:1955-1962.

Ikeda, S., Shiva, N., Ikeda, A., Smith, R. S., Nusinowitz, S., Yan, G., Lin, T. R., Chu, S., Heckenlively, J. R., North, M. A., Naggert, J. K., Nishina, P. M., and Duyao, M. P, 2000, Retinal degeneration but not obesity is observed in null mutants of the tubby-like protein 1 gene. *Hum. Mol. Genet.* **9**:155-163.

Ikeda, A., Zheng, Q. Y., Zuberi, A. R., Johnson, K. R., Naggert, J. K., and Nishina, P. M., 2002, Microtubule-associated protein 1A is a modifier of tubby hearing (moth1). *Nat. Genet.* **30**:401-405.

Meixner, A., Haverkamp, S., Wassle, H., Fuhrer, S., Thalhammer, J., Kropf, N., Bittner, R. E., Lassmann, H., Wiche, G., and Propst, F., 2000, MAP1B is required for axon guidance and is involved in the development of the central and peripheral nervous system. *J. Cell Biol.* **151**:1169-1178.

North, M. A., Naggert, J. K., Yan, Y., Noben-Trauth, K., and Nishina, P. M., 1997, Molecular characterization of TUB, TULP1, and TULP2, members of the novel tubby gene family and their possible relation to ocular diseases. *Proc. Natl. Acad. Sci. U.S.A.* **94**:3128-3133.

Ohlemiller, K. K., Hughes, R. M., Mosinger-Ogilvie, J., Speck, J. D., Grosof, D. H., and Silverman, M. S., 1995, Cochlear and retinal degeneration in the tubby mouse. *Neuroreport* **6**:845-849.

Santagata, S., Boggon, T. J., Baird, C. L., Gomez, C. A., Zhao, J., Shan, W. S., Myszka, D. G., and Shapiro, L., 2001, G-protein signaling through tubby proteins. *Science* **292**:2041-2050.

Tai, A. W., Chuang, J. Z., Bode, C., Wolfrum, U., and Sung, C. H., 1999, Rhodopsin's carboxy-terminal Cytoplasmic tail acts as a membrane receptor for cytoplasmic dynein by binding to the dynein light chain Tctex-1. *Cell* **97**:877-887.

West, K. A., Yan, L., Shadrach, K., Sun, J., Hasan, A., Miyagi, M., Crabb, J. S., Hollyfield, J. G., Marmorstein, A. D., and Crabb, J. W., 2003, Protein database, human retinal pigment epithelium. *Mol. Cell.* Proteomics (In Press).

TOWARDS UNDERSTANDING THE FUNCTION OF THE TUBBY GENE FAMILY IN THE RETINA

Sakae Ikeda, Akihiro Ikeda, Jürgen K. Naggert, and Patsy M. Nishina[1]

1. INTRODUCTION

The autosomal recessive mutation *tubby* (*tub*) occurred spontaneously in C57BL/6J (B6) mice (Coleman and Eicher, 1990). Tubby mice exhibit retinal degeneration (Heckenlively et al., 1995), cochlear degeneration (Ohlemiller et al., 1997) and maturity onset obesity with insulin resistance (Coleman and Eicher, 1990). The tripartite character of the tubby phenotypes shows striking similarity to human obesity syndromes, such as Alström (Alström et al., 1959) and Bardet-Biedl (Bray, 1995). The mutation was identified within a novel gene, *tub*, by positional cloning (Kleyn et al., 1996; Noben-Trauth et al., 1996). The mutation causes a splicing defect at the carboxy terminus, leading to a larger transcript which contains the unspliced intron. The TUB protein is predominantly hydrophilic and has no signal sequence, suggesting an intracellular localization. The sequence analysis of TUB did not reveal the function of this novel protein.

Tub is a member of a novel gene family with highly conserved carboxy and divergent amino termini. Three additional members, *TULP* (tubby-like protein) *1*, *TULP2* and *TULP3*, have been identified in humans (Nishina et al., 1998; North et al., 1997), and their mouse homologues have also been isolated (Nishina et al., 1998). A high degree of conservation across family members and through evolution suggests that they carry out some basic function in cells in which they are expressed (Ikeda et al., 2002a). This notion is supported by the disease phenotypes that result from mutations within the family members. Mutations in *TULP1* lead to retinitis pigmentosa (RP) (Banerjee et al., 1998; Gu et al., 1998; Hagstrom et al., 1998; Lewis et al., 1999; Paloma et al., 2000) and targeted mutants for *Tulp1* show retinal degeneration (Hagstrom et al., 1999; Ikeda et al., 2000). Disruption of the *Tulp3* gene results in embryonic lethality, with failure of neural tube closure (Ikeda et al., 2001). Taken together with tubby phenotypes, these findings suggest that the function of the tubby gene family is important for maintenance and proper function of neuronal cells. Determining the function of this

[1]The Jackson Laboratory, Bar Harbor, MA 04609, USA. Corresponding author: P. M. Nishina, E-mail: pmn@jax.org

family of proteins should enhance our understanding of the mechanisms through which the observed disease phenotypes are caused, and how they maintain the normal function of neuronal cells.

2. PROPOSED FUNCTIONS OF THE TUBBY GENE FAMILY

Several lines of evidences have suggested possible molecular functions of the tubby gene family. Kapeller et al. (1999) demonstrated that in chinese hamster ovary cells expressing insulin receptor (CHO-IR), transfected TUB was phosphorylated on a tyrosine residue in response to insulin and insulin-like growth factor-1. Furthermore, in vitro, TUB was phosphorylated by purified insulin receptor kinase as well as by ABL and JAK2 kinases. Phosphorylated TUB associated selectively with the SH2 domains of ABL, LCK, and phospholipase Cγ. These data suggested that TUB might function as an adaptor protein linking the insulin receptor to SH2-containing proteins. Since TUB and the insulin receptor colocalize in hypothalamic neurons in the brain, this signaling pathway is likely to be active in vivo.

X-ray crystallographic analyses of the C-terminal domain of TUB showed that it has a large groove of positively charged residues that would likely form protein-protein or protein-DNA binding sites (Boggon et al., 1999). Additional molecular biological experiments showed that the TUB C-terminal domain has the potential to bind double-stranded DNA oligomers and that the N-terminus of TUB and TULP1 can activate transcription when fused to the DNA binding domain of GAL4. Together, these results suggested that TUB might be a novel class of transcription factor.

Another proposed function of TUB and TULP1, which is specific for the retina, is described in greater detail below.

3. RETINAL DEGENERATION CAUSED BY MUTATIONS IN THE TUBBY GENE FAMILY AND THE POSSIBLE FUNCTION OF TULPs IN THE RETINA

In tubby mice, the earliest histological abnormality observed is the shortening and disorganization of inner and outer segments by 2 weeks of age. Loss of photoreceptor cells is apparent by 4 weeks of age (Heckenlively et al., 1995; Ohlemiller et al., 1997). The progressive reduction of the outer nuclear layer is observed through 8 months, when photoreceptor cells are diminished to a 1-2 nuclear layer thickness. Photoreceptor cells die through an apoptotic mechanism as demonstrated by TUNEL assay (Ikeda et al., 1999). Based on the nuclear morphology, cone and rod photoreceptor cells appear to die at a similar rate.

Mutations within the *TULP1* gene, which is expressed almost exclusively in the retina, were found in patients with retinitis pigmentosa (RP) (Banerjee et al., 1998; Gu et al., 1998; Hagstrom et al., 1998; Lewis et al., 1999; Paloma et al., 2000). As observed in tubby mice, these RP patients show severe, progressive loss of photoreceptor cells which often leads to blindness. *Tulp1* knockout mice, $Tulp1^{tm1Jax}$ ($Tulp1^{-/-}$), also show early-onset, progressive retinal degeneration (Hagstrom et al., 1999; Ikeda et al., 2000). These retinal degeneration phenotypes indicate that at least two of the tubby gene family members are necessary to maintain retinal integrity and function.

A unique feature observed in tubby and *Tulp1*$^{-/-}$ retina is the accumulation of extracellular vesicles in the interphotoreceptor matrix within the region of the inner segments of photoreceptor cells (Hagstrom et al., 1999; Heckenlively et al., 1995; Ikeda et al., 2000). In these mice, ectopic localization of rod and cone opsins is also observed (Hagstrom et al., 1999) (Ikeda et al., unpublished). Given the observation that the extracellular vesicles and the inner segment plasma membranes in *Tulp1*$^{-/-}$ mice were both immunoreactive for rhodopsin (Hagstrom et al., 1999), Hagstrom et al. (2001) proposed that these proteins normally function in intracellular vesicular trafficking of rhodopsin.

Another common feature of retinal degeneration in tubby and *Tulp1*$^{-/-}$ mice is that it is early-onset and affects both rods and cones. In *Tulp1*$^{-/-}$ mice, amplitudes of both rod and cone electroretinograms (ERGs) were markedly reduced by 19 days of age (Ikeda et al., 2000). We have recently found that ERGs of tubby mice are also abnormal at an early age (Figure 1). At 21 days of age, the rod a-wave was markedly reduced and the b-wave only slightly affected. In contrast, the cone b-wave was severely affected and almost diminished at this early timepoint. These abnormalities at the early age in both mutants suggest that TUB and TULP1 might have important functions during retinal, and particularly cone development. This notion was supported by the immunohistochemical studies which showed early expression of TULP1 by retinal neuroblasts in developing human retina. Differentiating cones were strongly TULP1-positive even before they expressed blue or red/green cone opsin (Milam et al., 2000).

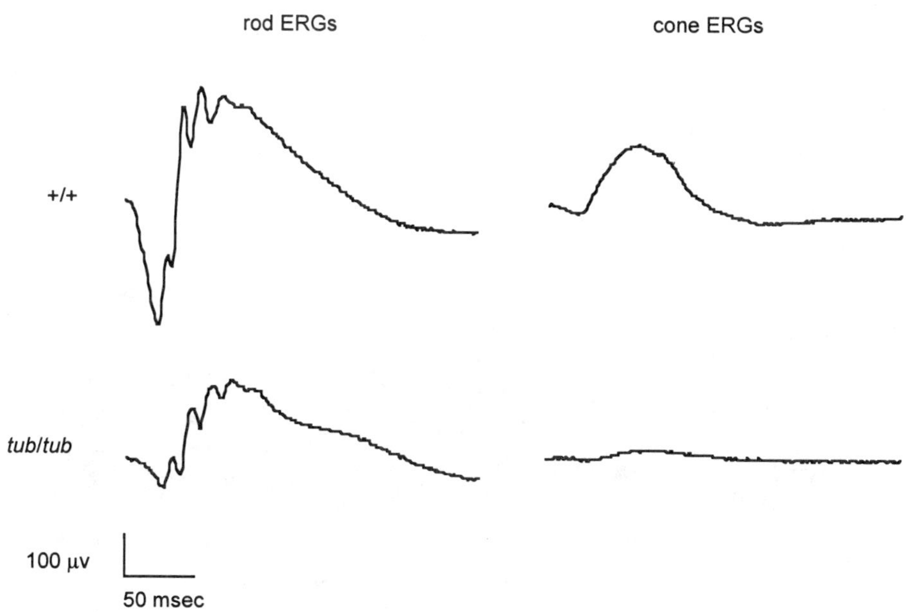

Figure 1. Rod and cone ERGs recorded from B6-+/+ and B6-*tub/tub* mice at 21 days of age. Both rod- and cone-mediated responses of B6-*tub/tub* are abnormal. Particularly, cone responses are non-detectable.

4. MODIFIER SCREENING AS A METHOD TO REVEAL FUNCTIONS OF THE TUBBY GENE FAMILY

An approach for revealing the functions of the tubby gene family is the use of genetic modifier screens. In these screens, modifier loci that affect the phenotypic outcome of a given genotype are localized in mapping crosses. These modifiers that interact with the disease causing gene may derive from same or parallel biological pathways (Haider et al., 2002). Therefore, identification of genetic modifiers for the tubby gene family will likely provide insights into the pathways through which they act, and through which they cause the associated disease phenotypes. Identification of a major quantitative trait loci (QTL) for hearing ability in tubby mice shows that this methodology is feasible and useful for identifying loci that interact with the tubby gene. This QTL, modifier of tubby hearing 1 (moth1), was recently identified as microtubule-associated protein 1A by positional cloning, and suggests an association of TUB with synaptic function or structure of neuronal cells (Ikeda et al., 2002b). Similar analyses were performed for the retinal degeneration phenotype of tubby mice. Offspring of the F2 intercross between (C57BL/6J-tub/tub and AKR/J-+/+) F1 hybrids homozygous for the tub mutation were phenotyped for the thickness of the outer nuclear layer (ONL)or the number of photoreceptor nuclei. A large variation in these phenotypes were observed, indicating a genetic background effect (Figure 2). A genome-wide scan revealed QTL modifiers with a significant linkage on Chromosome 11 (designated motr1) and suggestive linkages on Chromosome 2 and 8. Interestingly, the QTL on Chromosome 2 maps to the same region as the moth1 locus, suggesting the possibility that the same modifier gene is able
to suppress or delay cochlear and retinal degeneration in tubby mice. It is also important to note that no phenotypic variations were observed in F2-tub/+ and F2-+/+ mice from

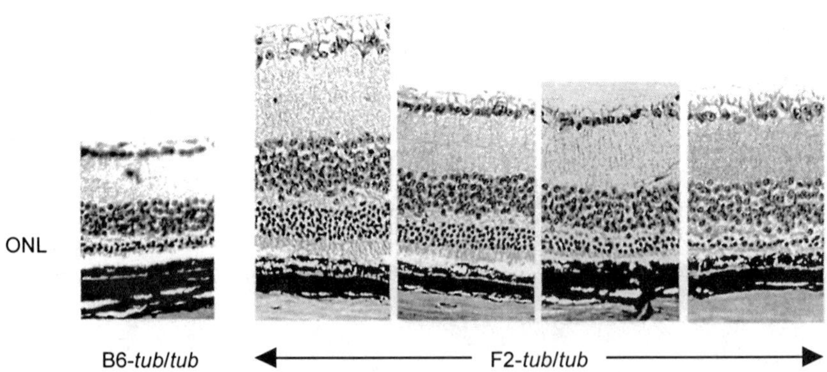

Figure 2. Segregation of the retinal degeneration phenotype in an intercross between B6-*tub/tub* and AKR/J mice. The thickness of ONL varies among F2--*tub/tub* mice.

the intercross. This suggests that the allelic difference between C57BL/6J and AKR/J are not crucial for maintaining the integrity of photoreceptor cells, but becomes significant only when the tubby gene is mutated and unable to function properly. Perhaps, the genes underlying these QTLs are involved in the retinal degeneration cascade induced by the tub mutation. Identification of these QTLs will, therefore, likely provide insights not only into the role of TUB in the retina, but also into the pathway through which retinal degeneration is caused in tubby mice. QTL analyses for retinal degeneration in Tulp1-/- mice are also underway.

5. ACKNOWLEDGEMENTS

We thank the National Eye Institute for the travel award provided to SI to attend this meeting.

6. REFERENCES

Alström, C. H., Hallgren, B., Nilsson, L. B., Åsander, H., 1959, Retinal degeneration combined with obesity, diabetes mellitus and neurogenous deafness. A specific syndrome (not hitherto described) distinct from Laurence-Moon-Biedl syndrome. A clinical endocrinological and genetic examination based on a large pedigree., *Acta Psychiatr Neurol Scand.* **34**: 1-35.

Banerjee, P., Kleyn, P. W., Knowles, J. A., Lewis, C. A., Ross, B. M., Parano, E., Kovats, S. G., Lee, J. J., Penchaszadeh, G. K., Ott, J., Jacobson, S. G., Gilliam, T. C., 1998, TULP1 mutation in two extended Dominican kindreds with autosomal recessive retinitis pigmentosa, *Nat Genet.* **18**: 177-9.

Boggon, T. J., Shan, W. S., Santagata, S., Myers, S. C., Shapiro, L., 1999, Implication of tubby proteins as transcription factors by structure-based functional analysis, *Science.* **286**: 2119-25.

Bray, G. A., 1995, Laurence, Moon, Bardet, and Biedl: reflections on a syndrome, *Obes Res.* **3**: 383-6.

Coleman, D. L., Eicher, E. M., 1990, Fat (fat) and tubby (tub): two autosomal recessive mutations causing obesity syndromes in the mouse, *J Hered.* **81**: 424-7.

Gu, S., Lennon, A., Li, Y., Lorenz, B., Fossarello, M., North, M., Gal, A., Wright, A., 1998, Tubby-like protein-1 mutations in autosomal recessive retinitis pigmentosa, *Lancet.* **351**: 1103-4.

Hagstrom, S. A., Duyao, M., North, M. A., Li, T., 1999, Retinal degeneration in tulp1-/- mice: vesicular accumulation in the interphotoreceptor matrix, *Invest Ophthalmol Vis Sci.* **40**: 2795-802.

Hagstrom, S. A., North, M. A., Nishina, P. L., Berson, E. L., Dryja, T. P., 1998, Recessive mutations in the gene encoding the tubby-like protein TULP1 in patients with retinitis pigmentosa, *Nat Genet.* **18**: 174-6.

Haider, N. B., Ikeda, A., Naggert, J. K., Nishina, P. M., 2002, Genetic modifiers of vision and hearing, *Hum Mol Genet.* **11**: 1195-206.

Heckenlively, J. R., Chang, B., Erway, L. C., Peng, C., Hawes, N. L., Hageman, G. S., Roderick, T. H., 1995, Mouse model for Usher syndrome: linkage mapping suggests homology to Usher type I reported at human chromosome 11p15, *Proc Natl Acad Sci U S A.* **92**: 11100-4.

Ikeda, A., Ikeda, S., Gridley, T., Nishina, P. M., Naggert, J. K., 2001, Neural tube defects and neuroepithelial cell death in Tulp3 knockout mice, *Hum Mol Genet.* **10**: 1325-34.

Ikeda, A., Nishina, P. M., Naggert, J. K., 2002a, The tubby-like proteins, a family with roles in neuronal development and function, *J Cell Sci.* **115**: 9-14.

Ikeda, A., Zheng, Q. Y., Zuberi, A. R., Johnson, K. R., Naggert, J. K., Nishina, P. M., 2002b, Microtubule-associated protein 1A is a modifier of tubby hearing (moth1), *Nat Genet.* **30**: 401-5.

Ikeda, S., He, W., Ikeda, A., Naggert, J. K., North, M. A., Nishina, P. M., 1999, Cell-specific expression of tubby gene family members (tub, Tulp1,2, and 3) in the retina, *Invest Ophthalmol Vis Sci.* **40**: 2706-12.

Ikeda, S., Shiva, N., Ikeda, A., Smith, R. S., Nusinowitz, S., Yan, G., Lin, T. R., Chu, S., Heckenlively, J. R., North, M. A., Naggert, J. K., Nishina, P. M., Duyao, M. P., 2000, Retinal degeneration but not obesity is observed in null mutants of the tubby-like protein 1 gene, *Hum Mol Genet.* **9**: 155-63.

Kleyn, P. W., Fan, W., Kovats, S. G., Lee, J. J., Pulido, J. C., Wu, Y., Berkemeier, L. R., Misumi, D. J., Holmgren, L., Charlat, O., Woolf, E. A., Tayber, O., Brody, T., Shu, P., Hawkins, F., Kennedy, B., Baldini, L., Ebeling, C., Alperin, G. D., Deeds, J., Lakey, N. D., Culpepper, J., Chen, H., Glucksmann-

Kuis, M. A., Moore, K. J., et al., 1996, Identification and characterization of the mouse obesity gene tubby: a member of a novel gene family, *Cell.* **85**: 281-90.

Lewis, C. A., Batlle, I. R., Batlle, K. G., Banerjee, P., Cideciyan, A. V., Huang, J., Aleman, T. S., Huang, Y., Ott, J., Gilliam, T. C., Knowles, J. A., Jacobson, S. G., 1999, Tubby-like protein 1 homozygous splice-site mutation causes early-onset severe retinal degeneration, *Invest Ophthalmol Vis Sci.* **40**: 2106-14.

Milam, A. H., Hendrickson, A. E., Xiao, M., Smith, J. E., Possin, D. E., John, S. K., Nishina, P. M., 2000, Localization of tubby-like protein 1 in developing and adult human retinas, *Invest Ophthalmol Vis Sci.* **41**: 2352-6.

Nishina, P. M., North, M. A., Ikeda, A., Yan, Y., Naggert, J. K., 1998, Molecular characterization of a novel tubby gene family member, TULP3, in mouse and humans, *Genomics.* **54**: 215-20.

Noben-Trauth, K., Naggert, J. K., North, M. A., Nishina, P. M., 1996, A candidate gene for the mouse mutation tubby, *Nature.* **380**: 534-8.

North, M. A., Naggert, J. K., Yan, Y., Noben-Trauth, K., Nishina, P. M., 1997, Molecular characterization of TUB, TULP1, and TULP2, members of the novel tubby gene family and their possible relation to ocular diseases, *Proc Natl Acad Sci U S A.* **94**: 3128-33.

Ohlemiller, K. K., Hughes, R. M., Lett, J. M., Ogilvie, J. M., Speck, J. D., Wright, J. S., Faddis, B. T., 1997, Progression of cochlear and retinal degeneration in the tubby (rd5) mouse, *Audiol Neurootol.* **2**: 175-85.

Paloma, E., Hjelmqvist, L., Bayes, M., Garcia-Sandoval, B., Ayuso, C., Balcells, S., Gonzalez-Duarte, R., 2000, Novel mutations in the TULP1 gene causing autosomal recessive retinitis pigmentosa, *Invest Ophthalmol Vis Sci.* **41**: 656-9.

IDENTIFICATION OF DOWNSTREAM MECHANISMS INVOLVED IN PEDF'S ACTIVITY IN THE RETINA BY LARGE SCALE GENE EXPRESSION PROFILING

Joyce Tombran-Tink[1], Nuria Lara[1], Arvind Chappa[1], Amber R. Tink[1], Lincoln V. Johnson[2], and Michelle Staples[2]

1. INTRODUCTION

PEDF is a neurotrophic and antiangiogenic protein in the retina. In 1987, Lincoln Johnson and I isolated and characterized the 50 kDa PEDF protein from fetal human RPE cells (Tombran-Tink et al., 1991) and showed that it was effective in switching Y79 retinoblastoma cells from an actively growing cell line to non-proliferating cells that attached and extended neurites. A role for PEDF in the development and differentiation of the retina was suggested then because of this activity.

We subsequently isolated, sequenced, and mapped the PEDF gene to human 17p13, a hot spot area for many retinal diseases (Steele et al., 1993; Tombran-Tink et al., 1996; Tombran-Tink et al., 1994) and found that the protein is a member of the serpin gene family, a group of serine protease inhibitors. Serpins are best known for their activity as inhibitors of serine proteinases, in particular, inhibition of members of the chymotrypsin gene family. A leucine in the exposed reactive serpin loop is a key residue that targets inhibition of chymotrypsin. Although the leucine residue is conserved in PEDF, the protein shows noninhibitory activity in assays performed so far (Becerra et al., 1995). Like ovalbumin and angiotensinogen, PEDF may have deviated from its original inhibitory function during evolutionary progression. Its biological activity, therefore, may be dependent on non-serpin like mechanisms.

Much of the described neurotrophic activity of PEDF comes from the ophthalmic field. PEDF strongly protects rat photoreceptors from degenerating after exposure to constant light (Cao et al., 2001). The effect is equally as potent as those reported for BDNF and CNTF, two well-characterized growth factors in the retina. Likewise, PEDF delays the progression of photoreceptor degeneration in *rd* mice and reduces apoptosis in photoreceptor cells in the *rds* mice, two mutant mouse models of retinal degeneration

[1]University of Missouri-Kansas City, Pharmaceutical Sciences, Kansas City, MO 64110, USA; [2]University of Califonria Santa Barabara, Neuroscience Research Institute, Santa Barbara, CA 93106, USA. Corresponding author: J. Tombran-Tink, E-mail: jttink@aol.com

Retinal Degenerations: Mechanisms and Experimental Therapy
Edited by LaVail *et al.*, Kluwer Academic/Plenum Publishers, 2003

(Cayouette et al., 1999). Cell death is significantly reduced in dissociated rat retinal cultures treated with PEDF after exposure to hydrogen peroxide toxicity (Cao et al., 1999). The protein is effective, as well, in preserving spatial organization, morphology, and functional integrity of photoreceptors after RPE detachment in a Xenopus model of retinal degeneration (Jablonski et al., 2000). In one study, PEDF is also shown to protect retinal neurons from damage caused by increased intraocular pressure from transient ischemic reperfusion (Ogata et al., 2001).

The neuroprotective effect of PEDF does not appear to be limited to photoreceptor neurons alone. It is apparent in other types of neurons as well. These include cerebellar granule cells (Taniwaki et al., 1995; 1997), hippocampal neurons (DeCoster et al., 1999), and spinal cord motor neurons (Bilak et al., 1999; Houenou et al., 1999) which are shown to survive the damaging effects of glutamate excitotoxicity after treatment with PEDF. In the eye, the protein may also play a significant protective role in more than one ocular cell type since its expression is reported in the vitreous, aqueous humor, RPE, cornea, iris/ciliary body, IPM, and the retina (Tombran-Tink et al., 1995; Ortego et al., 1996; Ogata et al., 2002; Behling et al., 2002)

A dual role for PEDF in the retina emerged when Dawson et al. (Dawson et al., 1999) published a landmark paper describing the effects of PEDF in angiogenesis. They reported that the secreted protein inhibits angiogenic processes and is more effective than the well-studied angiogenesis inhibitor, angiostatin. The finding is not entirely surprising, given the fact that PEDF is present in high concentrations in avascular regions of the eye, and that RPE cells secrete inhibitors of angiogenesis in cultured conditions. Those studies indicate that PEDF inhibits endothelial cell migration in a dose dependent manner in the presence of the angiogenic stimulators βFGF, VEGF, interleukin-8, αFGF, and lysophosphatic acid. Experiments of PEDF's antiangiogeneic effects were later described in a corneal model of neovascularization, in which recombinant PEDF inhibited blood vessel proliferation in the presence of βFGF (Mori et al., 2001). In addition to the antiangiogenic activity noted, the expression of endogenous PEDF can be manipulated experimentally by varying oxygen levels in several retinoblastoma cell lines. The protein is downregulated by hypoxia and is shown to increase under hyperoxic conditions (Dawson et al., 1999). Furthermore, the expression pattern of PEDF under hypoxic conditions is reciprocal to that seen for VEGF. Hypoxia is a strong stimulus of VEGF, one of the most potent mitogens of angiogenesis in the eye. An equilibrium shift has been shown between PEDF and VEGF during the uncontrolled growth of blood vessels in the eye (Holekamp et al., 2002; Ohno-Matsui et al., 2001; Ogata et al., 2001; Ogata et al., 2002; Spranger et al., 2001). Decreased levels of angiogenic inhibitors, such as PEDF, may be permissive to the angiogenic environment in the retina.

In an earlier study, we observed that the expression of PEDF is decreased in RPE cells during senescence in vitro (Tombran-Tink et al., 1995). A reduction in the physiological levels of PEDF in vivo may contribute to photoreceptor degeneration and choroidal neovascularization given the fact that PEDF is both neuroprotective and antiangiogenic. Another indication that PEDF plays a role in ocular angiogenesis is seen in studies showing that vitreous levels of PEDF are significantly lower in patients with active compared to inactive forms of diabetic retinopathy (Ogata et al., 2001). In addition, choroidal neovascular membranes are shown to contain increased amounts of PEDF in the proliferating RPE cells that cover the neovascular tissues after pan retinal laser photocoagulation. This finding suggests that the effects of pan retinal treatment in reducing neovascularization may be due, in part, to an upregulation of PEDF in proliferating RPE cells (Ogata et al., 2001).

Gene transfer technology is a recently developed strategy to test the effects of PEDF in the eye. Overexpression of the protein in the retina by viral-mediated gene transfer ameliorates retinal and choroidal neovascularization (Mori et al., 2002; Raisler et al., 2002). Intravitreal and subretinal applications of an adenoviral vector encoding PEDF can reduce the extent of vessel formation in an animal model of choroidal neovascularization arising from laser-induced damage to Bruch's membrane. The severity of vessel growth is reported to be diminished by sustained expression of PEDF in transgenic mice overexpressing VEGF in the photoreceptor cells and mice with oxygen induced ischemic retinopathy. Taken together, the data suggest that PEDF serves as an antiangiogenic switch in the retina and may play a central role in maintaining vascular quiescence in the eye. The studies have led to the development of protocols for a Phase 1 clinical trial of viral-mediated PEDF gene transfer to test the clinical feasibility of using PEDF intraocularly (Rasmussen et al., 2001) The potential use of PEDF gene therapy to enhance neuroprotective strategies or antagonize angiogenic stimuli is an exciting research direction for the treatment of ocular diseases, such as AMD and PDR, where both neuroprotection and antiangiogenic activities are important.

The studies are not conclusive but they point to a significant correlation between PEDF and the extent of neovascularization in the eye. They also raise the question of how PEDF inhibits angiogenic stimuli and protects neurons from degenerating in the retina. A possible signaling cascade of how PEDF protects cerebellar granule cells has been proposed recently. Phosphorylation of IkB, activation of NFkB, and upregulation of the antiapoptotic genes Bcl2, Bcl-x, and manganese superoxide dismutase are reported to be mechanisms triggered by PEDF. Although regulation of the NFkB pathway may alter the progression of apoptosis in cerebellar granule cells (Yabe et al., 2001) it is not known whether this is the signaling cascade that PEDF uses in protecting retinal neurons. The studies, however, clearly imply that PEDF plays a crucial role in neurodegenerative processes caused by apoptotic events. Identification of specific pathways triggered by PEDF in the retina may make it possible in the future to therapeutically modulate apoptotic signals that result in the loss of photoreceptors.

Our primary approach in this study was to identify potential signaling mechanisms of PEDF in two human retinal cell lines; ARPE19, a retinal pigment epithelial cell line, and the Y79 retinoblastoma cells. The regulation of four genes by PEDF is discussed: VEGRR3, which may be involved in angiogenic signaling; caspase 2, calpain 1, and Map kinase kinase 1, which are involved in elaborate cascades that control apoptosis, cell growth, and differentiation.

2. METHODS

The human ARPE19 and Y79 retinoblastoma cells were grown in MEM containing 10% FBS until they reached approximately 1 x 10^6 cells per ml. Then, the medium was replaced by serum-free medium 24 hrs before starting the experiment to eliminate the possibility of serum proteins interfering with gene expression. Subsequently, PEDF was added to the medium at a concentration of 100 ng/ml for 48 hours. Controls were kept in serum-free medium for the duration of the experiment and the experimental cultures were treated with or without 100 ng/ml PEDF for 24 hrs.

PEDF-treated and non-treated cells were harvested by centrifugation for RNA isolation, gene array analysis, and RT-PCR. The quality of RNA used for array analysis or RT-PCR was determined by capillary electrophoresis.

The RNA was prepared using Clontech's Atlas ™ Pure Total RNA labeling Kit and SMART™ Probe Amplification Kit. The quality and concentration of the purified RNA were determined by capillary electrophoresis. Probe synthesis was carried out by reverse transcription of 50 ug of total RNA and the probes labeled with cDNA Synthesis Primer Mix and α-^{32}P- dATP (Clontech). Radioactively labeled cDNA probes were purified by column chromatography, diluted in ExpressHyb solution, and used to hybridize 8K human gene arrays (Clontech) for 1 hr at 68° C. The hybridized membranes were washed and exposed to x-ray films or phosphoroimaging screens at varying time points.

Two software packages, the AtlasImage and AtlasNavigator, were used to plot expression signals and ratio differences between the control and experimental samples. Genes involved in angiogenic or apoptotic mechanisms that were regulated by PEDF by array analyses were selected to verify their expression using RT-PCR. These genes included: VEGF: isoforms 121, 165, and 189; VEGF receptors: VEGFR1, VEGFR2, VEGFR3; caspase 2, calpain 1, and map kinase kinase 1.

The genes selected for further studies were above a 2.5 fold change in expression after global normalization between genes and subtraction of background noise. cDNA generated by reverse transcription was quantitated and used at a constant concentration of 100ng/50 ul PCR reaction. Primer sequences, amplicon size, and PCR conditions for the selected genes are given below:

VEGF isoforms (Maximbio)
$VEGF_{121}$: (104 bp); $VEGF_{165}$: (234 bp); $VEGF_{189}$: (306 bp)
(PCR: 96° C, 1min, 1 cycle; 94° C, 1 min, 58° C, 1 min, 72° C, 1 min – 30 cycles)
Forward: 5'-CCAGCACATAGGAGAGATGAGCTTC-3'
Reverse: 5'-CACCGCCTCGGCTTGTCACAT-3'
VEGF receptors:
VEGFR1: (316 bp; PCR: 95° C, 1 min; 60° C, 1 min; 72° C, 1 min, 30 cycles)
Forward: 5'-GTAGCTGGCAAGCGGTCTTACCGGCTC-3'
Reverse:5'-GGATTTGTCTGCTGCCCAGTGGGTAGAGA-3'
VEGFR2: 550bp; 95° C, 1min, 57° C, 1 min, 72° C , 1 min; 30 cycles)
Forward: 5'-CCTTACCCTGGCTTACACAT-3'
Reverse: 5'-GCTGGAGGGCGAAATGGACA-3'
VEGFR3 (166 bp; PCR: 95⁰ C, 1 min; 59⁰ C, 1 min; 72⁰ C, 1 min; 30 cycles)
Forward: 5'-CAGGTGCTTCCCAGACACTGGCGTTACT-3'
Reverse: 5'-ACTCATATTACCAAGGAATAACTGGCGGGCA-3'
Caspase 2 (367 bp; PCR: 95⁰ C, 1min; 58⁰ C, 3min, 72⁰ C, 2min; 30 cycles)
Forward: 5'-TTTCAGCTCTTTGACAACGCCAACTGCC-3'
Reverse: 5'-CTTCCCGATCCTTGATAAGTGCGTTCAC-3'
Calpain 1 (338 bp; 95⁰ C, 1 min; 59 ⁰C, 1 min, 72⁰ C, 1 min; 30 cycles)
Forward: 5'-GATGACCAGATCCAGGCCAATCTCC-3'
Reverse: 5'-GCACTCATGCTGCCCGACTTGTCC-3'
MAP Kinase Kinase 1 (613 bp; 95⁰ C, 1 min; 59 C, 3 min, 72⁰ C, 2 min; 30 cycles)
Forward: 5'-GGTTCTGCATGAGTGCAAC-3'

Reverse: 5'-GGCTCGTTGACTATGTAATCC-3'
Actin: (540bp; 95 C, 1 min, 59 C, 1 min; 72 C, 1 min; 30 cycles)
Forward: 5'-GTGGGCCGCTCTAGGCACCAA
Reverse: 5'-CTCTTTGATGTCACGCACGATTTC-3'

3. RESULTS

Statistical analysis of large-scale gene profiling data allowed us to classify genes that are regulated by PEDF into specific functional clusters (Fig 1, Fig 2). PEDF regulates approximately 3% of the genes on the 8K arrays in human ARPE-19 and Y79 retinoblastoma cells. The two functional groups that showed consistent regulation by PEDF in both cell lines included genes involved in pathways leading to apoptosis and angiogenesis. Other clusters in the PEDF-induced transcription profile include: receptors, growth factors, cytokines, cell differentiation factors, and DNA binding proteins. A large percentage (47%) of genes that appear to be modulated by PEDF represents ribosomal proteins and genes with less well-characterized function or group association.

For this study, we selected six genes that are known to play an important role in events leading to angiogenesis and apoptosis so that we could verify their expression by PCR. These genes showed signal ratio differences of >2.5-fold changes between PEDF treated and non-treated cells on the array. They include:- the VEGF isoforms which play a key role in promoting angiogenesis: VEGF $_{121}$, VEGF $_{165}$, and VEGF $_{189}$; VEGF receptors: VEGFRI, VEGFR2, and VEGFR3, which are involved in signaling mechanisms that VEGF uses to promote angiogenesis; and three genes associated with initiation of apoptotic cascades in response to various stimuli, including neurotoxic or ischemic stimuli: Capsase 2, Calpain 1, and Map kinase kinase 1 (MKK1).

By RT-PCR analysis, none of the VEGF isoforms demonstrated changes after treatment with PEDF (Fig 3). We also did not observe significant modulation in the expression of VEGFR1 or VEGFR2 between PEDF treated and the control samples, even though the array data indicated that the expression of those genes decreased after treatment with 100 ng/ml PEDF (Fig 4, Fig 5). However, a significant down regulation of VEGR3, by approximately three folds, was shown in both the ARPE19 and Y79 cell lines by RT-PCR (Fig 4, Fig 5). The PCR result was consistent with the fold changes observed in the transcription profiling array experiments (Fig 8). Downregulation of caspase 2, calpain 1 (Fig 6), and map kinase kinase 1 (Fig 7), was evident in both the RPE and Y79 cell lines in the PCR experiments. The regulation of these genes was consistent with the data obtained from array analysis (Fig 8). Detectable changes in the expression of actin, were not observed between the control and experimental samples.

4. DISCUSSION

The data obtained from our experiments point to an increasing role for PEDF in regulating specific pathways as a means to possibly halt the progression of retinal degeneration. We show that PEDF targets key downstream regulators that may prevent apoptosis and redirect cell choices. Caspase 2 and Calpain 1 are intracellular cysteine proteases which are part of interrelated signal transduction cascades. Their activation can be triggered by calcium influx, oxidative stress, and ischemic injury and their

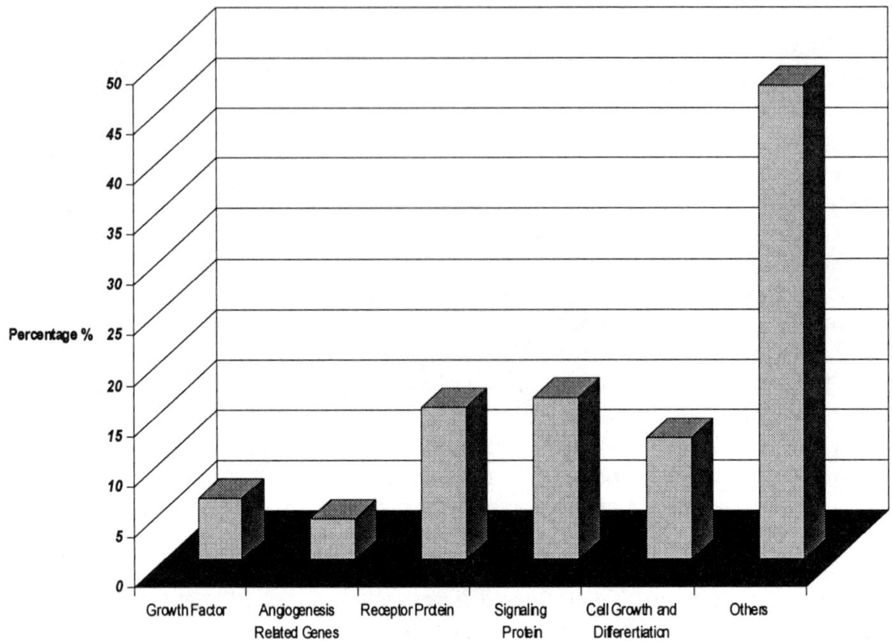

Fig. 1. Groups of genes regulated by PEDF.

overexpression linked to various necrotic and apoptotic conditions in acute neurodegenerative processes (Wang et al., 2000). The endpoint of these cascades appears to converge in the activation of transcription factors that regulate apoptosis, differentiation, and cell survival.

So far, two functions have been identified for PEDF: a direct protective effect of neuronal cells against apoptosis and the control of angiogenesis, possibly by mechanisms which induce apoptosis. These findings raise an important question-: How does PEDF protect neurons from multiple types of insults and induce apoptosis in endothelial cells? Regrettably, gaps in the fundamental understanding of basic molecular interactions of PEDF in the retina hinder systematic approaches that could exploit molecular targets of PEDF in retinal degeneration. A necessary first step in developing neuroprotective or antiangiogenic treatments that could be applied in acute settings is to dissect potential PEDF signaling pathways that modulate these processes. A few signaling targets have already been identified, but these may likely represent cell specific responses to PEDF. Our study demonstrates that multiple but interrelated pathways that may intersect are regulated by PEDF during its activity in the retina.

The remarkable consistency in regulation of caspase 2 and calpain 1 in two retinal cell lines treated with PEDF, by both array and PCR analysis, strongly imply that PEDF may exert a protective function by regulating Caspase 2/ Calpain 1 signaling cascades. Furthermore, the down regulation of MAPKK1 by PEDF also may be of significance in the control of apoptosis by PEDF. The MAP kinase superfamily of enzymes is a multipurpose, intermediate signal transducer system. It serves as a central switchboard that transmits signals that control cell growth and apoptosis. A growing body of evidence now suggests that MAP kinases become activated by MAPKKs following various injury stimuli, including focal cerebral ischemia. Their activation may play a key role in the control of neuronal apoptosis.

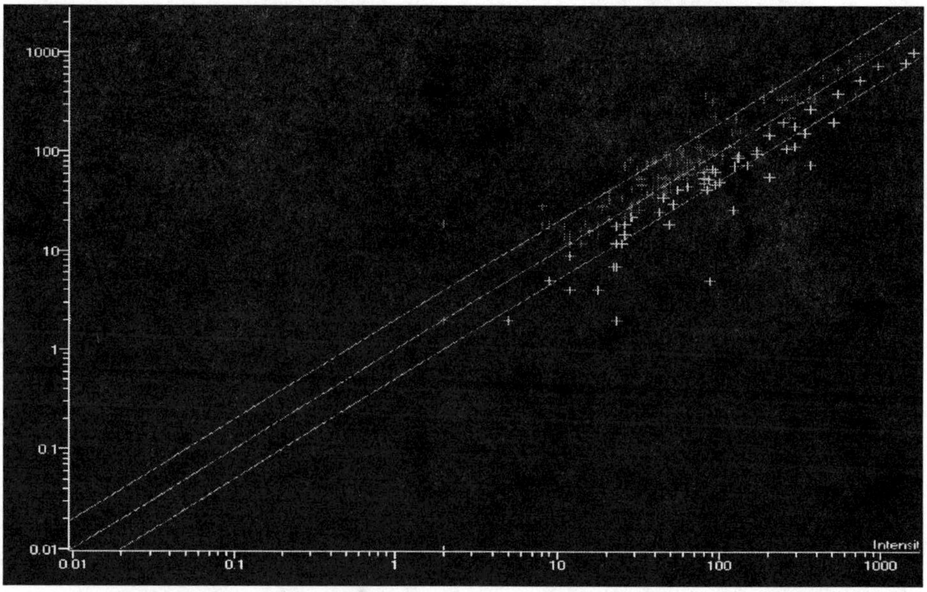

Fig 2. A Scatter plot showing relative gene expression intensities generated for ARPE19 and Y79 cells after PEDF treatment. The three diagonal lines are guides for comparing relative gene expression. The line in the center indicates equivalent gene expression. The upper line marks a 2-fold increase. The lower line marks a 2-fold decrease. Genes expressed by ARPE19 cells after treatment with PEDF are plotted on the x-axis. Genes expressed by Y79 cells after treatment with PEDF are plotted on the y-axis. The plot gives a comparison of PEDF-induced gene expression between the two cell lines.

We were rather surprised by the lack of regulation of VEGF isoforms by PEDF given the evidence supporting a distinct equilibrium shift between VEGF and PEDF in angiogenic eye models. The most pronounced expression in both cell lines was that of VEGF A 165 isoform. Only weak expression was seen for the VEGF A isoforms, 189 and 121 and no difference was observed in the expression profile for any of the isoforms. It is conceivable that because of the short duration of treatment with PEDF, most of the VEGF transcripts present in the cells before the experiment remained in the RNA pool used in the studies. Inhibition of VEGF may require longer exposure to PEDF or treatment with more effective doses of PEDF to mobilize a change in VEGF's expression.

A noteworthy finding was the expression and regulation of VEGFR3 in the RPE and Y79 cell lines. VEGFR3 is known to be expressed in lymphatic tissues and binds to VEGF C instead of VEGF A which is expressed in the retina. Although most adult tissues do not appear to express the VEGF R3 receptor, it may be important in angiogenic processes during development. In this regard, PEDF may control VEGF R3 mediated mitogenic signaling in undifferentiated developing tissues.

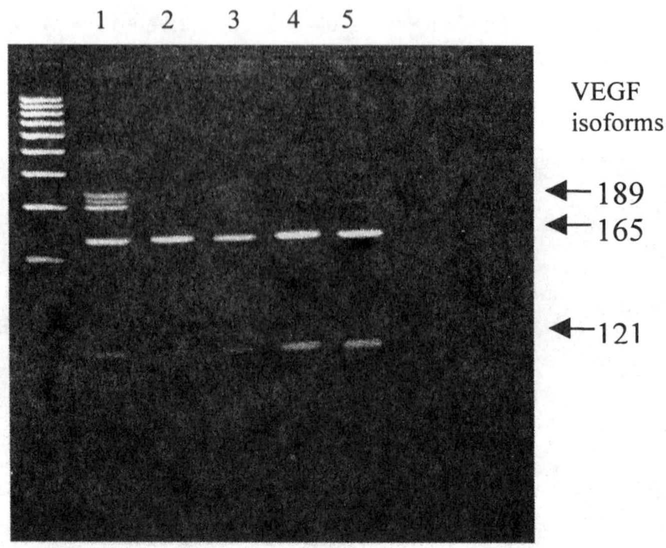

Fig. 3. RT-PCR: expression of VEGF isoforms in human ARPE19 and Y79 retinoblastoma cells treated with or without100 ng/ml PEDF for 48 hours. Lane 1: Expression of $VEGF_{121}$ (104 bp), $VEGF_{165}$ (234 bp), and $VEGF_{189}$ (306 bp) in whole brain cDNA (positive control). Lane 2: control, ARPE-19; Lane 3: ARPE19 treated with PEDF; Lane 4: control, Y79 cells; Lane 5: Y79 cells treated with PEDF. $VEGF_{165}$ is the most pronounced VEGF isoform in ARPE19 and Y79 cells. VEGF isoforms are not modulated by PEDF in the two cell lines.

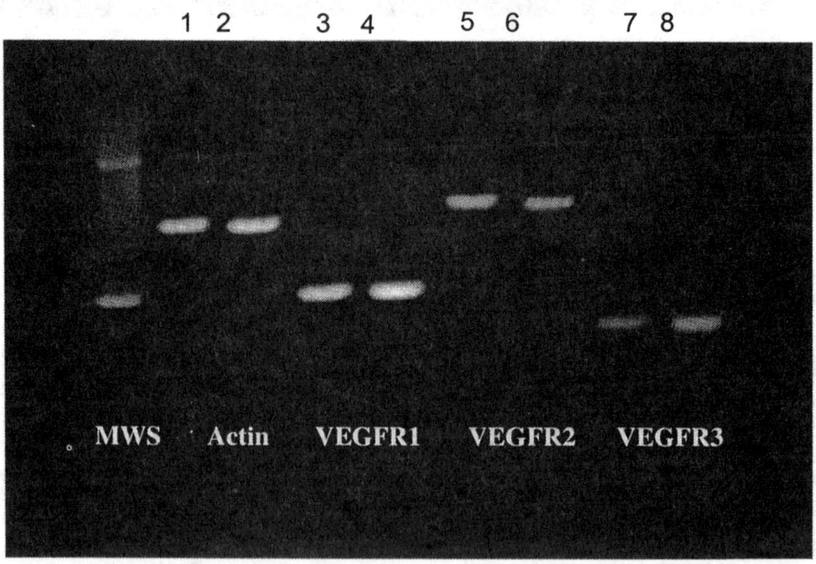

Fig 4. RT-PCR. Expression of VEGF receptors in human ARPE19 cells. Lanes: 1, 3, 5, and 7 represent cells treated with 100 ng/ml PEDF for 48 hours. Lanes: 2, 4, 6, and 8 represent non-treated cells. VEGFR3 is down regulated after treatment with PEDF (7). The expression of actin (1-2), VEGFR1 (3-4), and VEGFR2 (5-6) remained unchanged by PEDF.

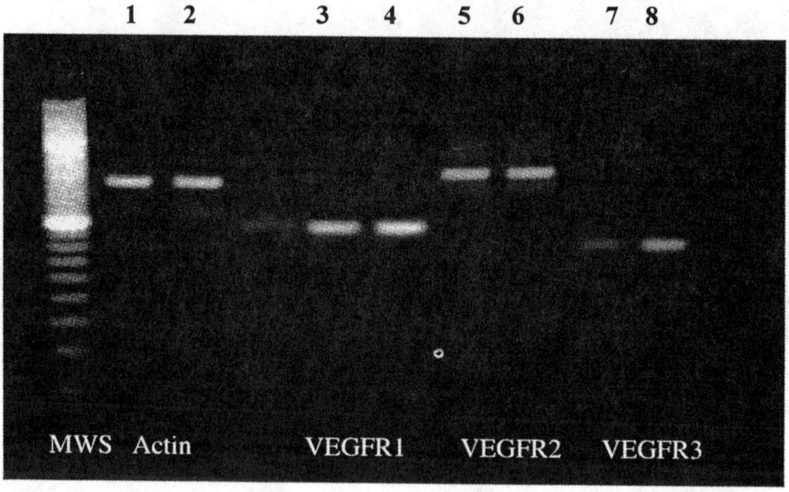

Fig 5. RT-PCR: Expression of VEGF receptors in human Y79 retinoblastoma cells. Lanes 1,3,5,and 7 represent cells treated with 100 ng/ml PEDF for 48 hrs; Lanes 2, 4, 6, and 8 represent non-treated cells. VEGFR3 is down regulated after treatment of the cells with PEDF (7). Expression of actin (1-2), VEGFR1 (3-4), and VEGFR2 (5-6) remained unchanged by PEDF.

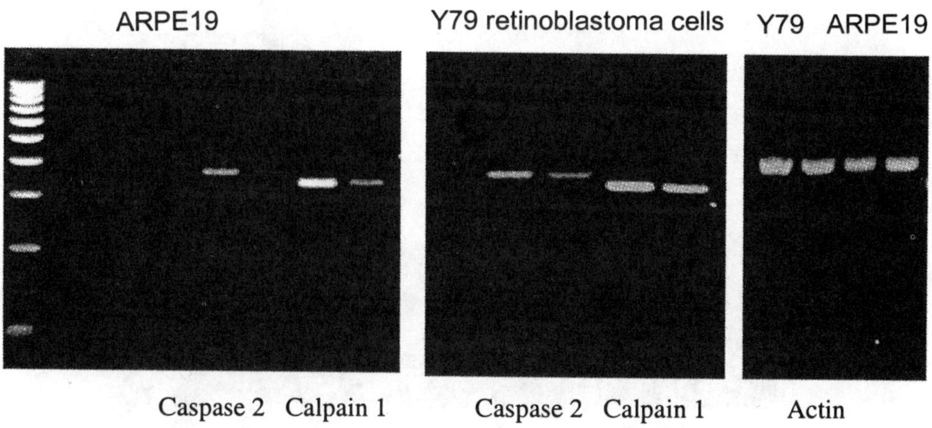

Fig 6. RT-PCR. Regulation of caspase 2 and calpain 1 in Y79 retinoblastoma and ARPE19 cells. The left lane in each group represents non-treated cells. The right lanes represent cells treated with 100 ng/ml PEDF for 48 hours. Caspase 2 and calpain 1 are down regulated by PEDF. MWS: 100bp ladder.

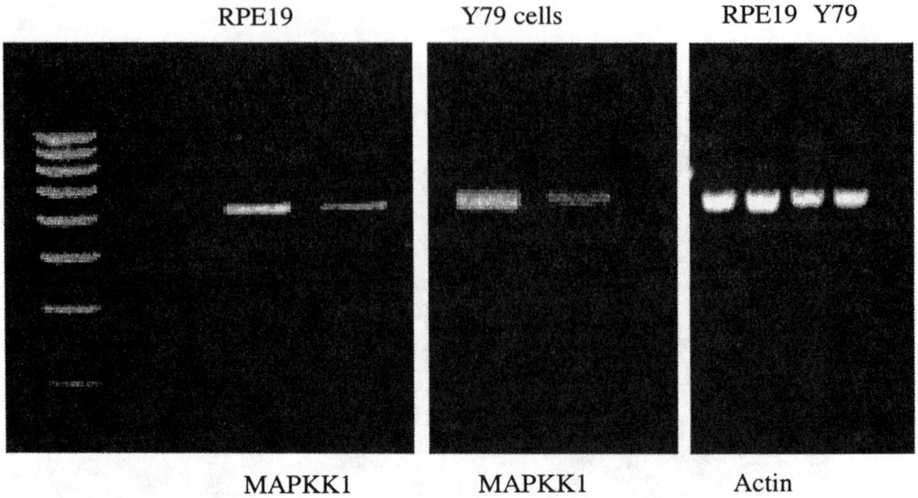

Fig 7. RT-PCR. Expression of Map kinase kinase 1 (MAPKK1) in human ARPE19 and Y79 retinoblastoma cell lines. Left lane in each group represents control; Right lanes represent cells treated with 100ng/ml PEDF for 48 hours. PEDF down regulates MAPKK1 in both cell lines. Actin expression remained constant between control and experimental samples. MWS: 100 bp ladder.

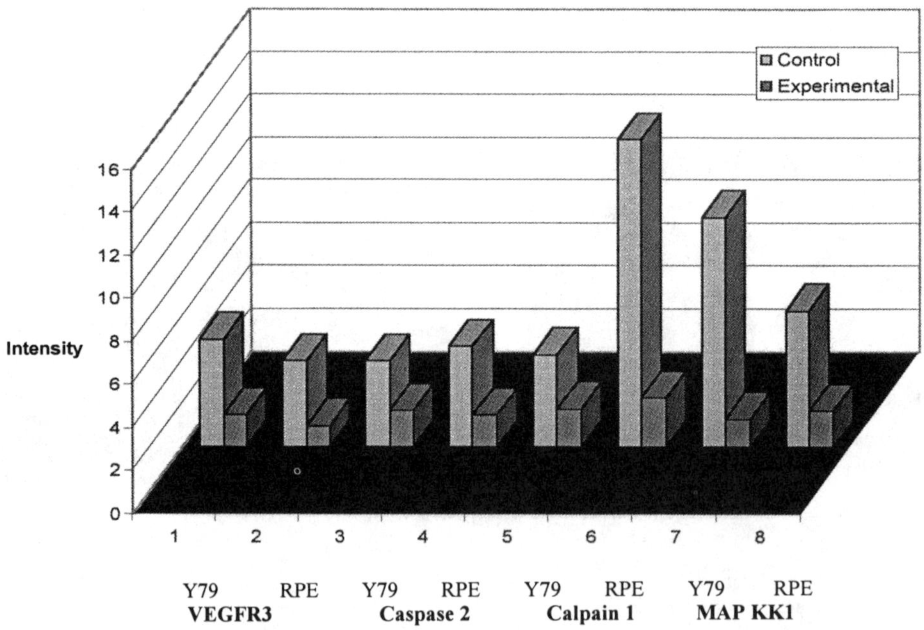

Fig. 8. The bar graph plots the signal intensity for VEGFR3, Caspase 2, Calpain 1, and MAP kinase kinase1 (MAPKK1) from gene arrays. The data represent an average of three gene array experiments and compare the signal intensity values between the control (left bars; non-treated samples) and experimental (right bars; PEDF-treated samples) data sets: VEGFR3 (1,2), Caspase 2 (3,4), Calpain 1 (5,6), and MAPKK1 (7,8). All three genes are significantly down regulated by PEDF.

In summary, our data strongly suggest a role for PEDF in regulating interrelated signaling cascades with endpoints in cell proliferation, protection, and death, processes that underlie neuroprotection and angiogenesis. Continuing studies may uncover distinct pathways and targets of PEDF that may offer a glimpse into the future use of PEDF in retinal therapeutics.

5. ACKNOWLEDGMENTS

We acknowledge Samit Shah for his technical assistance during the preparation of this manuscript. The study was supported by Inter-Cal and Pfizer Inc.

6. REFERENCES

Becerra, S.P., Sagasti, A., Spinella, P., and Notario, V., 1995, Pigment epithelium-derived factor behaves like a noninhibitory serpin. Neurotrophic activity does not require the serpin reactive loop. *J Biol Chem.* **270**:25992-9

Behling, K.C., Surace, E.M., and Bennett, J., 2002, Pigment epithelium-derived factor expression in the developing mouse eye. *Mol Vis.* **8**:449-54.

Bilak, M.M., Corse, A.M., Bilak, S.R., Lehar, M., Tombran-Tink, J., and Kuncl, R.W., 1999, Pigment epithelium-derived factor (PEDF) protects motor neurons from chronic glutamate-mediated neurodegeneration. *J Neuropathol Exp Neurol.* **58**:719-28.

Cao, W., Tombran-Tink, J., Chen, W., Mrazek, D., Elias, R., and McGinnis, J.F., 1999, Pigment epithelium-derived factor protects cultured retinal neurons against hydrogen peroxide-induced cell death. *J Neurosci Res.* **57**:789-800.

Cao, W., Tombran-Tink, J., Elias, R., Sezate, S., Mrazek, D., and McGinnis, J.F., 2001 In vivo protection of photoreceptors from light damage by pigment epithelium-derived factor. *Invest Ophthalmol Vis Sci.* **42**:1646-52.

Cayouette, M., Smith, S.B., Becerra, S.P., and Gravel, C., 1999, Pigment epithelium-derived factor delays the death of photoreceptors in mouse models of inherited retinal degenerations. *Neurobiol Dis.* **6**:523-32.

Dawson, D.W., Volpert, O.V., Gillis, P., Crawford, S.E., Xu, H., Benedict, W., and Bouck, N.P., 1999, Pigment epithelium-derived factor: a potent inhibitor of angiogenesis. *Science* **285**:245-8.

DeCoster, M.A., Schabelman, E., Tombran-Tink, J., and Bazan, N.G., 1999, Neuroprotection by pigment epithelial-derived factor against glutamate toxicity in developing primary hippocampal neurons. *J Neurosci Res.* **56**:604-10.

Holekamp, N.M., Bouck, N., and Volpert, O., 2002, Pigment epithelium-derived factor is deficient in the vitreous of patients with choroidal neovascularization due to age-related macular degeneration. *Am J Ophthalmol.* **134**:220-7.

Houenou, L.J., D'Costa, A.P., Li, L., Turgeon, V.L., Enyadike, C., Alberdi, E., and Becerra, S.P., 1999, Pigment epithelium-derived factor promotes the survival and differentiation of developing spinal motor neurons. *J Comp Neurol.* **412**:506-14.

Jablonski, M.M., Tombran-Tink, J., Mrazek, D.A., and Iannaccone, A., 2000, Pigment epithelium-derived factor supports normal development of photoreceptor neurons and opsin expression after retinal pigment epithelium removal. *J Neurosci.* **20**:7149-57.

Mori, K., Duh, E., Gehlbach, P., Ando, A., Takahashi, K., Pearlman, J., Mori, K., Yang, H.S., Zack, D.J., Ettyreddy, D., Brough, D.E., Wei, L.L., and Campochiaro, P.A., 2001, Pigment epithelium-derived factor inhibits retinal and choroidal neovascularization. *J Cell Physiol.* **188**:253-63.

Mori, K., Gehlbach, P., Ando, A., McVey, D., Wei, L., and Campochiaro, P.A., 2002, Regression of ocular neovascularization in response to increased expression of pigment epithelium-derived factor. *Invest Ophthalmol Vis Sci.* **43**:2428-34.

Ogata, N., Nishikawa, M., Nishimura, T., Mitsuma, Y., and Matsumura M., 2002, Unbalanced vitreous levels of pigment epithelium-derived factor and vascular endothelial growth factor in diabetic retinopathy. *Am J Ophthalmol.* **134**:348-53.

Ogata, N., Tombran-Tink, J., Jo, N., Mrazek, D., and Matsumura, M., 2001, Upregulation of pigment epithelium-derived factor after laser photocoagulation. *Am J Ophthalmol.* **132**:427-9.

Ogata, N., Tombran-Tink, J., Nishikawa, M., Nishimura, T., Mitsuma, Y., Sakamoto, T., and Matsumura, M. 2001, ,Pigment epithelium-derived factor in the vitreous is low in diabetic retinopathy and high in rhegmatogenous retinal detachment. Am J Ophthalmol. **132**:378-82.

Ogata, N., Wada, M., Otsuji, T., Jo, N., Tombran-Tink, J., and Matsumura, M., 2002, Expressio.n of pigment epithelium-derived factor in normal adult rat eye and experimental choroidal neovascularization. Invest Ophthalmol Vis Sci. **43**1168-75.

Ogata, N., Wang, L., Jo, N., Tombran-Tink, J., Takahashi, K., Mrazek, D., and Matsumura, M., 2001, Pigment epithelium derived factor as a neuroprotective agent against ischemic retinal injury. Curr Eye Res. **22**:245-52.

Ohno-Matsui, K., Morita, I., Tombran-Tink, J., Mrazek, D., Onodera, M., Uetama, T., Hayano, M., Murota, S.I., Mochizuki, M., 2001, Novel mechanism for age-related macular degeneration: an equilibrium shift between the angiogenesis factors VEGF and PEDF. J Cell Physiol. **189**:323-33.

Ortego, J., Escribano, J., Becerra, S.P., and Coca-Prados, M.,1996, Gene expression of the neurotrophic pigment epithelium-derived factor in the human ciliary epithelium. Synthesis and secretion into the aqueous humor. Invest Ophthalmol Vis Sci.**37**:2759-67.

Raisler, B.J., Berns, K.I., Grant, M.B., Beliaev, D., and Hauswirth, W.W., 2002, Adeno-associated virus type-2 expression of pigmented epithelium-derived factor or Kringles 1-3 of angiostatin reduce retinal neovascularization. Proc Natl Acad Sci U S A. 99:8909-14.

Rasmussen, H., Chu, K.W., Campochiaro, P., Gehlbach, P.L., Haller, J.A., Handa, J.T., Nguyen, Q.D., Sung, J.U.,2001, Clinical protocol. An open-label, phase I, single administration, dose-escalation study of ADGVPEDF.11D (ADPEDF) in neovascular age-related macular degeneration (AMD). Hum Gene Ther. **12**:2029-32.

Spranger, J., Osterhoff, M., Reimann, M., Mohlig, M., Ristow, M., Francis, M.K., Cristofalo, V., Hammes, H.P., Smith, G., Boulton, M., and Pfeiffer, A.F., 2001, Loss of the antiangiogenic pigment epithelium-derived factor in patients with angiogenic eye disease. Diabetes. **50**:2641-5.

Steele, F.R., Chader, G.J., Johnson, L.V., and Tombran-Tink, J.,1993, Pigment epithelium-derived factor: neurotrophic activity and identification as a member of the serine protease inhibitor gene family. Proc Natl Acad Sci U S A. **90**:1526-30.

Taniwaki, T., Becerra, S.P., Chader, G.J., and Schwartz, J.P., 1995, Pigment epithelium-derived factor is a survival factor for cerebellar granule cells in culture.J Neurochem. **64**:2509-17.

Taniwaki, T., Hirashima, N., Becerra, S.P., Chader, G.J., Etcheberrigaray, R., and Schwartz, J.P., 1997, Pigment epithelium-derived factor protects cultured cerebellar granule cells against glutamate-induced neurotoxicity. J Neurochem. **68**:26-32.

Tombran-Tink, J., Chader, G.G., and Johnson, L.V., 1991, PEDF: a pigment epithelium derived factor with potent neuronal differentiative activity. Exp Eye Res. **53**:411-4.

Tombran-Tink, J., Mazuruk, K., Rodriguez, I.R., Chung, D., Linker, T., Englander, E., and Chader, G.J., 1996, Organization, evolutionary conservation, expression and unusual Alu density of the human gene for pigment epithelium-derived factor, a unique neurotrophic serpin.Mol Vis. **4**:2:11.

Tombran-Tink, J., Pawar, H., Swaroop, A., Rodriguez, I., and Chader, G.J., 1994, Localization of the gene for pigment epithelium-derived factor (PEDF) to chromosome 17p13.1 and expression in cultured human retinoblastoma cells. Genomics. **19:**266-72.

Tombran-Tink, J., Shivaram ,S.M., Chader, G.J., Johnson, L.V., and Bok, D., 1995, Expression, secretion, and age-related downregulation of pigment epithelium-derived factor, a serpin with neurotrophic activity. J Neurosci. **15**:4992-5003.

Wang, K.K.,2000, Calpain and Caspases. Can you tell the difference? Trends Neurosci; **23**:20-60

Yabe, T., Wilson, D., and Schwartz, J.P., 2001, NFkappaB activation is required for the neuroprotective effects of pigment epithelium-derived factor (PEDF) on cerebellar granule neurons. J Biol Chem. **276**:43313-9.

THE *nob* MUTATION DOES NOT PROTECT AGAINST LIGHT-INDUCED RETINAL DEGENERATION

Machelle T. Pardue[1], Christian Grimm[2], Andreas Wenzel[2], and Charlotte E. Remé[2]

1. INTRODUCTION

Excessive exposure of the retina to light activates a series of events that lead to loss of photoreceptors by apoptosis (Remé et al., 1998). While the underlying mechanism of this pathway is not completely understood, mutant mice have proven to be useful tools for dissecting the processes that may be involved. For instance *c-fos* knockout mice are highly resistant to bright light exposure (Hafezi et al., 1998; Wenzel et al., 2000) while eliminating the α-subunit of transducin (*Gnat1-/-*) offers protection from low levels of light exposure but not brief periods of bright light exposure (Hao et al., 2002). The aim of this study is to examine any potential role of the inner retina in light-induced photoreceptor degeneration using a bright light-exposure model (Hafezi et al., 1997). The *nob* (*no b-wave*) mouse was used which has a defect in visual transmission to the inner retina. *nob* mice are characterized by an absent b-wave of the electroretinogram (ERG) but a normal a-wave (Pardue et al., 1998). This pattern indicates a defect in transmission from photoreceptors to bipolar cells. An 85 bp deletion in nyctalopin has been identified in the *nob* mouse (Gregg et al., 2003). While the exact location and function of nyctalopin has yet to be elucidated, it is clear that this protein is involved in the transmission of the visual signal to the inner retina. Thus, the *nob* mouse presents a model in which phototransduction is preserved in the photoreceptors, but the visual signal does not reach the inner retina.

---

[1]Rehab R&D, Atlanta VA Medical Center, Decatur, GA 30033, USA, and Department of Ophthalmology, Emory University, Atlanta, GA 30322, USA; [2]Laboratory of Retinal Cell Biology, Department of Ophthalmology, University Hospital Zürich, CH-8091 Zürich, Switzerland. Corresponding author: M. T. Pardue, E-mail: machelle.pardue@med.va.gov

2. METHODS

2.1. Animals

All procedures were carried out in accordance with the Association for Research in Vision and Ophthalmology statement concerning the use of animals in ophthalmic and vision research. We maintained pigmented lines of *nob* mice in a mixed background of C57Bl/6J and BALB/c. All animals used in this study were of the Rpe65 450Leu genotype which increases light damage susceptibility of the mice by enhancing the rate of rhodopsin regeneration after bleaching (Wenzel et al. 2001). Animals were raised in cyclic fluorescent light on a 12 h/12 h light/dark cycle with 60 lux within cages. Since *nob* is an X-linked trait, *nob* males were mated to carrier females to obtain affected (*nob*/Y) and wild-type(+/Y) males, affected females (*nob*/*nob*) and carrier females (*nob*/+). All genotypes were used in the experiments. However, due to the similarity in results between the wild-type males and carrier females, only the carrier female data is shown.

2.2. Light exposure

Experiments were conducted on mice aged 6-11 weeks. Mice were exposed to diffuse white fluorescent light in reflective cages. Average illuminance was measured on the cage floor, where food and water were provided. *nob* mice and carrier females were exposed to 5,800 lux cool, white, fluorescent light for 30, 60 or 120 minutes or 15,000 lux for 120 minutes. Temperature was maintained at about 25 °C during light exposure. Retinas were collected 24 hours after light treatment.

2.3. Retina histology and TUNEL staining

For morphological analysis of retinal tissue, enucleated eyes were fixed in 2.5% glutaraldehyde and embedded in Epon 812. Terminal transferase–mediated dUTP nick end labeling (TUNEL) was done essentially as described (Wenzel et al., 2000). Briefly, retinal tissue was fixed in 2% paraformaldehyde for 2 hours at 4°C and embedded in paraffin. The '*in situ* cell death detection kit' (Roche diagnostics, Rotkreuz, Switzerland) was used with minor modifications to perform staining on 5 µm sections.

3. RESULTS

Figure 1 illustrates the retinal morphology from *nob* and carrier mice for each light intensity. Mice raised in darkness showed no signs of retinal degeneration (left panels). However mice exposed to light at 5800 or 15000 lux showed disrupted photoreceptor inner and outer segments and reduced thickness of the outer nuclear layer. No differences were detected in the degree of degeneration seen in the *nob* mice compared to the carriers.

The number of photoreceptors undergoing apoptosis was detected using TUNEL –labeling (Fig. 2). No TUNEL labeling was detected in dark-reared carriers or *nob* mice while TUNEL labeling was abundant in the outer nuclear layer of both the *nob* and

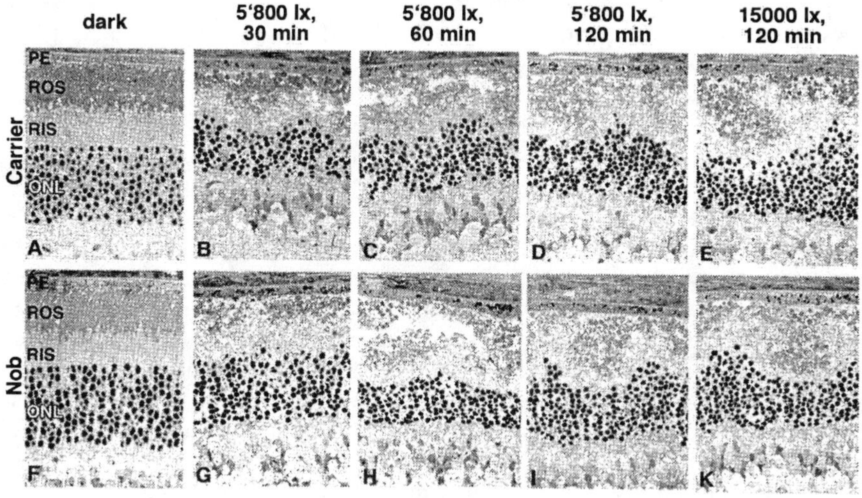

Figure 1. Retinal cross-sections of the retinal pigmented epithelium (PE), photoreceptor outer and inner segments (ROS, RIS), and outer nuclear layer (ONL) in *nob* mice and carriers showing similar susceptibility to light-induced photoreceptor degeneration.

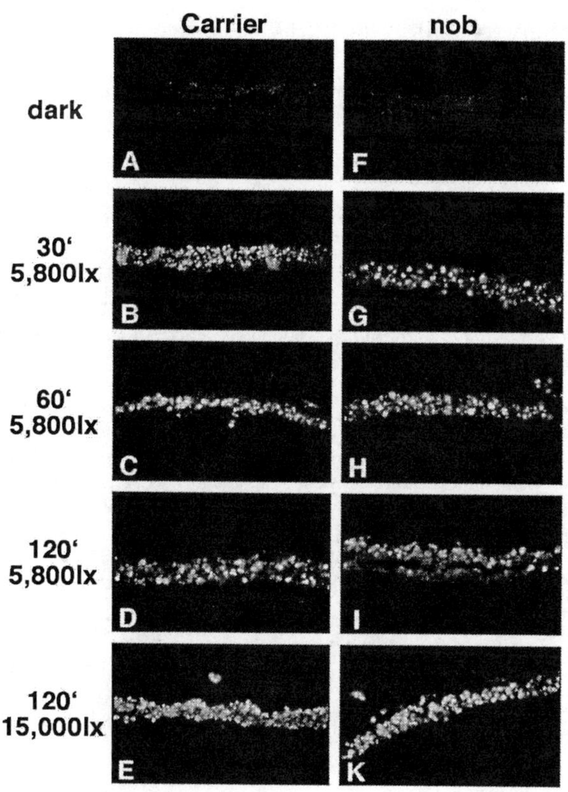

Figure 2. Retinal sections from carrier and affected *nob* mice showing comparable amounts of TUNEL labeling under 5,800 lux lighting for 30, 60, and 120 minutes and 15000 lux lighting for 120 minutes.

carrier mice after exposure to bright-intensity lights. No differences were detected in the amount of labeling seen between the *nob* mice and carriers suggesting equal susceptibility to light-induced photoreceptor degeneration.

4. DISCUSSION

These results show that *nob* mice are normally susceptible to bright light-induced retinal degeneration. While the defect causing the *nob* phenotype has been shown to be caused by a mutation in the gene encoding nyctalopin (Gregg et al., 2003), the exact function of nyctalopin is not known. However, the available evidence indicates that this nyctalopin mutation impairs visual transmission between photoreceptors and depolarizing bipolar cells (Pardue et al., 1998; Gregg et al., 2003). If the light-induced apoptotic pathway was influenced by inner retinal feedback we would expect to see full or partial protection from this type of environmental insult. This study indicates that inner retinal contributions from the ON-pathway do not contribute to bright-light induced retinal degeneration, a conclusion that is consistent with recent data obtained from other lines of mutant mice (Hao et al., 2002).

5. ACKNOWLEDGEMENTS

We thank Drs. Neal Peachey and Sherry Ball for insightful and critical reviews of the manuscript.

6. REFERENCES

Gregg, R.G., Mukhopadhyay, S., Candille, S.I., Ball, S.L., Pardue, M.T., McCall, M.A., and Peachey, N.S., 2003, Identification of the gene and the mutation responsible for the mouse *nob* phenotype, *Invest Ophthalmol Vis Sci.* (in press)

Hafezi, F., Abegg, M., Grimm, C., Wenzel, A., Munz, K., Strümer, J., Farber, D.B., and Remé, C.E., 1998, Retinal degeneration in the rd mouse in the absence of c-fos, *Invest Ophthalmol Vis Sci.* **39**:2239-2244.

Hafezi, F., Marti, A., Munz, K., and Remé, C.E., 1997, Light-induced apoptosis: Differential timing in the retina and pigment epithelium, *Exp Eye Res.* **64**:963-970.

Hao, W., Wenzel, A., Obin, M.S., Chen, C.K., Brill, E., Krasnoperova, N.V., Eversole-Cire, P., Kleyner, Y., Taylor, A., Simon, M.I., Grimm, C., Remé, C.E., and Lem, J., 2002, Evidence for two apoptotic pathways in light-induced retinal degeneration, *Nat Genetics* **32**:254-260.

Pardue, M.T., McCall, M.A., LaVail, M.M., Gregg, R.G., and Peachey, N.S., 1998, A naturally occurring mouse model of X-linked congenital stationary night blindness, *Invest Ophthalmol Vis Sci.* **39**:2443-2449.

Remé, C.E, Grimm, C., Hafezi, F., Marti, A., and Wenzel, A., 1998, Apoptotic cell death in retinal degenerations, *Progr Ret Eye Res.* **17**:443-464

Wenzel, A., Grimm, C., Marti, A., Kueng-Hitz, N., Hafezi, F., Niemeyer, G., and Remé, C.E., 2000, c-fos controls the "private pathway' of light-induced apoptosis of retinal photoreceptors, *J. Neurosci.* **20**, 81-88.

Wenzel, A., Remé, C.E., Williams, T.P., Hafezi, F., and Grimm, C., 2001, The Rpe65 Leu450Met variation increases retinal resistance against light-induced degeneration by slowing rhodopsin regeneration, *J Neurosci.* **21**:53-8.

THE PHAGOCYTOSIS OF OS IS MEDIATED BY THE PI3-KINASE LINKED TYROSINE KINASE RECEPTOR, MER, AND IS STIMULATED BY GAS6

Michael O. Hall[1], Brian J. Agnew[2], Toshka A. Abrams[1], and Barry L. Burgess[1]

1. INTRODUCTION

The phagocytosis of the shed tips of photoreceptor outer segments (OS) by the cells of the adjacent retinal pigment epithelium (RPE) is vital to the survival of the photoreceptor (PR) cells (Young and Bok, 1969). If this phagocytosis is inhibited, death of the PRs occurs. This situation is observed in the Royal College of Surgeons (RCS) strain of rat (Bourne et al., 1938; Dowling and Sidman, 1962). In this animal model, a recessive mutation in the receptor tyrosine kinase (RTK) gene, *mer*, causes the introduction of a stop codon early in the reading frame, with the result that functional Mer protein is never produced by the RPE cells (D'Cruz et al., 2000; Nandrot et al., 2000). This mutation affects the ability of the RPE cells to perform their normal function of phagocytosing the shed tips of the OS, resulting in degeneration of the underlying photoreceptor cells, and blindness (Bok and Hall, 1971; Mullen and LaVail, 1976; Edwards and Szamier, 1977).

Confirmation for the pivotal role of *mer* comes from recent studies in which the defect in OS phagocytosis has been reversed *in vitro* by transfection of RCS RPE cells with a plasmid carrying the normal *mer* cDNA (Vollrath et al., 2001), as well as *in vivo* by injection of an adenovirus vector also carrying native *mer* cDNA (Feng et al., 2002). *Mer* has also been identified as a gene that causes retinal dystrophy in humans (Gal et al., 2000; Thompson et al., 2002).

We have recently shown (Hall et al., 2001, 2002), that Gas6 (Manfioletti et al., 1993), the ligand for Mer (Nagata et al., 1996; Chen et al., 1997), is required for RPE

[1] Doris Stein Eye Research Center, UCLA Medical Center, 200 Stein Plaza, Los Angeles, CA 90095, USA; [2]Molecular Probes, Inc., 4849 Pitchford Avenue, Eugene, OR 97402-9165, USA. Corresponding author: M. O. Hall, E-mail: hall@jsei.ucla.edu

cells to phagocytise OS. Gas6 has been shown to activate Mer, resulting in autophosphorylation of the cytoplasmic tail of the receptor tyrosine kinase, transduction of an extracellular signal and the initiation of numerous and diverse physiological processes (Mark et al., 1996; Nagata et al., 1996; Chen et al., 1997). In the RPE cell, the major, and possibly the only signal transduced by Gas6 activation of Mer is that which results in the initiation of OS phagocytosis.

OS phagocytosis can proceed maximally in the presence of 10% FBS; however in the absence of serum, RPE cells are unable to ingest OS (Mayerson and Hall, 1986). When serum is replaced by Gas6, phagocytosis is completely restored (Hall et al., 2001, 2002). However, Gas6 cannot restore the phagocytic capability of RCS rat RPE cells, due to the absence of functional Mer.

After activation by extracellular ligand, many receptor tyrosine kinases transduce their signal to a phosphatidylinositol 3' kinase (PI3K), which phosphorylates phosphatidylinositol 4,5 diphosphate (PtdIns-4,5-P_2) to PtdIns-3,4,5-P_3. (Rameh and Cantley, 1999; Wymann and Pirola, 2000). Class I PI3Ks contain a regulatory 85kDa subunit and a catalytic 110kDa subunit.

In order to determine whether OS phagocytosis in RPE cells proceeds through a Mer-PI3K-linked signal pathway, we have investigated the effects of drugs that inhibit this pathway at various steps; genistein, a broad spectrum inhibitor of protein tyrosine and histidine kinases (Akiyama et al., 1987; Huang et al., 1992), and wortmannin and LY 294002 which are specific inhibitors of PI3K (Yano et al., 1993; Vlahos et al., 1994).

2. METHODS

2.1. Isolation and Growth of RPE Cells

Pigmented retinal dystrophy rats (rdy/rdy, p+/p+ [RCS]) (LaVail et al., 1975), and normal pigmented Long Evans rats (LE) were maintained on a 12L:12D cycle. Postnatal (P) day 12-15 rats were killed by CO_2 inhalation and the eyes were removed. RPE cells were isolated as previously described (Mayerson et al., 1985) and seeded on 13 mm glass coverslips at a density of 22,000 cells/75 μl of growth medium (GM) (MEM [Earle's salts] containing 10% FBS, 40 μg/ml gentamycin, 40 μg/ml kanamycin and 2 mM glutamine). The cells are used when they reach confluence after 6-7 days of incubation at 37°C in 95% air, 5% CO_2. The care and treatment of animals adhered to the NIH Guide for the Care and Use of Laboratory Animals, and to the UCLA Animal Research Committee Policies and Procedures.

2.2. Isolation and Feeding of OS

OS were isolated by sucrose density centrifugation of homogenized retinas from rats older than 30 days of age (Chaitin and Hall, 1983). The final pellet was suspended in GM, or in GM containing the inhibitor being investigated, at a concentration of 10^7 OS/ml. Glass coverslips with RPE cells were placed in individual wells of 24 well tissue culture plates. A 0.5 ml aliquot (0.5 x 10^7 OS) of the OS suspension was added to each well and incubated for 1 hr at 37°C in 95% air, 5% CO_2.

2.3. Double Immunofluorescence Labeling and Quantitation of OS Phagocytosis

After 1 hr of incubation with OS, cells were washed twice by vigorous agitation in PBS containing $CaCl_2$ and $MgSO_4$ to remove unbound OS and fixed in 3.7% formaldehyde in PBS, pH 7.2. Bound and ingested OS were differentially stained with rabbit anti-bovine OS antiserum using a double immunofluorescence method and quantitated as previously described (Chaitin and Hall, 1983; Mayerson and Hall, 1986). Each treatment is done in duplicate and at least five grid areas are counted on each disk. Each experiment was repeated at least twice. Results are expressed as the mean of the pooled data plus or minus the standard deviation.

2.4. Incubation with Inhibitors

All drugs were dissolved in DMSO. Genistein, genistin, and LY 294002 were added directly to complete GM, and preincubated with the cells for 30 min. Cells treated with wortmannin were preincubated with the drug for 45 minutes in GM in which the FBS concentration was reduced to 0.1%. In all cases the medium was withdrawn after preincubation and replaced with 0.5 ml of complete GM containing the same concentration of the drug, and 0.5 x 10^7 OS. Incubation was continued for 1 hr, after which the cells were washed, fixed and stained as described above. Preliminary experiments showed that the concentration of DMSO added to the cells, which was always less than 1%, had no effect on the cells or on their ability to phagocytise OS. Controls always contained the same concentration of DMSO as did the drug-treated cells.

3. RESULTS

To investigate whether OS ingestion requires PI 3-kinase activity, RPE cells were preincubated with the PI3K inhibitors wortmannin or LY 294002, and then challenged with OS. Pretreatment with PI3K inhibitors inhibited OS ingestion in a dose-dependent manner with an IC_{50} for wortmannin of 12 nM (Figure 1) and for LY 294002 of 18 μM (Figure 2). The broad spectrum tyrosine kinase inhibitor, genistein, also inhibited OS ingestion in a dose-dependent manner with an IC_{50} of 180 μM (Figure 3). This is well within the concentration reported to inhibit tyrosine kinases (Akiyama et al., 1987), as opposed to serine and threonine kinases, but could possibly inhibit histidine kinase (Huang et al., 1992). When RPE cells were incubated with the maximum indicated dose of each inhibitor, and allowed to recover in the absence of each inhibitor, they fully recovered their phagocytic ability (data not shown), in 12 hours (wortmannin), 1 hour (LY 294002), and less than 1 hour (genistein). Thus the inhibitors do not affect the viability of the cells over the concentration range used.

Figure 4 shows that medium conditioned with Gas6 completely replaces GM containing 10% FBS in supporting OS phagocytosis. In fact, at the highest levels of Gas6 used, OS ingestion exceeds that obtained with 10% FBS, which we have previously shown maximally supports OS phagocytosis (Mayerson and Hall, 1986).

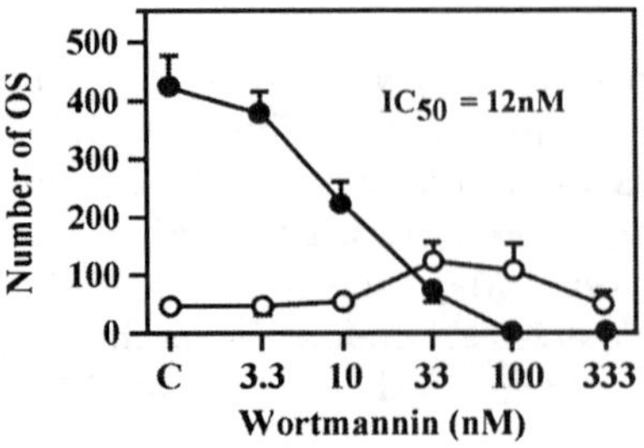

Figure 1. Wortmannin, a specific inhibitor of P13K, inhibits OS phagocytosis; ○ bound, ● ingested. C, control (no additions).

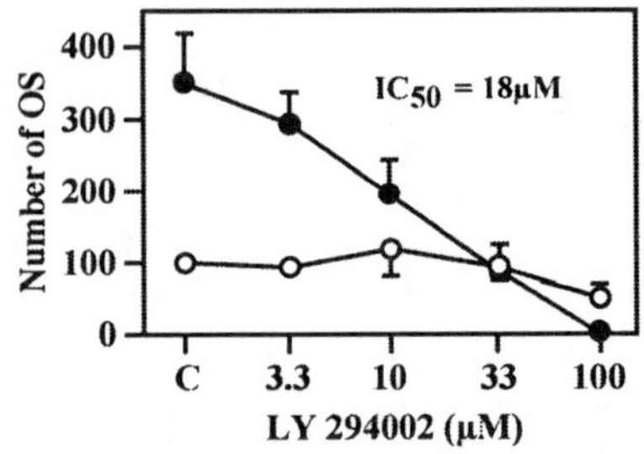

Figure 2. LY 294002, a specific inhibitor of P13K, inhibits OS phagocytosis; ○ bound, ● ingested. C, control (no additions).

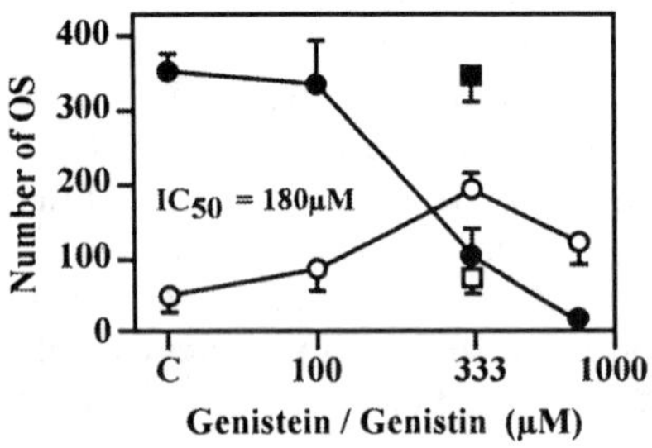

Figure 3. Genistein, a broad spectrum inhibitor of RTKs, inhibits OS phagocytosis; ○ bound, ● ingested. The inactive isomer, Genistin, has no effect; □ bound, ■ ingested. C, control (no additions).

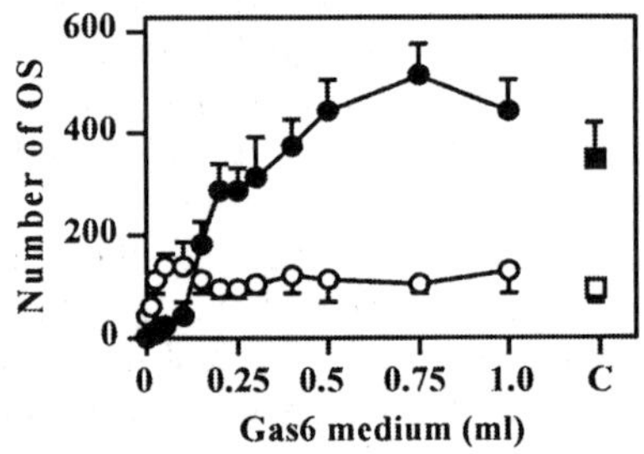

Figure 4. Gas6 stimulates OS phagocytosis; ○ bound, ● ingested. C, OS phagocytosis in 10% FBS; □ bound, ■ ingested.

4. DISCUSSION

Phagocytosis requires that external stimuli activate pathways leading to the accumulation of polymerized actin at sites of particle binding, prior to particle internalization. Diverse classes of integral membrane receptors can promote actin polymerization, some through PI3K (receptor tyrosine kinases, growth factor receptors), others through FAK (integrins), and still others through G-protein coupled receptors. The results presented here specifically implicate PI3K as participating in the signaling events related to OS phagocytosis. However they do not target Mer as the only surface receptor that is involved in this process. The αvβ5 integrin has also been implicated in the process of OS phagocytosis (Finneman et al., 1997; Finneman and Silverman, 2001).

It is well established that RCS rat RPE cells can bind OS normally but have a limited capacity to ingest them (Chaitin and Hall, 1983; Hall and Abrams, 1987). Since this mutant animal lacks Mer, this RTK alone cannot be responsible for both the binding of OS, and their subsequent ingestion — another molecule(s) must participate in the binding phase of this process. It has been suggested that αvβ5 integrin is the binding receptor, and that phagocytosis involves internalization of the OS-$\alpha v\beta 5$ complex (Finneman et al., 1997).

Although the key involvement of Mer in OS phagocytosis has been clearly proven, this process must involve a number of extracellular and intracellular molecules, including PI3K and $\alpha v\beta 5$ integrin. Since a mutation in any of the genes encoding the diverse players in this phagocytic pathway could result in blindness, experiments are underway to identify all of the molecules that are involved in this process.

5. ACKNOWLEDGEMENTS

Supported by grants from the National Institutes of Health to M. O. H. (EY 11967 and EY 00331).

6. REFERENCES

Akiyama, T., Ishida, J., Nakagawa, S., Ogawra, H., Watanabe, S., Itoh, N., Shibuya, M., and Fukami, Y., 1987, Genistein, a specific inhibitor of tyrosine-specific protein kinases, *J. Biol. Chem.* **262**:5592-5595.

Bok, D., and Hall, M. O., 1971, The role of the pigment epithelium in the etiology of inherited retinal degeneration, *J. Cell Biol.* **49**:664-682.

Bourne, M. C., Campbell, D. A., and Tansley, K., 1938, Hereditary degeneration of the rat retina, *Br. J. Ophthalmol.* **22**:613-622.

Chaitin, M. H., and Hall, M. O., 1983, Defective ingestion of rod outer segments by cultured dystrophic rat retinal pigment epithelium cells, *Invest. Ophthalmol. Vis. Sci.* **24**:812-820.

Chen, J., Carey, K., and Godowski, P. J., 1997, Identification of Gas6 as a ligand for Mer, a neural cell adhesion molecule related receptor tyrosine kinase implicated in cellular transformation, *Oncogene* **14**:2033-2039.

D'Cruz, P. M., Yasumura, D., Weir, J., Matthes, M. T., Abderrahim, H., LaVail, M. M., and Vollrath, D., 2000, Mutation of the receptor tyrosine kinase gene Mertk in the retinal dystrophic rat, *Hum. Mol. Gen.* **9**:645-651.

Dowling, J. E., and Sidman, R. L., 1962, Inherited retinal dystrophy in the rat, *J. Cell Biol.* **14**:73-109.

Edwards, R. B., and Szamier, R. B., 1977, Defective phagocytosis of isolated rod outer segments by RCS rat pigment epithelium in culture, *Science* **197**:1001-1003.

Feng, W., Yasamura, D., Matthes, M. T., LaVail, M. M., and Vollrath, D., 2002, Mertk triggers uptake of photoreceptor outer segments during phagocytosis by cultured retinal pigment epithelial cells, *J. Biol. Chem.* **277**:17016-17022.

Finneman, S. C., Bonilha, V. L., Marmorstein, A. D., and Rodriguez-Boulan, E., 1997, Phagocytosis of rod outer segments by retinal pigment epithelial cells requires $\alpha v\beta 5$ integrin for binding but not for internalization, *Proc. Natl. Acad. Sci. USA.* **94**:12932-12937.

Finneman, S.C., and Silverman, R. L., 2001, Differential roles of CD36 and $\alpha v\beta 5$ integrin in photoreceptor phagocytosis by the retinal pigment epithelium, *J. Exp. Med.* **194**:1289-1298.

Gal, A., Li, Y., Thompson, D. A., Weir, J., Orth, U., Jacobson, S. G., Apfelstedt-Sylla, D., and Vollrath, D., 2000, Mutations in Mertk, the human orthologue of the RCS rat retinal dystrophy gene cause retinitis pigmentosa, *Nature Genetics* **26**:270-271.

Hall, M. O., and Abrams, T., 1987, Kinetic studies of rod outer segment binding and ingestion by cultured rat RPE cells, *Exp. Eye Res.* **45**:907-922.

Hall, M. O., Prieto, A. L., Obin, M. S., Abrams, T. A., Burgess, B. L., Heeb, M. J., and Agnew, B. J., 2001, Outer segment phagocytosis by cultured retinal pigment epithelial cells requires Gas6, *Exp. Eye Res.* **73**:509-520.

Hall, M. O., Obin, M. S., Prieto, A. L., Burgess, B. L., and Abrams, T. A., 2002, Gas6 binding to photoreceptor outer segments requires gamma-carboxyglutamic acid (Gla) and Ca^{2+} and is required for OS phagocytosis by RPE cells in vitro, *Exp. Eye Res.* **75**:391-400.

Huang, J., Nasr, M., Kim, Y., and Matthews, H. R., 1992, Genistein inhibits protein histidine kinase, *J. Biol.Chem.* **267**:15511-15515.

LaVail, M. M., Sidman, R. L., and Gerhardt, C. O., 1975, Congenic strains of RCS rats with inherited retinal dystrophy, *J. Hered.* 66:242-244.

Manfioletti, G., Brancolini, C., Avanzi, G., and Schneider, C., 1993, The protein encoded by a growth arrest-specific gene (Gas6) is a new member of the vitamin K-dependent proteins related to protein S, a negative coregulator in the blood coagulation cascade, *Mol. Cell. Biol.* **13**:4976-4985.

Mark, M. R., Chen, J., Hammonds, R. G., Sadick, M., and Godowski, P. J., 1996, Characterization of Gas6, a member of the superfamily of G-domain-containing proteins, as a ligand for Rse and Axl, *J. Biol. Chem.* **271**:9785-9789.

Mayerson, P .L., Hall, M. O., Clark, V. M., and Abrams, T. A., 1985, An improved method for isolation and culture of retinal pigment epithelial cells, *Invest. Ophthalmol. Vis. Sci.* **26**:1599-1609.

Mayerson, P. L., and Hall, M. O., 1986, Rat retinal pigment epithelial cells show specificity of phagocytosis in vitro, *J. Cell Biol.* **103**:299-308.

Mullen, R. J., and LaVail, M. M., 1976, Inherited retinal dystrophy: primary defect in pigment epithelium determined with experimental rat chimeras, *Science* **192**:799-801.

Nagata, K., Ohashi, K., Nakano, T., Arita, H., Zong, C., Hanafusa, H., and Mizuno, K., 1996, Identification of the product of growth arrest-specific gene 6 as a common ligand for Axl, Sky and Mer receptor tyrosine kinase, *J. Biol. Chem.* **271**:30022-30027.

Nandrot, E., Dufour, E. M., Provost, A. C., Pequignot, M. O., Bonnel, S., Gogat, K., Marchant, D., Rouillac, C., Sepulchre de Conde, B., Bihoreau, M-T., Shaver, C., Dufier, J-L., Marsac, C., Lathrop, M., Menasche, M., and Abitbol, M. M., 2000, Homozygous deletion in the coding sequence of the c-mer gene in the RCS rat unravels general mechanisms of physiological cell adhesion and apoptosis, *Neurobiol. Dis.* **7**:586-599.

Rameh, L.E., and Cantley, L.C., 1999, The role of phosphoinositide 3-kinase lipid products in cell function, *J. Biol. Chem.* **274**:8347-835.

Thompson, D. A., McHenry, C. L., Li, Y., Richards, J. E., Othman, M. I., Schwinger, E., Vollrath, D., Jacobson, S. G., and Gal, A., 2002, Retinal dystrophy due to paternal isodisomy for chromosome 1 or chromosome 2, with homoallelism for mutations in RPE65 or Mertk, respectively, *Am. J. Hum. Genet.* **70**:224-229.

Vlahos, C. J., Matter, W. F., Hui, K. Y., and Brown, R. F., 1994, A specific inhibitor of phosphatidylinositol 3-kinase, 2-(4- morpholinyl)-8-phenyl-4H-1-benzopyran-4-one (LY 294002), *J. Biol. Chem.* **269**:5241-5248.

Vollrath, D., Feng, W., Duncan, J. L., Yasamura, D., D'Cruz, P. M., Chappelow, A., Matthes, M. T., Kay, M. A., and LaVail, M. M., 2001, Correction of the retinal dystrophy phenotype of the RCS rat by viral gene transfer of Mertk, *Proc. Natl. Acad. Sci. U.S.A.* **98**:12584-12589.

Wymann, M. P., and Pirola, L., 2000, Structure and function of phosphoinositide 3-kinases, *BBA, Lipids and Lipid Metabolism* **1436**:127-150.

Yano, H., Nakanishi, S., Kimura, K., Hanai, N., Saitoh, Y., Fukui, Y., Nonomura, Y., and Matsuda, Y., 1993, Inhibition of histamine secretion by wortmannin through the blockade of phosphatidylinositol 3-kinase in RBL-2H3 cells, *J. Biol. Chem.* **268**:25846-25856.

Young, R. W., and Bok, D., 1969, Participation of the retinal pigment epithelium in the rod outer segment renewal process, *J. Cell Biol.* **42**:392-402.

ROLE OF $\alpha v \beta 5$ INTEGRIN IN REGULATING PHAGOCYTOSIS BY THE RETINAL PIGMENT EPITHELIUM

Silvia C. Finnemann

1. INTRODUCTION

Daily phagocytosis of shed outer segment fragments (OS) by the retinal pigment epithelium (RPE) is an important element of the continuous photoreceptor rod and cone outer segment renewal in the retina. Shedding of outer segment distal tips every morning, synchronized by light and circadian rhythms, precedes a burst of phagocytic activity by the RPE that efficiently removes shed OS from the retina (Young and Bok, 1969; LaVail, 1976). This recurrent clearance of spent OS by the RPE is one of the highest capacity phagocytic systems known in nature.

The phagocytic mechanism of RPE for shed OS is a saturable, receptor-mediated process similar to the clearance mechanism of macrophages for apoptotic cells (Finnemann and Rodriguez-Boulan, 1999; Scott et al., 2001). Among the RPE surface proteins that participate in OS recognition and internalization are the tyrosine kinase receptor Mertk (D'Cruz et al., 2000, Nandrot et al., 2000), the scavenger receptor CD36 (Ryeom et al., 1996) and the integrin receptor $\alpha v \beta 5$ (Finnemann et al., 1997; Miceli et al., 1997). The relatively slow, highly synchronized interaction of OS and RPE surface receptors offers a unique opportunity to gain insight into how protein activities are coordinated to create and regulate a phagocytic machinery.

2. RPE CELLS ARE UNIQUE PHAGOCYTES

In the mammalian eye, one RPE cell serves approximately 30 photoreceptor cells, each of which sheds ~7% of their OS mass per day (Young, 1967; Young, 1971). Each RPE cell must completely dispose of several thousand OS disks before its next phagocytic challenge. Since RPE cells do not normally divide in the adult mammalian eye, individual RPE cells phagocytose more material over a life time than any other cell type in the body. Precise timing of clearance of spent OS is thus critical to the long-term

[1]Dyson Vision Research Institute, Departments of Ophthalmology and Cell and Developmental Biology, Weill Medical College of Cornell University, New York, NY 10021, USA. E-mail: sfinne@med.cornell.edu

maintenance of the retina. As OS shedding continues, even a minor delay in OS turnover during a 24 hour period will cause accumulation of undigested OS material resulting in photoreceptor damage and, eventually, blindness.

In the healthy eye, RPE cells are exclusively responsible for clearance of OS and OS account for most if not all of their phagocytic "meal" in a life time. However, the quantitative competition of apoptotic cells and OS for recognition by RPE cells as well as by monocyte-derived macrophages in culture suggests that the phagocytic machinery of RPE cells may not be uniquely or solely suited for OS clearance (Finnemann and Rodriguez-Boulan, 1999). Rather, the daily rhythm of OS phagocytosis may be the primary feature distinguishing phagocytosis by RPE from phagocytosis by other non-inflammatory phagocytes.

3. RPE CELLS USE PHAGOCYTIC RECEPTORS COMMON TO OTHER PHAGOCYTES

A deletion in the coding region of the gene for a receptor tyrosine kinase known as Mertk abolishes the RPE's ability to engulf (but not to bind) shed OS in the Royal College of Surgeons (RCS) rat model of hereditary retinal degeneration (Bok and Hall, 1971; Mullen and LaVail, 1976; Edwards and Szamier, 1977; Chaitin and Hall, 1983; D'Cruz et al., 2000; Nandrot et al., 2000). The rapid photoreceptor death in these animal models illustrates the vital importance of RPE phagocytosis of spent OS for the retina. Furthermore, Mertk deficiency causes retinal degeneration and a defect in the phagocytosis of apoptotic cells by macrophages in transgenic mice (Scott et al., 2001). Macrophages derived from RCS rats also exhibit phagocytic deficiencies (my unpublished observation). Re-expression of Mertk reconstitutes the phagocytic function of RCS rat RPE *in vitro* and *in vivo* (Vollrath et al., 2001; Feng et al., 2002). Despite this evidence that Mertk plays a direct and essential role in the phagocytic process, the precise function of Mertk in the phagocytic machinery is still unknown.

Like Mertk, the scavenger receptor CD36 does not act during the OS recognition/binding phase of RPE phagocytosis (Finnemann and Silverstein, 2001). Rather, recent experiments suggest that CD36 ligation by antibodies or other ligands, including lipoproteins or unidentified serum factors, regulates the rate with which RPE cells internalize surface-bound OS. As CD36 ligation at the basal surface of RPE is as effective as its ligation at the apical, phagocytic surface, CD36 likely functions primarily as a signaling molecule during RPE phagocytosis.

CD36 performs equivalent tasks in clearance phagocytosis by RPE and by macrophages. Its transfection into non-phagocytic Bowes melanoma cells renders them phagocytic towards apoptotic cells as well as towards OS (Ren et al., 1995; Ryeom et al., 1996). CD36 may act in cooperation with $\alpha v\beta 3$ integrin in macrophages to promote particle uptake (Savill et al., 1992). However, CD36 also participates in phagocytic mechanisms that are independent of integrin receptors (Fadok et al., 1998).

Developmental expression studies and *in vitro* functional assays strongly suggest a role for the integrin receptor $\alpha v\beta 5$ in OS phagocytosis by human and rodent RPE.

$\beta 5$ protein is absent from rat RPE at birth but reaches adult levels around postnatal day 11 (Finnemann et al., 1997). $\beta 5$ expression thus precisely coincides with the postnatal formation of photoreceptor OS and the onset of OS phagocytosis. In mature human and rodent retina, $\alpha v\beta 5$ is the sole integrin family adhesion receptor that localizes to the apical, phagocytic RPE surface facing interphotoreceptor matrix and photoreceptor OS (Anderson et al., 1995; Finnemann et al., 1997).

Several independent studies have demonstrated that adult rat and human RPE in primary culture as well as stable rat and human RPE derived cell lines employ αvβ5 during the recognition/binding phase of OS phagocytosis (Finnemann et al., 1997; Miceli et al., 1997; Lin and Clegg, 1998). αvβ5 inhibition using antibodies or RGD peptides reduces OS binding by differentiated RPE in culture by up to 84% but does not affect OS internalization (Finnemann et al., 1997). Like RPE, fibroblasts and dendritic cells use αvβ5 for clearance phagocytosis (Albert et al., 1998; Albert et al., 2000). Interestingly, the phagocytic capacity of β5 integrin transfected cells was shown to correlate directly with the levels of αvβ5 integrin (Albert et al., 2000). Monocyte-derived macrophages were first shown to employ αvβ3 integrin receptors which possess similar ligand binding properties as αvβ5 (Savill et al., 1990). More recent data demonstrate that these cells may also use αvβ5 depending on activation of αvβ5 receptors by cellular signaling pathways. Protein kinase C activation in monocytes and in RPE cells is required for αvβ5 cytoskeletal anchorage and OS (and apoptotic cell) binding mediated by αvβ5 (Finnemann and Rodriguez-Boulan, 1999). This illustrates the importance of receptor regulation for RPE phagocytosis.

Taken together, Mertk, CD36, and αvβ5 receptors have in common that they are also being utilized for the uptake of apoptotic cells by other professional and non-professional phagocytes, such as monocyte derived macrophages, dendritic cells, and fibroblasts. Furthermore, these receptors participate in cellular functions unrelated to phagocytic uptake. In RPE phagocytosis, αvβ5 acts in the recognition of OS while both CD36 and Mertk participate in the engulfment of bound OS (Finnemann and Silverstein, 2001; Feng et al., 2002). Regulatory mechanisms that affect the functions of CD36 or Mertk are not well understood and have not yet been reported in RPE. In contrast, much is known about the regulation of integrin receptor function.

4. REGULATION OF RPE PHAGOCYTOSIS: A ROLE FOR αvβ5 INTEGRIN

There is no evidence that RPE cells directly respond to circadian regulation themselves. Instead, circadian challenge with shed OS sets the timing of the RPE phagocytic rhythm every morning. At other times, the RPE phagocytic function is likely inhibited to prevent untimely attack of intact OS, which remain in immediate proximity to RPE apical projections at all times. Regulatory mechanisms that control the daily time course of the RPE phagocytic activity may include timely activation and inactivation of phagocytic receptors. Such mechanisms are not well understood on a molecular level.

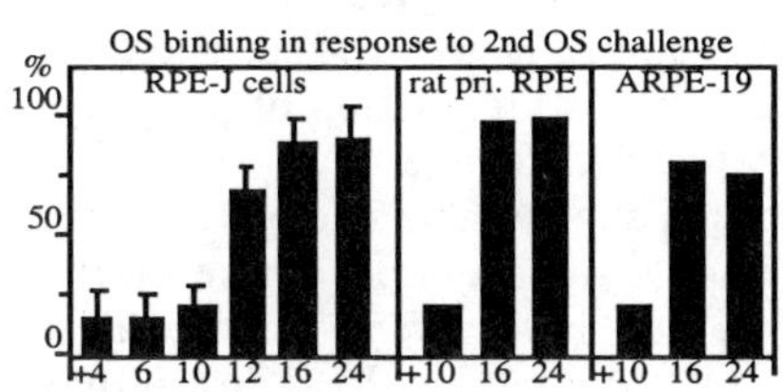

Figure 1. RPE cells in culture do not bind additional OS within ~ 10 hours of an initial phagocytic "meal". At different times up to 24 h following an initial OS challenge with unlabeled OS (as indicated), RPE-J cells, primary Long Evans rat RPE (rat pri. RPE), and ARPE-19 cells received a second phagocytic challenge with FITC-OS. Uptake values are given as percent of OS binding during the first OS challenge. Values are averages ± s.d., n=4, for RPE-J, and mean values, n=2, for rat primary RPE and ARPE-19.

When RPE cells in culture receive a second challenge with isolated OS within ~10 hours of an initial one, they bind only 10-25% as many OS as during the first challenge (Figure 1). Since bound OS are internalized after 5.5 hours (data not shown), inhibition of OS binding at later times is not due to surface receptor occupation. OS binding by RPE in culture, regardless of their origin, is completely restored at time points later than 12 hours. These re-challenge experiments demonstrate that the phagocytic activity of cultured RPE is regulated upon OS challenge through temporary inhibition of OS binding. Importantly, stable RPE cell lines retain this regulatory mechanism.

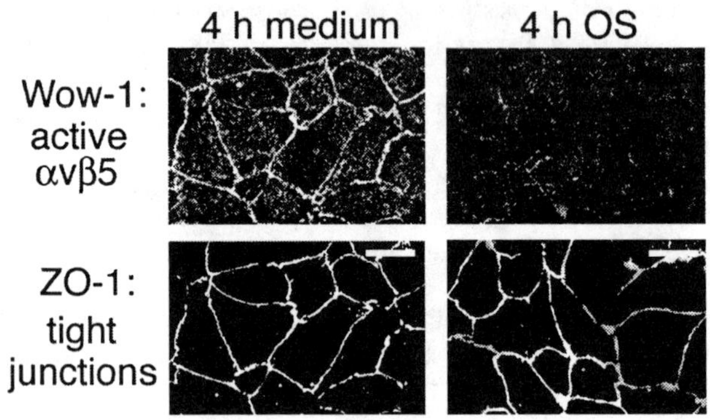

Figure 2. RPE cells downregulate high-affinity αvβ5 receptors in response to OS challenge. RPE-J cells received assay medium or OS as indicated before fixation and double labeling with Wow-1 and ZO-1. Scale bars: 10 μm.

The binding function of αvβ5 integrin receptors is subject to regulation by cellular signaling mechanisms. Regulatory mechanisms may include timed delivery to and retrieval from the cell surface and inactivation of receptors at the cell surface through conformational changes or changes in interactions with cytosolic integrin binding proteins such as linkers to cytoskeletal elements. Recently, Pampori and colleagues generated and tested a recombinant protein, termed Wow-1, that mimicks a monovalent ligand of αvβ3 and αvβ5 integrins (Pampori et al., 1999). Wow-1 reactivity can therefore distinguish inactive from activated integrin receptors. Strikingly, Wow-1 labeling shown in Figure 2 demonstrates that OS phagocytosis drastically reduces the levels of active αvβ5 integrin receptors at the apical, phagocytic surface of polarized RPE-J cells. This is the first direct demonstration of downregulation of phagocytic receptor activity by RPE cells in response specifically to OS challenge.

In conclusion, these results suggest that endogenous signaling pathways exist in the RPE to regulate the RPE phagocytic function. These mechanisms are –at least in part-retained by RPE in culture and involve the functional regulation of αvβ5 integrin receptors, which mediate OS binding by RPE in culture. Current experiments are addressing the molecular mechanisms of regulation that are responsible for the inactivation of αvβ5 integrin at the apical surface of RPE cells following phagocytic uptake.

5. PERSPECTIVE

It is well known that dysfunction in OS phagocytosis by the RPE causes retinal degeneration and blindness in experimental animals and may be an important contributor to retinal diseases in humans (Gal et al., 2000; Finnemann et al., 2002). Our complete understanding of the molecular mechanisms used by RPE cells to efficiently phagocytose shed OS once a day and every day is therefore important to gain insight into normal long-term retinal processes and forms of retinal degenerations, in which RPE phagocytosis may be impaired. Phagocytic receptors exclusive to RPE or OS specific ligands have not been identified thus far. Rather, all evidence suggests that RPE phagocytosis is one of a group of related clearance mechanisms that share common elements, among them the use of the phagocytic receptors Mertk, CD36, and αvβ3/5 integrins. Despite this major conceptual advance, it remains poorly understood precisely how any of these proteins contribute to the synchronized, timely, and repeated phagocytic clearance of OS by the RPE. Ongoing experiments in my lab aim to elucidate how signaling pathways in RPE regulate activities of RPE receptors and their cross-talk to promote the recurrent phagocytosis of OS unique to RPE.

6. ACKNOWLEDGEMENTS

This work was supported by a Career Development Award from Research To Prevent Blindness and NIH grant EY14184. I thank the National Eye Institute and the Foundation Fighting Blindness for the travel award provided to me to attend this meeting. I thank Dr. Sanford Shattil (Scripps Research Institute) who provided Wow-1 for these studies. I also thank Dr. Geri Kreitzer (Weill Medical College) for critically reading this manuscript.

7. REFERENCES

Albert, M. L., Kim, J. I., and Birge, R. B., 2000, αvβ5 integrin recruits the CrkII-Dock180-Rac1 complex for phagocytosis of apoptotic cells. *Nat. Cell Biol.*, **2**:899-905.

Albert, M. L., Pearce, S. F. A., Francisco, L. M., Sauter, B., Roy, P., Silverstein, R. L., and Bhardwaj, N., 1998, Immature dendritic cells phagocytose apoptotic cells via αvβ5 and CD36, and cross-present antigens to cytotoxic T lymphocytes. *J. Exp. Med.*, **188**:1359-1368.

Anderson, D. H., Johnson, L. V., and Hageman, G. S., 1995, Vitronectin receptor expression and distribution at the photoreceptor- retinal pigment epithelial interface. *J. Comp. Neurol.*, **360**:1-16.

Bok, D., and Hall, M. O., 1971, The role of the pigment epithelium in the etiology of inherited retinal dystrophy in the rat. *J. Cell Biol.*, **49**:664-682.

Chaitin, M. H., and Hall, M. O., 1983, Defective ingestion of rod outer segments by cultured dystrophic rat pigment epithelial cells. *Invest. Ophthalmol. Vis. Sci.*, **24**:812-820.

D'Cruz, P. M., Yasumura, D., Weir, J., Matthes, M. T., Abderrahim, H., LaVail, M. M., and Vollrath, D., 2000, Mutation of the receptor tyrosine kinase gene Mertk in the retinal dystrophic RCS rat. *Hum. Mol. Genet.*, **9**:645-651.

Edwards, R. B., and Szamier, R. B., 1977, Defective phagocytosis of isolated rod outer segments by RCS rat retinal pigment epithelium in culture. *Science*, **197**:1001-1003.

Fadok, V. A., Warner, M. L., Bratton, D. L. and Henson, P. M., 1998, CD36 is required for phagocytosis of apoptotic cells by human macrophages that use either a phosphatidylserine receptor or the vitronectin receptor (αvβ3). *J. Immunol.*, **161**:6250-6257.

Feng, W., Yasumura, D., Matthes, M. T., LaVail, M. M. and Vollrath, D., 2002, Mertk triggers uptake of photoreceptor outer segments during phagocytosis by cultured retinal pigment epithelial cells. *J. Biol. Chem.*, **277**:17016-17022.

Finnemann, S. C., Bonilha, V. L., Marmorstein, A. D., and Rodriguez-Boulan, E., 1997, Phagocytosis of rod outer segments by retinal pigment epithelial cells requires $\alpha v\beta 5$ integrin for binding but not for internalization. *Proc. Natl. Acad. Sci. U S A*, **94**:12932-12937.

Finnemann, S. C., and Rodriguez-Boulan, E., 1999, Macrophage and retinal pigment epithelium phagocytosis: apoptotic cells and photoreceptors compete for $\alpha v\beta 3$ and $\alpha v\beta 5$ integrins, and protein kinase C regulates $\alpha v\beta 5$ binding and cytoskeletal linkage. *J. Exp. Med.*, **190**:861-874.

Finnemann, S. C. and Silverstein, R. L., 2001, Differential roles of CD36 and $\alpha v\beta 5$ integrin in photoreceptor phagocytosis by the retinal pigment epithelium. *J. Exp. Med.*, **194**:1289-1298.

Finnemann, S. C., Leung, L. W., and Rodriguez-Boulan, E., 2002, The lipofuscin component A2E selectively inhibits phagolysosomal degradation of photoreceptor phospholipid by the retinal pigment epithelium. *Proc. Natl. Acad. Sci. U S A*, **99**:3842-3847.

Gal, A., Li, Y., Thompson, D. A., Weir, J., Orth, U., Jacobson, S. G., Apfelstedt-Sylla, E., and Vollrath, D., 2000, Mutations in MERTK, the human orthologue of the RCS rat retinal dystrophy gene, cause retinitis pigmentosa. *Nat. Genet.*, **26**:270-271.

LaVail, M. M., 1976, Rod outer segment disk shedding in rat retina: relationship to cyclic lighting. *Science*, **194**:1071-1074.

Lin, H., and Clegg, D. O., 1998, Integrin $\alpha v\beta 5$ participates in the binding of photoreceptor rod outer segments during phagocytosis by cultured human retinal pigment epithelium. *Invest. Ophthalmol. Vis. Sci.*, **39**:1703-1712.

Miceli, M. V., Newsome, D. A., and Tate, D. J., Jr, 1997, Vitronectin is responsible for serum-stimulated uptake of rod outer segments by cultured retinal pigment epithelial cells. *Invest. Ophthalmol. Vis. Sci.*, **38**:1588-1597.

Mullen, R. J., and LaVail, M. M., 1976, Inherited retinal dystrophy: primary defect in pigment epithelium determined with experimental rat chimeras. *Science*, **192**:799-801.

Nandrot, E., Dufour, E. M., Provost, A. C., Pequignot, M. O., Bonnel, S., Gogat, K., Marchant, D., Rouillac, C., Sepulchre de Conde, B., Bihoreau, M. T., Shaver, C., Dufier, J. L., Marsac, C., Lathrop, M., Menasche, M. and Abitbol, M. M., 2000, Homozygous deletion in the coding sequence of the c-mer gene in RCS rats unravels general mechanisms of physiological cell adhesion and apoptosis. *Neurobiol. Dis.*, **7**:586-599.

Pampori, N., Hato, T., Stupack, D. G., Aidoudi, S., Cheresh, D. A., Nemerow, G. R. and Shattil, S. J., 1999, Mechanisms and consequences of affinity modulation of integrin $\alpha v\beta 3$ detected with a novel patch-engineered monovalent ligand. *J. Biol. Chem.*, **274**:21609-21616.

Ren, Y., Silverstein, R. L., Allen, J., and Savill, J., 1995, CD36 gene transfer confers capacity for phagocytosis of cells undergoing apoptosis. *J. Exp. Med.*, **181**:1857-1862.

Ryeom, S. W., Sparrow, J. R., and Silverstein, R. L., 1996, CD36 participates in the phagocytosis of rod outer segments by retinal pigment epithelium. *J. Cell Sci.*, **109**:387-395.

Savill, J., Dransfield, I., Hogg, N., and Haslett, C., 1990, Vitronectin receptor-mediated phagocytosis of cells undergoing apoptosis. *Nature*, **343**:170-173.

Savill, J., Hogg, N., Ren, Y., and Haslett, C., 1992, Thrombospondin cooperates with CD36 and the vitronectin receptor in macrophage recognition of neutrophils undergoing apoptosis. *J. Clin. Invest.*, **90**:1513-1522.

Scott, R. S., McMahon, E. J., Pop, S. M., Reap, E. A., Caricchio, R., Cohen, P. L., Earp, H. S., and Matsushima, G. K., 2001, Phagocytosis and clearance of apoptotic cells is mediated by MER. *Nature*, **411**:207-211.

Vollrath, D., Feng, W., Duncan, J. L., Yasumura, D., D'Cruz, P. M., Chappelow, A., Matthes, M. T., Kay, M. A. and LaVail, M. M., 2001, Correction of the retinal dystrophy phenotype of the RCS rat by viral gene transfer of Mertk. *Proc. Natl. Acad. Sci. U S A*, **98**:12584-12589.

Young, R. W., 1967, The renewal of photoreceptor cell outer segments. *J. Cell Biol.*, **33**:61-72.

Young, R. W., and Bok, D., 1969, Participation of the retinal pigment epithelium in the rod outer segment renewal process. *J. Cell Biol.*, **42**:392-403.

Young, R. W., 1971, The renewal of rod and cone outer segments in the rhesus monkey. *J. Cell Biol.*, **49**:303-318.

COMPARATIVE STUDY OF CATHEPSIN D AND S IN RAT IPE AND RPE CELLS

Hiroshi Tomita, Eriko Sugano, Toshiaki Abe, Asahi Yamashita, and Makoto Tamai[1]

1. INTRODUCTION

We previously reported about the phagocytic activity of RPE and IPE cells (Sukuragi et al., 2001). We suggested that IPE cells have a phagocytic activity the same as RPE cells and that possibility IPE cells could be utilized for the treatment of patients with AMD as substitution for RPE transplantation (Abe et al., 1999; 2000). However, lysosomal function in IPE cells, which is the final process of phagocytosis, is not well documented.

In this study, we focused on the lysosomal function in phagocytosis. We investigated the activities, productions of cat D and cat S, and the expression of the mRNAs coding for these proteins in RPE and IPE cells.

2. MATERIALS AND METHODS

2.1. Cell preparation

IPE and RPE cells were prepared from 10-12-weeks-old Long Evans rats eyes. The cells were isolated as previously described cells (Sukuragi et al., 2001). The isolated IPE or RPE cells were transferred to a conical centrifuge tube containing the growth medium consist of Ham's F-12 (GIBCO BRL, Tokyo, Japan), 20% fetal bovine serum (FBS), and antibiotics (penicillin 100 units, streptomycin 0.1 mg, and amphotericin B 0.25 μg per ml of medium). The cell suspension was centrifuged at 600xg for 5 min. The cells were plated and incubated in a humidified incubator with 95% air and 5% CO_2

[1]Department of Ophthalmology, Tohoku University School of Medicine, 1-1 Seiryo Aoba Sendai Miyagi 980-8574 Japan. Corresponding author: H. Tomita, E-mail: h-tomita@oph.med.tohoku.ac.jp

2.2. mRNA purification

Expression of mRNA was investigated by a semi-quantitative RT-PCR method. The mRNA purification was carried using an mRNA purification kit (Amersham Pharmacia, Piscataway, NJ). The cDNA was synthesized using the First-Strand cDNA synthesis kit (Amersham Pharmacia, Piscataway, NJ). Cathepsin D-and S-specific primers were selected from rat cathepsin D and S sequences. PCR amplification was carried out using an Ex taq polymerase (TaKaRa, Kyoto, Japan). PCR products were quantified using SYBR Green I dye (BMA, Rockland, ME) and the fluorescein detectable specific thermal cycler (Smart cycler; TaKaRa, Kyoto, Japan).

2.3. Measurement of Cathepsin D and S activity

Cat D (Kageyama, 1995; Yasuda et al., 1999) and S (Bromme et al., 1989) activity were measured using the fluorogenic decapeptide substrate, MOCAc-Gly-Lys-Pro-Ile-Leu-Phe-Phe-Arg-Leu-Lys-(Dnp)-NH_2 (Peptide Institute, OSAKA, JAPAN) and the metylcumarylamide substrate Z-Val-Val-Arg-NHMec (Peptide Institute, OSAKA, Japan), respectively. Ascaris pepsin inhibitor was used for inhibiting the cat E activity (Kageyama, 1998; Abe-erreish and Peanasky, 1974).

3. RESULTS

Cat D expressions were observed both in cultured IPE and RPE cells. There was no significant difference in the expression levels of cat D among IPE and RPE cells (fig.1A). Likewise, the cat S expression level in IPE cells was 10% of that in RPE cells, which was significant low level compared to cultured RPE cells (fig.1B). Prolonged culture caused to decrease mRNA expression of cat D and cat S in both cells (data not shown).

Cat D activity in RPE cells was 37 times higher than that in IPE cells (fig.2A). Cat S activity in RPE cells was 4 times higher than that in IPE cells (fig.2B).

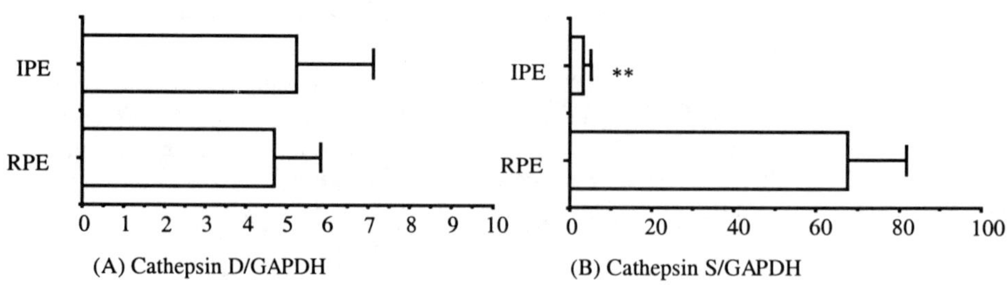

Figure 1. Semi-quantitative RT-PCR of cathepsin D (A) and S (B)_in cultured-IPE and –RPE cells. The levels of cat D expressions were indicated as the ratio of individual internal standard (GAPDH).Data shown mean ± SEM.

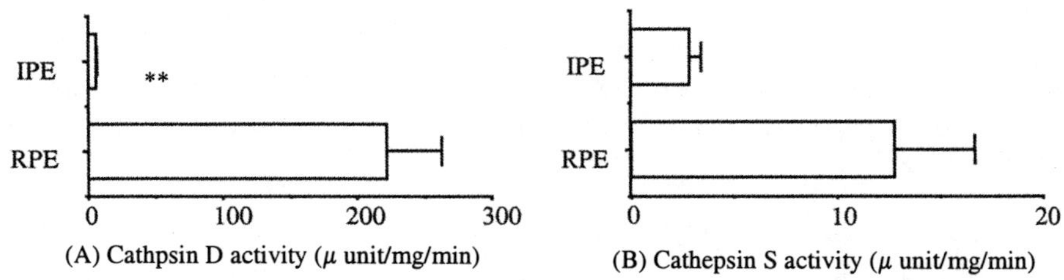

Figure 2. Cathepsin D (A) and S(B) activities were assayed by using a fluorogenic substrate. The enzyme activities were expressed as the unit of purified cat D or S bovine spleen.

As shown in Fig 3A, anti-cat D antibody recognizes all forms of cat D. Cultured IPE and RPE cells showed the presence of the inactive (43 kDa) and active (28 kDa) forms. Both of the signals (43 kDa and 28 kDa) intensities in RPE cells were grater than those in IPE cells. These data indicated that cat D activity in RPE cells is much than that in IPE cells. Cat S production was detected in both IPE and RPE cells at 30 kDa (Figure 3B). RPE cells showed high level of cat S protein production.

4. DISCUSSION

In this study, the mRNA expression of cat D in IPE was same level as that in RPE. However, cat D activites in IPE cells exhibited outstandingly low level compared to RPE cells. This controvacy could be explained by all forms of cat D such as pro-cat D was contained in the evaluated data in case of using semi-quantitative RT-PCR. The western blot analysis also showed that pro-cat D was same level both IPE and RPE cells. These results suggested that the procedures of processing from procat D to active cat D were different between IPE and RPE cells.

It has been proposed that the conversion of procat D into its active form involves an enzymatic cleavage step mediated by cysteine proteases (Rakoczy et al., 1994). It has been previously reported cat S downregulation influences the production of catD (Rakoczy et al., 1998). There might be some concernment in processing of catD between cat D and S. The role of catS is poorly understood, thus further studies are required to clarify the complex processes in this phenomenon.

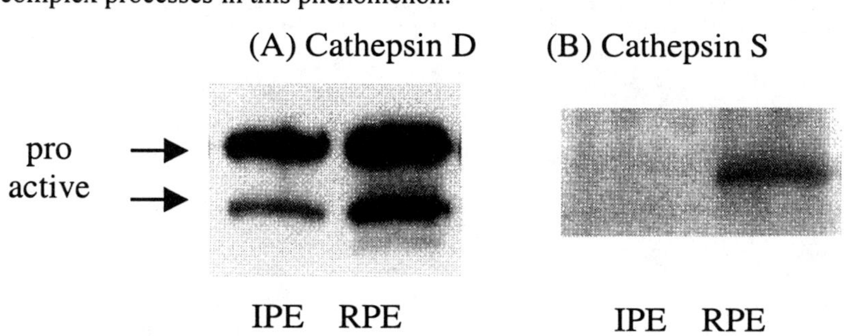

Figure 3. Western blot analysis of cathepsin D (A) and cathepsin S (B) in rat IPE and RPE cells. Arrows indicates pro-cathepsin D and active cathepsin S.

5. Conclusion

In this study, enzyme activities of cat D and S in IPE cells were low level compare to that in RPE cells. These results indicated that digestive activities of IPE cells were lower than that of RPE cells.

6. Acknowledgements

We thank the National Eye Institute (USA) for the travel award provided to H.T. to attend this meeting and also thank Dr. Takashi Kageyama at Kyoto University for the supply of Ascaris pepsin inhibitor. We also thank Dr. Yoshiyuki Yasuda and Dr. Kenji Yamamoto at Kyushu University Faculty of Dentistry for valuable help in the assay of cathepsin activity.

7. References

Abe T., Tomita H., Kano T., Yoshida M., Ohashi T., Nakamura Y., Nishikawa S., and Tamai M., 2000, Autologous iris pigment epithelial cell transplantation in monkey subretinal region. *Curr Eye Res.* **20**:268-275.

Abe T., Tomita H., Ohashi T., Yamada K., Takeda Y., Akaishi K., Yoshida M., Sato M., and Tamai M., 1999, Characterization of iris pigment epithelial cell for auto cell transplantation. *Cell Transplant.* **8**:501-510.

Abu-Erreish G.M., Peanasky R.J., 1974, Pepsin inhibitors from Ascaris lumbricoides. Isolation, purification, and some properties. *J Biol Chem.* **249**:1558-1565.

Bromme D., Steinert A., Friebe S., Fittkau S., Wiederanders B., and Kirschke H., 1989, The specificity of bovine spleen cathepsin S. A comparison with rat liver cathepsins L and B. *Biochem J.* **264**:475-481.

Kageyama T., 1998, Molecular cloning, expression and characterization of an Ascaris inhibitor for pepsin and cathepsin E. *Eur J Biochem.* **253**:804-809.

Kageyama T., 1995, Procathepsin E and cathepsin E. *Methods Enzymol.* **248**:120-136.

Rakoczy P.E., Mann K., Cavaney D.M., Robertson T., Papadimitreou J., and Constable I.J., 1994, Detection and possible functions of a cysteine protease involved in digestion of rod outer segments by retinal pigment epithelial cells. *Invest Ophthalmol Vis Sci.* **35**:4100-4108.

Rakoczy P.E., Lai M.C., Baines M.G., Spilsbury K., and Constable I.J., 1998, Expression of cathepsin S antisense transcripts by adenovirus in retinal pigment epithelial cells. *Invest Ophthalmol Vis Sci.* **39**:2095-2104.

Sakuragi M., Tomita H., Abe T., and Tamai M., 2001, Changes of phagocytic capacity in basic fibroblast growth factor-transfected iris pigment epithelial cells in rats. *Curr Eye Res.* **23**:185-191.

Yasuda Y., Kageyama T., Akamine A., 1999, Characterization of new fluorogenic substrates for the rapid and sensitive assay of cathepsin E and cathepsin D. *J Biochem (Tokyo).* **125**:1137-1143.

ISOLATION AND CULTURE OF PRIMARY MOUSE RETINAL PIGMENTED EPITHELIAL CELLS

Daniel Gibbs and David S. Williams

1. INTRODUCTION

A number of cases of retinal degeneration are caused by mutations in genes that are expressed in the retinal pigmented epithelial (RPE) cells. These genes include *RPE65* (Gu et al., 1997; Marlhens et al., 1997), *RGR* (Chen et al., 1996; Morimura et al., 1999), *MERTK* (Gal et al., 2000; Vollrath et al., 2001), *PEDF* (Goliath et al., 1996), and *MYO7A* (Weil et al., 1995; Liu et al., 1998b). To study the normal function of the products of these genes, the pathology caused by mutations in the genes, as well as methods of therapy, cultured RPE cells would provide an invaluable tool. Because passaged and immortalized RPE cells quickly lose some characteristics of polarized epithelial cells (Nabi et al., 1993; Davis et al., 1995), use of primary cultures of RPE cells is often a necessity.

Procedures for the primary culture of rat RPE cells are now well established (Hall, 1978; Irons and Kalnins, 1984; Mayerson et al., 1985; Chang et al., 1991). Indeed, due to a mutation in the orthologous *MERTK* gene in the RCS rat, primary RPE cell cultures from RCS rats have been used to advance our understanding of the Mertk protein as well as potential gene therapy (Feng et al., 2002). However, for most other gene mutations, the mouse is a better experimental animal. For this reason, we have sought to establish protocols for isolating mouse RPE cells and growing them in primary culture. Our interest is to study the role of *Myo7a*, the orthologue of the gene defective in Usher syndrome 1B (Weil et al., 1995). This gene is expressed in the RPE cells (Hasson et al., 1995; Liu et al., 1998a), and a number of different mutant mouse models are available (C. Lillo et al., this volume). Here, we describe a reproducible procedure for generating primary cultures of mouse RPE cells, and demonstrate their ability to phagocytose rod outer segments.

[1]Departments of Pharmacology and Neurosciences, UCSD School of Medicine, La Jolla, CA 92039-0983, USA. Corresponding author: D. S. Williams, E-mail: dswilliams@ucsd.edu

2. MOUSE PRIMARY RPE CELL ISOLATION AND CULTURE

Our method was developed from adaptations of an existing method developed for rat (Chang et al., 1991). It involves the isolation and enzymatic digestion of intact eyes from two week old animals, followed by rapid dissection and removal of the RPE as intact sheets of cells. These isolated sheets of RPE cells were minimally dissociated and grown in transwell culture upon uncoated polycarbonate filters to aid in the preservation of cellular polarity. The cells are viable and retain a high degree of morphological similarity to RPE cells *in vivo*, unlike immortalized cell lines under similar culture conditions, such as ARPE19 cells.

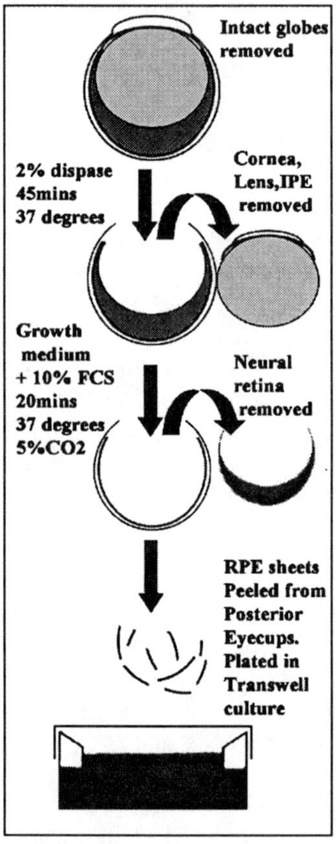

Figure 1. A schematic diagram of the RPE isolation and culture procedure described.

Our procedure for isolation of the RPE cells is illustrated in Fig. 1. Intact eyes were removed quickly from 10-15 day old mice after euthanasia by decapitation. The age of the mice used for RPE isolation was critical to the success of this procedure, as is the case for RPE isolation from rats, where 10 day old animals are most commonly used (Mayerson et al., 1985; Chang et al., 1991). Lots containing 10 eyes each were washed twice in 5 ml Dulbecco's modified eagle's medium (DMEM), containing high glucose (Invitrogen, CA), then incubated with 5 ml of sterile filtered 2% (w/v) Dispase (Invitrogen, CA), in DMEM for 45 min at 37^{0}C. The timing and temperature of the proteolytic dispase digestion and wash steps was critical and differs from the protocol for RPE isolation described for rat due to the smaller size of the eyes and the apparent tighter connection between the RPE and neural retina in the mouse.

Eyes were then washed twice in growth medium (GM) that consisted of DMEM (high glucose) plus 10% fetal calf serum (Invitrogen, CA), 1% penicillin/streptomycin (Invitrogen, CA), 2.5mM L-Glutamine and 1x MEM non-essential amino acids (Invitrogen, CA). Intact eyes were then transferred to a 90-mm culture dish with fresh GM for dissection.

Using microdissection scissors and an upright dissection microscope placed within a laminar airflow hood, a circular incision was made around the ora serrata of each eye. The anterior cornea, lens, capsule and associated iris pigmented epithelium were removed and discarded. The remaining posterior eyecups containing the neural retina loosely attached to the underlying RPE were placed in fresh GM in a 60-mm culture dish and incubated for 20 min at 37^{0}C/ 5% CO_2 in a humidified incubator to facilitate separation of the neural retina from the RPE. This separation may be a result of residual protease activity remaining after washing or simply a time-dependent breakdown of the normal adhesion between the apical RPE and the photoreceptor outer segments. As a result, the neural retina can be removed cleanly without damage to the apical surface of the RPE cells, and without excessive exposure to proteases that could greatly reduce cell viability and increase contamination.

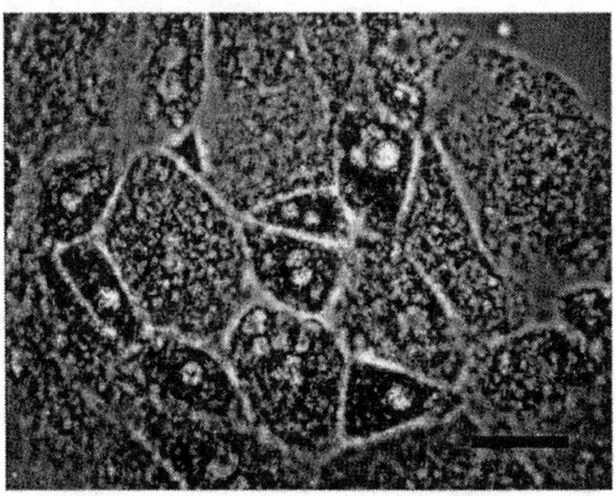

Figure 2. Micrograph of cultured primary mouse RPE cells 10 days after plating illustrating the preservation of pigmentation and epithelial morphology. Scale = 50 μm.

After removal of the neural retina, intact sheets of RPE cells were peeled from the underlying basement membrane using fine forceps. The eyecup was inverted and the forceps were used to gently pull the eyecup back and forth. Sheets of RPE then readily separate from the choroid and can be collected with minimal contamination and transferred to a sterile 60-mm culture dish, containing 5 ml of fresh GM. The sheets of

RPE were washed 3 times with GM. To gently dissociate these sheets without further enzymatic treatment, they were washed twice with Ca^{2+}- and Mg^{2+}-free Hank's balanced salt solution (5mM KCl, 0.5mM KH_2PO_4, 4mM $NaHCO_3$, 150mM NaCl, 3mM Na_2HPO_4 $7H_2O$, 5mM glucose, 0.05mM phenol red, pH 7.4), and then briefly triturated using a fine point Pasteur pipette. RPE cells were sedimented by centrifugation at 1500 rpm for 5 min and resuspended in GM to a final concentration of 50,000 cells per ml. Cells in suspension (0.5 ml) were added to the upper chamber of a 12-well plate, containing polycarbonate Transwell inserts, with a 3.0 μm pore diameter (Costar, MA). GM (1ml) was added to the lower well. Cells were cultured for 7 to 10 days until confluent colonies were present. The cells maintain a characteristic hexagonal morphology without the significant loss of pigmentation usually associated with the de-differentiation of RPE cells in culture (McLaren et al., 1993) (Fig. 2).

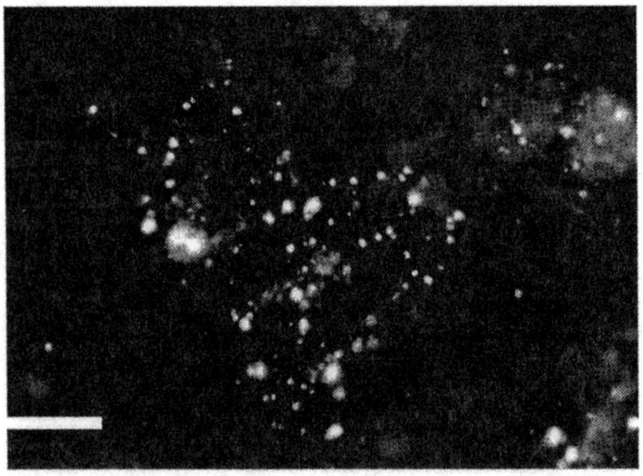

Figure 3. Micrograph showing the immunofluorescent labeling of ingested ROSs in cultured primary mouse RPE after a brief 5 min incubation. Scale bar = 25 μm.

3. PHAGOCYTOSIS OF ROD OUTER SEGMENTS BY CULTURED RPE

Mouse ROSs were isolated and purified from adult mouse retinas as described previously (Tsang et al., 1998). Purified ROSs were re-suspended in GM at a concentration of approximately 5 x10^7 ROS per ml. ROSs were then incubated with 7-10 day cultures of primary RPE cells by adding 0.5 ml of ROS suspension to the upper wells of a 12-well transwell plate. Note that this method rapidly exposes the RPE cells to a saturating number of ROSs, and thus differs from most published studies, using other RPE cell cultures, where fewer ROSs are added in a larger volume. After the addition of ROSs, cells were incubated at 37^0C, 5% CO_2 in a humidified incubator before fixation and immunocytochemical analysis. Double immunofluorescence labeling of bound and total (bound + ingested) ROSs was performed using a bovine opsin polyclonal antibody, pAb01 (Liu et al., 1999), following a previously described method (Colley and Hall, 1986).

The cultured RPE cells readily phagocytose purified mouse ROSs, demonstrating the utility of this culture system to study RPE phagocytosis. By incubating these cells with a saturating amount of ROSs in a small volume of medium we observed very rapid uptake of ROSs, without the lag due to the sedimentation time of the ROSs associated with lower concentrations and larger culture volumes (Fig. 3). Peak numbers of ingested ROSs were observed within 15 mins. In contrast, immortalized human or rat RPE cell lines, such as ARPE19 and RPEJ, require 3-5 hours before ingested ROS numbers reach a maximum (Finnemann et al., 1997).

In conclusion, our described method provides a rapid and reproducible procedure for obtaining primary cultures of mouse RPE cells. The cultured cells possess characteristic RPE cell morphology and phagocytic ability. This method should allow researchers to take advantage of the mouse models available for retinal degenerative disorders that result from RPE defects.

4. ACKNOWLEDGMENTS

We thank Dennis Defoe, Toshka Abrams, and Michael Hall for helpful discussions during our attempts to establish a procedure for the primary culture of mouse RPE cells. This research was supported by NIH R01 grant EY07042, NIH core grant EYEY12598, a grant from the Foundation Fighting Blindness. We thank the National Eye Institute for the travel award provided to DG to attend this meeting

5. REFERENCES

Chang, C. W., Roque, R. S., Defoe, D. M., and Caldwell, R. B., 1991, An improved method for isolation and culture of pigment epithelial cells from rat retina. *Curr. Eye Res.* **10**:1081-1086.

Chen, X. N., Korenberg, J. R., Jiang, M., Shen, D., and Fong, H. K., 1996, Localization of the human RGR opsin gene to chromosome 10q23. *Hum. Genet.* **97**:720-722.

Colley, N. J., and Hall, M. O., 1986, Phagocytosis of light- and dark-adapted rod outer segments by cultured RPE cells: a reassessment. *Exp. Eye Res.* **42**:323-329.

Davis, A. A., Bernstein, P. S., Bok, D., Turner, J., Nachtigal, M., and Hunt, R. C., 1995, A human retinal pigment epithelial cell line that retains epithelial characteristics after prolonged culture. *Invest. Ophthalmol. Vis. Sci.* **36**:955-964.

Feng, W., Yasumura, D., Matthes, M. T., LaVail, M. M., and Vollrath, D., 2002, Mertk triggers uptake of photoreceptor outer segments during phagocytosis by cultured retinal pigment epithelial cells. *J. Biol. Chem.* **277**:17016-17022.

Finnemann, S. C., Bonilha, V. L., Marmorstein, A. D., and Rodriguez-Boulan, E., 1997, Phagocytosis of rod outer segments by retinal pigment epithelial cells requires alpha(v)beta5 integrin for binding but not for internalization. *Proc. Natl. Acad. Sci. U S A* **94**:12932-12937.

Gal, A., Li, Y., Thompson, D. A., Weir, J., Orth, U., Jacobson, S. G., Apfelstedt-Sylla, E., and Vollrath, D., 2000, Mutations in MERTK, the human orthologue of the RCS rat retinal dystrophy gene, cause retinitis pigmentosa. *Nat. Genet.* **26**:270-271.

Goliath, R., Tombran-Tink, J., Rodriquez, I. R., Chader, G., Ramesar, R., and Greenberg, J., 1996, The gene for PEDF, a retinal growth factor is a prime candidate for retinitis pigmentosa and is tightly linked to the RP13 locus on chromosome 17p13.3. *Mol. Vis.* **2**:5.

Gu, S. M., Thompson, D. A., Srikumari, C. R., Lorenz, B., Finckh, U., Nicoletti, A., Murthy, K. R., Rathmann, M., Kumaramanickavel, G., Denton, M. J., and Gal, A., 1997, Mutations in RPE65 cause autosomal recessive childhood-onset severe retinal dystrophy. *Nat. Genet.* **17**:194-197.

Hall, M. O., 1978, Phagocytosis of light- and dark-adapted rod outer segments by cultured pigment epithelium. *Science* **202**:526-528.

Hasson, T., Heintzelman, M. B., Santos-Sacchi, J., Corey, D. P., and Mooseker, M. S., 1995, Expression in cochlea and retina of myosin VIIa, the gene product defective in Usher syndrome type 1B. *Proc. Natl. Acad. Sci. USA* **92**:9815-9819.

Irons, M. J., and Kalnins, V. I., 1984, Distribution of microtubules in cultured RPE cells from normal and dystrophic RCS rats. *Invest. Ophthalmol. Vis. Sci.* **25**:434-439.

Liu, X., Ondek, B., and Williams, D. S., 1998a, Mutant myosin VIIa causes defective melanosome distribution in the RPE of shaker-1 mice. *Nat. Genet.* **19**:117-118.

Liu, X., Udovichenko, I. P., Brown, S. D. M., Steel, K. P., and Williams, D. S., 1999, Myosin VIIa participates in opsin transport through the photoreceptor cilium. *J. Neurosci.* **19**:6267-6274.

Liu, X. Z., Hope, C., Walsh, J., Newton, V., Ke, X. M., Liang, C. Y., Xu, L. R., Zhou, J. M., Trump, D., Steel, K. P., Bundey, S., and Brown, S. D., 1998b, Mutations in the myosin VIIA gene cause a wide phenotypic spectrum, including atypical Usher syndrome. *Am. J. Hum. Genet.* **63**:909-912.

Marlhens, F., Bareil, C., Griffoin, J. M., Zrenner, E., Amalric, P., Eliaou, C., Liu, S. Y., Harris, E., Redmond, T. M., Arnaud, B., Claustres, M., and Hamel, C. P., 1997, Mutations in RPE65 cause Leber's congenital amaurosis. *Nat. Genet.* **17**:139-141.

Mayerson, P. L., Hall, M. O., Clark, V., and Abrams, T., 1985, An improved method for isolation and culture of rat retinal pigment epithelial cells. *Invest. Ophthalmol. Vis. Sci.* **26**:1599-1609.

McLaren, M. J., Sasabe, T., Li, C. Y., Brown, M. E., and Inana, G., 1993, Spontaneously arising immortal cell line of rat retinal pigmented epithelial cells. *Exp. Cell Res.* **204**:311-320.

Morimura, H., Saindelle-Ribeaudeau, F., Berson, E. L., and Dryja, T. P., 1999, Mutations in RGR, encoding a light-sensitive opsin homologue, in patients with retinitis pigmentosa. *Nat. Genet.* **23**:393-394.

Nabi, I. R., Mathews, A. P., Cohen-Gould, L., Gundersen, D., and Rodriguez-Boulan, E., 1993, Immortalization of polarized rat retinal pigment epithelium. *J. Cell Sci.* **104**:37-49.

Tsang, S. H., Burns, M. E., Calvert, P. D., Gouras, P., Baylor, D. A., Goff, S. P., and Arshavsky, V. Y., 1998, Role for the target enzyme in deactivation of photoreceptor G protein in vivo. *Science* **282**:117-121.

Vollrath, D., Feng, W., Duncan, J. L., Yasumura, D., D'Cruz, P. M., Chappelow, A., Matthes, M. T., Kay, M. A., and LaVail, M. M., 2001, Correction of the retinal dystrophy phenotype of the RCS rat by viral gene transfer of Mertk. *Proc. Natl. Acad. Sci. U S A* **98**:12584-12589.

Weil, D., Blanchard, S., Kaplan, J., Guilford, P., Gibson, F., Walsh, J., Mburu, P., Varela, A., Levilliers, J., Weston, M. D., Kelley, P. M., Kimberling, W. J., Wagenaar, M., Levi-Acobas, F., Larget-Piet, D., Munnich, A., Steel, K. P., Brown, S. D. M., and Petit, C., 1995, Defective myosin VIIA gene responsible for Usher syndrome type 1B. *Nature* **374**:60-61.

REGULATION OF THE VISUAL CYCLE: RETINOL DEHYDROGENASE AND RETINOL FLUORESCENCE MEASUREMENTS IN VERTEBRATE RETINA

M. Carter Cornwall[1], Efthymia Tsina[1], Rosalie K. Crouch[2], Barbara Wiggert[3], Chunhe Chen[4], and Yiannis Koutalos[4]

1. INTRODUCTION

The initial and only light-activated step in vision is the photoisomerization of the ligand of the visual pigment, 11-*cis* retinal to all-*trans* retinal, which occurs while being covalently bound to a lysine deep in the membrane region of the visual pigment. This event leads to the activation of the G-protein, transducin, which in turn activates c-GMP phosphodiesterase causing the destruction of c-GMP, the closure of cation channels in the plasma membrane of the photoreceptor, and eventually to the reduction in release of synaptic transmitter to other retinal neurons in the visual pathway. However, the visual pigment containing retinal in its all-*trans* form is now incapable of absorbing photons in the visual region of the spectrum and can no longer activate the transduction cascade. In order to do so, it must be regenerated into a form that contains11-*cis* retinal. The biochemical reactions by which this occurs are collectively called the visual cycle. This complex series of reactions is initiated in the photoreceptors themselves, by the reduction of all-*trans* retinal to all-*trans* retinol by all-*trans* retinol dehydrogenase (RDH) and the cofactor NADPH. All the subsequent steps in the regeneration of 11-*cis* retinal take place in the retinal pigment epithelium. Regenerated 11-*cis* retinal is then transported from the retinal pigment epithelium through the extracellular matrix back to the photoreceptor cells where it condenses with opsin to reform visual pigment.

The focus of this monograph is to review what is known presently regarding the mechanisms of the reduction of all-*trans* retinal to all-*trans* retinol, and its transport from the photoreceptor following bleaching. The reasons for focusing on this reaction are two. First, recent investigations of the biochemical reactions of the visual cycle have demonstrated that this step limits the rate of regeneration of retinoids in mouse retina

[1]Department of Physiology and Biophysics, Boston University School of Medicine, Room L714, 80 East Concord Street, Boston, MA, USA; [2]Department of Ophthalmology, Medical University of South Carolina, Charleston, SC, USA; [3]National Eye Institute, National Institutes of Health, Bethesda, MD, USA; [4]Department of Physiology and Biophysics, University of Colorado School of Medicine, Denver, CO, USA. Corresponding author: M. C. Cornwall, E-mail: cornwall@bu.edu

under bright-light adaptation conditions (Zimmerman et al., 1974; Saari et al., 1998). Second, all-*trans* retinal and its short-chain analogs can combine in vitro with opsin to form physiologically active complexes capable of interacting with the G-protein, transducin, resulting in activation of the transduction cascade (Fukada and Yoshizawa, 1981; Cohen et al., 1992; Sakmar et al., 1989; Hofmann et al., 1992; Palczewski et al., 1994; Surya et al., 1995) as well as promote rhodopsin phosphorylation by rhodopsin kinase (Hofmann et al., 1992; Palczewski et al., 1994) and arrestin binding (Jager et al., 1996). All of these biochemical reactions can occur in the absence of Schiff base formation (Surya et al., 1995; Jager et al., 1996; Melia et al., 1997). Taken together, these findings suggest that prolonged interactions between all-*trans* retinal and opsin following bleaching may be of crucial importance in the recovery of dark current in photoreceptors and therefore may determine the rate of dark adaptation following bright light exposure. Indeed, a large body of psychophysical and electrophysiological data now suggests strongly that this interaction is the source of the slow kinetics of response recovery as well as post-bleach induced noise in photoreceptors (Alpern et al., 1970; Lamb, 1980; Fain et al., 1996; Jin et al., 1993; Kefalov et al., 2001).

2. SUPPORTING DATA

2.1. RDH catalizes reduction all-*trans* retinal to all-*trans* retinol

Among the earliest descriptions of the existence and properties of all-*trans* retinal and its reduction in the retina following bleaching to all-*trans* retinol, were those made by Boll (1877) and Kuhne (1879). In Kuhne's review, published in 1879, he described that following bright light, the retina changed from a deep red to pale yellow, finally becoming colorless. For some minutes after becoming colorless, exposure to ultraviolet light resulted in the emanation of a greenish fluorescence, first described by Helmoltz and Setschenow (reviewed by Kuhne (1879)). To distinguish this fluorescent substance from visual purple (the visual pigment prior to exposure to light) and that produced immediately following light (visual yellow), Kuhne named this third substance visual white because it was not visible to the naked eye. He correctly surmised that this substance was produced in rod photoreceptors, and that it was a product of the bleaching of visual purple. However, he was unable to observe it in areas of the retina rich in cone photoreceptors, and concluded that this substance was exclusively found in rods. Subsequent work determined that the yellow substance present in the retina immediately following bleaching was all-*trans* retinal complexed with opsin and that the fluorescent substance present in the colorless retina was all-*trans* retinol (Wald, 1935).

RDH is responsible for converting all-retinal to all-*trans* retinol in photoreceptors. The enzyme was identified and partially characterized in early biochemical experiments (Blaner and Churchich, 1980; Ishiguro et al., 1991; Nicotra and Livrea, 1982), but only recently have experiments using biochemical and cloning techniques begun to elucidate this enzyme in detail (Palczewski et al., 1994; Haeseleer et al., 1998; Rattner et al., 2000). An RDH (prRDH) has been cloned from bovine tissue that has been found to contain approximately 310 amino acids and be a member of the short-chain dehydrogenase/reductase superfamily (Rattner et al., 2000). These studies demonstrated that bovine and human RDH are highly homologous, both are closely related to 17-βhydroxysteroid dehydrogenase, and in humans this enzyme is encoded on chromosome 19. Haeseleer and coworkers (Haeseleer et al., 1998) have reported the cloning of another short-chain dehydrogenase, retSDR1, from human and bovine retinal tissue. This enzyme localizes predominantly in cone photoreceptor cells. This protein is reported to contain 302 amino acids and the gene is located on chromosome 1. Cells infected with the retSDR1 sequence reduced all-*trans* retinal, but not 11-*cis* retinal; NADPH, but not NADH, was a required substrate.

Thus, two separate enzymes have been identified in bovine, mouse, and human retinal tissue. Both of these require NADPH as a cofactor and both are highly specific for all-*trans* retinal over 11-*cis* retinal (Haeseleer et al., 1998; Rattner et al., 2000). The fact that RetSDR1 and prRDH appear to localize predominantly or exclusively in rod and cone photoreceptors, respectively, suggests that each of these cell types may express a form of retinol reductase that suits a particular need, and may contribute to the widely different rates of reductase enzymatic activity that are thought to occur in these cell types. This assumption is based on the much more rapid rates at which the visual cycle regenerates visual pigment in cones compared to rods and the much more rapid rate of dark adaptation in cones. For instance, it is well known that under normal bright day light conditions, rod photoreceptors are unresponsive to light, and our visual system relies substantially, if not exclusively, on cones to transmit visual images to the brain. It has been demonstrated in psychophysical experiments on human subjects that the circulating current in cones recovered within about 3 sec following a 50% photopigment bleach, whereas 15 min was required for a similar extent of recovery in rods (Thomas and Lamb, 1999; Paupoo et al., 2000). Thus, human cones remain responsive under brightly lit conditions whereas rods are not. These authors attributed this 300-fold difference in rate of recovery to a difference in the way rods and cones handle the photoproducts of bleaching.

2.2. Microfluorometric measurements of photoreceptors

In addition to biochemical and molecular genetic experiments, microfluorometric measurements of retinal-to-retinol reduction have been made in retinal photoreceptors before and after bleaching (Denton, 1959; Liebman, 1969; Kemp, 1973; Kaplan and Liebman, 1977; Kaplan, 1985; Sears and Kaplan, 1989). These observations are based on the early observations referred to above, originally made by Kuhne (1879) and Boll (1877), that retinal rods exposed to near UV-light fluoresce weakly and bluish, but after bleaching they emit greenish fluorescent light. This occurs because all-*trans* retinol emits fluorescence more than 50-fold more intensely than either 11-*cis* retinal or all-*trans* retinal.

An example of this type of measurement is illustrated in Figure 1, which shows digital images of emitted fluorescent light measured before and following visual pigment bleaching in a solitary salamander rod. At the top left of Figure 1 is shown a bright-field image of the cell. It is oriented with the outer segment directed downward and to the left; the inner segment containing the ellipsoid, as well as the synaptic ending is at the top right. The fluorescence image in the middle panel was elicited when the cell was exposed to a 400 msec flash of 360 nm light as the cell was in the dark-adapted state. In this condition, all the visual pigment in the outer segment contained 11-*cis* retinal. It is apparent that little fluorescence can be detected from the outer segment. However, an intense fluorescence was observed to emanate from a region adjacent to the outer segment in the ellipsoid region of the cell containing mitochondria. The cell was then exposed to a bright 520 nm light that was calculated to have bleached in excess of 99% of

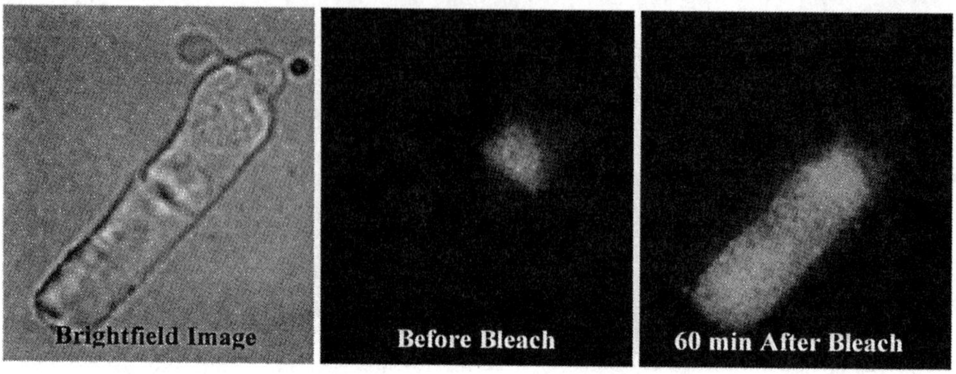

Figure 1. Measurement of all-*trans* retinol fluorescence in a salamander rod before and following rhodopsin bleaching. Left: bright-field image of a salamander red rod. Center: fluorescence of dark-adapted photoreceptor two min prior to exposure to a 520 nm light calculated to have bleached in excess of 99% of the visual pigment. Right: Fluorescence of the same cell measured 60 min after the bleach. See text for details.

the visual pigment. Following 60 min in darkness, the cell was again probed with the 400 msec, 360 nm fluorescence-inducing test flash. This image, which is shown in the right panel of Figure 1, shows that after 60 min, in addition to fluorescence in the ellipsoid region of the cell, the outer segment also emits an intense fluorescence. A large body of previous work has attributed this slowly developing bleach-induced fluorescence to the appearance of all-*trans* retinol. The evidence for this is summarized as follows. First, the fluorescence excitation spectrum is centered around 370 nm; the emission spectrum peaks at about 490 nm. These excitation and emission spectra are consistent with those of all-*trans* retinol, but are inconsistent with those of any of the 11-*cis*-retinal or all-*trans* retinal containing intermediate states associated with the visual pigment. Second, both the excitation and emission spectra are stronger for light polarized parallel rather than perpendicular to the long axis of the rod outer segment. This direction is orthogonal to the dichroism of the dark-adapted or light-activated (Meta II) rhodopsin states in the intact outer segment (Denton, 1959; Harosi and MacNichol, 1974; Liebman, 1969). Third, outer segment fluorescence after bleaching develops slowly. Denton (1959) determined that it approaches a maximum only after about 30 min; Kaplan and Liebman (1977) found that fluorescence in rods of *Rana pipiens* develops with a time constant of about 5-7 min at 22°C. Thus, post-bleach fluorescence develops much more slowly than any intermediate photoproducts of rhodopsin bleaching that are known to interact with proteins of the transduction cascade. Fourth, fluorescence appears to be substantially spatially restricted to regions of the outer segments that have experienced bleaching light (Kaplan, 1985; Sears and Kaplan, 1989). Finally, bleach-induced fluorescence can be elicited most strongly in outer segments of rods to which the ellipsoid is attached (Kaplan, 1985), suggesting that the energy required for reduction is derived, at least in part, from metabolic reactions that take place in the mitochondria-rich ellipsoid region located within the inner segment.

What is the metabolic source of reducing power that is required by retinol dehydrogenase to reduce all-*trans* retinal to all-*trans* retinol in photoreceptors? The in vitro biochemical experiments that have addressed this question have all been done with rod-rich preparations. In rods, energy in the form of NADPH and ATP that fuel this reduction has been shown to derive from anaerobic glycolytic as well as aerobic sources located in both the outer segment and the ellipsoid regions of rods. These pathways are illustrated diagrammatically in the cartoon in Figure 2. One route is via glycolysis and the hexose monophosphate pathways located in the outer segment of rods. The enzymes of these pathways that are required for the production of NADPH have all been detected within outer segments (Futterman, 1963; McConnell et al., 1969; Futterman et al., 1970; Schnetkamp and Daemen, 1981; Hsu and Molday, 1991). Furthermore, Hsu and Molday (1994) have demonstrated that much, if not all, of the reduction of retinal to retinol in bovine outer segments can be fueled by NADPH generated in this way. These reactions utilize glucose transported into the outer segment via a Glut 1 type glucose transporter from the extracellular space (Hsu and Molday, 1991; Lopez-Escalera et al., 1991; Li et al.

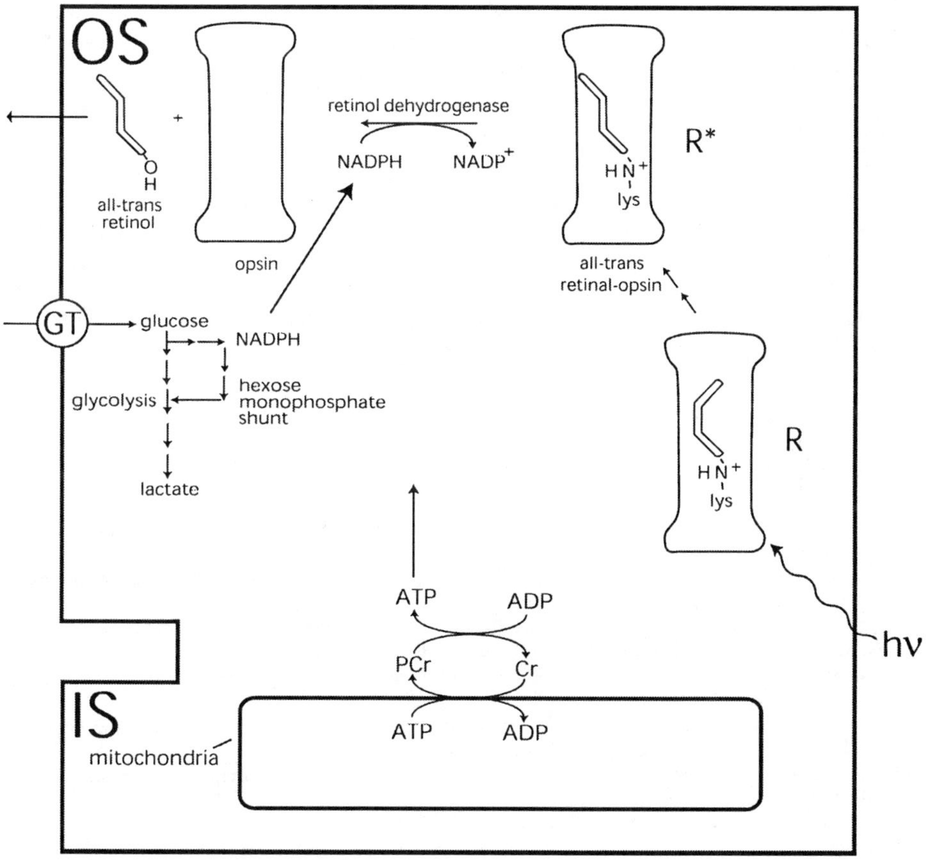

Figure 2. Cartoon of a single rod photoreceptor cell illustrating the principal mechanisms known for the production of NADPH and ATP necessary for reduction of all-*trans* retinal to all-*trans* retinol. Inner segment (IS) and outer segment (OS) are illustrated at the bottom and top of the cartoon, respectively. See text for details.

1994), and ATP derived from glycolysis within the outer segment or ATP that is produced by creatine kinase from creatine phosphate which has been transported from the inner segment to the outer segment (Wallimann et al., 1986; Hsu and Molday, 1991).

3. FUTURE STUDIES

A fruitful area of future investigation will be to combine single cell microfluorometric approaches that have been discussed above with simultaneous electrophysiological measurements of cell sensitivity. Using the combination of these techniques, the following important questions can be addressed. (1) How does the rate of reduction of all-*trans* retinal differ in rod and cone photoreceptors? (2) Are the large differences in the rates of recovery of sensitivity and light-suppressed dark current that have been observed in these two cell types following exposure to bright light due to differences in these respective rates? (3) Is the NADPH required for the reduction reaction produced solely by glycolysis, or are other reactions in the energy-rich ellipsoids of rods and cones also involved? (4) What are the mechanisms for removal of all-*trans* retinol from bleached photoreceptors? These and other important questions must be answered before we are to obtain a complete understanding of the relationship of the visual cycle to the recovery of photoreceptor sensitivity following bleaching.

4. ACKNOWLEDGMENTS

Funding was provided by grants from NIH (EY01157, EY04939 and EY11351), Foundation Fighting Blindness, and and an unrestricted grant to MUSC from Research to Prevent Blindness, Inc., New York, NY. The authors thank Luanna Bartholomew for editorial assistance.

5. REFERENCES

Alpern, M., Rushton, W. A. and Torii, S., 1970, The attenuation of rod signals by bleachings. *J. Physiol.* **207**(2):449-61.

Blaner, W. S. and Churchich, J. E., 1980, The membrane bound retinol dehydrogenase from bovine rod outer segments. *Biochem. Biophys. Res. Commun.* **94**(3):820-6.

Boll, F., 1877, Zur anatomie und physiologie der retina. *Arch. Anat. Physiol.* 4-36. [Translation: 1977, On the anatomy and physiology of the retina. *Vis. Res.* **17**(11-12):1249-65].

Cohen, G. B., Oprian, D. D., and Robinson, P. R., 1992, Mechanism of activation and inactivation of opsin: Role of Glu113 and Lys296. Biochemistry. **31**(50):12592-601.

Denton, E. J., 1959, The contributions of the oriented photosensitive and other molecules to the absorption of the whole retina. *Proc. R. Soc. (Lond.) Series B.* **150**:78-94.

Fain, G. L., Matthews, H. R., and Cornwall, M. C., 1996, Dark adaptation in vertebrate photoreceptors. *Trends Neurosci.* **19**(11):502-7.

Fukada Y., and Yoshizawa, T., 1981, Activation of phosphodiesterase in frog rod outer segment by an intermediate of rhodopsin photolysis. II. *Biochim. Biophys. Acta.* **675**(2):195-200.

Futterman, S., 1963, Metabolism of the retina. *J. Biol. Chem.* **238**:1145-50.

Futterman, S., Hendrickson, A., Bishop, P. E., Rollins, M. H., and Vacano, E., 1970, Metabolism of glucose and reduction of retinaldehyde in retinal photoreceptors. *J. Neurochem.* **17**(2):149-56.

Haeseleer, F., Huang, J., Lebioda, L., Saari, J. C., and Palczewski, K., 1998, Molecular characterization of a novel short-chain dehydrogenase/reductase that reduces all-*trans*-retinal. *J. Biol. Chem.* **273**(34):21790-9.

Harosi, F., I., and MacNichol Jr., E. F., 1974, Dichroic microspectrophotometer: A computer-assisted, rapid, wavelength-scanning photometer for measuring linear dichroism in single cells. *J. Opt. Soc. Am.* **64**(7):903-18.

Hofmann, K. P., Pulvermuller, A., Buczylko, J., Van Hooser, P., and Paczewski, K., 1992, The role of arrestin and retinoids in the regeneration pathway of rhodopsin. *J. Biol. Chem.* **267**(22):15701-6.

Hsu, S., C., and Molday, R., S., 1991, Glycolytic enzymes and a GLUT-1 glucose transporter in the outer segments of rod and cone photoreceptor cells. *J. Biol. Chem.* **266**(32):21745-52.

Hsu, S. C., and Molday, R. S., 1994, Glucose metabolism in photoreceptor outer segments. Its role in phototransduction and in NADPH-requiring reactions. *J. Biol. Chem.* **269**(27):17954-9.

Ishiguro, S., Suzuki, Y., Tamai, M., and Mizuno, K., 1991, Purification of retinol dehydrogenase from bovine retinal rod outer segments. *J. Biol. Chem.* **266**(23):15520-4.

Jager, S., Palczewski, K., and Hofmann, K. P., 1996, Opsin/all-*trans*-retinal complex activates transducin by different mechanisms than photolyzed rhodopsin. *Biochemistry.* **35**(9):2901-8.

Jin, J., Crouch, R. K., Corson, D. W., Katz, B. M., MacNichol, E. F., and Cornwall, M. C., 1993, Noncovalent occupancy of the retinal-binding pocket of opsin diminishes bleaching adaptation of retinal cones. *Neuron.* **11**(3):513-22.

Kaplan, M. W., Liebman, P. A., 1977, Slow bleach-induced birefringence changes in rod outer segments. *J. Physiol.* **265**(3):657-72.

Kaplan, M. W., 1985, Distribution and axial diffusion of retinol in bleached rod outer segments of frogs (*Rana pipiens*). *Exp. Eye Res.* **40**(5):721-9.

Kefalov, V. J., Crouch, R. K., and Cornwall, M. C., 2001, Role of noncovalent binding of 11-cis-retinal to opsin in dark adaptation of rod and cone photoreceptors. *Neuron.* **29**(3):749-55.

Kemp, C. M., 1973, Dichroism in rods during bleaching, *In*: Langer, H., editor, *Biochemistry and Physiology of Visual Pigments.* Berlin: Springer-Verlag. p. 307-12.

Kuhne W., 1879, Chemische vorgange in der netzhaut, (ed. Hermann, L.) Vol. 3 [Translation: Hubbard, R., Wald, G., 1977, Chemical processes in the retina. *Vis. Res.* **17**(11-12):1269-316].

Lamb, T. D., 1980, Spontaneous quantal events induced in toad rods by pigment bleaching. *Nature.* **287**(5780):349-51.

Li, X. B., Szerencsei, R. T., and Schnetkamp, P. P., 1994, The glucose transporter in the plasma membrane of the outer segments of bovine retinal rods. *Exp. Eye Res.* **59**(3):351-8.

Liebman, P. A., 1969, Microspectrophotometry of retinal cells. *Ann. N. Y. Acad. Sci.* **157**:250-64.

Lopez-Escalera, R., Li, X. B., and Szerencsei, P. P., 1991, Glycolysis and glucose uptake in intact outer segments isolated from bovine retinal rods. *Biochemistry.* **30**(37):8970-6.

McConnell, D. G., Ozga, G. W., and Solze, D. A., 1969, Evidence for glycolysis in bovine retinal microsomes and photoreceptor outer segments. *Biochim. Biophys. Acta.* **184**(1):11-28.

Melia Jr., T. J., Cowan, C. W., Angleson, J. K., and Wensel, T. G., 1997, A comparison of the efficiency of G protein activation by ligand-free and light-activated forms of rhodopsin. *Biophys. J.* **73**(6): 3182-91.

Nicotra, C., and Livrea, M. A., 1982, Retinol dehydrogenase from bovine retinal rod outer segments. Kinetic mechanism of the solubilized enzyme. *J. Biol. Chem.* **257**(19):11836-41.

Palczewski, K., Jager, S., Buczylko, J., Crouch, R. K., Bredberg, D. L., Hofmann, K. P., Asson-Batres, M. A., and Saari, J. C., 1994, Rod outer segment retinol dehydrogenase: Substrate specificity and role in phototransduction. *Biochemistry.* **33**(46):13741-50.

Paupoo, A. A., Mahroo, O. A., Friedburg, C., and Lamb, T. D., 2000, Human cone photoreceptor responses measured by the electroretinogram a-wave during and after exposure to intense illumination. *J. Physiol.* **529**(Pt2):469-82.

Rattner, A., Smallwood, P. M., and Nathans, J., 2000, Identification and characterization of all-*trans*-retinol dehydrogenase from photoreceptor outer segments, the visual cycle enzyme that reduces all-*trans*-retinal to all-*trans*-retinol. *J. Biol. Chem.* **275**(15):11034-43.

Saari, J. C., Garwin, G. G., Van Hooser, J. P., and Palczewski K., 1998, Reduction of all-*trans*-retinal limits regeneration of visual pigment in mice. *Vis. Res.* **38**(10):1325-33.

Sakmar, T. P., Franke, R. R., and Khorana, H. G., 1989, Glutamic acid-113 serves as the retinylidene Schiff base counterion in bovine rhodopsin. *Proc. Nat. Acad. Sci. U. S. A.* **86**(21):8309-13.

Schnetkamp, P. P., and Daemen, F. J., 1981, Transfer of high-energy phosphate in bovine rod outer segments. A nucleotide buffer system. *Biochim. Biophys. Acta.* **672**(3):307-12.

Sears, R. C., and Kaplan, M. W., 1989, Axial diffusion of retinol in isolated frog rod outer segments following substantial bleaches of visual pigment. *Vis. Res.* **29**(11):1485-92.

Surya, A., Foster, K. W., and Knox, B. E., 1995, Transducin activation by the bovine opsin apoprotein. *J. Biol. Chem.* **270**(10):5024-31.

Thomas, M. M., and Lamb, T. D., 1999, Light adaptation and dark adaptation of human rod photoreceptors measured from the a-wave of the electroretinogram. *J. Physiol.* **518**(Pt2): 479-96.

Wald, G., 1935, Carotenoids and the visual cycle. *J. Gen. Physiol.* **19**(2):351-71.

Wallimann, T., Wegmann, G., Moser, H., Huber, R., and Eppenberger, H. M., 1986, High content of creatine kinase in chicken retina: Compartmentalized localization of creatine kinase isoenzymes in photoreceptor cells. *Proc. Nat. Acad. Sci. U. S. A.* **83**(11):3816-9.

Zimmerman, W. F., Yost, M. T., and Daemen, F. J., 1974, Dynamics and function of vitamin A compounds in rat retina after a small bleach of rhodopsin. *Nature.* **250**(461):66-7.

MELATONIN ENHANCES RETINOIC ACID INDUCTION OF CONE ARRESTIN GENE EXPRESSION IN RETINOBLASTOMA CELLS

Aimin Li, Xuemei Zhu, Bruce Brown, and Cheryl M. Craft[1]

1. INTRODUCTION

Vitamin A plays a critical role in normal vision. Its acid form, retinoic acid (RA), is a potent modulator of hormonal transcription control and is critical for early eye and photoreceptor development and differentiation (Hyatt et al., 1996a; Hyatt et al., 1996b; Marsh-Armstrong et al., 1994; McCaffery et al., 1992; McCaffery et al., 1991). Melatonin (Mlt), rhythmically produced in pinealocytes of the pineal gland and photoreceptors of the retina with peak levels during darkness (Binkley et al., 1978; Craft, 1993; Craft et al., 1999; Ganguly et al., 2002), plays a crucial role in the circadian and seasonal rhythmic regulation of diverse organisms (Deguchi, 1979; Kasal et al., 1979; Klein and Moore, 1979; Wiechmann, 2002). Mlt also participates in modulating cyclic diurnal metabolic pathways in photoreceptors and retinal pigment epithelium (RPE).

Our laboratory has previously demonstrated that RA treatment of immortalized human retinoblastoma cells, Weri- Rb-1 and Y79, dramatically induces the mRNA and protein expression of a cone-specific gene, cone arrestin (CAR), but not its rod-specific counterpart, rod arrestin (Li et al., 2001; Li et al., 2002; Zhang et al., 2001). With the Affymetrix's oligonucleotide array-based expression profiling technology, we further demonstrated that the retinoblastoma cell line underwent G0/G1 cell cycle arrest and displayed cone–like genetic differentiation and selective cell apoptosis following RA treatment (Li et al., 2003).

In other systems, RA and Mlt individually inhibit the proliferation of the estrogen receptor-alpha (ER alpha)-positive MCF-7 breast cancer cell line (Blask et al., 1991; Cos and Sanchez-Barcelo, 2000; Czeczuga-Semeniuk et al., 2001; Dokmanovic et al., 2002). In addition, substantial evidence has accumulated demonstrating that RA, in combination with Mlt, is able to effectively and significantly suppress tumor development and growth and induce apoptosis of cancer cells (Czeczuga-Semeniuk et al., 2002; Eck-Enriquez et al., 2000; Nowfar et al., 2002).

[1]The Mary D. Allen Laboratory for Vision Research, Doheny Eye Institute and Department of Cell and Neurobiology, the Keck School of Medicine of the University of Southern California, BMT401, 1333 San Pablo Street, Los Angeles, CA 90089-9112, USA. Corresponding author: C. M. Craft, E-mail: ccraft@hsc.usc.edu

In this study, we sought to determine in childhood tumor-derived retinoblastoma cells whether Mlt alone or in combination with RA could affect CAR mRNA levels. The combined effects of Mlt and RA on CAR promoter activities were also monitored.

2. METHODS

2.1. Cell Culture and Hormone Treatment

Weri-Rb-1 retinoblastoma cells (American Type Tissue Culture, Rockville, MD) were maintained and treated with all-*trans* RA (RA), 9-cis RA, Mlt or RA plus Mlt, as previously described (Li et al., 2001; Li et al., 2002; Zhang et al., 2001). B27 (2%, GibcoBRL, http/www.lifetech.com) was substituted for the 10% fetal bovine serum in the medium. All tissue culture medium and supplements were obtained from Irvine Scientific (Santa Ana, CA). All chemicals used to treat the cells were obtained from Sigma (St. Louis, MO).

2.2. Northern Blot Analysis

Northern blot analysis was performed with 20 μg of total RNA isolated from Weri-Rb-1 cells as previously described (Li et al., 2002; Zhu and Craft, 2000). Each cDNA probe was specifically radiolabelled and was visualized by exposing the hybridized membrane to a phosphorimager screen (Molecular Dynamics, Sunnyvale, CA). The relative mRNA levels were normalized for loading variability and transfer efficiency by subsequently hybridizing the stripped membrane with an [α-^{32}P]dCTP-labeled β-actin probe or a 3'-terminal transferase labeled 18S oligonucleotide probe (Craft et al., 1990).

2.3. Immunoblot Analysis

Cells were treated as described above, and whole cell homogenates were prepared according to previously published procedures (Li et al., 2002; Zhu and Craft, 2000). Equal amounts of proteins were resolved on 11.5% sodium dodecylsulfate-polyacrylamide gel electrophoresis (SDS-PAGE) and were electrophoretically transferred to Immobilon-P membranes (Millipore, Bedford, MA) using previously published protocols (Craft et al., 1998). The immobilized proteins were detected with the anti-hCAR peptide polyclonal antibody (Luminaire Founders, LUMIF (Zhang et al., 2001), 1:10,000) and anti-rabbit secondary antibody (1:10,000, Bio Rad Laboratories), using an Enhanced Chemiluminescence Kit (Amersham, Arlington Heights, IL) (Craft et al., 1998).

2.4. Transient Transfection and Luciferase Assays

Both hCAR (ph388 and ph900) (Li et al., 2002) and mCAR (pm523 and pm1003) (Zhu et al., 2002) 5'-flanking fragments in the pRL-null vector were analyzed for responsiveness to Mlt or RA plus Mlt using transient transfection of Weri-Rb-1 cells as previously described (Li et al., 2002; Zhu et al., 2002). Briefly, Weri-Rb-1 cells grown in 24-well plates were transiently transfected with 0.8 μg of either mCAR or hCAR promoter–luciferase reporter construct or the promoterless pRL-null vector and 0.2 μg of pGL3-P plasmid (Promega) as an internal control of the transfection efficiency. After 3 hr incubation, either RA (1 μM), Mlt (1μM), RA and Mlt or the drug vehicle DMSO (0.1%) was added; the transfected cells were further incubated for 45 hrs before being harvested for luciferase reporter assay. *Renilla* luciferase activity was normalized to the firefly luciferase activity of the same sample and expressed as fold increase of the promoterless control.

3. RESULTS

3.1. Dose-dependent effects of Mlt on RA Induction of hCAR mRNA Expression

We have previously reported that the hCAR expression levels of the mRNA and the immunoreactive protein were highly induced in both Weri-Rb-1 and Y79 cells by RA with a peak effect observed at 10 μM (Li et al., 2001; Li et al., 2002; Zhang et al., 2001). The effect of Mlt on RA's ability to induce hCAR mRNA expression at various concentrations was tested. As shown in Figure 1A, treatment of Weri-Rb-1 cells with 1 μM of RA alone for 48 hrs has only minimum effect, and treatment with 1 μM of Mlt alone has no effect on hCAR mRNA expression. However, when the two reagents are applied together, hCAR mRNA expression is greatly enhanced. In contrast, 10 μM of RA has a dramatic effect on hCAR induction, but the addition of 10 μM Mlt does not further potentiate RA's effect on hCAR mRNA expression (Figure 1A).

We then determined the dose-response effects of Mlt on RA induction of hCAR mRNA expression using 1 μM of RA, in combination with Mlt at increasing concentrations (Figure 1B). Interestingly, Mlt's potentiation effect on RA induction of

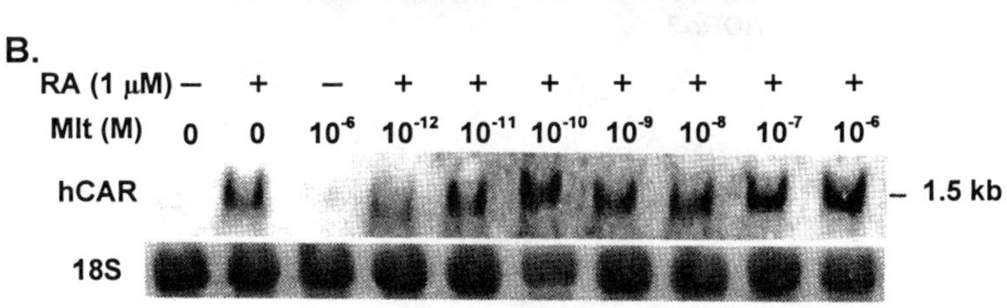

Figure 1. A. Northern blot analysis of hCAR expression in Weri-Rb-1 retinoblastoma cells treated with either 1 μM or 10 μM of Mlt and RA. **B.** Northern blot analysis of hCAR expression in Weri-Rb-1 cells treated with increasing concentrations of Mlt, in combination with 1 μM of RA. To quantitate gel loading and transfer efficiency, the membranes were striped and re-probed with a radioactively labeled 18S probe.

hCAR mRNA expression is not dose-dependent, with concentrations lower than 10^{-11} M having no effect and concentrations in the range from 10^{-11} M to 10^{-6} M (1 μM) having similar effect (Figure 1B).

3.2. Time-course effects of Mlt on RA Induction of hCAR mRNA Expression

To analyze the dynamics of the induction, we performed a time-course analysis of hCAR mRNA expression following the addition of 1 μM of Mlt and 1 μM of RA to Weri-Rb-1 cells. Northern blot analysis shows that hCAR mRNA levels are only slightly upregulated within 6 hrs of treatment with both RA and Mlt (Figure 2), similar to the time course of RA treatment alone (Li et al., 2002). However, Mlt significantly enhances the effect of RA on hCAR mRNA induction after 12 hrs of exposure compared to RA treatment alone. After 24 hrs of treatment with both RA and Mlt, the hCAR mRNA expression reaches an induction of 6-fold over RA treatment alone (Figure 2).

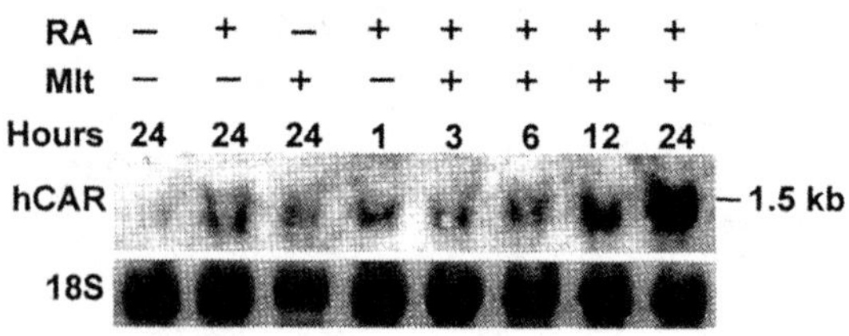

Figure 2. Time course of hCAR mRNA expression in Weri-Rb-1 cells following the addition of 1 μM of Mlt and/or 1 μM of RA.

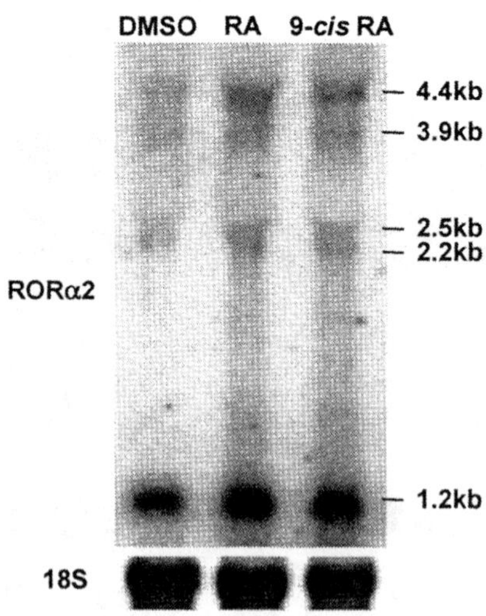

Figure 3. Northern blot analysis of RORα2 mRNA expression in Weri-Rb-1 cells treated with 10 μM of RA or 9-*cis* RA for 48 hrs. Sizes of the hybridized bands are indicated on the right.

3.3. RA induces retinoid-related orphan nuclear receptor alpha 2 (RORα2) Expression in Weri-Rb-1 cells

To elucidate the potential pathways through which Mlt potentiates RA induction of hCAR gene expression, we examined the mRNA expression of the Mlt receptor, RORα2, in Weri-Rb-1 cells. Northern blot analysis with a RORα2 cDNA probe reveals multiple RORα2 transcripts of 1.2, 2.2, 2.5, 3.9 and 4.4 kb. Treatment of the cells with either RA or 9-cis RA significantly increases the expression of RORα2 compared to the DMSO-treated control (Figure 3). This is consistent with our oligonucleotide microarray results showing that 48 hrs of RA treatment upregulated RORα2 mRNA expression by 4.2 fold in Weri-Rb-1 cells (Li et al., 2003).

3.4. Mlt Enhances the effect of RA on both hCAR and mCAR Promoter Activities in Weri-Rb-1 cells

RA stimulates hCAR mRNA expression through a direct action on a region between –852 and –702 of the hCAR 5' flanking region (Li et al., 2002). To examine if Mlt's potentiation of RA induction of hCAR expression in Weri-Rb-1 cells might be due to enhanced promoter activities, Weri-Rb-1 cells were transiently transfected with luciferase fusion constructs containing either the hCAR (Li et al., 2002) or the mCAR 5'-flanking fragments (Zhu et al., 2002).

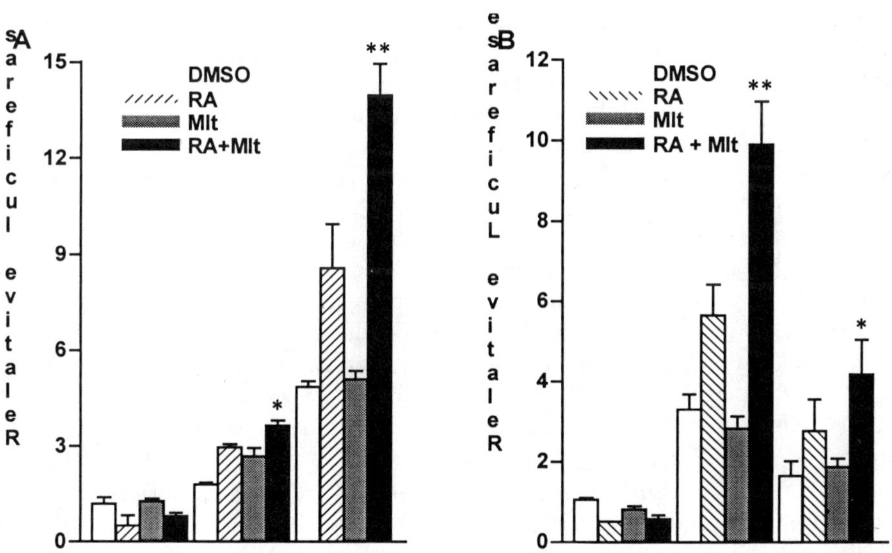

Figure 4. Mlt effects on RA stimulation of CAR promoter activities in Weri-Rb-1 retinoblastoma cells. Either hCAR (**A**) or mCAR (**B**) 5'-flanking fragments were inserted into the pRL-null Renilla luciferase reporter vector and transfected into Weri-Rb-1 cells followed by DMSO, RA, Mlt, or RA and Mlt treatment. The cells were processed and assayed for luciferase activity 48 hrs after transfection. The promoterless pRL-null vector was used as a control for basal luciferase activity of the cells and its luciferase activity was set at 1. Results represent means ± SEM (n = 3). * $P < 0.05$, ** $P < 0.01$ (the combinatorial treatment of Mlt and RA compared with RA alone).

As shown in Figure 4, RA alone stimulates both hCAR and mCAR promoter activities, as demonstrated by an increase in luminescence. In contrast, treatment of the transfected cells with Mlt alone has no effect on either the hCAR or the mCAR promoter. However, simultaneous treatment of the cells with both Mlt and RA significantly increases both the hCAR and mCAR promoter activities over those with RA treatment alone. The ph900 of the hCAR and the pm523 of the mCAR 5'-flanking fragments display the maximal response to the combined effect of Mlt and RA by 1.6- and 1.7-fold over that of RA alone, respectively.

4. DISCUSSION

For many years, vitamin A metabolites, including RA, have been known to have profound effects on development, cellular proliferation and differentiation, and tumor growth and invasion. The wide-ranging effects of RA on cellular proliferation and migration make this reagent a useful chemotherapeutic agent in the treatment of many types of cancer. Recent studies in our laboratory have demonstrated an extensive up-regulation of cone photoreceptor-specific genes and down-regulation of their rod counterparts by RA treatment of Weri-Rb-1 cells (Li et al., 2003). RA treatment also induces apoptosis and growth arrest in subpopulations of Weri-Rb-1 cells (Li et al., 2003), leading to implications for future retinoblastoma chemotherapy with RA.

One of the major drawbacks to the use of RA as a therapeutic agent is its toxicity at pharmacological doses (Eck-Enriquez et al., 2000). To explore if it would be possible to reduce the concentrations of retinoids without affecting their potential clinical efficacy by the combined use of Mlt and RA, we expanded our studies to determine if the addition of Mlt could allow the use of lower doses of RA while still maintaining its apoptotic and growth arrest effects, using hCAR induction as an indicator of retinobastoma cell differentiation. Our results demonstrate that the use of Mlt not only reduces the most effective concentration of RA from 10 μM to 1 μM, but also enhances its hCAR induction activity.

Results from our transfection experiments with both the hCAR and mCAR promoter constructs suggest that the additive/synergistic effects of RA and Mlt on CAR expression take place mainly at the transcriptional level by acting directly on the *cis*-acting elements within the 5'-flanking region of the CAR gene. Previous reports suggest that the actions of Mlt can be mediated either through G-protein coupled membrane receptors (Ebisawa et al., 1994; Reppert et al., 1995; Roth et al., 2001) or via retinoid orphan receptors (RORαs) (Steinhilber et al., 1995; Tini et al., 1995). Global gene expression profiling of RA-treated Weri-Rb-1 cells reveals upregulation of RORα2 by RA (Li et al., 2002). Northern blot analysis confirms a high basal level of RORα2 transcriptional activity, which is further stimulated by RA, in Weri-Rb-1 cells. It therefore points to the possibility that Mlt might bind to and activate the nuclear RORα2 receptor, which can cross-talk with the RAR/RXR pathway (Tini et al., 1995). Indeed, several potential RORα-responsive elements are present in both the hCAR and mCAR 5'flanking regions (data not shown), in addition to the potential RA-responsive elements (Li et al., 2002). Alternatively, Mlt may act through its G protein-coupled membrane receptors to potentiate the ability of RA to transactivate its receptors (Eck-Enriquez et al., 2000).

5. ACKNOWLEDGEMENTS

We wish to acknowledge Mary D. Allen for her continued generous support for our vision research. CMC is the Mary D. Allen Chair for Vision Research, Doheny Eye Institute. These studies were also supported, in part, from Grants EY00395 (CM Craft and RN Lolley), EY03042 Core Vision Research Center grant (Doheny Eye Institute), Cell and Neurobiology Microscopy Core and Neurogenetic Analysis Core, created through the Howard Hughes Medical Institute Research Resources Grant (CMC).

6. REFERENCES

Binkley, S.A., Riebman, J.B. and Reilly, K.B. 1978. The pineal gland: A biological clock in vitro. *Science*, **202,** 1198-20.

Blask, D.E., Pelletier, D.B., Hill, S.M., Lemus-Wilson, A., Grosso, D.S., Wilson, S.T. and Wise, M.E. 1991. Pineal melatonin inhibition of tumor promotion in the n-nitroso-n- methylurea model of mammary carcinogenesis: Potential involvement of antiestrogenic mechanisms in vivo. *J Cancer Res Clin Oncol*, **117,** 526-32.

Cos, S. and Sanchez-Barcelo, E.J. 2000. Melatonin, experimental basis for a possible application in breast cancer prevention and treatment. *Histol Histopathol*, **15,** 637-47.

Craft, C.M. 1993. Molecular biology of the pineal gland: Melatonin synthesizing enzymes. In *Melatonin biosynthesis, physiological effects and clinical applications*, H.S.Yu and R.J.Reiter (eds) pp. 17-38. CRC Press: Boca Raton.

Craft, C.M., Murage, J., Brown, B. and Zhan-Poe, X. 1999. Bovine arylalkylamine n-acetyltransferase activity correlated with mrna expression in pineal and retina. *Brain Res Mol Brain Res*, **65,** 44-51.

Craft, C.M., Whitmore, D.H. and Donoso, L.A. 1990. Differential expression of mrna and protein encoding retinal and pineal s-antigen during the light/dark cycle. *J Neurochem*, **55,** 1461-73.

Craft, C.M., Xu, J., Slepak, V.Z., Zhan-Poe, X., Zhu, X., Brown, B. and Lolley, R.N. 1998. Phlps and phlops in the phosducin family of g beta gamma binding proteins. *Biochemistry*, **37,** 15758-72.

Czeczuga-Semeniuk, E., Wolczynski, S., Anchim, T., Dzieciol, J., Dabrowska, M. and Pietruczuk, M. 2002. Effect of melatonin and all-trans retinoic acid on the proliferation and induction of the apoptotic pathway in the culture of human breast cancer cell line mcf-7. *Pol J Pathol*, **53,** 59-65.

Czeczuga-Semeniuk, E., Wolczynski, S., Dzieciol, J., Dabrowska, M., Anchim, T. and Tomaszewska, I. 2001. 13-cis retinoic acid and all-trans retinoic acid in the regulation of the proliferation and survival of human breast cancer cell line mcf-7. *Cell Mol Biol Lett*, **6,** 925-39.

Deguchi, T. 1979. Circadian rhythm of serotonin n-acetyltransferase activity in organ culture of chicken pineal gland. *Science*, **203,** 1245-7.

Dokmanovic, M., Chang, B.D., Fang, J. and Roninson, I.B. 2002. Retinoid-induced growth arrest of breast carcinoma cells involves co- activation of multiple growth-inhibitory genes. *Cancer Biol Ther*, **1,** 24-7.

Ebisawa, T., Karne, S., Lerner, M.R. and Reppert, S.M. 1994. Expression cloning of a high-affinity melatonin receptor from xenopus dermal melanophores. *Proc Natl Acad Sci U S A*, **91,** 6133-7.

Eck-Enriquez, K., Kiefer, T.L., Spriggs, L.L. and Hill, S.M. 2000. Pathways through which a regimen of melatonin and retinoic acid induces apoptosis in mcf-7 human breast cancer cells. *Breast Cancer Res Treat*, **61,** 229-39.

Ganguly, S., Coon, S.L. and Klein, D.C. 2002. Control of melatonin synthesis in the mammalian pineal gland: The critical role of serotonin acetylation. *Cell Tissue Res*, **309,** 127-37.

Hyatt, G.A., Schmitt, E.A., Fadool, J.M. and Dowling, J.E. 1996a. Retinoic acid alters photoreceptor development in vivo. *Proc Natl Acad Sci U S A*, **93,** 13298-303.

Hyatt, G.A., Schmitt, E.A., Marsh-Armstrong, N., McCaffery, P., Drager, U.C. and Dowling, J.E. 1996b. Retinoic acid establishes ventral retinal characteristics. *Development*, **122,** 195-204.

Kasal, C.A., Menaker, M. and Perez-Polo, J.R. 1979. Circadian clock in culture: N-acetyltransferase activity of chick pineal glands oscillates in vitro. *Science*, **203,** 656-8.

Klein, D.C. and Moore, R.Y. 1979. Pineal n-acetyltransferase and hydroxyindole-o-methyltransferase: Control by the retinohypothalamic tract and the suprachiasmatic nucleus. *Brain Res*, **174,** 245-62.

Li, A., Zhu, X. and Craft, C.M. 2001. Gene regulation of human cone arrestin. *Invest Ophthalmol Vis Sci.*, **42,** S352.

Li, A., Zhu, X. and Craft, C.M. 2002. Retinoic acid upregulates cone arrestin expression in retinoblastoma cells through a cis element in the distal promoter region. *Invest Ophthalmol Vis Sci*, **43,** 1375-83.

Li, A., Zhu, X. and Craft, C.M. 2003. Gene expression networks underlying retinoic acid-induced differentiation of human retinoblastoma cells. *Invest Ophthalmol Vis Sci.* In press.

Marsh-Armstrong, N., McCaffery, P., Gilbert, W., Dowling, J.E. and Drager, U.C. 1994. Retinoic acid is necessary for development of the ventral retina in zebrafish. *Proc Natl Acad Sci U S A*, **91,** 7286-90.

McCaffery, P., Lee, M.O., Wagner, M.A., Sladek, N.E. and Drager, U.C. 1992. Asymmetrical retinoic acid synthesis in the dorsoventral axis of the retina. *Development*, **115,** 371-82.

McCaffery, P., Tempst, P., Lara, G. and Drager, U.C. 1991. Aldehyde dehydrogenase is a positional marker in the retina. *Development*, **112,** 693-702.

Nowfar, S., Teplitzky, S.R., Melancon, K., Kiefer, T.L., Cheng, Q., Dwived, P.D., Bischoff, E.D., Moro, K., Anderson, M.B., Dai, J., Lai, L., Yuan, L. and Hill, S.M. 2002. Tumor prevention by 9-cis-retinoic acid in the n-nitroso-n-methylurea model of mammary carcinogenesis is potentiated by the pineal hormone melatonin. *Breast Cancer Res Treat*, **72,** 33-43.

Reppert, S.M., Godson, C., Mahle, C.D., Weaver, D.R., Slaugenhaupt, S.A. and Gusella, J.F. 1995. Molecular characterization of a second melatonin receptor expressed in human retina and brain: The mel1b melatonin receptor. *Proc Natl Acad Sci U S A*, **92,** 8734-8.

Roth, J.A., Rosenblatt, T., Lis, A. and Bucelli, R. 2001. Melatonin-induced suppression of pc12 cell growth is mediated by its gi coupled transmembrane receptors. *Brain Res*, **919,** 139-46.

Steinhilber, D., Brungs, M., Werz, O., Wiesenberg, I., Danielsson, C., Kahlen, J.P., Nayeri, S., Schrader, M. and Carlberg, C. 1995. The nuclear receptor for melatonin represses 5-lipoxygenase gene expression in human b lymphocytes. *J Biol Chem*, **270,** 7037-40.

Tini, M., Fraser, R.A. and Giguere, V. 1995. Functional interactions between retinoic acid receptor-related orphan nuclear receptor (ror alpha) and the retinoic acid receptors in the regulation of the gamma f-crystallin promoter. *J Biol Chem*, **270,** 20156-61.

Wiechmann, A.F. 2002. Regulation of gene expression by melatonin: A microarray survey of the rat retina. *J Pineal Res*, **33,** 178-85.

Zhang, Y., Li, A., Zhu, X., Wong, C.H., Brown, B. and Craft, C.M. 2001. Cone arrestin expression and induction in retinoblastoma cells. In *Proceceedings of the ninth international symposium on retinal degeneration* pp. 309-317. Kluwer Academic/Plenum Publishers: New York.

Zhu, X. and Craft, C.M. 2000. Modulation of crx transactivation activity by phosducin isoforms. *Mol Cell Biol*, **20,** 5216-26.

Zhu, X., Ma, B., Babu, S., Murage, J., Knox, B.E. and Craft, C.M. 2002. Mouse cone arrestin gene characterization: Promoter targets expression to cone photoreceptors. *FEBS Lett*, **524,** 116-22.

REGULATION OF RETINAL PHOSPHOINOSITIDE 3-KINASE ACTIVITY IN P85α-SUBUNIT KNOCKOUT MICE

Raju V.S. Rajala[1,3], Mark E. McClellan[1,3], John D. Ash[1,3], and Robert E. Anderson[1,2,3]

1. INTRODUCTION

A variety of growth factors and hormones mediate their cellular affects via interactions with cell surface receptors that possess protein kinase activity (Williams, 1989; Ullrich and Schlessinger, 1990). The interaction of most of these ligands with their receptors induces tyrosine kinase activation and autophosphorylation of the receptor, resulting in physical association of the receptors with several cytoplasmic substrates having SH2 domains (Otsu et al., 1991; Skolnik et al., 1991). Phosphoinositide 3-kinase (PI3K) has been identified through its ability to associate with cellular protein kinases, including numerous growth factor receptors and oncogene products (Otsu et al., 1991; Skolnik et al., 1991).

PI3K consists of an ~85 kDa regulatory subunit (p85) and a ~110 kDa catalytic subunit (p110), the latter being responsible for the phosphorylation of phosphoinositides at the D3 position and of serine residues in proteins (Skolnik et al., 1991; Dhand et al., 1994). The p85 subunit contains an SH3 domain capable of binding to proline-rich sequences, a p110 binding domain, and two SH2 domains. Many activated receptors with tyrosine kinase activity interact with an SH2 domain in the regulatory subunit through a phosphorylated

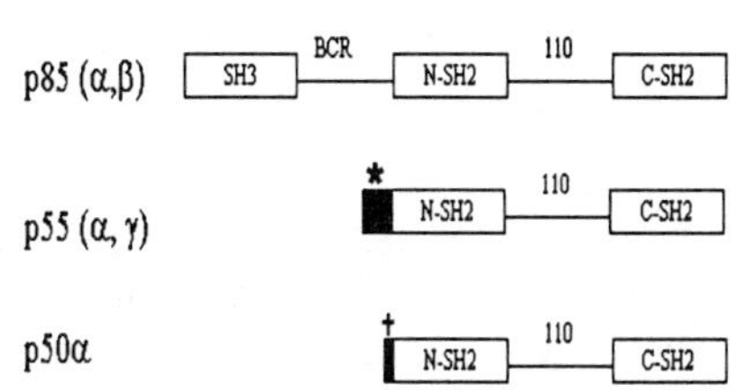

Figure 1. Schematic comparison of the three groups of PI3K regulatory subunits. *Unique 34 amino acid and †unique 6 amino acid extensions.

Dept. of [1]Ophthalmology and [2]Cell Biology, Univ. of Oklahoma Health Sciences Center; [3]Dean A. McGee Eye Inst., Oklahoma City, OK 73104, USA. Corresponding author: R. V. S. Rajala, E-mail: raju-rajala@ouhsc.edu

YXXM motif in the receptors themselves (Songyang et al., 1993), resulting in the activation or recruitment of PI3K (White and Khan, 1994). To date, there are five regulatory subunit isoforms in mammals, including two 85 kDa proteins (p85α, p85β), two 55 kDa proteins (p55α, p55γ), and one 50 kDa protein (p50α) (Inukai et al., 1996; Antonetti et al., 1996; Inukai et al., 1997). The domain organization of various p85 isoforms is given in Figure 1. It was thought that the 50-55 kDa proteins were degradation products of p85α and p85β, so they have thus attracted little attention. However, it has been demonstrated that the p85α gene generates three isoforms of regulatory subunits from PI3K: p50α, p55α, and p85α, which have different PI3K activities in response to insulin stimulation (Inukai et al., 1997).

PI3K activity increases in response to PDGF binding to its receptor, in large part because the p85/p110 complex is translocated from the cytosol to the plasma membrane, by the direct binding of the p85 SH2 domain to tyrosine phosphorylated sites on the receptor (Hu et al., 1993; Klippel et al., 1992). PI3K activity is also regulated by activated receptor-tyrosine kinase substrates such as the insulin receptor substrate-1 (IRS-1) (McGlade et al., 1992). Thus, substantial evidence exists suggesting that the Class I p85/p110 complex of PI3K is a common element of numerous signaling pathways involving a large number of tyrosine kinases (Fruman et al., 1998) and that a variety of stimuli can trigger distinct and specific biological responses in different cell types.

We have reported that bovine rod outer segments (ROS) contain a Class I p85/p110 enzyme complex that is more active in light-adapted retinas *in vitro* (Guo et al., 1997) and can be activated by tyrosine phosphorylation of proteins in these membranes (Guo et al., 2000). We have identified a mechanism for regulation of PI3K activity in bovine photoreceptor cells through tyrosine phosphorylation of the insulin receptor β-subunit (IRβ) *in vitro* (Rajala and Anderson, 2001). The activation is mediated through direct binding of the N-terminal SH2 domain of p85 subunit of PI3K to the tyrosine-phosphorylated C-terminus of the insulin receptor (Rajala and Anderson, 2001). Recently we have shown that PI3K in retina is regulated *in vivo* through light activation of the insulin receptor β-subunit (Rajala et al., 2002).

Mice with disruption of the p85α subunit of PI3K have impaired B cell development (Suzuki et al., 1999). Disruption of mouse gene encoding PI3K adapter subunit p85α and its splice variants p55α and p50α had a lethal phenotype of impaired B cell development and proliferation (Fruman et al., 1999). In this communication, an attempt was made to examine the phenotypic changes of retina and retinal PI3K regulation in p85α knockout mice that retain the ability to express p50 and p55.

2. METHODS

2.1. Materials

P85α subunit knockout and wild-type control mice (Balb/CANTac, 5-6 weeks old) were purchased from Taconic Farms, Inc (Germantown, NY). The animal care strictly confirmed to the ARVO Statement for the Use of Animals in Ophthalmic and Vision Research and the University of Oklahoma Health Sciences Center (OUHSC) Guidelines for Animals in Research. All protocols were reviewed and approved by the Institutional Animal Care and Utilization Committees of the OUHSC and the Dean A. McGee Eye Institute. Polyclonal antisera to the p85α regulatory subunit of PI3K was from Upstate

Biotechnology (Lake Placid, NY). Polyclonal anti-IRβ was from Santa Cruz Biotechnology (Santa Cruz, CA). Mouse anti-opsin was a gift of Dr. Colin Barnstable (Yale University). Anti-Akt and pAkt (Ser 473) were purchased from Cell Signaling Technology (Beverly, MA). [γ^{32}P]ATP was from New England Nuclear (Boston, MA). Echelon Research Laboratories (Salt Lake City, UT) provided D-myo-phosphatidylinositol-4,5-bisphosphate (PI-4,5-P_2). All other reagents were of analytical grade from Sigma (St. Louis, MO).

2.2. Immunoprecipitation (IP)

Retinal lysates were solubilized for 30 min at 4 °C in a lysis buffer containing 1% Triton X-100, 137 mM NaCl, 20 mM Tris-HCl (pH 8.0), 10% glycerol, 1 mM EGTA, 1 mM $MgCl_2$, 1 mM PMSF, 0.2 mM Na_3VO_4, 10 μg/ml leupeptin, and 1 μg/ml aprotinin. Insoluble material was removed by centrifugation at 17,000 x g for 20 min and the solubilized proteins were pre-cleared by incubation with 40 μl protein A-Sepharose for 1 hour at 4 ϒC with mixing. The supernatant was incubated with anti-IRβ (4 μg) overnight at 4 °C and subsequently with 40 μl of protein A-Sepharose for 2 hr at 4 °C. Immune complexes were washed twice with modified solubilization buffer.

2.3. SDS-PAGE and Western Blot Analysis

Retinal proteins were resolved by 10% SDS-PAGE and transferred onto nitrocellulose membranes, and the blots were washed 2 times for 10 min with TTBS [20 mM Tris-HCl (pH 7.4), 100 mM NaCl, and 0.1% Tween-20] and blocked with 10% bovine serum albumin in TTBS overnight at 4 ϒC. Blots were then incubated with anti-p85α (1:4000), anti-Akt (1:1000), anti-pAkt (Ser 473), or anti-opsin (1:20,000) antibodies for 2 hr at room temperature. Following primary antibody incubations, immunoblots were incubated with HRP-linked secondary antibodies (either anti-rabbit, anti-mouse, or anti-Goat IgG) and developed by ECL according to the manufacturer's instructions.

2.4. PI 3-kinase Assay

Enzyme assays were carried out essentially as previously described (Kaplan et al., 1987). Briefly, assays were performed directly on anti-IRβ immunoprecipitates in 50 μl of a reaction mixture containing 0.2 mg/ml PI-4,5-P_2, 50 μM ATP, 0.2 μCi [γ^{32}P]ATP, 5 mM $MgCl_2$, and 10 mM HEPES buffer (pH 7.5). The reactions were carried out for 15 min at room temperature and stopped by the addition of 100 μl of 1 N HCl followed by 200 μl of chloroform/methanol (1/1, v/v). Lipids were extracted and resolved on oxalate-coated TLC plates (silica gel 60) with a solvent system of 2-propanol/2 M acetic acid (65/35, v/v). The plates were coated with 1% (w/v) potassium oxalate in 50% methanol (v/v) and then baked in oven at 100 °C for 1 hr prior to use. TLC plates were exposed to X-ray film overnight at –70 °C and radioactive lipids were scraped and quantified by liquid scintillation counting.

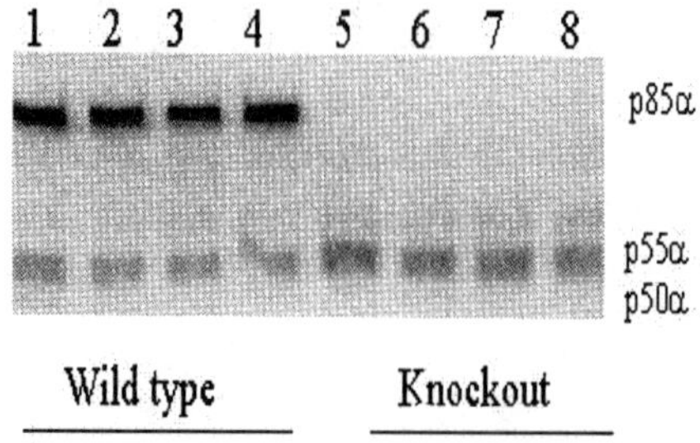

Figure 2. Western blot of wild type (lane 1-4) and p85α knockout (lanes 5-8) mouse retina lysates probed with anti-p85α antibody. This antibody cross reacts with the p55α and p50α alternative splice variants of p85α. Equal amounts of protein were applied to each lane (25 μg).

2.5. Histology

Mice were euthanized with CO_2 and eyes were excised immediately, placed in fixative (4% paraformaldehyde, 2% trichloroacetic acid, 20% isopropyl alcohol, 2% aqueous zinc chloride, and 72% distilled water), and embedded in paraffin. Sections of 5 μm were cut along the vertical meridian through the optic nerve. After staining with hematoxylin and eosin, slides were cover-slipped and visualized on a Nikon Eclipse E800 digital microscope. Images were captured using Adobe Photoshop 5.5.

3. RESULTS

3.1. Expression of p85 in p85α-subunit knockout and wild type mice

Wild type and knockout mouse retina lysates probed with anti-p85 antibody indicated the presence of p85 immunoreactivity only in wild type mouse retinas. However, the immunoreactivity of anti-p85 towards p55 and p50 were found in wild type and p85 knockout mouse retinas (Figure 2), and appeared to be more abundant in the latter.

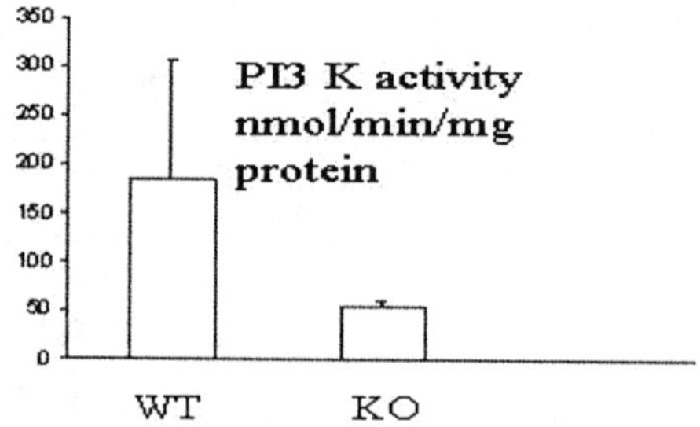

Figure 3. PI3K activities were measured in anti-IRβ IPs of retina lysates from light-adapted wild type (WT) and p85α (KO) knockout mice. The radioactive spots of PI-3,4,5-P_3 were scarped from the TLC plate and counted. Results are mean ± SD (n=3).

3.2. Association of PI3K activity with insulin receptor β-subunit

Immunoprecipitation of PI3K activity with anti-IRβ antibody indicated a 3.3-*fold* decrease in the PI3K activity associated with IRβ in light-adapted p85α-subunit knockout mice compared to light-adapted wild type mice (Figure 3). These results suggest that the reduction of PI3K activity is due the absence of p85 in the retinal homogenates.

3.3. Effect of p85 α-subunit knockout on Akt and pAkt levels in retina

The catalytic activity of PI3K is controlled through the regulatory subunit of p85 and this complex is required for the generation of PI-3,4,5-P_3 and the subsequent activation of Akt. Therefore, we examined the expression of Akt and pAkt in wild type and knockout mouse retina lysates. The results indicate that the levels of Akt and pAkt were not different in wild type and knockout mouse retinas (Figure 4). Opsin expression was also examined in wild type and knockout mouse retina lysates (Figure 4) and found to be same in both groups.

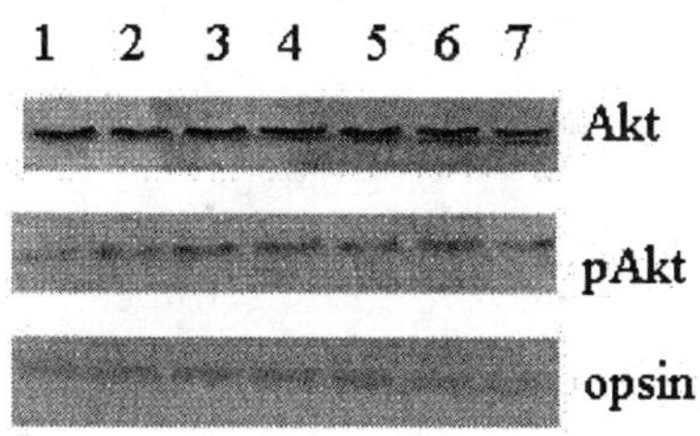

Figure 4. Wild type (lanes 1-3) and knockout (lanes 4-7) mouse retina lysates probed with anti-Akt, anti-pAkt, or anti-opsin antibodies.

3.4. Effect of p85 α-subunit knockout on retina morphology

No gross differences were observed in the morphology of wild type and knockout mice. The photoreceptor cells were preserved in the knockout mice. There was no significant difference in ONL thickness in wild type and p85 α-subunit knockout mice (Figure 5). The ROS of the knockout mice appeared to be shorter than those of the wild type.

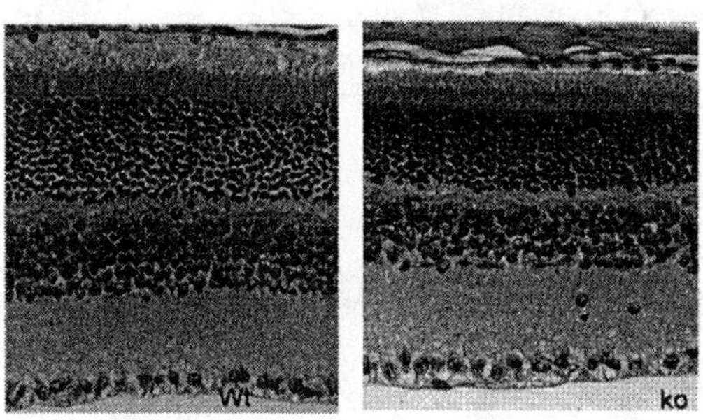

Figure 5. Representative micrographs of retina from wild type and p85α knockout mice.

4. DISCUSSION

We did not observe gross phenotype changes in retinas of the knockout mice, although there was decreased PI3K activity in the anti-IRβ immunoprecipitates of the knockout animals. The absence of a lethal retinal phenotype could be due to the complementation of p55α and p50α for the p85α disruption. The regulation of retinal PI3K is mediated through the N-terminal SH2 domain in the p85 regulatory subunit (Rajala and Anderson, 2001), which is retained in the alternative splice variants p55α and p50α (Inukai et al., 1996). Disruption of the mouse gene encoding PI3K adapter subunit p85α and its splice variants p55α and p50α had a lethal phenotype of impaired B cell development and proliferation (Fruman et al.,1999). Since global knockout of p85α and its alternative splice variants is lethal, these mice were not available for study. To address this issue, we have developed mice expressing Cre recombinase under the control of the rod opsin promoter to allow conditional knockout of the PI3K gene in rod photoreceptor cells (Le and Anderson, unpublished results).

Recently we reported the light-induced tyrosine phosphorylation of IRβ in the regulation of PI3K in retina (Rajala et al., 2002). Light-dependent activation of PI3K is unique to photoreceptor cells. The functional consequence of light-dependent activation of the IRβ and PI3K in photoreceptor cells is currently not known. Activation of the IR signaling pathway has been shown to have complex physiological roles in a variety of cell types (Saltiel and Pessin, 2002). In rods, IRβ activity is regulated by light, which suggests its function could be related to rod specific activities that are driven by light such as shedding of photoreceptor tips, biogenesis of new ROS membranes through addition of newly synthesized membranes at the base of the ROS, or light adaptation. Alternatively, it is possible that light-induced PI3K activity may mediate an innate self-protection mechanism. In some neuronal cell types, such as cerebellar granular neurons (D'Mello et al., 1997) and PC-12 cells (Spear et al., 1997), receptor activation of PI3K has been shown to protect these cells from stress-induced neurodegeneration. It has been well established that PI3K/pAkt is crucial in mediating cell survival. Bright light stimulation can result in the death of rod and cone photoreceptor cells (Noell, 1980). Activation of the PI3K/pAkt pathway by relatively modest light levels may serve to prevent death of photoreceptors. Consistent with this hypothesis, we have recently observed that biotinylated-PIP_3 (phosphatidylinositol-3,4,5-triphosphate) could pull down Akt from bovine rod outer segments (unpublished observation), suggesting that light-induced PI3K activity could regulate downstream effector molecules. Understanding the regulation of PI3K in photoreceptor cells will greatly facilitate studies on the regulation of downstream targets that may reveal a role for PI3K in protection from retinal degeneration. Whether PI3K activation protects the retina from stress is currently under study in our laboratory.

5. ACKNOWLEDGEMENTS

Raju V.S. Rajala is a recipient of RD2002 Travel Award from the National Eye Institute. This work was supported by grants from the National Institutes of Health (EY00871, EY04149, EY12190, and RR17703), Research to Prevent Blindness Inc., New York, NY; The Foundation Fighting Blindness, Baltimore, MD; Samuel Roberts

Nobel Foundation, Inc., Ardmore, OK; and Presbyterian Health Foundation, Oklahoma City, OK; The Oklahoma Center for Advancement of Science and Technology, Oklahoma City, OK; and the Knights Templar Eye Foundation, Chicago, IL.

6. REFERENCES

Antonetti DA, Algenstaedt P, Kahn CR., 1996, Insulin receptor substrate 1 binds two novel splice variants of the regulatory subunit of phosphatidylinositol 3-kinase in muscle and brain. *Mol Cell Biol.*, **16**: 2195-2203.

D'Mello RD, Borodez K, Solto SP., 1997, Insulin like growth factor and potassium depolarization maintain neuronal survival by distinct pathways: possible involvement of PI 3-kinase in IGF-1 signaling. *J Neurosci.*, **17**:1548-1560.

Dhand R, Hiles I, Panayotou G, Roche S, Fry MJ, Gout I, Totty NF, Truong O, Vicendo P, Yonezawa K, Kasuga M, Courtneidge SA, Waterfield MD., 1994, PI3-kinase is a dual specificity enzyme: autoregulation by an intrinsic protein-serine kinase activity. *EMBO J.*, **13**: 522-533.

Fruman DA, Meyers RE, Cantley LC., 1998, Phosphoinositide kinases. *Ann Rev Biochem.*, **67**:481-507.

Fruman DA, Snapper SB, Yballe CM, Davidson L, Yu JY, Alt FW, Cantley LC., 1999, Impaired B cell development and proliferation in absence of phosphoinositide 3-kinase p85α. *Science.*, **283**:393-397.

Guo XX, Ghalayini AJ, Chen HM, Anderson RE., 1997, Phosphatidylinositol 3-kinase in bovine photoreceptor rod outer segments. *Invest Ophthalmol Vis Sci.*, **38**: 1873-1882.

Guo XX, Huang Z, Bell MW, Chen H, Anderson RE., 2000, Tyrosine phosphorylation is involved in phosphatidylinositol 3-kinase activation in bovine rod outer segments. *Mol Vis.*, **6**:216-221.

Hu P, Mondino A, Skolnik EY, Schlessinger J., 1993, Cloning of a novel, ubiquitously expressed human phosphatidylinositol 3-kinase and identification of its binding site on p85. *Mol Cell Biol.*, **13**:7677-7688.

Inukai K, Anai M, Van Breda E, Hosaka T, Katagiri H, Funaki M, Fukushima Y, Ogihara T, Yazaki Y, Kikuchi, Oka Y, Asano T., 1996, A novel 55-kDa regulatory subunit for phosphatidylinositol 3-kinase structurally similar to p55PIK is generated by alternative splicing of the p85α gene. *J. Biol Chem.*, **271**:5317-5320.

Inukai K, Funaki M, Ogihara T, Katagiri H, Kanda A, Anai M, Fukushima Y, Hosaka T, Suzuki M, Shin BC, Takata K, Yazaki Y, Kikuchi M, Oka Y, Asano T., 1997, p85alpha gene generates three isoforms of regulatory subunit for phosphatidylinositol 3-kinase (PI 3-Kinase), p50α, p55α, and p85α with different PI 3-kinase activity elevating responses to insulin. *J Biol. Chem.*, **272**:7873-7882.

Kaplan DR, Whitman M, Schaffhausen B, Pallas DC, White M. Cantley LC, Roberts TM., 1987, Common elements in growth factor stimulation and oncogenic transformation: 85 kd phosphoprotein and phosphatidylinositol kinase activity. *Cell.*, **50**: 1021-1029.

Klippel A, Escobedo JA, Fantl WJ, Williams LT., 1992, The C-terminal SH2 domain of p85 accounts for the high affinity and specificity of the binding of phosphatidylinositol 3-kinase to phosphorylated platelet-derived growth factor-β receptor. *Mol Cell Biol.*, **12**: 1451-1459.

McGlade CJ, Ellis C, Reedijk M, Anderson D, Reith AD, Panayotou G, End P, Bernstein A, Kazlauskas A, Waterfiled MD, Paswon T., 1992, SH2 domains of the p85α subunit of phosphatidylinositol 3-kinase regulate binding to growth factor receptors. *Mol Cell Biol.*, **12**:991-997.

Noell WK., 1980, Possible mechanisms of photoreceptor damage by light in mammalian eyes. *Vis Res.*, **20**:1163-1171.

Otsu M, Hiles I, Gout I, Fry MJ, Ruiz-Larrea F, Panayotou G, Thompson A, Dhand R, Hsuan J, Totty N., 1991, Characterization of two 85 kd proteins that associate with receptor tyrosine kinases, middle-T/pp60c-src complexes, and PI 3-kinase. *Cell.*, **65**:91-104.

Rajala RVS, Anderson RE., 2001, Interaction of the insulin receptor β-subunit with phosphatidylinositol 3-kinase in bovine ROS. *Invest Ophthalmol Vis Sci.*, **42**: 3110-3117.

Rajala RVS, McClellan ME, Ash JD, Anderson RE., 2002, In vivo regulation of phosphoinositide 3-kinase in retina through light-induced tyrosine phosphorylation of the insulin receptor β-subunit. *J Biol. Chem.*, **277**:43319-43326.

Saltiel AR, Pessin JE., 2002, Insulin signaling pathways in time and space. *Trends Cell Biol.*, **12**:65-71.

Skolnik, EY, Margolis B, Mohammadi M, Lowenstein E, Fischer R, Drepps A, Ullrich, A, Schlessinger J., 1991, Cloning of PI3K-associated p85 utilizing a novel method for expression/cloning of target proteins for receptor tyrosine kinases. *Cell.*, **65**: 83-90.

Songyang Z, Shoelson SE, Chaudhuri M, Gish G, Pawson T, Haser WG, King F, Robserts T, Ratnofsky S, Lechleider RJ, Neel BG, Birge RB, Fajardo JE, Chou MM, Hanafusa H, Schaffhausen B, Cantley LC., 1993, SH2 domains recognize specific phosphopeptide sequences. *Cell.*, **72**: 767-778.

Spear N, Estevez AG, Barbeito L, Beckman JS, Johnson GV., 1997, Nerve growth factor protects PC12 cells against peroxynitrite-induced apoptosis via a mechanism dependent on phosphatidylinositol 3-kinase. *J Neurochem.*, **69**:53-59.

Suzuki H, Terauchi Y, Fujiwara M, Aizawa S, Yazaki Y, Kadowaki T, Koyasu S., 1999, Xid-like immunodeficiency in mice with disruption of the p85α subunit of phosphoinositide 3-kinase. *Science.*, **283**:390-392.

Ullrich A, Schlessinger J., 1990, Signal transduction by receptors with tyrosine kinase activity. *Cell.*, **61**:203-212.

White MF, Khan CR., 1994, The insulin signaling system. *J Biol Chem.*, **269**: 1-4.

Williams LT.,1989, Signal transduction by the platelet-derived growth factor receptor. *Science.*, **243**:1564-1570.

PERIPHERIN/*RDS* IN SKATE RETINA

Muna I. Naash[1], Xi-Qin Ding[1], Chibo Li[2], John O'Brien[3], and Muayyad R. Al-Ubaidi[1]

1. INTRODUCTION

Peripherin/*rds*, a transmembrane glycoprotein, has been localized to the rim region of mature disks and the basal region adjacent to the cilia of rod and cone outer segments where disk morphogenesis occurs (Arikawa *et al.*, 1992). The importance of peripherin/*rds* to photoreceptor outer segment structure is inferred by its involvement in the *retinal degeneration slow* (*rds*) mutant mouse line (Van Nie *et al.*, 1978). Outer segment membrane biogenesis does occur in the homozygous *rds* mutant, but the membrane is unable to fold into the proper disk structure, suggesting an important role for peripherin/*rds* in the assembly, orientation, and physical stability of the disks (Travis and Bok, 1993; Molday, 1994). The peripherin/*rds* cDNA has been isolated and sequenced from human (Travis *et al.*, 1991), cow (Connell and Molday, 1990), dog (Moghrabi *et al.*, 1995), cat (Gorin *et al.*, 1993), rat (Begy and Bridges, 1990), mouse (Travis *et al.*, 1989), chicken (Weng *et al.*, 1998), and *Xenopus* (Kedzierski *et al.*, 1996) and found to code for proteins containing 346 to 364 residues, depending on the species. Studies have shown a noncovalent association between peripherin/*rds* and its homolog ROM-1 to form homo and heterotetramer and higher oligomers (Goldberg and Molday, 1996).

Interest in peripherin/*rds* has increased since the discovery of its association with different forms of human retinal diseases. Over 70 different pathogenic mutations have been identified that are associated with retinitis pigmentosa (RP) and different forms of macular dystrophies (Kajiwara *et al.*, 1991 and 1993; 1993; Weleber *et al.*, 1993; Wells *et al.*, 1993; Lam *et al.*, 1995; Richards and Creel, 1995; Fishman, *et al.*, 1997; Grover *et al.*, 2002). These mutations include base substitutions causing missense mutations or premature termination, and in-frame insertion/deletion mutations that change the reading frame. The majority of these mutations are located in the large intradiscal loop, emphasizing the important role played by this region in the function of peripherin/*rds*. In

[1]Cell Biology, University of Oklahoma Health Sciences Center, 940 Stanton L. Young Blvd., Oklahoma City, OK 73104, USA; [2]Ophthalmology and Visual Sciences, University of Illinois at Chicago, Chicago, IL 60612, USA; [3]Ophthalmology & Visual Science, University of Texas School of Medicine, 6431 Fannin MSB 7.024, Houston, TX 77030, USA. Corresponding author: M. I. Naash, E-mail: muna-naash@ouhsc.edu

the present study, we analyzed and compared peripherin/rds from the skate retina to its homologs and orthologs.

2. MATERIALS AND METHODS

2.1. Light microscopy and immunocytochemistry

Skates (*Raja erinacea*) were netted in the deep waters of Vineyard Sound and maintained at ~14 °C in opaque, aerated seawater tanks. Eyes were enucleated from animals anesthetized with 0.1% MS 222 (3-aminobenzoic acid ethyl ester, tricaine). The eyes were hemisected, the vitreous humor was removed with Kimwipes, and the globe was fixed for 1 hr in 0.1 M phosphate buffered saline (pH 7.4) (PBS) containing 4% formaldehyde. The anterior segment was then discarded, and the posterior half of the globe was placed in cold fixative for an additional 12 hr. After multiple rinses in cold PBS, the tissues were incubated in 10% sucrose in PBS overnight followed by cryoprotection with 30% sucrose in the same buffer. The eyecups were then mounted in Tissue-Tek O.C.T. (Miles Laboratories, Naperville, IL), sectioned on a cryostat (-20 °C) at 6 μm, and collected on gelatin-coated slides. The sections were rinsed twice in PBS at room temperature, immersed in freshly prepared pre-incubation medium (0.1% Triton X-100 and 10% normal serum in PBS) for 30 minutes at room temperature to saturate non-specific binding sites and then washed three times with PBS for 5 minutes each. Sections were incubated with various concentrations (1:2, 1:10, 1:50 and 1:100) of a monoclonal antibody prepared against cow peripherin/*rds* (Per-5H2), kindly provided by Dr. Robert Molday (University of British Columbia, Canada), and allowed to incubate overnight at room temperature. Sections were then rinsed three times for 5 minutes each with PBS and incubated with FITC-labeled secondary antibodies (1:100) for 30 minutes at room temperature. The sections were washed three times (10 min each) in PBS, cover-slipped with Vectashield (Vector Laboratory Inc., Burlingame, CA), and photographed in a Zeiss universal microscope equipped for incident-light fluorescence. Control preparations (omission of the primary antibody) gave no fluorescent signals.

2.2. RNA blot analysis

Total RNA was isolated from skate retina, brain, heart, stomach, kidney, spleen, liver, skeletal muscle, skin, and lens. A total of 8 μg RNA from each source was run on a 1% agarose gel containing 6.7% formaldehyde. The RNA was transferred to membrane and probed with a ^{32}P-labeled skate peripherin/*rds* cDNA clone. The probe was labeled by nick translation, and hybridized to the blot under moderate stringency conditions. The hybridization solution contained 50% formamide, 5 × SSC, 1 × Denhart's solution, 0.2% SDS, and 0.25 mg/ml salmon sperm DNA. Hybridization was performed at 42 °C overnight.

2.3. Immunoblot Analysis

Polyclonal antibody against residues 336-351 from the carboxyl terminus of mouse ROM-1 (rom-1-C-term Ab) was generated by immunizing rabbits with the corresponding peptides coupled to keyhole limpet hemocyanin. Dissected retinas were homogenized on ice in solubilization buffer containing 50 mM Tris-HCl, pH 7.5, 100 mM NaCl, 5 mM EDTA, 1% Triton X-100, 0.05% SDS, 2.5% glycerol, and 1.0 mM phenylmethylsulfonyl fluoride. After solubilization at 4 °C for 1 hr the homogenates were centrifuged at 100,000 g for 30 min, and supernatants were collected for protein concentration and Western blot analysis. For Western blotting, retinal extracts were mixed with 2x Laemmli buffer containing 100 mM Tris-HCl, 4% SDS, 200 mM DTT and 20% glycerol, separated on 10% SDS PAGE, and then transferred onto PVDF membrane in transfer buffer containing 25 mM Tris-base, 190 mM glycine, 20% methanol and 0.03% SDS. The blots were preblocked with 10% nonfat dry milk in Tris-based saline buffer containing 0.1% Tween 20 (TBST) at 4 °C for overnight before incubation with rom-1-C-term Ab at a dilution ratio of 1:1000 for 2 hr at room temperature. Blots were then washed for several times with TBST and incubated with the secondary antibody at a dilution ratio of 1:25,000 for 1 hr at room temperature. Super Signal chemiluminescent substrate (Pierce, Rockford, IL) was used for detection.

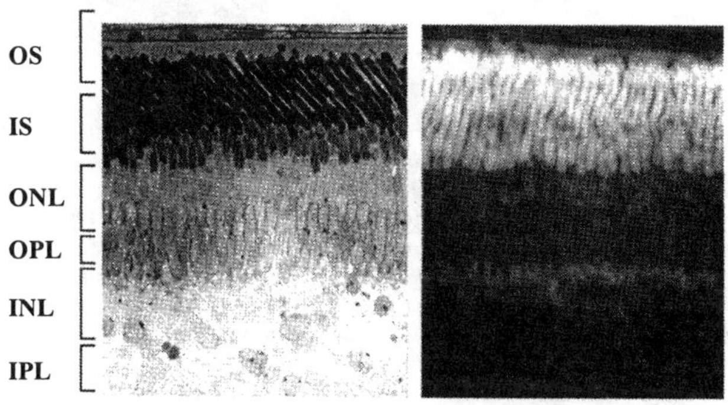

Figure 1. Localization of peripherin/*rds* in the skate retina. (A) A light micrograph of a skate retinal section stained with methylene blue. (B) Immunofluorescent labeling of a frozen section from a skate retina with a monoclonal antibody to bovine peripherin/*rds*. Immunofluorescence was confined almost exclusively to the outer segments and inner segments of the rod photoreceptors. The fluorescent intensity appeared to be greatest in the outer segments. Note also the less intense fluorescence in the outer plexiform layer. OS: outer segment; IS: inner segment; ONL: outer nuclear layer; OPL: outer plexiform layer; INL: inner nuclear layer; IPL: inner plexiform layer.

3. RESULTS

3.1. Localization of peripherin/*rds* in the skate retina

An anti-bovine peripherin/*rds* monoclonal antibody was used to localize peripherin/*rds* in the skate retina (Fig. 1). Photoreceptor outer and inner segments, darkly stained with methylene blue in Fig. 1A, were strongly labeled by the antibody (Fig. 1B). The immunofluorescence was confined almost exclusively to the outer and inner segments, although a faint signal appeared at the photoreceptor synaptic terminals (Fig. 1B). This pattern of immunoreactivity did not significantly differ from that observed in other species (Molday *et al.*, 1987; Arikawa *et al.*, 1992). These results show that, as in other vertebrates, peripherin/*rds* is a component of skate photoreceptors. The immunoreactivity seen in the skate inner segment is rather unusual when compared to mammalian retinas.

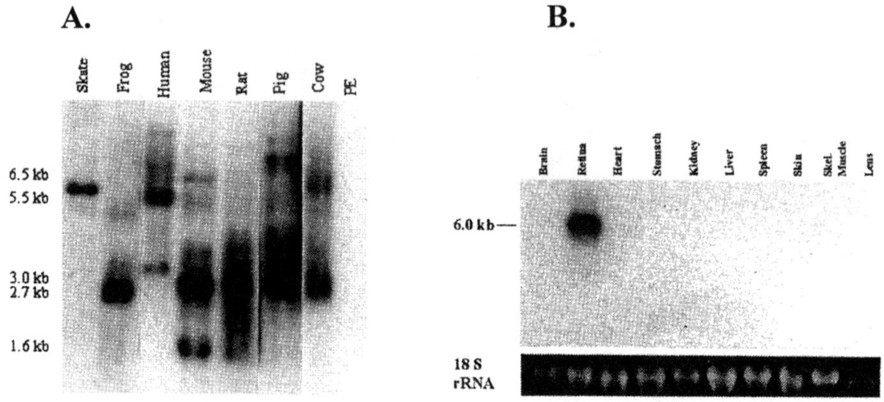

Figure 2 Conservation of peripherin/*rds* transcripts. (A) Northern blot analysis of total RNA isolated from skate, frog, human, mouse, rat, pig, and cow retinas. Cow pigment epithelium was used as a control. The blot was probed with the full-length bovine cDNA clone. More than one transcript was observed for all species studied, with the exception of the skate, for which only a single band of ~ 6 kb was found. No signal was detected in RNA isolated from cow pigment epithelium. The molecular weights of RNA size standards are indicated to the right in kb. (B) Northern blot analysis of total RNA isolated from various skate tissues probed with the skate peripherin/*rds* cDNA clone (Upper panel). A single transcript was detected only in the retinal RNA sample. Ethidium bromide staining of the gel reveals approximately equal loading of RNA samples in all lanes (Lower panel). The 18S rRNA band is shown.

3.2. Conservation of the skate peripherin/rds gene

To determine the screening conditions for the isolation of skate peripherin/*rds*, an RNA blot containing total retinal RNA from skate as well as from frog, human, mouse, rat, pig, and cow was probed with a cow peripherin/*rds* cDNA clone. As shown in Fig. 2 (A), the retinas of all species tested expressed more than one transcript, with the exception of the skate, which gave rise to a single message ~ 6 kb in size. Five transcripts were detected in mouse retinal RNA samples (1.6, 2.7, 4.0, 5.5, and ~ 6.5 kb), four in human (3.0, 5.8, 7.0 and ~ 8.0 kb), two in rat (2.7 and 4.2 kb), two in cow (2.7 and 6.5 kb), four in pig (2.7, 4.5, 7.0 and ~ 7.5 kb), and two in frog (2.5 and 5 kb). No transcript was observed in RNA isolated from cow pigment epithelium. The size of the major transcript obtained from mouse, rat, pig, and cow retinal RNA was similar, suggesting that this 2.7 kb transcript is a common message for peripherin/*rds*. However, the major transcript from human and skate retinal RNA was ~ 6 kb. Fig. 2 (B) shows Northern blot analysis of total RNA isolated from skate retina, brain, heart, stomach, kidney, spleen, liver, skeletal muscle, skin, and lens. Consistent with peripherin/*rds* being a retina-specific protein, the transcript was detected only in retinal RNA.

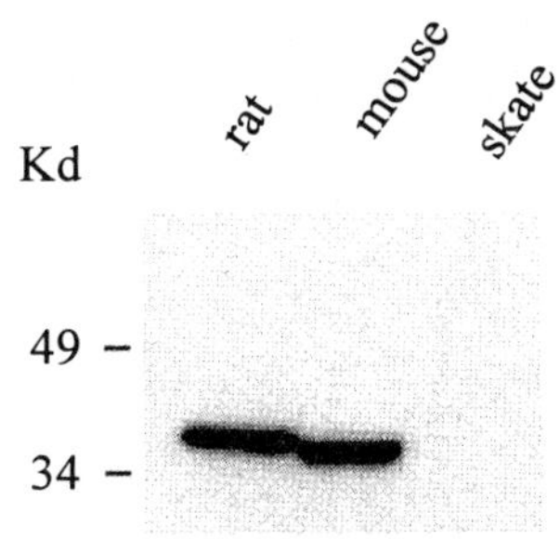

Figure 3. Lack of expression of ROM-1 in the skate retina. Western blot analysis of ROM-1 in skate, mouse, and rat retinal extracts was performed using anti-ROM-1 C-terminus antibody as described under "Material and Methods". ROM-1 signal was detected in mouse and rat retina, but not in skate retina. Note that rom-1 detected in rat retina migrated at a position slightly higher than the mouse homolog.

3.3. The absence of rom-1 in the skate retina

Peripherin/*rds* is known to associate non-covalently with ROM-1 to form heterotetramers and higher oligomers. To better understand the involvement of the skate peripherin/*rds* in the morphogenesis of the skate rim region, we looked for the presence of *rom-1* or peripherin/*rds* like sequences in the skate retina. We first screened the skate retinal cDNA library with the full-length bovine *rom-1* cDNA probe. With several screening attempts we failed to detect any positive clones. Similarly, it was reported that *rom-1* is not present in *Xenopus* or the chicken (Kedzierski *et al.*, 1996; Weng *et al.*, 1998). However, both of these species have peripherin/*rds*-like proteins in addition to peripherin/*rds*. Furthermore, Western blot analysis of skate retinal extracts immunoreacted with anti-mouse ROM-1 specific to the C-terminus region could not detect ROM-1 signal in the skate retina, while this protein was clearly detected in mouse and rat retinas (Fig. 3). We recognize that the data shown in Fig.3 on its own does not conclude the absence of ROM-1 in the skate retina. However, in combination with the absence of positive clones from the multiple low stringency screening we have performed on the skate cDNA library in search for *rom-1* and rds-like clones, we can conclude that the skate retina very likely lacks ROM-1 and rds-like proteins. Of note, ROM-1 from rat retina migrated at a position slightly higher than that of the mouse (Fig. 3).

4. DISCUSSION

This study was undertaken to examine the evolutionary conservation of peripherin/*rds* among vertebrates. The skate, a member of the class including sharks and rays, is the most primitive vertebrate in which this protein has been studied. Furthermore, the skate retina is an all rod retina in which many of the functions of both rod and cone photoreceptors are subserved by the rods (Dowling and Ripps, 1970; Dowling and Ripps, 1991; Ripps and Dowling, 1991). The outer segment organization of skate photoreceptors is similar to that of other vertebrate rods, and immunocytochemistry with anti-bovine peripherin/*rds* antibody showed strong labeling in the skate outer segments. Therefore, we expect the peripherin/*rds* protein to perform the same function in the skate as it does in mammals.

The high degree of conservation of the peripherin/*rds* group suggests that this protein has a very important function. The large number of mutations in this gene that cause different forms of retinal degeneration attests to this fact. Since the skate does not have rom-1 or other rds-like proteins, we conclude that these proteins are not essential for the morphogenesis and maintenance of outer segment structure in the skate. The divergence of sequence among the ROM-1 and rds-like proteins provides further evidence that their function is not highly conserved. In keeping with this concept is the observation that knockout of *rom-1* in mice results in a far less severe retinal phenotype compared to the peripherin/*rds* knockout (*rds/rds* mice) (Clarke *et al.*, 2000).

5. ACKNOWLEDGMENT

This research was funded by National Eye Institute (NEI) Grant EY-10609 (MIN), Department of Ophthalmology NEI Core grant (EY-01792), the Foundation Fighting Blindness (MIN, MRA). Dr. Naash is a recipient of the Research to Prevent Blindness James S. Adams Scholar award.

6. REFERENCES

Arikawa, K., Molday, L.L., Molday, R.S., and Williams, D.S., 1992, Localization of peripherin/*rds* in the disk membranes of cone and rod photoreceptors: relationship to disk membrane morphogenesis and retinal degeneration, *J.Cell Biol.* **116:**659-667.

Begy, C. and Bridges, C.D.B., 1990, Nucleotide and predicted protein sequence of rat retinal degeneration slow (rds), *Nucl.Acids Res.* **18:**3058-3058.

Clarke, G., Goldberg, A.F., Vidgen, D., Collins, L., Ploder, L., Schwarz, L., Molday, L.L., Rossant, J., Szel, A., Molday, R.S., Birch, D.G., McInnes, R.R., 2000, Rom-1 is required for rod photoreceptor viability and the regulation of disk morphogenesis, *Nat Genet.* 25:67-73.

Connell, G.J. and Molday, R.S., 1990, Molecular cloning, primary structure, and orientation of the vertebrate photoreceptor cell protein peripherin in the rod outer segment disc membrane, *Biochem.* **29:**4691-4698.

Dowling, J.E. and Ripps, H., 1970, Visual adaption in the retina of the skate, *J.Gen.Physiol.* **56**:491-520.

Dowling, J.E. and Ripps, H., 1991, On the duplex nature of the skate retina, *J.Exp.Zool.Supp.* **5**:55-65.

Fishman, G.A., Stone, E.M., Alexander, K.R., Gilbert, L.D., Derlacki, D.J., Butler, N.S., 1997, Serine-27-phenylalanine mutation within the peripherin/rds gene in a family with cone dystrophy, *Ophthalmology*, **104**:299-306

Goldberg, A.F.X. and Molday, R.S., 1996, Subunit composition of the peripherin/rds-rom-1 disk rim complex from rod photoreceptors: hydrodynamic evidence for a tetrameric quaternary structure, *Biochem.* **35**:6144-6149.

Gorin, M.B., Snyder, S., To, A., Narfstrom, K., and Curtis, R., 1993, The cat rds transcript: candidate gene analysis and phylogenetic sequence analysis. *Mamm.Genome* **4**, 544-548.

Grover S, Fishman GA, Stone EM., 2002, Atypical presentation of pattern dystrophy in two families with peripherin/rds mutations. *Ophthalmology*, **109**:1110-7.

Kajiwara, K., Hahn, L.B., Mukai, S., Travis, G.H., Berson, E.L., and Dryja, T.P., 1991, Mutations in the human retinal degeneration slow gene in autosomal dominant retinitis pigmentosa, *Nature* **354**:480-483.

Kajiwara, K., Sandberg, M.A., Berson, E.L., and Dryja, T.P., 1993, A null mutation in the human peripherin/RDS gene in a family with autosomal dominant retinitis punctata albescens, *Nature Genet.* **3**:208-212.

Kedzierski, W., Moghrabi, W.N., Allen, A.C., Jablonski-Stiemke, M.M., Azarian, S.M., Bok, D., and Travis, G.H., 1996, Three homologs of rds/peripherin in *xenopus leavis* photoreceptors that exhibit covalent and non-covalent interactions, *J.Cell Sci.* **109**:2551-2560.

Lam, B.L., Vandenburgh, K., Sheffield, V.C., and Stone, E.M., 1995, Retinitis pigmentosa associated with a dominant mutation in codon 46 of the peripherin/rds gene (arginine-46-stop), *Am.J.Ophthalmol.* **119**:65-71.

Moghrabi, W.N., Kedzierski, W., and Travis, G.H., 1995, Canine homolog and exclusion of retinal degeneration slow (rds) as the gene for early retinal degeneration (erd) in the dog, *Exp.Eye Res.* **61**:641-643.

Molday, R.S., 1994, Peripherin/rds and rom-1: molecular properties and role in photoreceptor cell degeneration, in: *Progress in Retinal and Eye Research*, N.N. Osborne and G.J. Chader, ed., Pergamon Press Ltd., Great Britain, pp. 271-299.

Molday, R.S., Hicks, D., and Molday, L., 1987, Peripherin: a rim-specific membrane protein of rod outer segment discs, *Invest.Ophthalmol.Vis.Sci.* **28**:50-61.

Richards, S.C. and Creel, D.J., 1995, Pattern dystrophy and retinitis pigmentosa caused by a peripherin/rds mutation, *Retina* **15**:68-72.

Ripps, H. and Dowling, J.E., 1991, Structural features and adaptive properties of photoreceptors in the skate retina, *J.Exp.Zool.Supp.* **5**:46-54.

Travis, G.H. and Bok, D., 1993, A molecular characterization of the *retinal degeneration slow* (*rds*) mouse mutation. in: *Retinal Degeneration: Clinical and Laboratory Applications*,. J.G Hollyfield, R.E. Anderson, and M.M LaVail, ed., Plenum Press, New York & London, pp. 219-230.

Travis, G.H., Brennan, M.B., Daneilson, P.E., Kozak, C.A., and Sutcliffe, J.G., 1989, Identification of a photoreceptor-specific mRNA encoded by the gene responsible for retinal degeneration slow (*rds*), *Nature* **338**:70-73.

Travis, G.H., Christerson, L., Danielson, P.E., Klisak, I., Sparkes, R.S., Hahn, L.B., Dryja, T.P., and Sutcliffe, J.G., 1991, The human retinal degeneration slow (rds) gene: chromosome assignment and structure of the mRNA, *Genomics* **10**:733-739.

Van Nie, R., Ivanyi, D., and Demant, P., 1978, A new h-2-linked mutation rds, causing retinal degeneration in the mouse, *Tissue Antigens* **12**:106-108.

Weleber, R., Carr, R.E., Murphey, W., Sheffield, V., and Stone, E., 1993, Phenotypic variation including retinitis pigmentosa, pattern dystrophy, and fundus flavimaculatus in a single family with a deletion of codon 153 or 154 of the peripherin/rds gene, *Arch.Ophthalmol.* **111**:1531-1542.

Wells, J., Wroblewski, J., Keen, J., Inglehearn, C., Jubb, C., Eckstein, A., Jay, M., Arden, G.B., Battacharya, S.S., Fitzke, F., and Bird, A.C., 1993 Mutations in the human retinal degeneration slow (rds) gene can cause either retinitis pigmentosa or macular dystrophy, *Nature Genet.* **3**:213-218.

Weng, J., Belecky-Adams, T., Adler, R., and Travis, G.H., 1998, Identification of two rds/peripherin homologs in the chick retina, *Invest.Ophthalmol.Vis.Sci.* **39**:440-443.

XIAP PROTECTS PHOTORECEPTORS FROM N-METHYL-N-NITROSOUREA-INDUCED RETINAL DEGENERATION

Dino Petrin[1], Adam Baker[1], Jennifer Brousseau[2], Stuart Coupland[2], Peter Liston[1], William W. Hauswirth[3], Robert G. Korneluk[1], Catherine Tsilfidis[2]

1. INTRODUCTION

Retinitis Pigmentosa (RP) is a genetically heterogeneous disease that affects 1 in 3500 people globally (Pagon, 1988). Regardless of the causative mutation, the endpoint is the same; photoreceptor cell death occurs via apoptosis. The mechanisms that cause program cell death (PCD) in photoreceptors remain elusive. Anti-apoptotic strategies have shown promise as a therapeutic approach in the treatment of neurodegenerative diseases in animal models. Since the retina is of neuronal origin, it may be possible to treat RP by blocking the apoptotic cascade in photoreceptors, irrespective of the mutation involved, thereby delaying, if not preventing altogether, progression of the disease.

A number of anti-apoptotic genes have been identified, including the Bcl-2 family, small stress proteins (Mehlen et al., 1996) and the inhibitor of apoptosis (IAP) gene family (Deveraux et al., 2001). The IAPs prevent PCD by binding to and inhibiting initiator and effector caspases, the executioners of cell death (Salvesen, 1999). Of the eight mammalian IAPs identified, the X-linked inhibitor of apoptosis (XIAP) is the most potent. It has three **B**aculoviral **I**nhibitor of Apoptosis **R**epeat (BIR) domains and a C-terminal RING zinc finger (Takahashi et al., 1998). The BIRs and the linker regions between them are responsible for XIAP's anti-apoptotic function by binding directly to and inhibiting initiator caspase-9 (Deveraux et al., 1998) and the effector caspases-3 and 7 (Suzuki et al., 2001). The RING zinc finger confers an E3 ubiquitin ligase activity for XIAP (MacFarlane et al., 2002; Yang & Li, 2000). In animal models of neuro-degeneration, adenoviruses encoding XIAP (Ad-XIAP) have demonstrated not only structural protection but also functional protection as demonstrated by the preservation of

[1]Children's Hospital of Eastern Ontario Research Institute, 401 Smyth Rd., Ottawa, ON, Canada, K1H 8L1; [2]University of Ottawa Eye Institute, 501 Smyth Rd., Ottawa, ON, Canada, K1H 8L6; [3]University of Florida College of Medicine, J. Hillis Miller Health Center, Gainesville, FL 32610-0284, USA. Corresponding author: D. Petrin, E-mail: dino@mgcheo.med.uottawa.ca

Retinal Degenerations: Mechanisms and Experimental Therapy
Edited by LaVail *et al.*, Kluwer Academic/Plenum Publishers, 2003

higher order functions (Eberhardt et al., 2000; Robertson et al., 2000; Xu et al., 1999). Recently, adeno-associated virus (AAV) encoding XIAP has been shown to protect optic nerve axons in a hypertensive rat model of glaucoma (McKinnon et al., 2001).

Since XIAP is able to protect neurons in a variety of neurodegeneration models, it is possible that XIAP could protect the retina's photoreceptors from a variety of cell death triggers. This chapter examines the neuro-protective effects of XIAP in an acute chemical model of retinal degeneration.

In addition to being a potent carcinogen and teratogen, N-methyl-N-nitrosourea (MNU) has been used extensively to study photoreceptor apoptosis (Herrold, 1967; Nakajima et al., 1996; Nambu et al., 1997). MNU is an alkylating agent that methylates DNA directly at the O^6 position of guanine causing O^6 methylguanine (O^6-meGua) lesions to form (Christmann & Kaina, 2000). Subsequently, O^6-meGua pairs with thymine instead of cytosine leading to GC→AT transition mutations in photoreceptor cell DNA that initiate PCD (Yoshizawa et al., 1999).

In this study, we show that XIAP can protect photoreceptor cells both structurally and functionally from MNU-induced degeneration. Since XIAP is able to prevent apoptosis from such an acute trigger, it holds promise as a therapeutic agent for less severe and slower retinal degenerations that are found in retinitis pigmentosa.

2. MATERIAL AND METHODS

2.1. Animals

Male Sprague-Dawley rats were purchased from Charles River Laboratories (Wilmington, MA) and cared for under the guidelines outlined in the ARVO Statement for the Use of Animals in Ophthalmic and Vision Research. Rats were reared on a diurnal cycle at 22°C with 60% relative humidity and had free access to water and food throughout the duration of the experiment.

2.2. Construction of Recombinant AAV Vectors Expressing XIAP or GFP

A Notl fragment containing the entire human XIAP ORF with an N-terminal HA-tag was cloned into the pTRUF2 vector. pTRUF2 contains the mouse opsin minimal promoter (MOP500) which is exclusively expressed in rod photoreceptors. The final construct MOP500-HA-XIAP, and MOP500-GFP were used in the construction of serotype 2 recombinant AAV (rAAV2) as previously described (Flannery et al., 1997). Vectors were packaged, purified, concentrated and titered as described previously (Hauswirth et al., 2000). Titres of the rAAV2 (rAAV-XIAP and rAAV-GFP) were adjusted to $4x10^{12}$ physical particles/mL. Ratios of physical-to-infectious particles were all less than 100. For production of these vectors, a mini-adenovirus helper plasmid, pDG, (Grimm et al., 1998) was used to produce rAAV vectors with no detectable adenovirus or wild-type AAV contamination. rAAV vectors were purified using iodixanol gradient/heparin-affinity chromatography and were found to be greater than 99% pure as determined by silver stained PAGE. Contaminating helper adenovirus and wild-type AAV were below detection levels as assayed by serial dilution cytopathic effect or infectious centre assay respectively.

2.3. Sub-retinal Injections

Rats were anaesthetized by halothane gas inhalation for the duration of the procedure. Rats were sub-retinally injected with rAAV-XIAP or rAAV-GFP using a transcorneal approach as previously described (Timmers et al., 2001). Only the right eye of each rat was injected. A volume of 2µL of virus or BSS (Balanced Salt Solution) with fluorescein as a tracer was injected slowly into the sub-retinal matrix of the rat, inducing a localized retinal detachment. The fluorescein tracer indicated whether the injection procedure was successful. The needle was slowly withdrawn and Maxitrol (Alcon Canada, Mississauga, ON) was applied on the cornea to prevent infection from occurring.

2.4. Experimental Procedure

MNU (Sigma) was dissolved in physiological saline to a final concentration of 5mg/mL immediately prior to use and kept on ice. A dose response to different concentrations of MNU (20-60mg/kg) was performed in wild-type rats which did not receive a sub-retinal injection. For each animal, the average trace at each intensity was scored and latencies and amplitudes were exported to excel worksheets. The b-wave amplitudes were then imported into SigmaPlot 8.0. Here, a Naka-Rushton equation (Velten et al., 2001) was used to fit the response versus luminance:

$$f = (R_{max}*(L^n))/((L^n)+(K^n))$$

where R_{max} is the maximum b-wave amplitude attained, L is the luminance, K is the light intensity where the b-wave amplitude is _ R_{max} and n is related to the slope of the response function. For the gene-therapy study, six weeks after sub-retinal injection of rAAV vectors, rats received a 60mg/kg dose of MNU via intraperitoneal injection.

2.5. Electroretinography

Rats were weighed and dark-adapted overnight prior to ERG recordings. Under safelight conditions, rats were anaesthetized using an 80mg/kg ketamine and 6.4mg/kg xylazine cocktail. The rats were placed on a heated pad to prevent heat loss. Pupils were dilated with 1% tropicamide and anaesthetized using 0.5% proparacaine. DTL micro conductive fibre electrodes were placed on each cornea using a sponge applicator. A saline-moistened gold mini-disc electrode was placed on the tongue as a reference electrode. Gonak (Akorn, Buffalo) was applied to the corneas to keep them hydrated during ERG recordings. A ground needle electrode was placed subcutaneously in the tail. The rat was positioned inside a ganzfeld (GS2000, Nicolet Inc.) stimulator facing the rear of the globe. ERGs were recorded using a Nicolet Pathfinder II with a band pass filter of 0.3 to 300 Hz. ERGs were amplified 50,000 times and twenty ERG traces were averaged over a luminance range of –3.0 to 1.4 log cd.s/m^2. ERGs were recorded pre-treatment and at 24, 48, 72 hours and 1 week post-MNU injection.

2.6. Tissue Fixation and Processing

At each time point, post-ERG, rats were given a lethal dose of sodium pentobarbital and transcardially perfused with 4% paraformaldehyde (PFA). The eyes were scored prior to removal using a white-hot 18-gauge needle for orientation. Eyes were placed in 4% PFA at 4°C overnight. Eyes were washed in PBS, the lenses were dissected out, and the eyecups were placed in 30% sucrose at 4°C until saturated. The eyes were then placed in a 1:1 mixture of 30% sucrose : OCT compound and allowed to equilibriate for 1 h at 4°C. Eyes were embedded and cryosectioned at 16μm. The slides were air-dried for three hours and stored at –20°C with dessicant. Hematoxylin and Eosin staining was carried out using standard procedures.

2.7. mRNA Expression Levels

Three rats were sacrificed 12 weeks post sub-retinal injection of rAAV-XIAP. The retinas were dissected out, processed using Trizol reagent (Gibco) and purified further using a Qiagen RNAeasy column as per the manufacturers instructions. The purified RNA was quantitated and Taqman analysis was carried out on XIAP-injected retinas and uninjected controls with XIAP-specific primers.

2.8. Protein Expression Levels

Six weeks after the rAAV-XIAP injection, three rats that were sub-retinally injected with rAAV-XIAP were euthanized. The retinas were dissected out and placed on dry ice. Each retina was sonicated in 200μL of RIPA buffer (150mM NaCl, 50mM Tris-Cl, pH 7.4, 1mM EDTA, 1% NP-40, 0.25% Na-deoxycholate, 1mM NaF, 1mM Na-vanadate, 1mM PMSF, and 10_g/mL each of aprotinin and leupeptin). Thirty μg of protein from XIAP-injected retinas and uninjected contralateral controls were electrophoresed on a 10% polyacrylamide gel under denaturing conditions and transferred onto PVDF membrane (Immobilon P, Millipore). Blots were probed with antibodies to RIAP3 (rat XIAP) or to the HA-epitope tag (Boehringer-Mannheim).

3. RESULTS

3.1. Statistical Analysis of the Dose Response

Using the Naka-Rushton equation, R_{max} values were obtained for the four animals tested in each dose group (20, 30, 40, 50, 60mg/kg MNU). Using an analysis of variance (ANOVA) test, we found that no significant difference in R_{max} was found between the 72h and 1-week time points, indicating that maximal damage to the retina occurs by 72 hours. In addition, we found that a significant difference in the 72h response begins to occur at the 40mg/kg dose. This suggests that a 40mg/kg dose is sufficient to screen for neuro-protective compounds that can prevent retinal degeneration from occurring.

3.2. XIAP Over-expression in the Rat Retina

The rAAV-XIAP vector contains an N-terminal HA-tagged full-length human XIAP cDNA under the control of the mouse opsin promoter. We performed real-time PCR using Taqman analysis (ABI systems) with primers specific to human XIAP but which also cross-react with endogenous rat IAP. RNA was isolated from retinal tissue at 12 weeks following injection to assess the expression of XIAP mRNA (Fig. 1A). To check

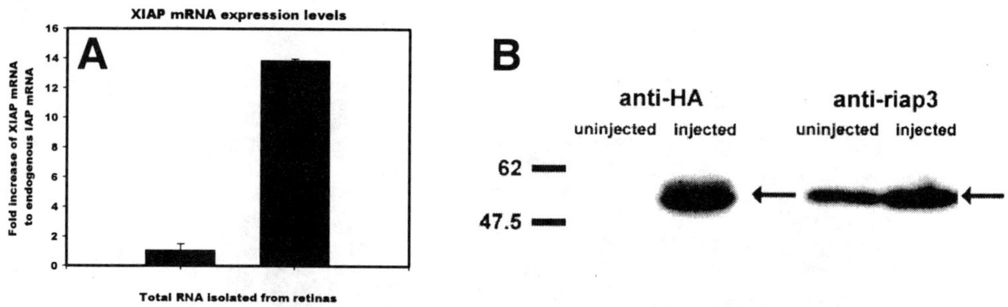

Figure 1. A) Taqman™ analysis of retinal samples at 12 weeks following injection of rAAV-XIAP. Expression of XIAP mRNA is 13 fold greater in the eyes that received rAAV-XIAP than in the uninjected controls. **B)** Western blot analysis of XIAP expression in a rat that received an rAAV-XIAP injection using anti-HA and anti-RIAP3 antibodies.

the integrity of the mRNA, GAPDH was also amplified (data not shown). Expression of XIAP mRNA was found to be at least 10 fold greater in the right eye injected with rAAV-XIAP in relation to endogenous IAP levels detected in the uninjected left eye. Western blot analysis confirmed that XIAP protein was also upregulated (Fig. 1B) and the HA-tagged XIAP protein is only found in the injected eye. Probing the Western blot with RIAP3, which cross-reacts to rat as well as human IAP, clearly shows that the level of IAP is higher in the injected eye in comparison to the uninjected control.

3.3. Structural Protection of Photoreceptors

Hematoxylin and Eosin staining of retinal sections shows that at 24h post-MNU, the retinal morphologies of the injected and uninjected eyes are indistinguishable (Fig. 2C,D). At 48h (Fig. 2E,F) and 72h (Fig. 2G,H) there is wide spread destruction of the photoreceptors in the uninjected eye whereas the rAAV-XIAP injected eye shows preservation of the outer nuclear layer (ONL). At one week, the photoreceptors of the control eye have been essentially wiped out. The XIAP treated retina still shows structural integrity of the photoreceptors in the area of injection (Fig. 2I,J) in 4 out of 6 animals.

3.4. Functional Protection of Photoreceptors from MNU-Induced Retinal Degeneration

ERG recordings were taken to ascertain if functional protection correlated with histological rescue. At 24h post-MNU (Fig. 3B), both the XIAP-treated and the control eye show a reduction in both the a- and b-wave amplitudes. This trend continues at the 48 and 72h time points (data not shown). However, at 1-week post-MNU, the left eye shows a flat-lined ERG whereas the XIAP-treated eye shows a reduced but clearly present ERG response in 2 out of the 4 animals which showed histological protection (Fig. 3C). The 1-week ERGs from the XIAP treated eyes have retained between 5-15% of the pre-MNU baseline b-wave amplitudes.

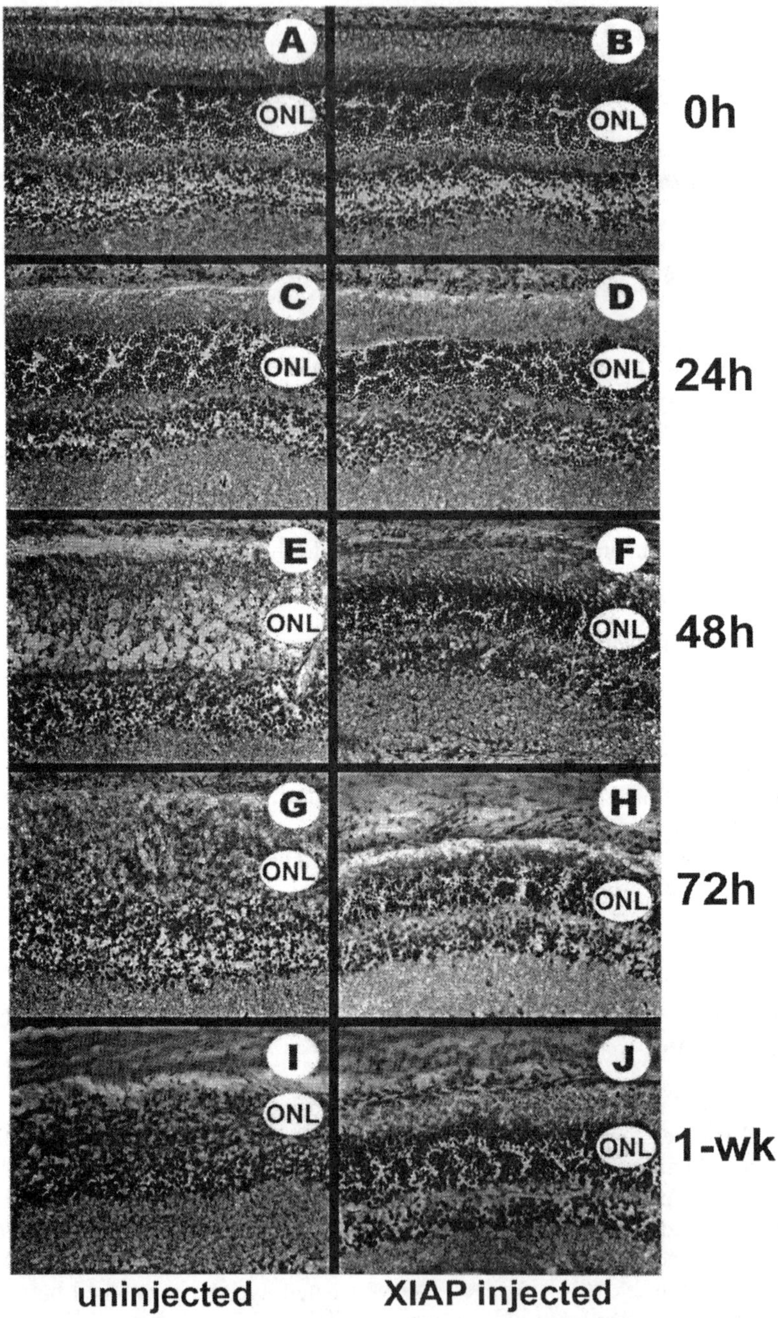

Figure 2. H&E sections of rat retinas at set time points pre, panel **A** and **B**, and post-MNU, panels **C-J**. The outer-nuclear layer (ONL) is indicated to show the differences in structural integrity between the animals that received rAAV-XIAP, panels on the right, to those that did not receive any virus, panels on the left.

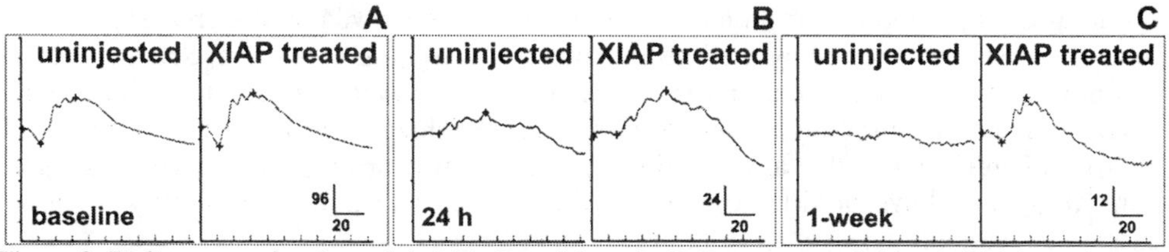

Figure 3. ERG traces at time 0 panel **A,** 24 h, **B,** and 1-week, **C.** Y-axes are scaled in μV/division and the X-axes are all 20ms/division. The luminance intensity is the same for all time points taken; 0.0 log cd-s/m^2.

4. DISCUSSION

Over 150 genes, a large number of which are expressed in the photoreceptors, are implicated in causing retinal dystrophies when mutated (Daiger et al., 1998). A mutation in any one of these genes seems to impede the cells ability to function optimally and to repair themselves when subjected to injury due to extrinsic (intense light) (Wenzel et al., 2000) or intrinsic (genetic) factors (Portera-Cailliau et al., 1994). For a comprehensive list of retinal dystrophies visit http://www.sph.uth.tmc.edu/Retnet/. Every chromosome has a least one gene which when mutated causes RP. The end point of RP, regardless of the mutation involved, is the death of the photoreceptor cells via apoptosis. In RP, all the photoreceptors in an affected individual have the mutation, yet they function normally for many years prior to the onset of disease. Photoreceptor cell death occurs slowly over a long period of time. If the apoptotic threshold in diseased photoreceptors could be raised, the cells could survive and function, thereby preventing disease progression in RP patients.

Survival-promoting agents such as cytokines, neurotrophins or growth factors have been examined as therapeutic agents in treating retinal degenerations (LaVail et al., 1992). Bcl-2, an anti-apoptotic protein and the caspase-3 inhibitor Ac-DEVD-CHO have been used recently with the hope of preventing retinal degenerations in various animal models (Nir et al., 2000; Yoshizawa et al., 2000). Though morphological protection was observed in the two anti-apoptotic approaches listed above, no functional protection was reported. In this study, we examined whether XIAP, a potent apoptotic inhibitor, could rescue photoreceptors structurally and functionally in an acute model of retinal degeneration induced by MNU.

XIAP was delivered sub-retinally and over-expression was confirmed in the photoreceptors using both real-time PCR and Western blot analysis (Fig.1). Histological analysis (Fig. 2A-J) shows remarkable protection of the ONL in the injected eye at all time points studied. This is noteworthy given that MNU is such a quick acting chemotoxin and DNA adducts are detected 12h post-treatment. The apoptotic cascade is initiated only after sufficient DNA adducts form in the photoreceptor cell (Yoshizawa et al., 1999). Though XIAP is known to be a potent apoptotic inhibitor, it may also function in other anti-death pathways that have yet to be elucidated. The ERG amplitudes in the two of the four animals, which showed histological protection at 1-week post-MNU, are approximately 5-15% of the baseline values. There are two reasons for this dramatic decrease in signal. Firstly, the 2μL of virus that is injected is estimated to cover 10–30 % of the rat retina and thus there are treated and untreated areas adjacent to each other. The full field ERG records a response from the entire retina. Therefore, any localized activity resulting from photoreceptor protection would be overshadowed by the greater part of the retina that was not covered by the injection and is undergoing apoptosis. Secondly, MNU is a powerful alkylating agent and is expected to cause severe damage to the cells DNA even prior to the initiation of apoptotic signals (Yoshizawa et al., 1999; Yuge et al., 1996). It is remarkable that any functional protection was observed at all, which suggests that XIAP may have additional protective effects in addition to its ability to inhibit caspase activation.

MNU is a potent inducer of photoreceptor apoptosis rendering animals blind within 72h of its administration. In this study we have shown that XIAP not only rescues photoreceptor cells at the histological level, but at the functional level as well. This is the first time that any compound has shown functional protection from MNU. Since XIAP is able to confer functional protection of the photoreceptors in this severe model of retinal degeneration, it holds promise as a therapeutic in slower retinal degenerations such as retinitis pigmentosa.

5. ACKNOWLEDGEMENTS

We thank the National Eye Institute for the travel award provided to DP to attend this meeting.

6. REFERENCES

Christmann, M. & Kaina, B., 2000, Nuclear translocation of mismatch repair proteins MSH2 and MSH6 as a response of cells to alkylating agents, *J Biol Chem.* **275**:36256-62.

Daiger, S., Rossiter, B., Greengerg, J., Christoffels, A. & Hide, W., 1998, Ret Net: data services and software for identifying genes and mutations causing retinal degeneration [ARVO Abstract], *Invest Ophthalmol Vis Sci.* **39**:S295 Abstract nr 1352.

Deveraux, Q. L., Roy, N., Stennicke, H. R., Van Arsdale, T., Zhou, Q., Srinivasula, S. M., Alnemri, E. S., Salvesen, G. S. & Reed, J. C., 1998, IAPs block apoptotic events induced by caspase-8 and cytochrome c by direct inhibition of distinct caspases, *Embo J.* **17**:2215-23.

Deveraux, Q. L., Schendel, S. L. & Reed, J. C., 2001, Antiapoptotic proteins. The bcl-2 and inhibitor of apoptosis protein families, *Cardiol Clin.* **19**:57-74.

Eberhardt, O., Coelln, R. V., Kugler, S., Lindenau, J., Rathke-Hartlieb, S., Gerhardt, E., Haid, S., Isenmann, S., Gravel, C., Srinivasan, A., Bahr, M., Weller, M., Dichgans, J. & Schulz, J. B., 2000, Protection by synergistic effects of adenovirus-mediated X-chromosome- linked inhibitor of apoptosis and glial cell line-derived neurotrophic factor gene transfer in the 1-methyl-4-phenyl-1,2,3,6- tetrahydropyridine model of Parkinson's disease, *J Neurosci.* **20**: 9126-34.

Flannery, J. G., Zolotukhin, S., Vaquero, M. I., LaVail, M. M., Muzyczka, N. & Hauswirth, W. W., 1997, Efficient photoreceptor-targeted gene expression in vivo by recombinant adeno-associated virus, *Proc Natl Acad Sci USA.* **94**:6916-21.

Grimm, D., Kern, A., Rittner, K. & Kleinschmidt, J. A., 1998, Novel tools for production and purification of recombinant adenoassociated virus vectors, *Hum Gene Ther.* **9**:2745-60.

Hauswirth, W. W., LaVail, M. M., Flannery, J. G. & Lewin, A. S., 2000, Ribozyme gene therapy for autosomal dominant retinal disease, *Clin Chem Lab Med.* **38**:147-53.

Herrold, K. M., 1967, Pigmentary degeneration of the retina induced by N-methyl-N-nitrosourea. An experimental study in syrian hamsters, *Arch Ophthalmol.* **78**:650-3.

LaVail, M. M., Unoki, K., Yasumura, D., Matthes, M. T., Yancopoulos, G. D. & Steinberg, R. H., 1992, Multiple growth factors, cytokines, and neurotrophins rescue photoreceptors from the damaging effects of constant light, *Proc Natl Acad Sci USA.* **89**:11249-53.

MacFarlane, M., Merrison, W., Bratton, S. B. & Cohen, G. M., 2002, Proteasome-mediated degradation of Smac during apoptosis: XIAP promotes Smac ubiquitination in vitro, *J Biol Chem.* **277**:36611-6.

McKinnon, S. J., Ransom, N. L., Tahzib, N. G., Lehman, D. M. & Hauswirth, W. W., 2001, Intravitreal gene therapy using adeno-associated virus (AAV) delivery of X-linked inhibitor of apoptosis protein (XIAP) protects optic nerve axons in a hypertensive rat model of glaucoma (abstract), *Invest Ophathal Vis Sci.* **42**:2209.

Mehlen, P., Schulze-Osthoff, K. & Arrigo, A. P., 1996, Small stress proteins as novel regulators of apoptosis. Heat shock protein 27 blocks Fas/APO-1- and staurosporine-induced cell death, *J Biol Chem.* **271**:16510.

Nakajima, M., Yuge, K., Senzaki, H., Shikata, N., Miki, H., Uyama, M. & Tsubura, A., 1996, Photoreceptor apoptosis induced by a single systemic administration of N-methyl-N-nitrosourea in the rat retina, *Am J Pathol.* **148**:631-41.

Nambu, H., Yuge, K., Nakajima, M., Shikata, N., Takahashi, K., Miki, H., Uyama, M. & Tsubura, A., 1997, Morphologic characteristics of N-methyl-N-nitrosourea-induced retinal degeneration in C57BL mice, *Pathol Int.* **47**:377-83.

Nir, I., Kedzierski, W., Chen, J. & Travis, G. H., 2000, Expression of Bcl-2 protects against photoreceptor degeneration in retinal degeneration slow (rds) mice, *J Neurosci.* **20**:2150-4.

Pagon, R. A., 1988, Retinitis pigmentosa, *Surv Ophthalmol.* **33**:137-77.

Portera-Cailliau, C., Sung, C. H., Nathans, J. & Adler, R., 1994, Apoptotic photoreceptor cell death in mouse models of retinitis pigmentosa, *Proc Natl Acad Sci USA.* **91**:974-8.

Robertson, G. S., Crocker, S. J., Nicholson, D. W. & Schulz, J. B., 2000, Neuroprotection by the inhibition of apoptosis, *Brain Pathol.* **10**:283-92.

Salvesen, G. S., 1999, Programmed cell death and the caspases, *Apmis* **107**, :73-9.

Suzuki, Y., Nakabayashi, Y., Nakata, K., Reed, J. C. & Takahashi, R., 2001, X-linked inhibitor of apoptosis protein (XIAP) inhibits caspase-3 and -7 in distinct modes, *J Biol Chem.* **276**:27058-63.

Takahashi, R., Deveraux, Q., Tamm, I., Welsh, K., Assa-Munt, N., Salvesen, G. S. & Reed, J. C., 1998, A single BIR domain of XIAP sufficient for inhibiting caspases, *J Biol Chem.* **273**:7787-90.

Timmers, A. M., Zhang, H., Squitieri, A. & Gonzalez-Pola, C., 2001, Subretinal injections in rodent eyes: effects on electrophysiology and histology of rat retina, *Mol Vis.* **7**:131-7.

Velten, I. M., Horn, F. K., Korth, M. & Velten, K., 2001, The b-wave of the dark adapted flash electroretinogram in patients with advanced asymmetrical glaucoma and normal subjects, *Br J Ophthalmol.* **85**:403-9.

Wenzel, A., Grimm, C., Marti, A., Kueng-Hitz, N., Hafezi, F., Niemeyer, G. & Reme, C. E., 2000, c-fos controls the "private pathway" of light-induced apoptosis of retinal photoreceptors, *J Neurosci.* **20**:81-8.

Xu, D., Bureau, Y., McIntyre, D. C., Nicholson, D. W., Liston, P., Zhu, Y., Fong, W. G., Crocker, S. J., Korneluk, R. G. & Robertson, G. S., 1999, Attenuation of ischemia-induced cellular and behavioral deficits by X chromosome-linked inhibitor of apoptosis protein overexpression in the rat hippocampus, *J Neurosci.* **19**:5026-33.

Yang, Y. L. & Li, X. M., 2000, The IAP family: endogenous caspase inhibitors with multiple biological activities, *Cell Res.* **10**: 169-77.

Yoshizawa, K., Nambu, H., Yang, J., Oishi, Y., Senzaki, H., Shikata, N., Miki, H. & Tsubura, A., 1999, Mechanisms of photoreceptor cell apoptosis induced by N-methyl-N-nitrosourea in Sprague-Dawley rats, *Lab Invest.* **79**:1359-67.

Yoshizawa, K., Yang, J., Senzaki, H., Uemura, Y., Kiyozuka, Y., Shikata, N., Oishi, Y., Miki, H. & Tsubura, A., 2000, Caspase-3 inhibitor rescues N -methyl- N -nitrosourea-induced retinal degeneration in Sprague-Dawley rats, *Exp Eye Res.* **71**:629-35.

Yuge, K., Nambu, H., Senzaki, H., Nakao, I., Miki, H., Uyama, M. & Tsubura, A., 1996, N-methyl-N-nitrosourea-induced photoreceptor apoptosis in the mouse retina, *In Vivo* **10**:483-8.

NEUROPROTECTIVE EFFECT OF ESTROGEN UPON RETINAL NEURONS IN VITRO

Wei Cao, Raju V.S. Rajala, Feng Li, Robert E. Anderson, Na Wei, Charles E. Soliman, and James F. McGinnis[1]

1. INTRODUCTION

The female sex hormone estrogen has a variety of metabolic activities including numerous effects on neurons. Therefore, the picture has changed from estrogen being a pure female sexual hormone important for sex differentiation and maturation to estrogen as a neuromodulator and neuroprotectant. It appears that estrogen specifically maintains verbal memory in women and may prevent the deterioration in short- and long-term memory that is associated with normal aging (Sherwin, 1999). Several epidemiological studies reported that estrogen replacement therapy of postmenopausal women may prevent neurodegenerative disorders including Alzheimer's disease (Henderson, 1997). In addition, estrogen is not restricted to females since the male sex hormone testosterone (and other steroids with a 19 carbon atom structure, so-called C-19 steroid) can be chemically converted to estradiol in various tissues, including the brain, by an aromatase P450 enzyme (Schumacher et al., 2002). Many lines of evidence suggest that estrogen has neurotrophic and neuroprotective properties (Wise, 2002; Garcia-Segura et al., 2001). 17 ß-estradiol (ßE2) is a steroid hormone produced mainly in the ovaries, by the placenta during pregnancy, and to a lesser extent in the adrenal cortices, testes, and peripheral tissues (Schumacher et al., 2002). It is synthesized enzymatically from acetate, cholesterol, progesterone, and testosterone (Schumacher et al., 2002). For many years, it has been known that ßE2 promotes viability and survival of neurons in primary neuronal cultures. Addition of ßE2 to defined culture media increased the viability, survival and differentiation of primary neuronal cultures from different brain areas including amygdala (Lorenzo et al., 1992), hypothalamus (Cambiasso and Carrer, 2001), neocortex (Vedder et al., 1999). However, the role of estrogen in the protection of retinal neurons is unknown. We have previously established H_2O_2-induced cell death of cultured retinal neurons as a model system for studying stress-induced degeneration, and used this model

[1]Department of Ophthalmology, University of Oklahoma Health Sciences Center, Dean A. McGee Eye Institute, 608 Stanton L. Young Blvd., Oklahoma City, OK 73104, USA. Corresponding author: W. Cao, E-mail: wei-cao@ouhsc.edu

Retinal Degenerations: Mechanisms and Experimental Therapy
Edited by LaVail *et al.*, Kluwer Academic/Plenum Publishers, 2003

system to study the role of several neurotrophic factors, such as pigment epithelium-derived factor, in retinal neuronal protection (Cao et al., 1999). In the present study, we used our cultured retinal neurons as a model system to determine whether estrogen protects retinal neurons.

2. METHODS

2.1. Primary Culture of Retinal Neurons

Timed pregnant Sprague-Dawley rats were ordered each week and the retinas of 10-15 pups, 0-2 days old, were removed with the aid of a dissecting microscope under sterile conditions in a tissue culture hood. The retinas were suspended in 25 ml of DMED-F12 plus 10% fetal calf serum in a plastic bag and mechanically dissociated for 2 minutes using a Stomacher set on low power. The suspension was first filtered through a 230μm sieve which was then rinsed with medium, and the filtrate was then passed through a 140 μm sieve followed by a rinse with undiluted fetal calf serum. The filtered suspension was centrifuged at 800 rpm in a clinical centrifuge for 5 minutes, the supernatant decanted, and the cell pellets resuspended in 25 ml of media using a sterile 5 ml pipette. The concentration of cells was determined with a cell counter or hemocytometer and the suspension diluted with medium to $1x10^5$ cells per ml. The cells (1 ml) were plated in 24 well tissue culture plates on 12 mm coverslips that had been pre-treated overnight with poly-D-lysine (10 μg/ml). The cells were maintained in either DMEM-F12 with 2% fetal calf serum, or the synthetic serum-free media. The cultures were used in experiments at 10-14 days after plating.

2.2. MTT Assay

MTT (3,(4,5-dimethylthiazol-2-yl)2,5-diphenyl-tetrazolium bromide) (Sigma, St. Louis, MO) was dissolved at a concentration of 5 mg/ml in PBS. Lysing buffer was prepared as follows: 20% w/v of SDS (Sigma, St. Louis, MO) was dissolved at 37°C in a solution of 50% of each DMF (N,N-dimethyl formamide) (Sigma, St. Louis, MO) and deionized water. 25 μl of the 5 mg/ml stock solution of MTT was added to each well and, after 2 hours of incubation at 37°C, 100 μl of the lysing buffer was added. After an overnight incubation at 37°C, absorbane of the samples was read at 562 nm using a microtiter plate ELISA reader.

2.3. TUNEL Assay

TUNEL (TdT-mediated digoxigenin-dUTP nick-end labeling) method was carried out using a commercially available *in situ* apoptosis detection kit and as described previously (Cao et al., 1999). Briefly, staining for the TUNEL technique was performed according to the manufacturer's protocol. TUNEL positive cells were visualized using diaminobenzidine as the substrate for the kit's horseradish peroxidase. The results were viewed with a Nikon Eclipse 800 microscope. The pictures were transferred by a digital camera and stored in the computer.

2.4. SDS-PAGE and Western Blot Analysis

Protein samples were resolved by 7.5%, 10%, or 15% SDS-PAGE and transferred on to nitrocellulose membranes. The blots were washed two times for 5 minutes with TTBS [100 mM NaCl, 20 mM Tris-HCl (pH 7.4), and 0.1% Tween-20] and blocked with 10% non-fat dry milk in TTBS overnight at 4°C. The blots were then incubated with anti-p85 subunit of phosphatidylinositol 3-kinase (PI3K) (1:1000), anti-Akt (1:500), anti-pAkt s473 (1:2500), anti-Insulin receptor β–subunit (IRβ) (1:1000) and anti-opsin (R1D4, 1:5000) antibodies for 2 hours at room temperature. Following primary antibody incubations, the blots were incubated with horseradish peroxidase (HRP)-linked secondary antibodies (anti-rabbit, anti-mouse, or anti-goat IgG) and developed by enhanced chemiluminescence (ECL), according to the manufacturer's instructions.

2.5. PI 3-Kinase Assay

Enzyme assays were carried out essentially as described previously (Rajala and Anderson, 2001). Briefly, assays were performed directly on total cells in 50 μl of the reaction mixture containing 0.2 mg/ml PI-4,5-P_2, 50 μM ATP, 0.2 μCi [γ^{32} P]ATP, 5 mM $MgCl_2$, and 10 mM HEPES buffer (pH 7.5). The reactions were performed for 15 minutes at room temperature and stopped by the addition of 100 μl of 1 N HCL followed by 200 μl of chloroform-methanol (1:1, v/v). Lipids were extracted and resolved on oxalate-coated TLC plates (silica gel 60) with a solvent system of 2-propanol/2 M acetic acid (65:35, v/v). The TLC plates were prepared by placing in 1% (w/v) potassium oxalate in 50% methanol (v/v) and baked in an oven at 100°C for 1 hour before use. TLC plates were exposed to x-ray film overnight at -70°C and radioactive lipids were scraped and quantified by liquid scintillation counting.

2.6. GST-p85 Fusion Proteins and Pull-Down Experiments

GST-p85 N-SH2 (314-446 amino acids) fusion proteins were generated by PCR amplification of the p85α cDNA and cloned into a GST-vector (pGEX2T; Amersham Pharmacia Biotech). The sequence of each clone was verified by DNA sequencing. All inductions yielded proteins of the expected size, as judged by Coomassie blue staining. Pull-down experiments were performed as described (Raju et al., 2001), using 5 μg GST-fusion proteins that had been adsorbed onto GST-Sepharose 4B matrix. Cultured retinal neurons were incubated with GST/GST-p85 fusion proteins with continuous mixing at 4°C for 1.5 hours. The Sepharose beads were washed three times in 500 μl HNTG buffer (20 mM HEPES [pH 7.5], 150 mM NaCl, 0.1% Triton X-100, and 10% glycerol) and centrifuged at 5000 rpm for 1 min at 4°C. Bound proteins to GST-p85 was eluted by boiling in 2x SDS sample buffer and subjected to SDS-PAGE. The gels were transferred to nitrocellulose membranes followed by Western blot analysis with anti-IRβ antibody.

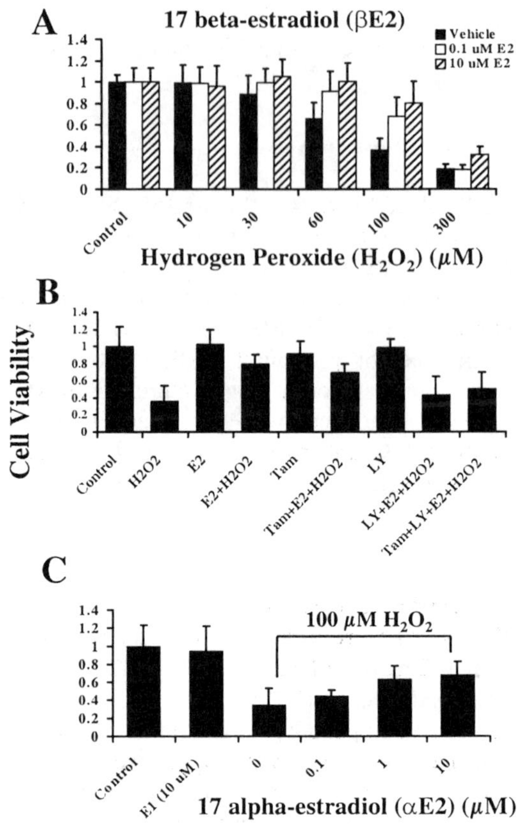

Figure 1. Estradiols attenuate H_2O_2-induced cytotoxicity in cultured retinal neurons. (A) Neuroprotective effect of 17 β-estradiol (βE2) in different concentrations on H_2O_2-induced cytotoxicity. (B) Tamoxifen, estrogen receptor antagonist, did not block neuroprotective effect of βE2, whereas LY249002, a specific PI3K inhibitor, blocks neuroprotective effect of βE2. (C) Dose-dependent neuroprotective effect of 17 α-estradiol (αE2) on H_2O_2-induced cell death.

3. RESULTS

3.1. Estradiols Attenuate H_2O_2-Induced Cytotoxicity

To evaluate the potential of 17 β-estradiol (βE2) to protect retinal neurons from H_2O_2-induced cytotoxicity, we have pretreated retinal neuron cultures with two different concentrations of βE2 (0.1 μM and 10 μM) 30 minutes prior to exposure to different doses of H_2O_2 (10, 30, 60, 100 and 300 μM) for 24 hours. Cell viability was determined by the MTT assay as described in methods. Figure 1A shows that βE2 at 0.1 μM significantly inhibits cytotoxicity induced by either 60 μM or 100 μM H_2O_2, whereas this concentration of βE2 did not inhibit cytotoxicity when the cultures were exposed to 300 μM H_2O_2. Pretreatment with 10 μM βE2 completely inhibited cytotoxicity induced by 60 μM H_2O_2 and also significantly attenuated 100 μM H_2O_2-induced cytotoxicity. There was no protective effect of 10 μM βE2 on cytotoxicity induced by 300 μM H_2O_2.

Figure 1B shows that additions of tamoxifen, an estrogen receptor blocker, at 5 μM or LY294002, a PI3K inhibitor, at 10 μM for 24 hours did not significantly affect cell viability. Tamoxifen pretreatment did not significantly block the neuroprotective effect of βE2, whereas LY294002 greatly inhibited the βE2 effect in this study.

17 α-estradiol (αE2) is inactive at the estrogen receptor site. The effect of αE2 on H_2O_2-induced cell death was also examined in this study. Retinal neuron cultures were treated with different concentration of αE2 (0.1, 1 and 10 μM) 30 minutes prior to exposure to 100 μM H_2O_2 for 24 hours. Figure 1C shows that αE2 at concentrations of either 1 or 10 μM significantly inhibits cytotoxicity induced by 100 μM H_2O_2, but ineffective at 0.1 μM.

3.2. Estradiols Inhibit Apoptosis Induced by H_2O_2

To determine whether estradiols can inhibit apoptosis induced cell death by H_2O_2, the TUNEL assay was performed. A few positive staining cells were noted in control cultures (Figure 2-V-A), whereas cultures treated with 100 μM H_2O_2 for 24 hours resulted in a large numbers of cells undergoing apoptotic cell death (Figure 2-V-B). However, pretreatment of retinal neurons with βE2 at the concentration of 10 μM 30 minutes prior to H_2O_2 (100 μM) exposure exhibited a dramatic decrease in the numbers of apoptotic cells (Figure 2-V-C). αE2 also significantly reduced the number of apoptotic cells although to a lesser extent than βE2 (Figure 2-V-D).

3.3. Involvement of Insulin Receptor and PI 3-Kinase/Akt signal Pathway

Western blot analysis demonstrated that insulin receptor beta subunit expression was clearly found to be present in these cultured rat retinal neurons (Figure 2-I). We further employed GST-p85 fusion proteins study the receptor phosphorylation (activation) in pull-down experiments. Our results indicate that both βE2 and αE2 activate insulin receptor, but progesterone does not (Figure 2-II). To investigate whether PI3K activity can be activated by addition of βE2 to the culture, retinal neuron cultures were treated with 10 μM βE2 for different times (0.5, 1, 3, 6 and 12 hours). As shown in Fig. 2-III (top panel), PI3K activity was increased by a 30-minute treatment, and reached a peak at 1 hr and 3 hrs. PI3K activity declined after 6 hrs of treatment and returned to control level after 12 hrs. We also examined the expression of PI3K p85 alpha subunit in response to βE2 treatment using Western blot analysis. There were not significant differences at any time-point examined compared to control group (Figure 2-III, low panel). Addition of βE2 (10 μM) to the culture for 30 min significantly increased phosphorylated Akt expression as shown in Western blotting (Figure 2-IV upper panel). This increase was maintained at least for 3 hrs. The expression of non-phosphorylated Akt is shown as a control (Figure 2-IV, low panel).

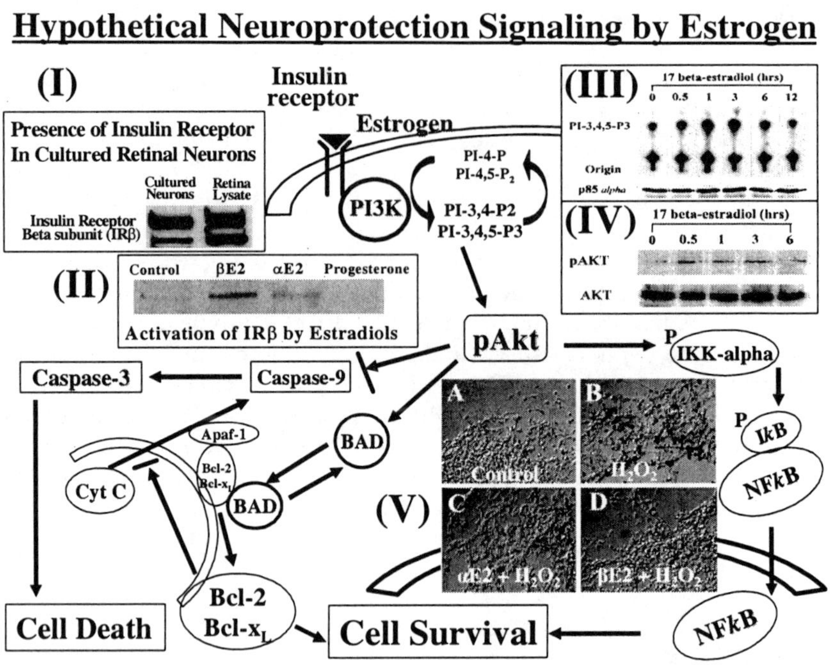

Figure 2. Involvement of insulin/PI3K/Akt signal pathway in estrogen-mediated neuroprotection. (I) Presence of insulin receptor beta subunit in cultured rat retinal neurons. (II) Both 17 β-estradiol (βE2) and 17 β-estradiol (βE2) but not progesterone activate insulin receptor. (III) TLC autoradiogram showing a transient increase in PI3K activity by βE2. (IV) βE2 activates Akt in cultured retinal neurons. (V) Estradiols inhibit H_2O_2-induced apoptotic cell death: (A) H_2O_2 untreated control. (B) H_2O_2 treated (100 μM, 24 hrs) cells. (C) Cultured retinal neurons pretreated with 17 α-estradiol (αE2, 10 μM) 30 min before H_2O_2 exposure. (D) Cultured retinal neurons pretreated with 17 β-estradiol (βE2, 10 μM) 30 min before H_2O_2 exposure.

4. DISCUSSION

Although it has been known that βE2 promotes viability and survival of other primary neuronal culture systems, such as cortex (Honda et al., 2001), hippocampus (Gallaher et al., 2001), hypothalamus (Cambiasso and Carrer, 2001), this is the first observation in a primary retinal neuronal culture demonstrating that βE2 attenuated H_2O_2-induced cytotoxicity, inhibited apoptosis, and that insulin receptor and PI3K/Akt signal pathway may be involved in the protective effect of βE2. The MTT assay is a colorimetric assay for the nonradioactive quantitation of cell viability and is based on the cleavage of the yellow tetrazolium salt MTT to purple formazan crystals by metabolically active cells. The formation of formazan is thought to take place via intact mitochondria. A decrease in the number of living cells results in a decrease in the total metabolic activity in the samples. This assay has been widely used to assess H_2O_2-induced cytotoxicity and neuroprotection (Cao et al., 1999). We used this assay demonstrating that βE2 increased cell viability of retinal neurons in response to H_2O_2, suggesting that the inhibition of mitochondrial function by H_2O_2 can be attenuated by βE2. Inhibition of mitochondrial activity has been shown to result in apoptosis (Green et al., 2001). Our data also showed that βE2 greatly reduced H_2O_2-induced apoptotic cells, suggesting the inhibition of mitochondrial activity by H_2O_2 can be attenuated or blocked by βE2.

Estradiol can bind intracellular-specific estrogen receptors and the complex binds to a specific site on genome cDNA and controls its transcription; thus, estradiol neuroproection may be mediated by the estrogen receptor and protein synthesis (Wise, 2002; Dhandapani and Brann, 2002). Indeed, βE2 was reported to provide neuroprotection mediated by estrogen receptors in cultured cortical neurons (Singer et al., 1996) and in hippocampal-derived cell line (Fitzpatrick et al., 2002). In the present study we observed that a competitive estrogen receptor antagonist, tamoxifen, did not significantly attenuate the protection provided by βE2. Although the absence of any effects by tamoxifen did not completely exclude the possibility that the neuroprotection was mediated by estrogen receptors, 17 α-estradiol, a biologically inactive stereoisomer which has little effect as a female sex steroid hormone, also provides neuroprotection in retinal neurons against H_2O_2-induced apoptotic cell death. This suggests that neuroprotective effects of βE2 in this culture system may not act through activation of estrogen receptors.

In addition to the activation of genome transcription mediated by estrogen receptors, estrogens have been reported to activate IGF-I receptor kinase resulting in enhanced binding of p85 (PI3K) to insulin receptor substrate-1 (IRS-1) and IRS-2 in the mouse uterus (Richards et al., 1998). As shown in Fig 2, in the present study, we demonstrated that estradiol activates the insulin receptor but not the IGF-I receptor in cultured retinal neurons. The insulin receptor is present in neuronal retina (Gosher et al., 2000) and the interaction of the insulin receptor β-subunit with PI3K has been shown (Rajala and anderson, 2001; Barber et al., 2001). In the present study, we demonstrate that estradiol significantly increased PI3K activity and actived Akt. We have also demonstrated the absence of a neuroprotective effect of βE2 in the presence of PI3K inhibitor. These observations led to the hypothesis that estrogen mediated insulin receptor signaling could be one of the protective mechanisms in retinal neurons. Further detailed studies, however, are required to understand the molecular mechanism underlying estrogen-mediated retinal neuronal protection.

5. ACKNOWLEDGEMENTS

Supported by National Center for Research Sources grant P20 RR17703, and by grants from the National Institutes of Health (EY014427, EY13050, EY00871, EY04149, EY12190, EY06973), and by a Jules and Doris Stein Professorship from Research to Prevent Blindness (RPB) and an unrestricted grant from RPB to the Department of Ophthalmology; Samuel Roberts Nobel Foundation, Inc., Ardmore, OK; and Presbyterian Health Foundation, Oklahoma City, OK. We thank the National Eye Institute and the Foundation Fighting Blindness for the travel award provided to WC to attend RD2002 meeting.

6. REFERENCES

Sherwin, B. B., 1999, Can estrogen keep you smart? Evidence from clinical studies, *J Psychiatry Neurosci* **24**, 315-321.

Henderson, V. W., 1997, The epidemiology of estrogen replacement therapy and Alzheimer's disease, *Neurology* **48**, 27-35.

Schumacher, M., Akwa, Y., Guennoun, R., Robert, F., Labombarda, F., Desarnaud, F., Robel, P., Nicola, A. F., and Baulieu, E. E., 2002, Steroid synthesis and metabolism in the nervous system: trophic and protective effects, *J Neurocytol* **29**, 307-26.

Wise, P. M., 2002, Estrogens and neuroprotection, *Trends Endocrinol Metab* **13**, 229-230.

Garcia-Segura, L. M., Azcoitia, I., and DonCarlos, L. L., 2001, Neuroprotection by estradiol, *Prog Neurobiol* **63**, 29-60.

Lorenzo, A., Diaz, H., Carrer, H., and Caceres, A., 1992, Amygdala neurons in vitro: neurite growth and effects of estradiol, *J Neurosci Res* **33**, 418-435.

Cambiasso, M. J., and Carrer, H. F., 2001, Nongenomic mechanism mediates estradiol stimulation of axon growth in male rat hypothalamic neurons in vitro, *J Neurosci Res* **266**, 475-481.

Vedder, H., Anthes, N., Stumm, G., Wurz, C., Behl, C., and Krieg, J. C., 1999, Estrogen hormones reduce lipid peroxidation in cells and tissues of the central nervous system, *J Neurochem* **72**(6), 2531-2538.

Cao, W., Tombran-Tink, J., Chen, W., Mrazek, D., Elias, R., and McGinnis, J. F., 1999, Pigment epithelium-derived factor protects cultured retinal neurons against hydrogen peroxide-induced cell death, *J Neurosci Res* **57**, 789-800.

Rajala, R. V., and Anderson, R. E., 2001, Interaction of the insulin receptor beta-subunit with phosphatidylinositol 3-kinase in bovine ROS, *Invest Ophthalmol Vis Sci* **42**, 3110-3117.

Honda, K., Shimohama, S., Sawada, H., Kihara, T., Nakamizo, T., Shibasaki, H., and Akaike, A., 2001, Nongenomic antiapoptotic signal transduction by estrogen in cultured cortical neurons, *J Neurosci Res* **64**, 466-475.

Gallaher, B. W., Hille, R., Raile, K., and Kiess, W., 2001, Apoptosis: live or die--hard work either way, *Horm Metab Res* **33**, 511-519.

Green, P. S., Yang, S. H., Nilsson, K. R., Kumar, A. S., Covey, D. F., and Simpkins, J. W., 2001, The nonfeminizing enantiomer of 17beta-estradiol exerts protective effects in neuronal cultures and a rat model of cerebral ischemia, *Endocrinology* **142**, 400-406.

Dhandapani, K. M., and Brann, D. W., 2002, Protective effects of estrogen and selective estrogen receptor modulators in the brain, *Biol Reprod* **67**, 1379-1385.

Singer, C. A., Rogers, K. L., Strickland, T. M., and Dorsa, D. M., 1996, Estrogen protects primary cortical neurons from glutamate toxicity, *Neurosci Lett* **212**, 13-16.

Fitzpatrick, J. L., Mize, A. L., Wade, C. B., Harris, J. A., Shapiro, R. A., and Dorsa, D. M., 2002, Estrogen-mediated neuroprotection against beta-amyloid toxicity requires expression of estrogen receptor alpha or beta and activation of the MAPK pathway, *J Neurochem* **82**, 674-682.

Richards, R. G., Walker, M. P., Sebastian, J., and DiAugustine, R. P., 1998, Insulin-like growth factor-1 (IGF-1) receptor-insulin receptor substrate complexes in the uterus. Altered signaling response to estradiol in the IGF-1(m/m) mouse, *J Biol Chem* **273**, 11962-11969.

Gosbell, A. D., Favilla, I., Baxter, K. M., Jablonski, P., 2000, Insulin receptor and insulin receptor substrate-I in rat retinae, *Clin Experiment Ophthalmol* **28**, 212-215.

Barber, A. J., Nakamura, M., Wolpert, E. B., Reiter, C. E., Seigel, G. M., Antonetti, D. A., and Gardner, T. W., 2001, Insulin rescues retinal neurons from apoptosis by a phosphatidylinositol 3-kinase/Akt-mediated mechanism that reduces the activation of caspase-3, *J Biol Chem* **276**, 32814-32821.

TRANSGENIC *RD* MICE HARBORING AXOKINE GENE BY *RPE65* GENE PROMOTER DOES NOT RESCUE PHOTORECEPTOR DEGENERATION

Toshiaki Abe*, Yuki Yoshioka, Hiroshi Tomita, Yoko Saigo, Tetuso Udono, Shigeki Shibahara, and Makoto Tamai

1. INTRODUCTION

Neurotrophic factors have reported to rescue photoreceptor degeneration in dystrophic rats (Faktorovich et al., 1990), and variable strains of mice showing hereditary photoreceptor degeneration (LaVail et al., 1998). Axokine is a modified form of human CNTF to enhance its specific activity (Panayotatos et al., 1993) and its signaling pathway is reported in retinal neurons and glia via CNTF receptor (CNTFRα) (Peterson et al., 2000). In this study, we introduced Axokine gene in *rd/rd* mice, a mice reported to have mutation in ß subunit of phosphodiesterase (PDE) gene (Pittler and Baehr, 1991) and examined its effect for retinal degeneration. The Axokine gene was introduced into retinal pigment epithelium (RPE) by *Rpr65* promoter, which has acitivity of RPE specific expression (Boulanger et al, 2000).

2. MATERIALS AND METHODS

2.1. Construction of the Transgenic Gene and Microinjection Into *Rd* Mice

Axokine cDNA was generously supplied by Takeda pharmaceutical company. *Rpe65* promoter region (-655 to +52), which region was reported to have RPE specific activity (Boulanger et al., 2000) was amplified from the mouse genomic DNA. Then the promoter region was combined with the Axokine cDNA and enhanced green fluorescein protein (EGFP) gene with internal ribosome entry site (IRES). The constructs were were microinjected into mouse embryos and implanted into *rd* mother (C3H/HeNCrj)(CHARLES RIVER JAPAN, Inc., Yokohama, Japan).

[1]Department of Opthalmology, Tohoku University, School of Medicine, 1-1 Seiryomachi Aobaku, Sendai Miyagi, 980-8574, Japan. Corresponding author: T. Abe, E-mail:toshi@oph.med.tohoku.ac.jp

2.2. Genomic Southern Blot Analysis

Copy number of the transgene construct was estimated by Southern blot analysis of BamHI-digested genomic DNA, which were hybidized by Axokine cDNA with the known quantity of linearized transgene.

2.3. Reverse Transcriptase-polymerase Chain Reaction (RT-PCR)

mRNA was extracted from each transgenic (Tg) and non-transgenic (non-Tg) mice using oligo dT cellulose (Pharmacia Biotech Inc., Uppsala, Sweden) and PCR was carried out by the standard methods.

2.4. Primers

For *Rpe65* promoter:
5'-CCCAAGCTTGCAATGGTGAAGACAGTGA-3' and
5'-CCCAAGCTTCTTCCAGTGAAGATTAGAG-3'
For Axokine:
5'-TTCACAGAGCATTCACCGC-3' and 5'-GAAGGTTCTCTTGGAGTCGC-3'
For ß-actin:
5'-CTACAATGAGCTGCGTGTGG-3' and 5'-CGGTGAGGATCTTCATGAGG-3'.

2.5. Histological Examination

The eyes were sectioned and examined at postnatal days of 6, 13, 36, and 45. We measured the thickness of outer nuclear layer (ONL) at the points of 400 μm from the optic nerve in both Tg and non-Tg using IPLab program (Scanalytics, Inc. Fairfax VA) of Leica DC 100 (Leica AG, Heerbrugg, Switzerland).

Cryo sections (4 days postnatal) were also made and photographed (Leica DMIRE2, Leica Microsystems Imaging Solutions Ltd., Cambridge UK).

All procedures were preformed in accordance with the ARVO Statement for the Use of Animals in Ophthalmic and Vision Research.

2.6. Statistical Analysis

Statistical significance was determined by Mann-Whitney's U test.

3. Results

Southern blotting analysis, RT-PCR and fluorescein microscopy of the cryo-section of Tg showed transgene expression. Axokine and EGFP genes were expressed only in transgenic mice but not non-transgenic mice.

Histological sections of Tg and non-Tg were examined. We examined the thickness of the ONL at posterior region (400 μm from optic disc) in both Tg and non-Tg. As shown in the figure, statistical analysis was not significant between Tg and non-Tg of the thickness of ONL.

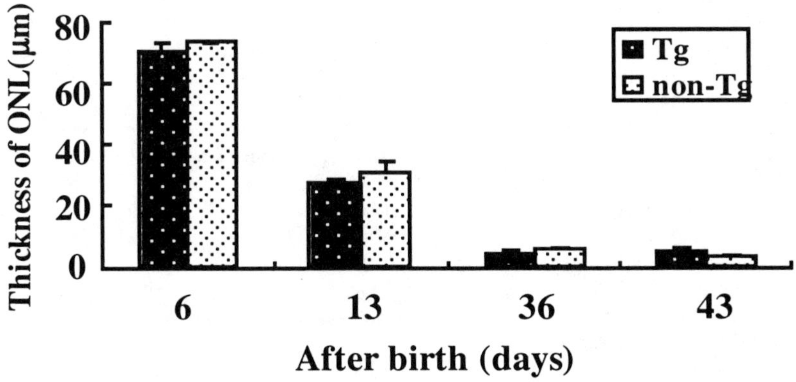

Figure 1. Outer nuclear layer (ONL) thickness in Tg and non-Tg *rd/rd* mice.

4. DISCUSSION

In spite of the Axokine gene expression in the RPE, we could not find any photoreceptor rescue in the *rd/rd* mice. Although rat CNTF rescued photoreceptor degeneration in *rd/rd* mice, it is well known that CNTF has species-specific activity and human CNTF showed no or far less effect, if applied intravitreally (LaVail et al., 1998). Axokine is an analogue of human CNTF (Panayotatos et al., 1993) and may be less effect for rescue of retinal degeneration of *rd/rd* mice (LaVail et al., 1998).

CNTF binds to a CNTF-specific component (CNTFRα) and promotes its binding to another components of gp130 and leukemia inhibitory factor receptor (LIFR)ß (Stahl and Yancopoulos, 1994). The CNTFRα expression has reported on horizontal, amacrine, and ganglion cells in normal adult rat retina (Krisch and Hofmann, 1994)). When we examined the CNTFRα expression, it is distributed widely in all layer of the retina at postnatal day 8. However, after starting photoreceptor degeneration, the expression of CNTFRα was focussed at the photoreceptor layer. Further the CNTFRα expression was suspected to be down-regulated in the Tg by real-time PCR (data not shown). The aberrant expression of CNTFRα may influence the Axokine function to photoreceptor degeneration, even though the Axokine expression was continuous and abundant. Continuous expression of Axokine from the early days using transgenic mice may be modified systemically.

5. ACKNOWLEDGMENT

This work was supported in part by grants from Grant-in-Aid for Scientific Research 10671630 (Dr. T. Abe); from the Ministry of Education and Culture of the Japanease Government, Tokyo (Dr. Tamai). We are indebted to the families studied for their cooperation and support.

6. REFERENCES

Boulanger A, Liu S, Henningsgaard AA, Yu S Redmond TM, 2000, The upstream region of the Rpe65 gene confers retinal pigment epithelium-specific expression in vivo and in vitro and contains critical octamer and E-box binding sites. *J Biol Chem.*, **275**:31274-31282.

Faktorovich EG, Steinberg RH, Yasumura D, Matthes MT and LaVail MM, 1990, Photoreceptor degeneration in inherited retinal dystrophy delayed by basic fibroblast growth factor. *Nature,* **347**:83-86.

Kirsch M and Hofmann HD, 1994, Expression of ciliary neurotrophic factor receptor mRNA and protein in the early postnatal and adult rat nervous system. *Neurosci Lett.*, **180**:163-166

LaVail MW, Yasumura D, and Matthes MT, et al., 1998, Protection of mouse photoreceptors by survival factors in retinal degenerations. *Invest Ophthalmol Vis Sci.*, **39**:592-602.

Panayotatos N, Radziejewska E, Acheson A, Pearsall D, Thadani A and Wong V., 1993, Exchange of a single amino acid interconverts the specific activity and gel mobility of human and rat ciliary neurotrophic factors. *J Biol Chem.*, **268**:19000-19003.

Peterson WM, Wang Q, Tzekova R and Wiegand SJ., 2000, Ciliary neurotrophic factor and stress stimuli activate the Jak-STAT pathway in retinal neurons and glia. *J Neurosci.*, **20**:4081-4090.

Pittler SJ and Baehr W., 1991, Identification of a nonsense mutation in the rod photoreceptor cGMP phosphodiesterase beta-subunit gene of the rd mouse. *Proc Natl Acad Sci U S A,* **88**:8322-8326.

Stahl N and Yancopoulos GD., 1994, The tripartite CNTF receptor complex: activation and signaling involves components shared with other cytokines. *J Neurobiol.*, **25**:1454-1466.

MARROW STROMAL CELLS (MSC): A SPECIES COMPARISON

MSCs: A species comparison

Anthony Kicic[1], Adam C. Shanley[1,2], Christine M. Hall[1,2], and P. Elizabeth Rakoczy[2]

1. INTRODUCTION

1.1. Retinal Degenerations

Retinal degenerations and dystrophies are the major causes of inherited blindness in the developed world comprising a variety of disparate genetically determined conditions that differ from one another in their mode of inheritance, severity, topographic pattern of visual loss, and ophthalmoscopic appearance. Onset of most genetically inherited retinal degenerations and dystrophies usually occurs early-on in life, and thus they have a great socio-economic impact on the community. Since these retinal disorders are phenotypically heterogeneous, early clinical and molecular diagnosis is rare and as a result, no treatment methods currently exist for these conditions. Histologically, they are characterized by the apoptotic death of one subset of cells in the retina, namely the photoreceptors (Lolley et al., 1994). However, some function may still exist in the remaining retina and in the axons connecting the retina to the brain. Hence photoreceptor replacement as part of a gene or cell-based therapy may aid in the restoration of some degree of vision (Ali et al., 2000).

1.2. Embryonic and Adult-derived Stem Cells

Most tissues contain unique cells that originate from a small subset of highly undifferentiated, self-renewing elements called "stem cells" that have the potential to

[1]Stem Cell Unit, Department of Molecular Ophthalmology, Lions Eye Institute affiliated with the Centre of Ophthalmology and Visual Science, University of Western Australia, Nedlands, 6009, Western Australia, Australia; [2]Centre of Ophthalmology and Visual Science, The University of Western Australia, Nedlands, 6009, West. Australia, Australia. Corresponding author: P. E. Rakoczy, E-mail: rakoczy@cyllene.uwa.edu.au

Retinal Degenerations: Mechanisms and Experimental Therapy
Edited by LaVail *et al.*, Kluwer Academic/Plenum Publishers, 2003

persist throughout adulthood (Morrison et al., 1999). Thus during normal development these stem cells generate intermediate progenitor cells that are capable of proliferating and differentiating into multiple yet distinct cell lineages. Embryonically derived stem cells are capable of differentiating into all somatic cell types (Kopen et al., 1999), whereas those derived from adult tissue were initially thought to produce only cell lineages characteristic of the tissue of isolation (Morrison et al., 1999). Embryonic stem cells have been successfully isolated from both animal and human fetal tissue (Reh and Kljavin, 1989; Rubio et al., 2000), but despite their enormous therapeutic potential, the ethical and moral dilemmas raised by their use as well as the fact that recipients of embryonic stem cell transplantations must still overcome the problem of rejection, could limit their application in cell therapy.

1.3. Bone Marrow Stromal Cells

The bone marrow (consisting of hematopoietic stem cells (HSC) and marrow stromal cells (MSC)) remains the primary site where self-renewal and differentiation of stem cells occurs (Dexter et al., 1977). Marrow stromal cells have been shown to possess greater plasticity than first thought. Initially, MSCs were believed to differentiate only into osteocytes, chondrocytes and adipocytes, all of which are located in tissues immediately surrounding the bone marrow (Kopen et al., 1999). This multilinear potential has been expanded with a number of recent studies showing successful differentiation of MSCs into numerous cell types (Kopen et al., 1999).

There are several advantages in using MSCs as opposed to HSCs. Firstly, MSCs are easy to isolate and culture (Dexter et al., 1977; Phinney et al., 1999) compared to HSCs, which have been observed to be quite difficult to expand *in vitro*. Secondly, only small volumes of MSCs are required at harvesting whilst larger volumes are needed to obtain adequate numbers of HSCs (Prockop, 1997). MSCs harvested from the recipient themselves bypasses the risk of tissue rejection and transmission of prion proteins (Zhang and Bok, 1998). Due to their autologous nature, MSCs are considered advantageous over other therapies such as retinal transplantation.

1.4. Stem Cell Therapy of Ocular Disease

To date most studies that have investigated the potential of stem cells for the treatment of ocular diseases have focussed on the isolation of progenitor cells from the tissue of choice, in this case from both embryonic and adult retina (Reh and Kljavin, 1989; Ahmad et al., 1999, 2000). Isolation has involved the excision of both the optic nerve and mesenchymal tissue, the separation of the retina from the retinal pigment epithelium and the removal of the central portion of the retina connected to the optic nerve (Reh and Kljavin, 1989; Ahmad et al., 1999, 2000). Retinal progenitor cells are then dissociated into single cell suspensions and cultured under standard aseptic conditions. Retinal stem cells have been successfully differentiated into ganglion cells (Davis et al., 2000), amacrine cells (Ahmad et al., 2000) and both rod and cone photoreceptors (Reh and Kljavin, 1989). Despite the potential of retinal stem cells, the contra-lateral nature of the disease itself, the difficult microsurgical procedures involved and the availability of only a limited number of pluripotent cells makes this a non-viable option for treating these inherited ocular conditions.

1.5. Variation in stem cell characteristics

Marrow stromal cells have been successfully isolated and subcultured from a variety of species, including mouse, rat and human (Phinney et al., 1999; Azizi et al., 1998; Lu et al., 2001). Although the plasticity of MSCs between species has been demonstrated, slight variations have been noted to exist between MSC characteristics from various species. For example, despite a similar heterogeneity in MSC populations being observed in both rat and human, Javazon and colleagues (2001), found that rat MSCs were far more sensitive to low plating densities, yet they expanded more readily after low-density plating to give rise to confluent cultures.

Interestingly, an investigation conducted earlier by Phinney and co-workers (1999), found that despite the MSC population being heterogeneous in nature, progenitor cells isolated from five different strains of mice exhibited marked variations in their growth kinetics. However, despite these variations, it was also observed that these cells could be induced into desired cell types such as the osteocytes and adipocytes (Phinney et al., 1999). Knowing that variations may exist in MSCs isolated from different sources, a study was conducted to directly compare and characterize and plasticity of these cells from both rodent and murine species.

2. MATERIALS AND METHODS

2.1. Reagents

Alpha minimal essential medium (α-MEM), fetal bovine serum (FBS), normal rabbit serum, penicillin G, streptomycin sulfate, trypsin-EDTA and L-glutamine, were purchased from Invitrogen (Melbourne, Australia). Insulin, 5,8,11,14-eicosatetraynoic acid, dexamethasone, bovine serum albumin (BSA), activin A, opsin antibody and all secondary antibodies were obtained from Sigma Chemical Company (St Louis, Missouri). BD Pharmingen (San Diego, California) supplied all primary antibodies and the streptavidin-allophycocyanin conjugate (SAv-APC). All tissue culture plasticware was obtained from Becton Dickinson Labware (Franklin Lakes, New Jersey).

2.2. Isolation and Culture of MSCs

Primary cultures of rat (*RCS-rdy$^+$-p$^+$*) and mouse (*C57BL/6J*) MSCs were performed using standard techniques previously described (Phinney et al., 1999) and conducted in accordance with the ARVO Statement for the Use of Animals on Ophthalmic Vision Research.

2.3. Stromal Cell Characterization and Purification

Marow stromal cell population profiles from both animal models were compared using flow cytometry (FACSCalibur®, Becton Dickinson, San Jose, California). Briefly, harvested cells were fixed with 70% methanol, washed in PBS and incubated with fluorescence-conjugated monoclonal antibodies(1:100; 1h on ice) directed against the following cell surface markers (CD11b, CD45 and CD90. After washing, secondary antibody (SAv-APC; 1:100) was added and the cells incubated again on ice for 1h. Cells were washed a further two times prior to analysis. For purification the cells were sorted using a FACSVantage Cell Sorter (Becton Dickinson, San Jose, California) following the fluorescence-labeling procedure mentioned above. MSCs solely expressing CD90 were then aseptically collected, recultured and used in all subsequent experiments.

2.4. Photoreceptor Induction *in vitro*

Twenty four hours post seeding, media to $CD90^+$ cultures were replaced with α-MEM supplemented with FBS (1% v/v), penicillin (100U/mL), streptomycin (100μg/mL) and the inducing agent, activin A (100ng/mL). Cells were maintained for a further 8-10 days, fixed and then immunocytochemically analyzed.

2.5. Immunocytochemistry

$CD90^+$ MSCs were fixed, washed and then blocked in 5% BSA, 10% FBS (v/v), 0.1% Triton X-100 in 1 x PBS, for 1h at room temperature. Cells were incubated with various primary antibodies for 24h at $4^{o}C$ followed by secondary antibodies (FITC conjugated or Strepavidin biotin) for a similar period, and the antibody complex visualized using a fluorescence microscope (Olympus, Tokyo Japan). Primary antibodies included: microtubule-associated protein 2 (MAP2, 1:500), glial fibrillary acid protein (GFAP; 1:500), rhodopsin (RHOS, 1:100), protein kinase C (PKC, 1:100), cellular retinoic acid-binding protein (CRABP-1, 1:100), and nestin (1:1000). The number of immunoreactive cells was then quantitated and expressed as a percentage of the total number of cells counted.

2.6. Transduction of $CD90^+$ MSCs

$CD90^+$ MSCs were transduced with either adeno-associated virus (rAAV.GFP) or adenovirus (Ad.CMV.GFP) in α-MEM using the method of Lai et al., (2001).

3. RESULTS & DISCUSSION

3.1. MSC Population Profiles

Analysis of passage 1 rat MSC cultures revealed that very few MSCs expressed CD11b, with none expressing CD45, indicating that these cultures were predominantly free of MSCs commited to any of the lymphohematopoeitic lineages (Table 1). However, they were found to predominantly express CD90, consistent with their undifferentiated state (Table 1) and these observations confirmed results obtained by Woodbury and co-workers (2000), who had obtained a similar surface marker expression profile for rat MSCs. With continual passage it was observed that the number of cells expressing CD11b decreased, but the percentage of cells solely expressing CD90 remained high (~70% up to passage 8), indicating an easily accessible pool of undifferentiated cells.

In contrast, mouse MSCs were found to exhibit a markedly different profile. At early passage, mouse MSCs were heterogenoeus in nature, expressing multiple lineage markers in addition to the undifferentiated marker CD90. Over time, mouse MSCs became less heterogenous with most of the population expressing both CD90 and CD45 (Table 1). The significant observation was that the number of solely expressing CD90 cells increased to approximately 40% by passage 8, indicating that these dedifferentiated when cultured in basal medium over time (Table 1). A similar heterogeneity in the cell surface marker expression profile for mouse MSCs was found by Phinney and colleagues[20] who showed that over 80% of their murine MSC cultures coexpressed both CD45 and CD11b at early passage and that this expression pattern was maintained over multiple passages.

Table 1. CD90 expression comparison between normal rat and mice.

Species	Passage	90/45/11b	90/45	90/11b	90	45	11b
		% Cell Surface marker expression					
Rat	1	-	-	-	84.0 ± 3.8	-	4.15 ± 1.7
	4	-	-	-	80.3 ± 4.4	-	2.11 ± 0.8
	8	-	-	-	68.9 ± 6.1	-	2.00 ± 0.6
Mouse	1	99.8 ± 0.1	0.14 ± 0.01	0.03 ± 0.02	0.55 ± 0.007	-	-
	4	99.9 ± 0.5	0.01 ± 0.005	0.04 ± 0.02	0.05 ± 0.007	-	-
	8	9.6 ± 0.2	52.3 ± 0.2	0.83 ± 0.2	37.3 ± 0.5	-	-

When the sole expression of CD90 was compared between mouse and rat MSCs, a distinct difference was observed (Fig. 1). At early passage, rat MSCs predominantly express CD90, which decreases with subsequent passage to <20%. However, a statistically significant decrease ($p<0.001$) in number of $CD90^+$ cells only occurs after 8 passages, indicating a readily accessible pool of undifferentiated cells (Fig. 1). In contrast, very few mouse MSCs (<1%) are observed to solely express CD90 at early passage, but this pool increases markedly in size (~40%) after passage 8 ($p<0.001$). These results show that rat and mouse MSC populations are heterogeneous with respect to their cell surface expression markers. In addition, the pool size of undifferentiated cells also varies and needs to be considered when assessing their potential application to treat disease. Therefore, standard methods cannot be applied to the isolation and culturing of MSCs across species and further investigations will be required to examine whether this variation is also evident in human MSC cell populations.

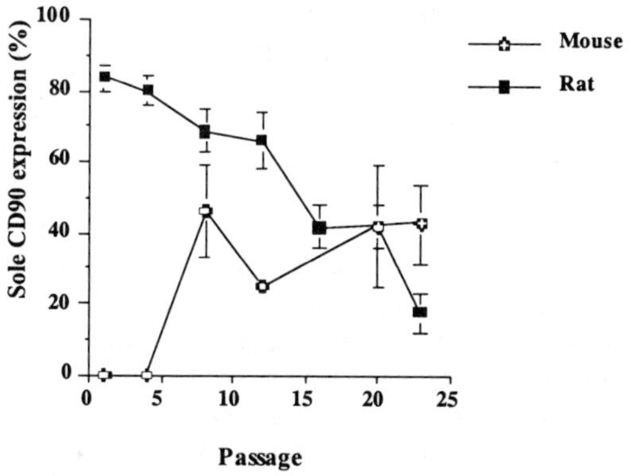

Figure 1. Comparison of the percentage of CD90 expression between mouse and rat MSCs.

3.2. Rat vs Mouse MSC Plasticity

Once profiled, undifferentiated $CD90^+$ cells from both rat (passage 2) and mouse (passage 18) origin were sorted and compared for their ability to differentiate into any of the retinal cell lineages. Here, CD90$^+$ cells were incubated with an inducing agent; activin A, which has previously been shown to be essential in retinal cell development and differentiation[9]. Following induction with activin A, ~ 30% of the rat stem cell population (Fig. 2A) were found to express rhodopsin (RHOS), a photoreceptor specific cell marker, whereas only ~2% of mouse CD90$^+$ MSCs were induced to express this marker (Fig. 2B).

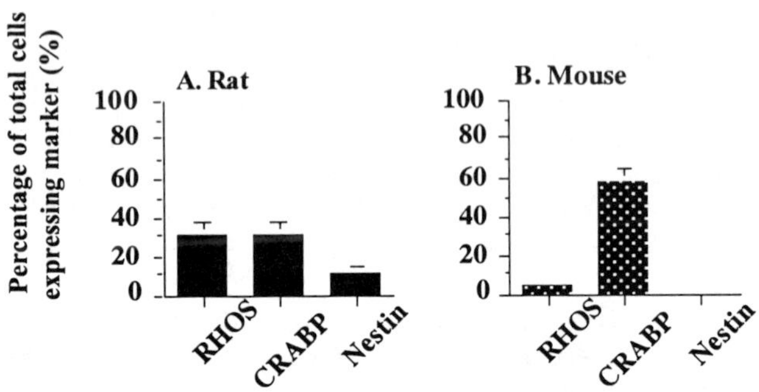

Figure 2. Effect of activin A on the differentiation of rat (A) and mouse (B) CD90$^+$ MSCs.

In addition, ~30% of rat CD90$^+$ MSCs were immunoreactive for cellular retinoic acid binding protein-1 (CRABP-1), which in the retina is primarily expressed by the amacrine cells. When mouse CD90$^+$ MSCs were incubated under similar conditions, over half (55%) of the cells were found to stain positively for this retinal specific lineage marker (Fig. 2A & B). Finally, activin A was also shown to induce a small number of rat CD90$^+$ MSCs (5-9%) to express nestin, a typical neural stem cell marker, whereas mouse CD90$^+$ cells did not stain for this marker (Fig. 2A & B). These results showed that rat and mouse MSCs varied in their capacity to differentiate into retinal specific lineages and that the rate of induction into these lineages also differed markedly between the two models. It remains uncertain whether this reduced capacity to differentiate in the mouse MSC population is a result of the extended culture period required to obtain a larger pool of undifferentiated MSCs, or whether other factors are involved which affect their ability to differentiate into retinal specific lineages.

3.3. Viral Tranduction Comparisons

The major limitation to the autologous transplantation of MSCs is that they themselves will contain the genetic abnormality responsible for the development of the retinal degeneration. Thus, permanent genetic correction in these cells is required in order to produce normal functioning photoreceptors and there are numerous vehicles with which this aim can be achieved. A set of experiments were conducted to determine whether a foreign gene (green fluorescent protein; GFP) could be introduced into these cells using either adenovirus (Ad.CMV. GFP) or adeno-associated virus (rAAV.GFP) as the vehicle for gene transfer. Results showed that CD90$^+$ MSC could be successfully transduced (Table 2) with both Ad.CMV.GFP and rAAV.GFP, although efficiencies did vary with different multiplicities of infection (MOI).

The Ad.CMV.GFP vehicle resulted in a higher efficiency of transduction in both rat and mouse CD90$^+$ MSCs, meaning that cells incubated with the adenoviral vehicle were successfully transduced at lower MOIs in both species. In contrast, higher MOIs were required to successfully transduce both rodent and murine CD90$^+$ MSC when utlizing rAAV.GFP. A potential problem to the clinical use of Ad.CMV.GFP, however, is that it has been shown to cause retinal degeneration in the rat when used in ocular gene therapy (Shen et al., 2001) and does not result in long-term transgene expression in MSCs (Schwarz et al., 1999). In addition, Cognet and Minguell (2000) observed that MSCs transduced with Ad.CMV.GFP could not differentiate into adipocytes suggesting that this vehicle may impair their trans-differentiation ability. The effects of rAAV.GFP transduction on MSCs have not been investigated, and thus experiments assessing their transgene expression and influence on pluripotency needs to be conducted for a direct comparison to be made. The fact that the transduction efficiency is extremely low highlights the need for either other viral vehicles such as retroviruses (Zhang et al., 2002) or other methods, such as non-viral DNA plasmids (Hamm et al., 2002) to be assessed.

Table 2. Transduction efficiencies of MSCs

Species	Vehicle	MOI	Efficiency (%)
Rat	rAAV.GFP	10:1	~1%
		100:1	~7%
	Ad.CMV.GFP	10:1	~28%
		100:1	~38%
Mouse	rAAV.GFP	10:1	-%
		100:1	~1%
	Ad.CMV.GFP	10:1	~1%
		100:1	~45%

4. CONCLUSION

The results show that MSCs have the potential to treat genetically inherited retinal degenerations and dystrophies. It was observed that bone marrow stem cell profiles differ between the rat and mouse models, and that they change over time, indicating that one standard method of characterization cannot be applied to cells from different species. In addition, it was also shown that pools of undifferentiated MSCs from both models were capable of being induced into retinal specific lineages, although the extent of their plasticity did vary. Both rat and mouse MSCs could be transduced using various viral vehicles, but the efficiencies of transduction did vary at different MOIs as well as between the different vehicles used. Finally, the adenoviral vehicle was found to be cytotoxic to both rat and mouse CD90$^+$ cells (data not shown) even at relatively low MOIs, indicating that these vehicles must be screened and biologically assessed before being used in conjunction with stem cell therapies.

5. ACKNOWLEDGEMENTS

The National Eye Institute (NEI) and the Foundation Fighting Blindness (FFB) are sincerely thanked for providing the first author with a travel award to attend this meeting.

6. REFERENCES

Ahmad, I., Dooley, C.M., Thoreson, W.B., Rogers, J.A. and Afiat, S., 1999, In vitro analysis of a mammalian retinal progenitor that gives rise to neurons and glia. *Brain Res.* **831**:1-10.

Ahmad, I., Tang, L., and Pham, H., 2000, Identification of neural progenitors in the adult mammalian eye. *Biochem Biophys Res Commun.* **270**:517-521.

Ali, R.R., Sarra, G.M., Stephens, C., Alwis, M.D., Bainbridge, J.W., Munro, P.M., Fauser, S., Reichel, M.B., Kinnon, C., Hunt, D.M., Bhattacharya, S.S., and Thrasher, A.J., 2000, Restoration of photoreceptor ultrastructure and function in retinal degeneration slow mice by gene therapy. *Nat Genet.* **25**:306-310.

Azizi, S.A., Stokes, D., Augelli, B.J., DiGirolamo, C., and Prockop, D.J., 1998, Engraftment and migration of human bone marrow stromal cells implanted in the brains of albino rats--similarities to astrocyte grafts. *Proc Natl Acad Sci USA.* **95**:3908-3913.

Conget, P.A. and Minguell, J.J., 2000, Adenoviral-mediated gene transfer into ex vivo expanded human bone marrow mesenchymal progenitor cells. *Exp Hematol.* **28**:382-390.

Davis, A.A., Matzuk, M.M., and Reh, T.A., 2000, Activin A promotes progenitor differentiation into photoreceptors in rodent retina. *Mol Cell Neurosci.* **15**:11-21.

Dexter, T.M, Allen, T.D., and Lajtha, L.G., 1977, Conditions controlling the prolifration of hematopoietic stem cells in vitro. *J Cell Physiol.* **91**:335-344.

Hamm, A., Krott, N., Breibach, I., Blindt, R., and Bosserhoff, A.K., 2002, Efficient transfection method for primary cells. *Tissue Eng.* **8**:235-245.

Javazon, E.H., Colter, D.C., Schwarz, E.J., and Prockop, D.J., 2001, Rat marrow stromal cells are more sensitive to plating density and expand more rapidly from single-cell-derived colonies than human marrow stromal cells. *Stem Cells* **19**:219-225.

Kopen, G.C., Prockop, D.J., and Phinney, D.G., 1999, Marrow stromal cells migrate throughout forebrain and cerebellum, and they differentiate into astrocytes after injection into neonatal mouse brains. *Proc Natl Acad Sci USA.* **96**:10711-10716.

Lai, C.M., Brankov, M., Zaknich, T., Lai, Y.K., Shen, W.Y., Constable, I.J., Kovesdi, I., and Rakoczy, P.E., 2001, Inhibition of angiogenesis by adenovirus-mediated sFlt-1 expression in a rat model of corneal neovascularization. *Hum Gene Ther.* **12**:1299-1310.

Lolley, R.N., Rong, H., and Craft, C.M., 1994, Linkage of photoreceptor degeneration by apoptosis with inherited defect in phototransduction. *Invest Ophthalmol Vis Sci.* **35**:358-362.

Lu, D., Mahmood, A., Wang, L., Li, Y., Lu, M., and Chopp, M., 2001, Adult bone marrow stromal cells administered intravenously to rats after traumatic brain injury migrate into brain and improve neurological outcome. *Neuroreport* **12**:559-563.

Morrison, S.J., White, P.M., Zock, C., and Anderson, D.J., 1999, Prospective identification, isolation by flow cytometry, and in vivo self-renewal of multipotent mammalian neural crest stem cells. *Cell* **96**:737-749.

Phinney, D.G., Kopen, G., Isaacson, R.L., and Prockop, D.J., 1999, Plastic adherent stromal cells from the bone marrow of commonly used strains of inbred mice: variations in yield, growth, and differentiation. *J Cell Biochem.* **72**:570-585.

Prockop, D.J., 1997, Marrow stromal cells as stem cells for nonhematopoietic tissues. *Science* **276**:71-74.

Reh, T.A. and Kljavin, I.J., 1989, Age of differentiation determines rat retinal germinal cell phenotype: induction of differentiation by dissociation. *J Neurosci.* **9**:4179-4189.

Rubio, F.J., Bueno, C., Villa, A., Navarro, B., and Martinez-Serrano, A., 2000, Genetically perpetuated human neural stem cells engraft and differentiate into the adult mammalian brain. *Mol Cell Neurosci.* **16**:1-13.

Schwarz, E.J., Alexander, G.M., Prockop, D.J., and Azizi, S.A., 1999, Multipotential marrow stromal cells transduced to produce L-DOPA: engraftment in a rat model of Parkinson disease. *Hum Gene Ther.* **10**:2539-2549.

Shen, W.Y., Lai, M.C., Beilby, J., Barnett, N.L., Liu, J., Constable, I.J., and Rakoczy, P.E., 2001, Combined effect of cyclosporine and sirolimus on improving the longevity of recombinant adenovirus-mediated transgene expression in the retina. *Arch Ophthalmol.* **119**:1033-1043.

Woodbury, D., Schwarz, E.J., Prockop, D.J., and Black, I.B., 2000, Adult rat and human bone marrow stromal cells differentiate into neurons. *J Neurosci Res.* **61**:364-370.

Zhang, X. and Bok, D., 1998, Transplantation of retinal pigment epithelial cells and immune response in the subretinal space. *Invest Ophthalmol Vis Sci.* **39**:1021-1027.

Zhang, X.Y., La Russa, V.F., Bao, L., Kolls, J., Schwarzenberger, P., and Reiser, J., 2002, Lentiviral vectors for sustained transgene expression in human bone marrow-derived stromal cells. *Mol Ther.* **5**:555-565.

GENE THERAPY FOR LEBER CONGENITAL AMAUROSIS

Nadine S. Dejneka, Enrico M. Surace, and Jean Bennett[1]

1. LEBER CONGENITAL AMAUROSIS (LCA)

Leber congenital amaurosis (LCA) accounts for 5% of all inherited retinal dystrophies (Foxman et al., 1985; Kaplan et al., 1990), causing blindness in infants (Leber, 1869). Children are diagnosed before one year of age and present with impaired vision or total blindness, a normal fundus exam and a severely depressed or absent electroretinogram (ERG) (Foxman et al., 1985). Nystagmus, eye poking and hyperopia are also often observed (Franceschetti and Forni, 1958; Wagner et al., 1985). Over time, visual acuity, visual field and the ERG generally remain constant; however, fundus abnormalities begin to develop. These can include salt-and-pepper pigmentation, bone spicule pigment, an attenuation of retinal vessels and atrophy of the retinal pigment epithelium (RPE) and choriocapillaris (Perrault et al., 1999).

2. THE GENETIC BASIS OF LCA

LCA is a genetically heterogeneous disorder that primarily has a recessive mode of transmittance; however, dominant disease has been reported (Cremers et al., 2002). Several genes can cause the disease and are thought to account for 50% of all LCA cases (Cremers et al., 2002). Mutations in *GUCY2D* (*RETGC1*) (Perrault et al., 1996), *RPE65* (Gu et al., 1997; Marlhens et al., 1997; Morimura et al., 1998), *CRX* (Freund et al., 1998; Swaroop et al., 1999), *AIPL1* (Sohocki et al., 2000; Sohocki et al., 2000), *RPGRIP1* (Dryja et al., 2001; Gerber et al., 2001) and *CRB1* (den Hollander et al., 2001; Lotery et al., 2001) have been identified in patients. Each of these genes is expressed in the retina and has a unique function.

[1]University of Pennsylvania, Scheie Eye Institute, Philadelphia, PA 19104-6069, USA. Corresponding author: N. S. Dejneka, E-mail: ndejneka@mail.med.upenn.edu

The first gene identified, *GUCY2D,* appears to be the major LCA causative gene (Perrault et al., 2000) and is primarily localized in cone and rod outer segments (Liu et al., 1994). It codes for guanylate cyclase 2D, which plays a critical role in the phototransduction cascade during the recovery of the dark state following excitation of photoreceptors by light (Perrault et al., 1999). Photoexcitation results in the hydrolysis of cGMP and a decrease in intracellular free Ca^{2+}. Ca^{2+} levels stimulate the production of guanylate cyclase, and this enzyme then catalyzes the conversion of GTP to 3'5'cGMP, restoring cGMP levels in the cell. *In vitro* studies suggest that missense mutations in the catalytic domain of *GUCY2D* destroy its ability to hydrolyze GTP, while mutations in the extracellular domain affect protein folding, resulting in enzyme degradation (Duda et al., 1999; Rozet et al., 2001).

The second gene identified, *RPE65*, accounts for 7-16% of all known cases of LCA (Marlhens et al., 1997; Morimura et al., 1998). This gene codes for a unique, highly conserved, 61kDa protein (533 amino acids) (Boulanger et al., 2001; Nicoletti et al., 1995) that is developmentally regulated and primarily expressed in the microsomal membrane fraction of the RPE (Tsilou et al., 1997). In addition, expression has also been confirmed in mammalian cone photoreceptors (Znoiko et al., 2002) and transformed kidney and RPE cell lines (Ma et al., 1999). The function of RPE65 is not clear; however, studies in *Rpe65-/-* mice reveal that the protein is a crucial component of the visual cascade. *Rpe65-/-* mice are incapable of producing detectable amounts of 11-*cis*-retinal and all-*trans*-retinyl esters accumulate in the RPE (Redmond et al., 1998). There is also an accumulation of lipid droplets and a considerable reduction in lipofuscin levels in the RPE (Katz and Redmond, 2001). Free opsin is readily available in photoreceptors, but rhodopsin regeneration is impaired and animals possess disorganized outer segment discs and severely depressed rod ERGs (Redmond et al., 1998).

CRX (cone-rod homeobox) was the third gene identified to cause LCA. It encodes a homeobox transcription factor that plays an important role in photoreceptor cell differentiation and maintenance (Chen et al., 1997). Localized to photoreceptors and pinealocytes, it is expressed during development and is required for the elongation of photoreceptor outer segments. Furthermore, it works synergistically with the transcription factor NRL (neural leucine-zipper) to regulate the expression of photoreceptor-specific genes (Chen et al., 1997).

AIPL1 (aryl hydrocarbon receptor-interacting protein-like 1), the fourth causative gene identified, has also been localized to photoreceptors and pinealocytes (Sohocki et al., 2000; van der Spuy et al., 2002), and studies suggest it may cause approximately 9-11% of LCA (Sohocki et al., 2001). The protein's function is not clear; however, it is structurally similar to the FK506-binding proteins, which act as molecular chaperones (Sohocki et al., 2001).

The functions of CRB1 (crumbs homolog 1) and RPGRIP1 (retinitis pigmentosa GTPase regulator-interacting protein 1) are also not well defined. *CRB1* encodes a protein that is homologous to the *Drosophila* crumbs protein, which plays a role in photoreceptor morphogenesis (Izaddoost et al., 2002; Pellikka et al., 2002). CRB1 may have an analogous role in mammalian photoreceptors. *RPGRIP1* encodes a protein that interacts with the retinitis pigmentosa GTPase regulator (RPGR). It is localized to rod outer segments (Roepman et al., 2000) and connecting cilium of photoreceptor cells (Hong et al., 2001). It may be involved in vesicular trafficking or anchoring RPGR.

3. GENE THERAPY FOR RPE65 MUTATIONS

It is clear that the genes responsible for LCA are very diverse, and little is still known about many of their functions. In order to develop treatments for patients with these gene mutations, it will be necessary not only to learn more about the proteins they encode, but also to develop or identify animal models that mimic the human disease.

Two animal models with *RPE65* mutations exist, and they have allowed for testing of potential therapeutics for LCA. The first is the genetically engineered animal model, the *Rpe65-/-* mouse (Redmond et al., 1998), described above. It has been used to test whether oral administration of 9-*cis* retinal can restore the LCA phenotype (Van Hooser et al., 2000; Van Hooser et al., 2002). Pharmacological intervention resulted in the formation of rod photopigment and an improvement in rod physiology. The effects have persisted for more than 6 months, suggesting that this compound may be useful in humans with *RPE65* mutations.

The second model is the Swedish Briard dog. This animal has a naturally occurring homozygous 4-bp deletion in the *RPE65* gene that causes an early and severe visual impairment that is similar to humans (Veske et al., 1999). Viral-mediated delivery of canine *RPE65* has been explored in these animals and shows great promise for future clinical trials. An adeno-associated virus (AAV) carrying the canine *RPE65* cDNA, under the regulation of chicken-β-actin promoter, was delivered to these animals unilaterally via a subretinal approach (Acland et al., 2001). Three months post-injection, animals were analyzed for functional improvement using ERG analysis and pupillometry. The results demonstrated that retinal function was definitively restored in eyes treated with AAV-*RPE65* and that retinal activity was being transmitted to higher visual pathways in the brain. Follow-up analysis 13 months post-treatment showed that the effect persisted (Acland et al., 2002). In another group of studies using a cytomegalovirus (CMV) promoter to drive AAV-*RPE65* expression, it was reported that treatment not only restored function, but also partially restored retinal morphology (Bragadottir et al., 2002; Katz et al., 2002; Narfstrom et al., 2002). Characteristic lipid droplets present in the RPE disappeared in the areas transduced by virus; however, there were no changes in outer segment morphology. The studies also concluded that there were no adverse systemic effects following AAV-mediated therapy; however, there was a risk of developing uveitis (Caro et al., 2002).

4. FUTURE THERAPEUTIC DIRECTIONS

It is clear that gene therapy holds much promise for the treatment of LCA; however, numerous issues must be addressed before such a therapy is administered to humans. Long-term safety and efficacy are critical. Ideally, one would like to develop a viral vector that exclusively targets the affected tissue, avoiding systemic complications.

Advances in vector biology are allowing researchers to develop vectors with very specific targeting abilities. Eight different AAV serotypes have been identified and the exchange of surface proteins impacts the viral vector's cellular specificity and transduction characteristics. Recombinant AAV is produced by a triple transfection procedure. One plasmid encodes the AAV ITRS flanking the therapeutic gene of interest. A second encodes the AAV Rep and Cap proteins, and a third encodes helper replication sequences of the helper virus (Monahan and Samulski, 2000). By cloning each of the serotype capsid coding domains into a common vector backbone containing AAV type 2 replication genes, it is possible to cross-package a single AAV2 vector genome into multiple AAV serotypes (Auricchio et al., 2001; Rabinowitz et al., 2002; Yang et al., 2002). For example, AAV2/5 refers to an AAV serotype 2 genome packaged in an AAV serotype 5 capsid. In the mouse retina, AAV2/2 primarily targets photoreceptor cells and a smaller number of RPE cells following subretinal administration (Ali et al., 1996; Bennett et al., 1997; Flannery et al., 1997). In contrast, AAV2/5 efficiently transduces both cell types (Auricchio et al., 2001), while AAV2/1 and 2/6 efficiently target RPE (Auricchio et al., 2001; Yang et al., 2002).

Lentiviruses have also been shown to successfully transduce the retina (Miyoshi et al., 1997). These complex, enveloped retroviruses may be modulated by pseudotyping the envelope protein (Auricchio et al., 2001; Miyoshi et al., 1997; Naldini et al., 1996; Naldini et al., 1996). The vesicular stomatitis virus glycoprotein (VSV-G) envelope allows for transduction of RPE and photoreceptors following subretinal delivery, while the Mokola envelope restricts transgene expression to the RPE (Auricchio et al., 2001).

Another aspect of vector design, which may improve the efficacy of treatment, includes the use of tissue- and species-specific constructs. The *RPE65* cDNA is highly conserved between species (Boulanger et al., 2001). The mouse and human genes are nearly 94% homologous, while the canine and human genes are 98% homologous, suggesting that any of the three translated products would function well between species. Nevertheless, in many cases a gene may not be highly conserved and so it will be important to deliver the endogenous transgene to the species of interest.

Additionally, it may be efficacious to regulate expression with a highly tissue-specific promoter. Analysis of the upstream region of the mouse *RPE65* gene revealed that the 5' flanking region, containing bases –655 to +52, was sufficient to direct high RPE-specific expression in transgenic mice (Boulanger et al., 2000). *In vitro* analysis revealed that the region between –83 and +39 was responsible for basal human promoter activity (Nicoletti et al., 1998). Furthermore, analysis of the human promoter indicated the 5' flanking region containing bases –800 to +21 supported AAV-mediated gene expression in rodent RPE cells, with less efficient expression in photoreceptor cells (Li et al., 2002).

A final consideration for gene therapy is the timing of viral vector delivery in order to have a therapeutic effect. Viral administration in adult rodents has been performed in many different mouse models. In addition, transgene expression can be achieved following post-natal administration. The success of the treatment is determined not only by the ability of the virus to transduce cells efficiently, but is also influenced by the therapeutic window of the disease.

It is clear that in the canine LCA model virus may be delivered as late as 11 months of age (Narfstrom et al., 2002) to achieve a therapeutic effect. It is not clear if treatment in a child of equivalent age will be as successful or if earlier delivery may be more advantageous or necessary. Patients with rapidly degenerative diseases may benefit from fetal delivery of virus, and such an approach may ultimately be useful in human LCA since degeneration may commence during gestation (Porto et al., 2002). Recently a technique has been developed to deliver virus *in utero* which results in efficient transduction of retinal cells (Surace et al., 2002). Injections of AAV-EGFP were performed in pregnant mice on embryonic days 13 or 14. The injected pups were delivered by Caesarian section on gestational day 19 and nursed by a foster mother.

EGFP expression could be detected histologically on days 8, 3 and 1 following injection of AAV2/2, AAV2/5 and AAV2/1 respectively. In all cases, expression was first observed in RPE. Shortly afterwards, expression was observed in neural retina, and upon retinal differentiation, high levels of transgene expression were observed in photoreceptors, with strong, stable expression persisting throughout the entire retina.

5. CONCLUSIONS

Gene therapy for LCA may soon become a reality for patients with *RPE65* mutations. Experiments are currently underway to develop an optimal viral-vector delivery vehicle, determine the appropriate time for treatment and assess the safety of the therapy. Studies in the Swedish Briard dog are the first of their kind to demonstrate long-term rescue of retinal function in a large animal model and provide hope for a similar therapeutic effect in humans.

6. ACKNOWLEDGMENTS

We thank the National Eye Institute and Foundation Fighting Blindness for the travel award provided to NSD to attend this meeting.

7. REFERENCES

Acland, G. M., Aguirre, G. D., Aleman, T. S., Cideciyan, A. V., Maguire, A. M., Jacobson, S. G., Hauswirth, W. W., and Bennett, J., 2002, Continuing evaluation of gene therapy in the Rpe65 mutant dog, ARVO Annual Meeting, Ft. Lauderdale, FL, 4593.

Acland, G. M., Aguirre, G. D., Ray, J., Zhang, Q., Aleman, T. S., Cideciyan, A. V., Pearce-Kelling, S. E., Anand, V., Zeng, Y., Maguire, A. M., Jacobson, S. G., Hauswirth, W. W., and Bennett, J., 2001, Gene therapy restores vision in a canine model of childhood blindness, *Nat Genet.* **28**:92-5.

Ali, R. R., Reichel, M. B., Thrasher, A. J., Levinsky, R. J., Kinnon, C., Kanuga, N., Hunt, D. M., and Bhattacharya, S. S., 1996, Gene transfer into the mouse retina mediated by an adeno-associated viral vector, *Hum Mol Genet.* **5**:591-4.

Auricchio, A., Kobinger, G., Anand, V., Hildinger, M., O'Connor, E., Maguire, A. M., Wilson, J. M., and Bennett, J., 2001, Exchange of surface proteins impacts on viral vector cellular specificity and transduction characteristics: the retina as a model, *Hum Mol Genet.* **10**:3075-81.

Bennett, J., Duan, D., Engelhardt, J. F., and Maguire, A. M., 1997, Real-time, noninvasive in vivo assessment of adeno-associated virus- mediated retinal transduction, *Invest Ophthalmol Vis Sci.* **38**:2857-63.

Boulanger, A., Liu, S., Henningsgaard, A. A., Yu, S., and Redmond, T. M., 2000, The upstream region of the Rpe65 gene confers retinal pigment epithelium-specific expression in vivo and in vitro and contains critical octamer and E-box binding sites, *J Biol Chem.* **275**:31274-82.

Boulanger, A., Liu, S., Yu, S., and Redmond, T. M., 2001, Sequence and structure of the mouse gene for RPE65, *Mol Vis.* **7**:283-7.

Bragadottir, R., Lei, B., and Narfstrom, K., 2002, Electroretinographic monitoring of gene therapy treated RPE65 mull mutation dog, a model of human Leber's Congenital Amaurosis, ARVO Annual Meeting, Ft. Lauderdale, FL, 4596.

Caro, M. L., Estes, D. M., Cohn, L. A., and Narfstrom, K., 2002, The systemic and ocular immune response after recombinant adeno-associated virus gene transfer of RPE65 in 6 dogs affected with the RPE65 null mutation, ARVO Annual Meeting, Ft. Lauderdale, FL, 4594.

Chen, S., Wang, Q. L., Nie, Z., Sun, H., Lennon, G., Copeland, N. G., Gilbert, D. J., Jenkins, N. A., and Zack, D. J., 1997, Crx, a novel Otx-like paired-homeodomain protein, binds to and transactivates photoreceptor cell-specific genes, *Neuron.* **19**:1017-30.

Cremers, F. P., Van Den Hurk, J. A., and Den Hollander, A. I., 2002, Molecular genetics of Leber congenital amaurosis, *Hum Mol Genet.* **11**:1169-76.

den Hollander, A. I., Heckenlively, J. R., van den Born, L. I., de Kok, Y. J., van der Velde-Visser, S. D., Kellner, U., Jurklies, B., van Schooneveld, M. J., Blankenagel, A., Rohrschneider, K., Wissinger, B., Cruysberg, J. R., Deutman, A. F., Brunner, H. G., Apfelstedt-Sylla, E., Hoyng, C. B., and Cremers, F. P., 2001, Leber congenital amaurosis and retinitis pigmentosa with Coats-like exudative vasculopathy are associated with mutations in the crumbs homologue 1 (CRB1) gene, *Am J Hum Genet.* **69**:198-203.

Dryja, T. P., Adams, S. M., Grimsby, J. L., McGee, T. L., Hong, D. H., Li, T., Andreasson, S., and Berson, E. L., 2001, Null RPGRIP1 alleles in patients with Leber congenital amaurosis, *Am J Hum Genet.* **68**:1295-8.

Duda, T., Venkataraman, V., Goraczniak, R., Lange, C., Koch, K. W., and Sharma, R. K., 1999, Functional consequences of a rod outer segment membrane guanylate cyclase (ROS-GC1) gene mutation linked with Leber's congenital amaurosis, *Biochemistry.* **38**:509-15.

Flannery, J. G., Zolotukhin, S., Vaquero, M. I., LaVail, M. M., Muzyczka, N., and Hauswirth, W. W., 1997, Efficient photoreceptor-targeted gene expression in vivo by recombinant adeno-associated virus, *Proc Natl Acad Sci U S A.* **94**:6916-21.

Foxman, S. G., Heckenlively, J. R., Bateman, J. B., and Wirtschafter, J. D., 1985, Classification of congenital and early onset retinitis pigmentosa, *Arch Ophthalmol.* **103**:1502-6.

Franceschetti, A., and Forni, S., 1958, Degenerescence Tapeto-retinienne (type Leber) avec aspect marbre du fond d'oeil peripherique, *Ophthalmologica.* **135**:610-616.

Freund, C. L., Wang, Q. L., Chen, S., Muskat, B. L., Wiles, C. D., Sheffield, V. C., Jacobson, S. G., McInnes, R. R., Zack, D. J., and Stone, E. M., 1998, De novo mutations in the CRX homeobox gene associated with Leber congenital amaurosis, *Nat Genet.* **18**:311-2.

Gerber, S., Perrault, I., Hanein, S., Barbet, F., Ducroq, D., Ghazi, I., Martin-Coignard, D., Leowski, C., Homfray, T., Dufier, J. L., Munnich, A., Kaplan, J., and Rozet, J. M., 2001, Complete exon-intron structure of the RPGR-interacting protein (RPGRIP1) gene allows the identification of mutations underlying Leber congenital amaurosis, *Eur J Hum Genet.* **9**:561-71.

Gu, S. M., Thompson, D. A., Srikumari, C. R., Lorenz, B., Finckh, U., Nicoletti, A., Murthy, K. R., Rathmann, M., Kumaramanickavel, G., Denton, M. J., and Gal, A., 1997, Mutations in RPE65 cause autosomal recessive childhood-onset severe retinal dystrophy, *Nat Genet.* **17**:194-7.

Hong, D. H., Yue, G., Adamian, M., and Li, T., 2001, Retinitis pigmentosa GTPase regulator (RPGRr)-interacting protein is stably associated with the photoreceptor ciliary axoneme and anchors RPGR to the connecting cilium, *J Biol Chem.* **276**:12091-9.

Izaddoost, S., Nam, S. C., Bhat, M. A., Bellen, H. J., and Choi, K. W., 2002, Drosophila Crumbs is a positional cue in photoreceptor adherens junctions and rhabdomeres, *Nature.* **416**:178-83.

Kaplan, J., Bonneau, D., Frezal, J., Munnich, A., and Dufier, J. L., 1990, Clinical and genetic heterogeneity in retinitis pigmentosa, *Hum Genet.* **85**:635-42.

Katz, M. L., and Redmond, T. M., 2001, Effect of Rpe65 knockout on accumulation of lipofuscin fluorophores in the retinal pigment epithelium, *Invest Ophthalmol Vis Sci.* **42**:3023-30.

Katz, M. L., Redmond, T. M., and Narfstrom, K., 2002, Adeno-associated viral mediated gene therapy for RPE65 gene defect in a canine model - partial correction of morphological phenotype, ARVO Annual Meeting, Ft. Lauderdale, FL, 4592.

Leber, T., 1869, Uber retinitis pigmentosa and angeborene amaurose, *Graefes Arch Klin Exp Ophthalmol.* **15**:13-20.

Li, Q., Glushakova, L., Doyle, T. J., Timmers, A. M., Beliaev, D., and Hauswirth, W. W., 2002, Only a specific region of the human RPE65 promoter supports AAV-vectored reporter gene expression in rodent RPE cells, ARVO Annual Meeting, Ft. Lauderdale, FL, 3640.

Liu, X., Seno, K., Nishizawa, Y., Hayashi, F., Yamazaki, A., Matsumoto, H., Wakabayashi, T., and Usukura, J., 1994, Ultrastructural localization of retinal guanylate cyclase in human and monkey retinas, *Exp Eye Res.* **59**:761-8.

Lotery, A. J., Jacobson, S. G., Fishman, G. A., Weleber, R. G., Fulton, A. B., Namperumalsamy, P., Heon, E., Levin, A. V., Grover, S., Rosenow, J. R., Kopp, K. K., Sheffield, V. C., and Stone, E. M., 2001, Mutations in the CRB1 gene cause Leber congenital amaurosis, *Arch Ophthalmol.* **119**:415-20.

Ma, J. X., Zhang, D., Laser, M., Brownlee, N. A., Re, G. G., Hazen-Martin, D. J., Redmond, T. M., and Crouch, R. K., 1999, Identification of RPE65 in transformed kidney cells, *FEBS Lett.* **452**:199-204.

Marlhens, F., Bareil, C., Griffoin, J. M., Zrenner, E., Amalric, P., Eliaou, C., Liu, S. Y., Harris, E., Redmond, T. M., Arnaud, B., Claustres, M., and Hamel, C. P., 1997, Mutations in RPE65 cause Leber's congenital amaurosis, *Nat Genet.* **17**:139-41.

Miyoshi, H., Takahashi, M., Gage, F. H., and Verma, I. M., 1997, Stable and efficient gene transfer into the retina using an HIV-based lentiviral vector, *Proc Natl Acad Sci U S A.* **94**:10319-23.

Monahan, P. E., and Samulski, R. J., 2000, Adeno-associated virus vectors for gene therapy: more pros than cons?, *Mol Med Today.* **6**:433-40.

Morimura, H., Fishman, G. A., Grover, S. A., Fulton, A. B., Berson, E. L., and Dryja, T. P., 1998, Mutations in the RPE65 gene in patients with autosomal recessive retinitis pigmentosa or leber congenital amaurosis, *Proc Natl Acad Sci U S A.* **95**:3088-93.

Naldini, L., Blomer, U., Gage, F. H., Trono, D., and Verma, I. M., 1996, Efficient transfer, integration, and sustained long-term expression of the transgene in adult rat brains injected with a lentiviral vector, *Proc Natl Acad Sci U S A.* **93**:11382-8.

Naldini, L., Blomer, U., Gallay, P., Ory, D., Mulligan, R., Gage, F. H., Verma, I. M., and Trono, D., 1996, In vivo gene delivery and stable transduction of nondividing cells by a lentiviral vector, *Science.* **272**:263-7.

Narfstrom, K., Bragadottir, R., Redmond, T. M., Katz, M. L., Lei, B., Lai, C. M., and Rakoczy, E. P., 2002, Gene therapy in 6 dogs with RPE65 null mutation improves visual function: a short-term study using clinical observations, electrophysiology and morphology, ARVO Annual Meeting, Ft. Lauderdale, FL, 4601.

Nicoletti, A., Kawase, K., and Thompson, D. A., 1998, Promoter analysis of RPE65, the gene encoding a 61-kDa retinal pigment epithelium-specific protein, *Invest Ophthalmol Vis Sci.* **39**:637-44.

Nicoletti, A., Wong, D. J., Kawase, K., Gibson, L. H., Yang-Feng, T. L., Richards, J. E., and Thompson, D. A., 1995, Molecular characterization of the human gene encoding an abundant 61 kDa protein specific to the retinal pigment epithelium, *Hum Mol Genet.* **4**:641-9.

Pellikka, M., Tanentzapf, G., Pinto, M., Smith, C., McGlade, C. J., Ready, D. F., and Tepass, U., 2002, Crumbs, the Drosophila homologue of human CRB1/RP12, is essential for photoreceptor morphogenesis, *Nature.* **416**:143-9.

Perrault, I., Rozet, J. M., Calvas, P., Gerber, S., Camuzat, A., Dollfus, H., Chatelin, S., Souied, E., Ghazi, I., Leowski, C., Bonnemaison, M., Le Paslier, D., Frezal, J., Dufier, J. L., Pittler, S., Munnich, A., and Kaplan, J., 1996, Retinal-specific guanylate cyclase gene mutations in Leber's congenital amaurosis, *Nat Genet.* **14**:461-4.

Perrault, I., Rozet, J. M., Gerber, S., Ghazi, I., Ducroq, D., Souied, E., Leowski, C., Bonnemaison, M., Dufier, J. L., Munnich, A., and Kaplan, J., 2000, Spectrum of retGC1 mutations in Leber's congenital amaurosis, *Eur J Hum Genet.* **8**:578-82.

Perrault, I., Rozet, J. M., Gerber, S., Ghazi, I., Leowski, C., Ducroq, D., Souied, E., Dufier, J. L., Munnich, A., and Kaplan, J., 1999, Leber congenital amaurosis, *Mol Genet Metab.* **68**:200-8.

Porto, F. B., Perrault, I., Hicks, D., Rozet, J. M., Hanoteau, N., Hanein, S., Kaplan, J., and Sahel, J. A., 2002, Prenatal human ocular degeneration occurs in Leber's congenital amaurosis (LCA2), *J Gene Med.* **4**:390-6.

Rabinowitz, J. E., Rolling, F., Li, C., Conrath, H., Xiao, W., Xiao, X., and Samulski, R. J., 2002, Cross-packaging of a single adeno-associated virus (AAV) type 2 vector genome into multiple AAV serotypes enables transduction with broad specificity, *J Virol.* **76**:791-801.

Redmond, T. M., Yu, S., Lee, E., Bok, D., Hamasaki, D., Chen, N., Goletz, P., Ma, J. X., Crouch, R. K., and Pfeifer, K., 1998, Rpe65 is necessary for production of 11-cis-vitamin A in the retinal visual cycle, *Nat Genet.* **20**:344-51.

Roepman, R., Bernoud-Hubac, N., Schick, D. E., Maugeri, A., Berger, W., Ropers, H. H., Cremers, F. P., and Ferreira, P. A., 2000, The retinitis pigmentosa GTPase regulator (RPGR) interacts with novel transport-like proteins in the outer segments of rod photoreceptors, *Hum Mol Genet.* **9**:2095-105.

Rozet, J. M., Perrault, I., Gerber, S., Hanein, S., Barbet, F., Ducroq, D., Souied, E., Munnich, A., and Kaplan, J., 2001, Complete abolition of the retinal-specific guanylyl cyclase (retGC-1) catalytic ability consistently leads to leber congenital amaurosis (LCA), *Invest Ophthalmol Vis Sci.* **42**:1190-2.

Sohocki, M. M., Bowne, S. J., Sullivan, L. S., Blackshaw, S., Cepko, C. L., Payne, A. M., Bhattacharya, S. S., Khaliq, S., Qasim Mehdi, S., Birch, D. G., Harrison, W. R., Elder, F. F., Heckenlively, J. R., and Daiger, S. P., 2000, Mutations in a new photoreceptor-pineal gene on 17p cause Leber congenital amaurosis, *Nat Genet.* **24**:79-83.

Sohocki, M. M., Perrault, I., Leroy, B. P., Payne, A. M., Dharmaraj, S., Bhattacharya, S. S., Kaplan, J., Maumenee, I. H., Koenekoop, R., Meire, F. M., Birch, D. G., Heckenlively, J. R., and Daiger, S. P., 2000, Prevalence of AIPL1 mutations in inherited retinal degenerative disease, *Mol Genet Metab.* **70**:142-50.

Sohocki, M. M., Sullivan, L. S., Tirpak, D. L., and Daiger, S. P., 2001, Comparative analysis of aryl-hydrocarbon receptor interacting protein- like 1 (Aipl1), a gene associated with inherited retinal disease in humans, *Mamm Genome.* **12**:566-8.

Surace, E. M., Auricchio, A., Reich, S. J., Pineles, S., O'Connor, E., Wilson, J. M., and Bennett, J., 2002, In utero AAV vector-mediated gene delivery to the retina, ARVO Annual Meeting, Ft. Lauderdale, FL, 4617.

Swaroop, A., Wang, Q. L., Wu, W., Cook, J., Coats, C., Xu, S., Chen, S., Zack, D. J., and Sieving, P. A., 1999, Leber congenital amaurosis caused by a homozygous mutation (R90W) in the homeodomain of the retinal transcription factor CRX: direct evidence for the involvement of CRX in the development of photoreceptor function, *Hum Mol Genet.* **8**:299-305.

Tsilou, E., Hamel, C. P., Yu, S., and Redmond, T. M., 1997, RPE65, the major retinal pigment epithelium microsomal membrane protein, associates with phospholipid liposomes, *Arch Biochem Biophys.* **346**:21-7.

van der Spuy, J., Chapple, J. P., Clark, B. J., Luthert, P. J., Sethi, C. S., and Cheetham, M. E., 2002, The Leber congenital amaurosis gene product AIPL1 is localized exclusively in rod photoreceptors of the adult human retina, *Hum Mol Genet.* **11**:823-31.

Van Hooser, J. P., Aleman, T. S., He, Y. G., Cideciyan, A. V., Kuksa, V., Pittler, S. J., Stone, E. M., Jacobson, S. G., and Palczewski, K., 2000, Rapid restoration of visual pigment and function with oral retinoid in a mouse model of childhood blindness, *Proc Natl Acad Sci U S A.* **97**:8623-8.

Van Hooser, J. P., Liang, Y., Maeda, T., Kuksa, V., Jang, G. F., He, Y. G., Rieke, F., Fong, H. K., Detwiler, P. B., and Palczewski, K., 2002, Recovery of visual functions in a mouse model of Leber congenital amaurosis, *J Biol Chem.* **277**:19173-82.

Veske, A., Nilsson, S. E., Narfstrom, K., and Gal, A., 1999, Retinal dystrophy of Swedish briard/briard-beagle dogs is due to a 4-bp deletion in RPE65, *Genomics.* **57**:57-61.

Wagner, R. S., Caputo, A. R., Nelson, L. B., and Zanoni, D., 1985, High hyperopia in Leber's congenital amaurosis, *Arch Ophthalmol.* **103**:1507-9.

Yang, G. S., Schmidt, M., Yan, Z., Lindbloom, J. D., Harding, T. C., Donahue, B. A., Engelhardt, J. F., Kotin, R., and Davidson, B. L., 2002, Virus-mediated transduction of murine retina with adeno-associated virus: effects of viral capsid and genome size, *J Virol.* **76**:7651-60.

Znoiko, S. L., Crouch, R. K., Moiseyev, G., and Ma, J. X., 2002, Identification of the RPE65 protein in mammalian cone photoreceptors, *Invest Ophthalmol Vis Sci.* **43**:1604-9.

FUNCTIONAL AND STRUCTURAL EVALUATION AFTER AAV.RPE65 GENE TRANSFER IN THE CANINE MODEL OF LEBER'S CONGENITAL AMAUROSIS

Kristina Narfström[1], Ragnheidur Bragadóttir[2], T. Michael Redmond[3], P. Elizabeth Rakoczy[4], Theo van Veen[5], and Anitha Bruun[5]

1. INTRODUCTION

The Briard dog is affected by an autosomal recessively inherited retinal disease with severe, early-onset visual impairment or blindness. The external structures of the eye appear normal but most affected dogs have a rapid, quivering nystagmus and a wider resting pupillary diameter than normal dogs (Narfstrom et al., 1994). Electroretinography (ERG) is diagnostic in puppies, with non-recordable dark adapted responses and either non-recordable or low amplitude photopic ERG responses. The internal structures of the eye are normal appearing up to several years of age. In older animals changes such as attenuation of retinal vasculature and paling of the tapetal fundus are most often present. In 3-4 year-old affected dogs, yellow specks or spots may also be observed, mainly in the central parts of the fundus, that increase and spread peripherally with age.

The genetic defect is a four base-pair deletion in the RPE65 gene that results in a frameshift with a null-mutation in homozygous dogs (Veske et al., 1999; Aguirre et al., 1998). RPE65 is a 61-kDa protein that is preferentially expressed in the retinal pigment epithelium (RPE) (Hamel et al., 1993 a, b). RPE65 is a 61-kDa protein that is preferentially expressed in the retinal pigment epithelium (RPE). In the visual cycle, the protein is part of an enzyme complex necessary for the isomerization of all-trans retinol to 11-cis retinal. When the protein is absent, retinyl esters accumulate and rhodopsin is not formed, as shown in the Rpe65 knock-out mouse by Redmond and collaborators (1998) (Redmond et al., 1998). The accumulation of stored retinoids is also a hallmark of affected RPE65 dogs, whereby lipoid-like inclusions are seen by light and electron microscopy (Wrigstad et al, 1992). Mutations in the RPE65 gene in humans cause severe forms of early onset retinal dystrophy, including Leber's congenital amaurosis (LCA) (Gu et al., 1997; Thompson et al., 2000).

[1]University of Missouri-Columbia, MO, USA; [2]University Hospital, Oslo, Norway; [3]National Eye Institute, National Institutes of Health, Bethesda, MD, USA; [4]Lions Eye Institute, University of Western Australia, Perth, Australia; [5]Department of Ophthalmology, University Hospital of Lund, Sweden. Corresponding author: K. Narfstrom, E-mail: NarfstromK@missouri.edu

Gene therapy is a promising tool for the treatment of inherited genetic disorder in which the gene defect is known. In the eye, this therapy appears exceptionally promising and has already been shown to be effective in RPE65 null mutation dogs by two different groups (Acland et al., 2001; Narfstrom et al., 2003a). We now report on results from functional and structural studies using ERG and immunocytochemistry in treated dogs, performed 5-9 months after gene transfer surgery.

2. MATERIALS AND METHODS

2.1. Animals and Constructs

A large group of dogs were initially used for our gene transfer studies and the details regarding the procedures and short-term follow-up have been reported (Narfstrom et al., 2003a). In short normal RPE65 cDNA from a heterozygous (RPE65+/-) Briard-Beagle dog, born and bred in our laboratory, was introduced into a recombinant adenoassociated virus with a cytomegalovirus promotor. Control vector with Green Fluorescent Protein (GFP) was also produced. The titres of the two batches of constructs were $2x10^{12}$ particles/ml for the rAAV.RPE65 and $2x10^{10}$ transducing units/ml for the rAAV.GFP, respectively. Surgery was performed, as previously described, on dogs at the ages between 4 and 30 months. In general rAAV.RPE65 was injected subretinally into the right eye and rAAV.GFP into the left eye. Control dogs were sham injected subretinally into one eye and injected with rAAV.GFP subretinally into the fellow eye. For the present publication morphologic follow-up is described for one of the young dogs (Buddy; 4 months old at the time of surgery), and electrophysiologic long-term follow-up for the oldest dog (Bonus; 2.5 years old at surgery). The older dog had received 100 microliters of the construct subretinally, and treated as the majority of dogs above, while the young dog was treated by 60 microliters of the gene construct subretinally into one eye and its contralateral eye was left untouched.

2.2. Electroretinography

ERGs were performed on the dogs pre- and post-operatively, starting 4-6 weeks after surgery and then regularly during a 5 and 9 month follow-up period for the young and older dog, respectively. Retinal function was assessed using simultaneous bilateral full-field flash ERGs. A commercial Ganzfeld ERG system (ERG System TOR, Scanditronix Uppsala, Sweden) was utilized. Recording contact lens electrodes (Jet lenses; Universo Plastique, Switzerland) were cushioned on the cornea with methylcellulose (Methocel 2%, Ciba Vision, München, Germany). Platinum reference needle electrodes (Grass Instrumental Division, Astro-Med. Inc., West Warwick, USA) were placed subcutaneously at the base of each ear and the ground at the occiput. After 2 hours of dark adaptation, ERG recordings were made with the dogs in sternal recumbency. Stay sutures were used to position the globes and eyelids. Low (–2.0 log cd sec/m2) and high (0.6 log cd sec/m2) intensity scotopic ERGs were recorded. After 10 min of light adaptation using white background light (37 cd/m2), photopic ERG recordings were obtained at 5.1 and 30.1 Hz using white light stimuli at 0.0 log cd sec/m2.

2.3. Tissue preparation and Immunohistochemistry

For morphological studies, the younger dog Buddy was euthanized with an intravenous overdose of barbiturates. The eyes were enucleated and transsected at the ora ciliaris retinae. The eyecups were immersion fixed in freshly prepared 4% phosphate-buffered formaldehyde for 4 hours at 4° C. The tissue was rinsed and cryoprotected by transferring it stepwise through solutions of increasing concentrations of sucrose in Sorensen's buffer. The eyecups were divided into two parts by a horizontal incision from the nasal to the temporal peripheral margins comprising the inferior part of the tapetal fundus and the optic disc. The tissues were embedded in Yazulla media (30% egg albumen and 3% gelatine in water) and mounted for sectioning. 12 µm thick sections were cut in a cryostat, from the subretinal injection site in the vertical plane, comprised a total width of approximately 2 mm. Similarly, the same area in the contralateral eyecup was sectioned. Sections were collected on chrome aluminium-coated slides, air dried, and stored at –20 ° C until used.

The sections were rapidly thawed at room temperature and washed in phosphate-buffered saline at pH 7.2 (PBS) with 0.25% Triton-X100 (PBST, Sigma Immuno Chemicals, St Louis, Mo, USA). Bovine serum albumin (1%) was added to the PBST for the purpose of diluting the primary and secondary antibodies. The sections were then incubated in the appropriate dilution of the primary antibody (Table 1) for 16-18 hours at 4 °C. After 1 hour at room temperature, the slides were rinsed in PBST and further incubated for 45 min in darkness with the appropriate fluorescein isothiocyanate (FITC)

Table 1. Antibodies used in the present study

Antibody	Antigen	Target	Poly/ monoclonal	Source
Calbindin	Purified chicken gut calbindin-D	Horizontal cells, cone bipolar cells	Mono	Sigma Immuno Chem. St. Louis, MO, USA
GFAP (glial fibrillary acidic protein)	Purified GFAP	Glial cells	Mono	Chemicon Internatl.
OS	Purified blue pigment peptide	Blue cones	Rabbit	Theirry Leveillard, France
PNA	Fluorescein-peanut agglutinin	Cones		Vector Labs Burlingame, USA
RPE65	RPE65	Retinal pigment epithelium	Mono	Dr. Debra A. Thompson, Univ. Michigan, Ann Arbor
Synaptophysin	Purified rat retina synaptophysin	Neurons	Mono	Chemicon Internatl.
Vimentin	Purified vimentin	Muller cells	Mono	Chemicon Internatl.

or Texas-red conjugated secondary antibodies (1:200) (Jackson Chemicals). After rinsing, the slides were mounted in antifading mounting medium (Vectashield, Vector Lab., Inc., Burlingame, CA, USA). Negative controls were performed by omission of the primary antiserum.

The slides were examined using a fluorescence microscope (Nikon Eclipse 800, Bergström Instruments, Sweden) optimized for FITC or Texas Red.

3. RESULTS

3.1. Clinical Findings including ERG

Bonus had always been a very shy dog and reluctant to move around especially in unknown surroundings. He bumped into obstacles in dim lighting conditions but less so in daylight. Both eyes were normal appearing externally although the resting diameter of his pupils were wider than in normal dogs. At the time of surgery minor changes in his fundi included slight generalized vascular attenuation and some small white specks in the non-tapetal central fundus, as reported previously (Narfstrom et al., 2003a). ERGs recorded pre-operatively showed non- or barely recordable scotopic responses and only low-amplitude photopic responses (Fig. 1 B, Table 2).

No complications occurred at the time of surgery. On the second post-operative day Bonus developed a slight inflammatory response (uveitis) in the AAV.RPE65 treated eye. Three days later, following symptomatic topical and systemic anti-inflammatory treatment with corticosteroids, the uveitis subsided. The contralateral eye that was treated with AAV.GFP never developed any inflammatory response.

One month following surgery Bonus behavior changed. He became more alert, used his eyes to navigate his surroundings, and did not bump into objects as frequently as before surgery. Three months after surgery his visual behavior continued to markedly improve. In addition his ERG recordings revealed both scotopic and photopic responses, with the latter increased over those obtained preoperatively, in the eye treated with the transgene. In contrast, the contralateral eye, injected with AAV.GFP, revealed similar responses to those obtained pre-operatively. At six months post-operatively continued increases in scotopic high ERG amplitudes in the RPE65 gene transfer eye were recorded and there was an increase in photopic b-wave 30 Hz flicker amplitudes. Surprisingly, an increase in scotopic high intensity amplitudes and in photopic b-wave responses were noted in the contralateral eye. The increase in response amplitudes was sustained at the 9 month post-surgery follow-up in both eyes. (Fig 1, table 2).

Buddy appeared both day and night blind when first evaluated at 5 weeks of age. The external and internal structures of his eyes were normal appearing, however, by 6 weeks of age, demonstrated a quick, bilateral, quivering nystagmus. Preoperative scotopic ERGs were non-recordable and only low amplitude 30 Hz flicker responses were recorded.

Buddy did not have any adverse reactions in the eye treated with the gene transfer. Postoperatively he continued to bump into objects in dim light but appeared to function better in day light. As has been described earlier for all dogs treated with subretinal injection of the gene construct, his nystagmus disappeared bilaterally approximately 10 weeks after surgery (Narfstrom et al., 2003a). Follow-up ERG recordings showed both low amplitude scotopic responses and increased photopic responses in the eye treated with the gene construct only (data not shown).

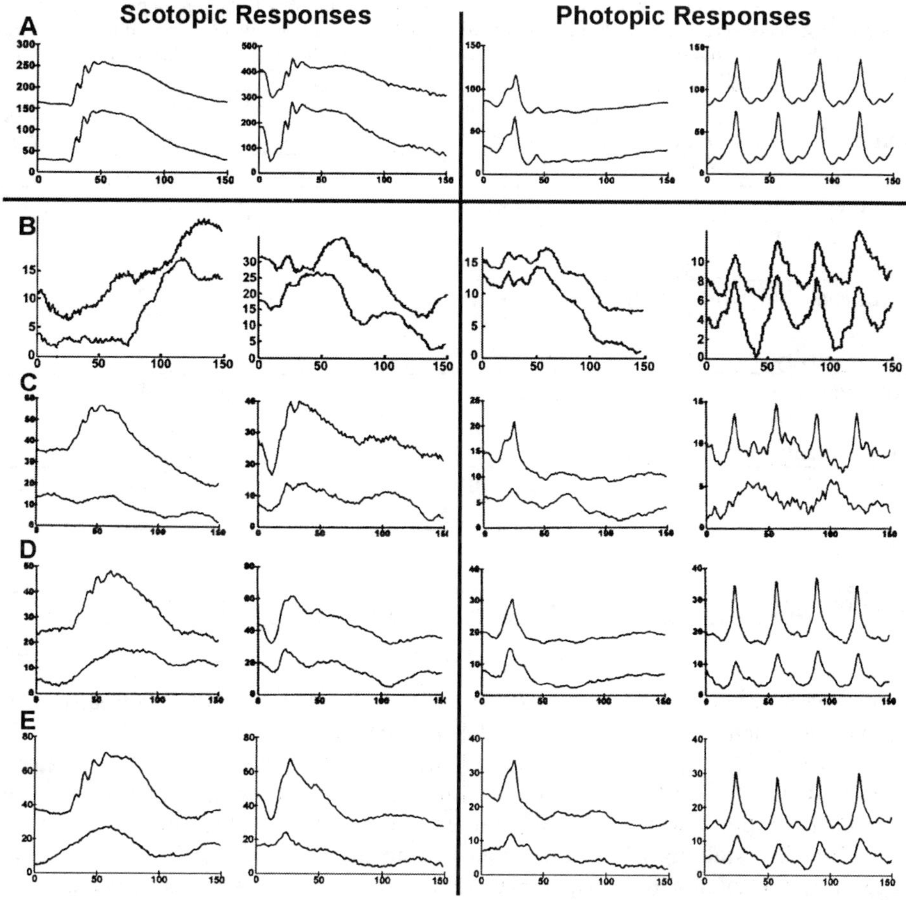

Fig. 1 A-E: **A**: Actual bilateral full-field ERGs from a 2.5-year-old control dog, that obtained subretinal injections of AAV.GFP into the right eye (OD) and saline into the left eye (OS) and from an RPE65 null mutation dog (Bonus, 2.5-years-old) treated by the AAV.RPE65 gene construct subretinally in OD and AAV.GFP subretinally in OS. Upper recording in each tracing is from OD and lower recording is from OS, **B**: preoperative, **C**: 3 months follow-up, **D**: 6 months follow-up and **E**: 9 month follow-up ERGs of Bonus. Horizontal calibration = msec and vertical calibration in each tracing = microvolts.

3.2. Immunocytochemistry

Marked changes of the retina were identified in the untouched eye compared to the subretinally rAAV.RPE65 treated eye of Buddy. The RPE layer was filled with autofluorescent vacuoles in the untouched eye while, in the AAV.RPE65 treated eye, there were only minor traces of autofluorescence observed in the RPE cells in the region of the subretinal injection (Fig. 2 A,B). Another major difference between the 2 eyes was

Table 2. ERG a- and b-wave amplitudes of the RPE65 null mutation dog Bonus, pre-operatively, and at 3, 6 and 9 month follow-up recordings. Normal values are given for age-matched control dogs of the same breed and age in our laboratory. OD: right eye, treated with rAAV.RPE65; OS: left eye, treated with rAAV.GFP nr: non-recordable.

	ERG amplitude values ± SEM (μV)								
Bonus	Peroperative		3 months		6 months		9 months		Normal
	OD	OS	OD	OS	OD	OS	OD	OS	
Scotopic b-wave (low intensity	nr	nr	24	nr	25	nr	36	nr	93 ± 39
Scotopic a-wave (high intensity)	nr	nr	10	2	13	4	14	2	85 ± 21
Scotopic b-wave (high intensity)	3	2	24	8	32	14	36	8	149 ± 26
Photopic a-wave	nr	nr	2	nr	3	2	5	nr	14 ± 2
Photopic b-wave	2	3	12	3	12	9	12	5	18 ± 4
30 Hz flicker	4	4	8	4	15	10	16	8	26 ± 6

the marked immunolabeling using RPE65 antibody in the rAAV.RPE65 treated eye (Fig. 2B) with no such labeling of RPE in the untouched eye (Fig. 2 A). Immunolabeling using PNA showed positive staining similarly in both treated and un-treated eyes however, in the AAV.RPE65 injected eye, the cone population appeared more organized and robust compared to the more disorganized appearance to the untouched eye (Fig. 2 D,C). Specific staining of the blue cones using the OS antibody resulted in a similar appearance of treated and untreated eyes (Fig. 2 C-F). No differences between eyes were observed using calbindin (Fig 2 A,B), PKC (Fig. 2 G, H), synaptophysin, vimentin or GFAP. There was no difference in either the transgene treated and the untouched retina between the width of the outer nuclear layer (Fig. 2 E,F) or the number of photoreceptor nuclei in the outer nuclear layer (data not shown).

4. DISCUSSION

We have described the long-term effects of gene transfer in two previously blind RPE65 null mutation dogs using follow-up ERGs and immunocytochemistry. In both the 2.5-year-old and the 4-month-old dogs, gene transfer resulted in a noticeable behavioral improvement. Since birth both dogs had been extremely careful; reluctant to move around in new surroundings and regularly bumped into objects. Only 4 weeks after surgery, their visual capacity appeared to increase. There were also improvements in their retinal function, evaluated objectively by ERG. In the older dog, Bonus, there was a slow improvement of ERG parameters up to at least six months post-surgery. A slight positive

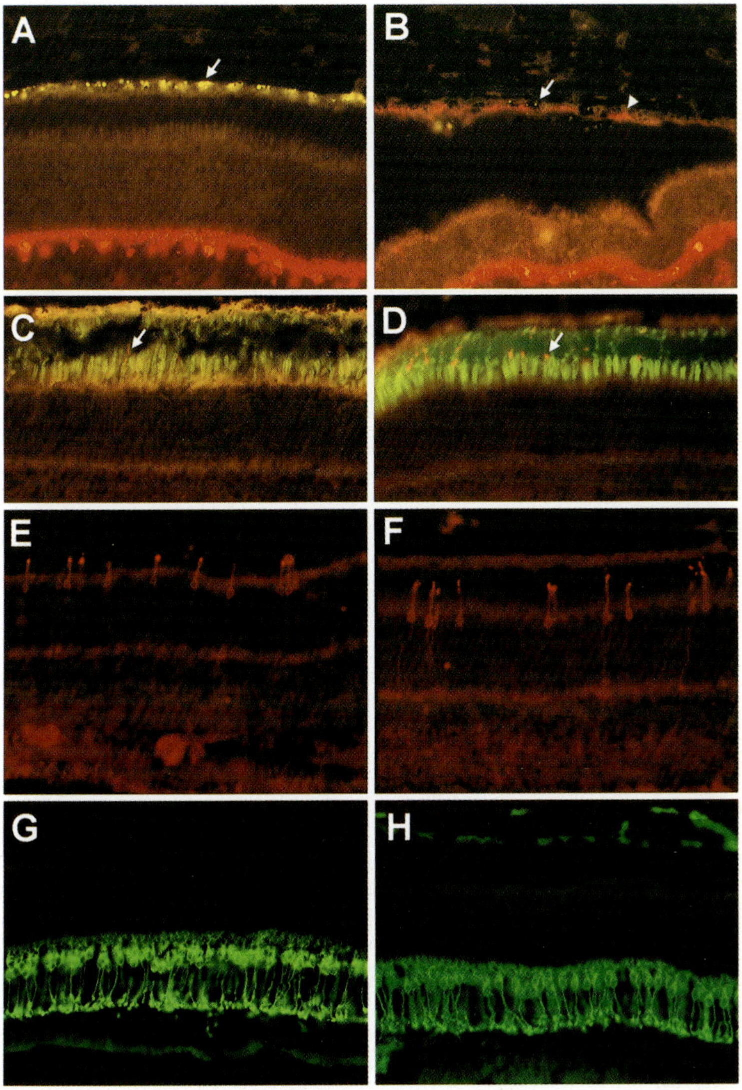

Fig. 2 A-H: Montage of fluorescence micrographs following immunolabeling of retina from untreated eye (**A,C,E,G**) and from retina treated by gene transfer using AAV.RPE65 (B,D,F,H). No RPE65 immunostaining could be observed in the RPE cells in A, while there was a continuous band of immunostaining in RPE of B, (arrowhead). Further, there was an obvious difference in the amount of lipoid inclusions observed in the RPE of A (arrow) versus **B** (arrow). PNA positive staining of cone sheaths was seen in both **C** and **D** as well as OS positive staining for blue cones (arrow). Single staining of OS showed blue cones in both **E** and **F**. Single staining using PKC demonstrated a normal configuration of rod bipolar cells in both **G** and **H**.

improvement, mainly on the light adapted ERGs, was also identified in the contralateral eye, the eye that was not treated with the transgene. At nine months follow-up the ERG amplitudes recorded in the gene construct treated eye of Bonus appeared to stabilize, as did the effect on the contralateral eye.

Earlier studies on the larger group of subretinally treated RPE65 null mutation dogs (n=11) have included a short term (3 months) follow-up mainly using clinical studies, but also including electrophysiology and morphology (Narfstrom et al., 2003a). Further the long-term electrophysiologic effects of the gene transfer treatment was evaluated in a subgroup consisting of 5 of the treated animals, that had received similar high amounts of the transgene subretinally (Narfstrom, et al., 2003b). The local inflammatory response that occurred in the gene transfer treated eye of Bonus was observed in 75% of the eyes treated with the AAV.RPE65 gene construct in the short term study. It was hypothesized that the RPE65 is immunopathogenic, such as has been described for other ocular-specific molecules (Gery et al., 1986) and which has been shown for RPE65 injected rats (Ham et al. 2002). Further, dark adapted ERG b-wave amplitudes were found to recover to an average of 28% of normal dogs and light adapted b-waves to 32% of normal dogs three months post-surgery when all homozygous affected dogs treated subretinally by the transgene were taken into account. This functional improvement was surprising as no structural improvement was observed at 3 and 6 months post-surgery using light and electron microscopy. The present study was initiated to more thoroughly investigate retinal organization and viability following the gene transfer, using immunocytochemistry. Antibodies directed against cone photoreceptors were included since they appeared to gain function (Seeliger et al., 2001) after the treatment by gene transfer.

We have verified through the 5 month follow-up study using the dog Buddy, that there is a reduction of the lipoid-like inclusions found in the RPE associated with treatment. In the normal visual cycle recent work suggests that there is a constant turnover of retinoids in the RPE (Qtaishat et al., 2003). In the RPE65 null mutation dog, this mechanism appears defective, which results in a build-up of retinoids in the RPE. Gene transfer appears to switch on the local turn-over of these retinoids in the area of treatment. The precise mechanism behind this process is unknown.

Antibodies against cone sheaths, utilizing PNA, showed the regular configuration of the cones in the treated eye compared to the untreated eye, where they appeared more disorganized. Specific antibodies directed against the blue cones showed a similar appearance of these photoreceptors in both treated and untreated eyes. Thus, there appears to be a structural stabilization of cone photoreceptors after gene transfer but further long-term ultrastructural studies are needed for verification of such an effect.

We were pleased to find such a marked improvement in visual behavior and in ERG recordings, even long-term for the dog Bonus, that was comparably old at the time of surgery. This gives hope that there is a relatively wide time window for surgery to be effective in human patients with Leber's congenital amaurosis. At the present time we have no explanation for the small but distinct increase in the long-term photopic ERG responses observed in the fellow eye of Bonus. Although treatable in most cases, the resultant uveitis in relation to subretinal injection of the transgene is a serious complication. Further immunological, molecular and ultrastructural studies, not only of the treated eye, but also of the contralateral "control" eye are needed for clarification of these important issues.

5. ACKNOWLEDGEMENTS

The authors wish to thank Ginny Dodam and Jenny Garland for excellent technical assistance. We would also like to acknowledge the support of the Foundation Fighting Blindness.

6. REFERENCES

Acland, G.M., Aguirre, G.D., Ray, J. et al., 2001, Gene therapy restores vision in a canine model of childhood blindness, *Nat Genet.* **28**:92-95.

Aguirre,G., Baldwin,V., Pearce-Kelling,S. et al.,1994, Congenital stationary night blindness in the dog: common mutation in the RPE65 gene indicates founder effect, *Mol Vis.* **4**:23.

Gery, I., Mochizuki, M., and Nussenblatt, R.B., 1986, Retinal specific antigens and immunopathogenic processes they provoke. In:Progress in Retinal Research, Osborne N, Chader,G, eds. Pergamon Press, Oxford. 75-109.

Gu,. S.-G., Thompson, D.A., Srikumari, C.R.S., et al., 1997, Mutations in the RPE65 cause autosomal recessive childhood-onset severe retinal dystrophy. *Nat Genet.* **17**:194-197.

Ham, D.I., Gentleman, S., Chan, C., et al., 2002, RPE65 is highly uveitogenic in rats. *Invest Ophthalmol. Vis Sci.* **43**:2258-2263.

Hamel, C.P., Tsilou, E., Harris, E., et al., 1993a, A developmentally regulated microsomal protein specific for the pigment epithelium of the vertebrate retina. *J Neurosci Res.* **34**:414-25.

Hamel, C.P., Tsilou, E., Pfeffer, B.A., Hooks, J.J., Detrick, B., and Redmond, T.M., 1993b, Molecular cloning and expression of RPE65, a novel retinal pigment epithelium-specific microsomal protein that is post-transcriptionally regulated in vitro. *J of Biol Chem.* **268**:15751-15757.

Narfström, K., Katz, M.L., Bragadotti,r R., Seeliger, M., Boulanger, A., Redmond, T.M., Caro, L., Lai, C., and Rakoczy, E., 2003a, Functional and structural recovery of the retina after gene therapy in the RPE65 null mutation dog. *Invest Ophthalmol Vis Sci*, In press.

Narfström, K., Katz, M.L., Ford, M., Redmond, T.M., Rakoczy, E., and Bragadottir, R., 2003b, In Vivo gene therapy in young and adult RPE65-/- dogs produces long-term visual improvement. *J of Hered.* In press.

Narfström, K., Wrigstad, A., Ekesten, B., and Nilsson,S.E.G., 1994, Hereditary retinal dystrophy in the Briard dog: Clinical and hereditary characteristics, *Prog. in Vet. and Comp. Ophthalmol.* **4**,85-92.

Redmond, T.M., Yu, S., Lee, E., Bok, D., Hamasaki, D., Chen, N., et al., 1998, RPE65 is necessary for production of 11-cis-vitamin A in the retinal visual cycle. *Nat Genet.* **20**:344-351.

Seeliger, M.W., Grimm, C., Stahlberg, F., et al., 2001, New views on RPE65 deficiency: the rod system is the source of vision in a mouse model of Leber Congenital amaurosis. *Nat Genet.* **29**:70-74.

Thompson, D.A., Gyurus, P., Fleischer, L.L., et al., 2000, Genetics and phenotypes of RPE65 mutations in inherited retinal degeneration. *Invest Ophthalmol Vis Sci.* **41**:4293-4299.

Veske, A., Nilsson, S.E.G., Narfstrom, K., and Gal, A., 1999, Retinal dystrophy of Swedish Briard/Beagle dogs is due to a 4-bp deletion in RPE65. *Genomics.* **57**:57-61.

Wrigstad, A., Nilsson, S.E.G., and Narfström, K., 1992, Ultrastructural changes of the retina in Briard dogs with hereditary congenital night blindness and partial day blindness. *Exp Eye Res.* **55**:805-818.

Qtaishat, N., Redmond, T.M., and Pepperberg, D.R., 2003,Acute radiolabeling of retinoids in eye tissues of normal andRPE65-deficient mice. *Invest Ophthalmol Vis Sci.* In press.

ASSESSMENT OF rAAV-MEDIATED GENE THERAPY IN THE *Rpe65*$^{-/-}$ MOUSE

P. Elizabeth Rakoczy[1], Chooi-May Lai[1], Meaghan J.T. Yu[2], Dru M. Daniels[1], Meliha Brankov[2], Ben C. Rae[2], Chris W. Stoddart[2], Nigel L. Barnett[3], Matthew T. Martin-Iverson[4], T. Michael Redmond[5], Kristina Narfstrom[6], Xiaohuai Zhou[7], and Ian J. Constable[1]

1. INTRODUCTION

There have been several reports demonstrating the potential of virus-mediated gene delivery for use in treating congenital retinal degenerations (Bennett and Maguire, 2000; Acland et al., 2001; Campochiaro et al., 2002; Narfstrom et al., 2002). A recent report on the restoration of vision to Briard dogs marked the first successful gene therapy for blindness in a large animal model, bringing the possibility of a human retinal gene therapy trial closer (Bennett and Maguire, 2000; Narfstrom et al., 2002). We wished to pursue this potential, by developing a successful gene therapy protocol for the treatment of retinal degeneration in a small animal model. Once established, this small animal model could be used to assess the clinical benefits and limitations of using gene therapy in treating congenital retinal degeneration.

The small animal model of choice for this project was the *Rpe65*$^{-/-}$ knockout mouse strain (Redmond et al., 1998). The *Rpe65*$^{-/-}$ mice were produced via a targeted disruption of RPE65, a retinal pigment epithelial (RPE)-specific gene that in humans leads to severe, early onset retinal degeneration (Marlhens et al., 1997; Lorenz et al., 2000). In

[1]Centre for Ophthalmology and Visual Science, University of Western Australia (UWA), Perth, Western Australia, Australia; [2]Lions Eye Institute affiliated with UWA, Perth, Western Australia; [3]Vision Touch and Hearing Research Centre, University of Queensland, Queensland, Australia; [4]Department of Pharmacology, UWA, Perth, Western Australia, Australia; [5]Laboratory of Retinal Cell and Molecular Biology, National Eye Institute, National Institutes of Health, MD, USA; [6]Department of Veterinary Medicine and Surgery, University of Missouri-Columbia, Columbia, MO, USA; [7]Virus Vector Core Facility, Gene Therapy Centre, University of North Carolina, Chapel Hill, NC, USA. Corresponding author: P. E. Rakoczy, Email: rakoczy@ cyllene.uwa.edu.au

the mouse model, disruption of the RPE65 gene causes inhibition of the visual cycle, reduction in electroretinography (ERG) signals, lipid accumulation in the RPE cells and, in the older animals, degeneration of the photoreceptors (Redmond et al., 1998; Seeliger et al., 2001).

The gene therapy trial with the *Rpe65*$^{-/-}$ mice involved replacing the disrupted RPE65 gene with normal, functional mouse RPE65 cDNA. The RPE65 cDNA delivery was achieved using a recombinant adenoassociated virus (rAAV), which are known to provide efficient delivery and expression of transgenes in RPE cells *in vivo* (Grant et al., 1997; Rolling et al., 1999; Acland et al., 2001). The effects of the rAAV-mediated RPE65 cDNA delivery were then determined, in order to assess the presence and/or extent of any therapeutic effect.

2. METHODS

2.1. Production, Delivery and Expression of rAAV.RPE65

An rAAV construct, carrying a human cytomegalovirus promoter to drive the mouse RPE65 cDNA expression (rAAV.RPE65), was generated by standard gene technology procedures (Rolling and Samulski, 1995; Zolotukhin et al., 1999). The large scale production and purification of rAAV.RPE65 was performed by the Vector Core Facility (University of North Carolina, NC, USA), with the resulting viral stock having a titer of 6×10^{13} particles/ml. A control rAAV construct, in which the cDNA for Green Fluorescent Protein (GFP) was used to substitute the mouse RPE65 cDNA, was also produced (rAAV.GFP, 2×10^{10} tu/ml). A 1 µl aliquot of rAAV.RPE65, rAAV.GFP or PBS was delivered to the subretinal space of *Rpe65*$^{-/-}$ and control C57BL/6J mice. The subretinal injection was performed on mice of 3 weeks of age, using procedures that have been described previously (Rolling et al., 1999). Injected eyes were monitored for up to 18 months post injection using indirect ophthalmoscopy (rAAV.RPE65 and rAAV.GFP) and fluorescence fundus photography (rAAV.GFP). At 2 weeks and 2, 6, 12 and 18 months post injection mice were euthanased, their eyes enucleated, and the retina isolated and separated into the neural retina and RPE-choroid components. The two components were then whole mounted and used to assess RPE65 expression (immunohistochemistry, rabbit anti-bovine RPE65, Redmond and Hamel, 2000), or GFP expression (fluorescence microscopy).

All procedures were approved by the University of Western Australia Animal Experimentation Committee, and were in compliance with the Association for Research in Vision and Ophthalmology Statement for the Use of Animals in Ophthalmic and Vision Research.

2.2. Functional Analysis of the Effects of Subretinal rAAV.RPE65 Delivery

The effects of subretinal rAAV.RPE65 injection were analyzed functionally using ERG and behavioral studies. ERG was performed on rAAV.RPE65-injected (n = 6) and PBS-injected *Rpe65*$^{-/-}$ (n= 6) mice at 3 and 12 months. The mice were anaesthetized and dark adapted for 2 hours. ERG responses were recorded using a xenon strobe light (0.1 Hz) and a series of neutral density (ND) filters (maximum flash intensity = 1.52 cd/m^2).

Four consecutive flashes were recorded at each density, and the resulting curve was then used to calculate the a- and b-wave responses for that intensity. The mice were light adapted for 10 minutes, before the same procedure was repeated to obtain photopic responses (background light 30 cd/m^2).

For behavioral analyses, individual mice (7 rAAV.RPE65-injected *Rpe65*$^{-/-}$, 10 PBS-injected *Rpe65*$^{-/-}$, 10 C57BL/6J) were housed in custom built plexiglass cages fitted with infrared photocell beams. The photocells were linked to a computer, that monitored the level of mouse activity as the number of photocell beam interruptions over time (counts/hr)(Martin-Inverson et al., 1988). The mice were then placed in normal animal housing rooms, with the lighting set on a 12 hr light:12 hr dark cycle. After 6 days, the lighting pattern was reversed 180° (12 hr dark:12 hr light) and the data recorded for a further 12 days. The counts/hr were analyzed in blocks of 3 days (72 hrs), with particular attention being placed on the shifting time of acrophase.

2.3. Morphological Analysis of the Effects of Subretinal rAAV.RPE65 Delivery

rAAV.RPE65-injected (n = 3) and PBS-injected (n = 3) *Rpe65*$^{-/-}$ mice were euthanased at 2 weeks, and 3, 12, 18 and 21 months and their eyes enucleated, fixed and embedded in either resin or paraffin. The eyes were then serial sectioned and stained with toluidine blue or haematoxylin and eosin for analysis by light microscopy.

3. RESULTS

3.1. Delivery and Expression of rAAV.RPE65 to the RPE of *Rpe65*$^{-/-}$ Mice

The ability to deliver and express transgenes in the RPE cells of *Rpe65*$^{-/-}$ mice was assessed using rAAV.GFP. Subretinal injection of rAAV.GFP resulted in an area of GFP

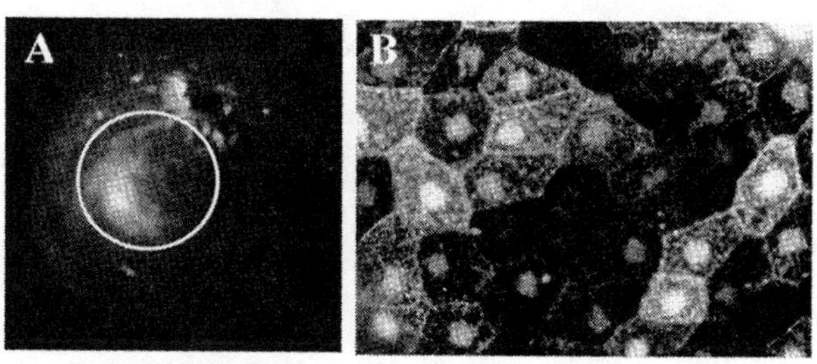

Figure 1. GFP expression in rAAV.GFP-injected *Rpe65*$^{-/-}$ mouse eyes. (A) Fluorescent fundus photography at 2 months post injection showing GFP localized to an area covering approximately 30% of the retina (circle). (B) RPE-choroid whole-mount of an rAAV.GFP-injected eye showing GFP expression within the RPE cells. Photomicrographs were taken at 2 months post injection.

expression that covered approximately 30% of the injected mouse's retina (Fig.1A). The signal was detectable from as early as 2 weeks post injection, and was still visible at 18 months post injection although at the later time points (15 months) the signal was reduced slightly. Enucleation and whole-mounting of the rAAV.GFP-injected *Rpe65*$^{-/-}$ eyes showed that the GFP signal was localized to the RPE cells (Fig.1B), with some GFP expression also being seen in the neural retina (data not shown).

Once the ability to transduce the RPE of *Rpe65*$^{-/-}$ mice had been confirmed, rAAV.RPE65 was delivered and its affects *in vivo* investigated. Enucleation, whole-mounting and RPE65-immunohistochemistry on the RPE-choroidal layers of rAAV.RPE65-injected eyes revealed the presence of an area of RPE65 immunoreactivity which, like rAAV.GFP, covered an area of approximately 30% of the retina. The immunoreactivity was localized to the RPE cells, and was still visible at 18 months post injection (Fig.2A). Whole-mounting and immunohistochemistry of the neural retina of rAAV.RPE65-treated eyes showed some RPE65 immunoreactivity which, by its location, appeared to be within the photoreceptor layers (data not shown).

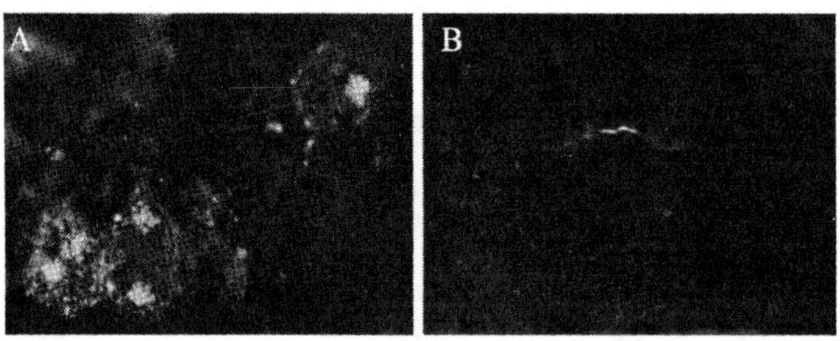

Figure 2. RPE65 expression in the retinas of rAAV.RPE65-injected *Rpe65*$^{-/-}$ mice. Fluoromicrograph showing (A) the presence of RPE65 immunoreactivity in the RPE cells of an *Rpe65*$^{-/-}$ 2 months post-rAAV.RPE65 injection and (B) the absence of RPE-65 immunoreactivity in PBS-injected *Rpe65*$^{-/-}$ mice.

3.2. Functional Assessment of the Effects of rAAV.RPE65 Delivery in *Rpe65*$^{-/-}$ Mice

The functional effects of subretinal rAAV.RPE65 injection were assessed by ERG and behavioral studies. In an ERG analysis, rAAV.RPE65 delivery demonstrated a modest, but measurable, improvement compared to PBS-injected controls. The increased signal was seen in the b-wave component of the ERG response under both scotopic (Fig.3) and photopic conditions (data not shown). Both rAAV.RPE65-injected and PBS-injected *Rpe65*$^{-/-}$ mouse eyes gave negligible a-wave responses, under both scotopic and photopic conditions (data not shown).

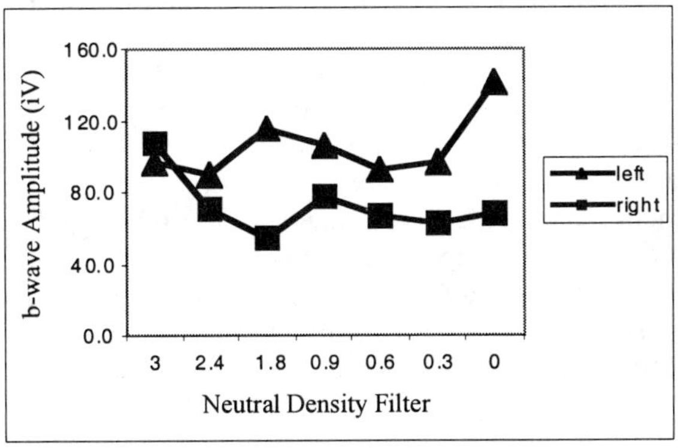

Figure 3. Improvement in ERG response after rAAV.RPE65 injection. Diagram shows a typical intensity response curve from an rAAV.RPE65-injected (left _) and PBS-injected (right _) eyes of an *Rpe65*$^{-/-}$ mouse. Maximum stimulus intensity (ND0) was 1.52 log cd/m^2. ERG was performed at 3 months post injection.

The improvements seen in the ERG responses of rAAV.RPE65-injected *Rpe65*$^{-/-}$ mice were also reflected in a behavioral study. In this study, the ability of rAAV.RPE65-injected mice to perceive light was assessed by measuring their ability to re-adjust their circadian rhythm to a 180° reversal of the light/dark cycle. The ability of the rAAV.RPE65-injected mice to re-adjust was intermediary, and fell in between that of PBS-injected *Rpe65*$^{-/-}$ mice and control C57BL/6J mice, as measured by their shifting time of acrophase (daily circadian time of maximal activity, Fig.4). There was a high degree of variability between the individual injected animal's ability to re-adjust to the light reversal, with 3 of the 7 animals tested showing particularly promising results.

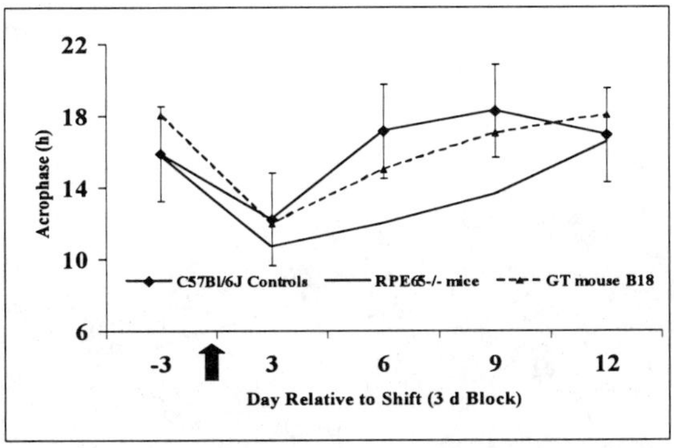

Figure 4. Re-adjustment of the circadian rhythm of rAAV.RPE65-injected *Rpe65*$^{-/-}$ mice following light reversal. Graph shows the shift in acrophase over time of an rAAV.RPE65-injected *Rpe65*$^{-/-}$ mouse (_) following a 180^0 light reversal. For comparison, the mean acrophase from PBS-injected *Rpe65*$^{-/-}$ (_, n = 10) and normal C57BL/6J (_, n = 10) are also shown. Arrow indicates the time of light reversal. Error bars show critical differences of pair-wise comparisons (multiple F-test $\alpha = 0.05$).

3.3. Morphological Assessment of the Effects of rAAV.RPE65 Delivery

An assessment of the effects of subretinal rAAV.RPE65 injection on the RPE and photoreceptor cell morphology of *Rpe65*$^{-/-}$ mice was performed using light microscopy. The difference between the RPE cell morphology of rAAV.RPE65-injected and PBS-injected *Rpe65*$^{-/-}$ eyes was striking. rAAV.RPE65-injected eyes showed a significant reduction in the amount of vacuole-like structures (Fig.5), believed to be lipid inclusions (Redmond et al., 1998), when compared to PBS- or rAAV.GFP-injected eyes, where the lipid inclusions were abundant and formed a thick layer along the basement membrane of the RPE cells (Fig.5B, arrows). In rAAV.RPE65-injected eyes, the lipid inclusions were sparse and only accounted for a small proportion of the total cell area (Fig.5A, arrows).

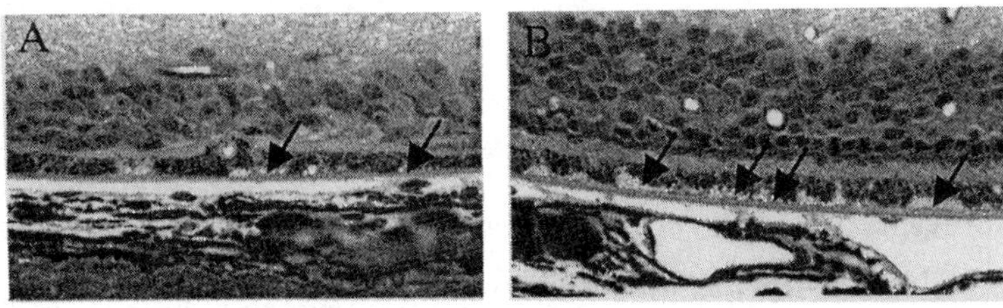

Figure 5. Effect of rAAV.RPE65 injection on RPE morphology. (A) Photomicrograph shows the small number of lipid inclusions (arrows) present in rAAV.RPE65-injected *Rpe65*$^{-/-}$ mouse eyes. (B) In contrast, photomicrograph of PBS-injected Rpe65$^{-/-}$ mouse eye shows a large number of lipid inclusions (arrows). Photomicrographs were taken at 21 months post injection.

rAAV.RPE65-mediated delivery of RPE65 into *Rpe65*$^{-/-}$ mouse retina did not appear to slow the rate of photoreceptor degeneration when compared to PBS-injected mouse eyes (data not shown). This lack of difference in photoreceptor number between rAAV.RPE65-injected and PBS-injected eyes was seen at all time points examined, from 2 weeks (Fig.6A) to 18 months post injection (Fig.6B).

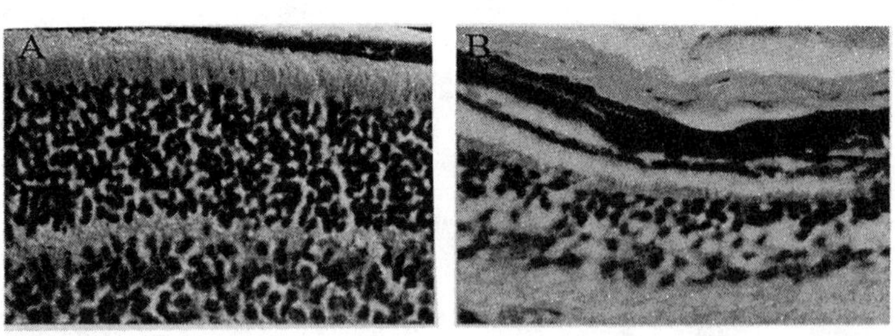

Figure 6. Lack of effect of rAAV.RPE65 injection on photoreceptor morphology. Photomicrograph of an rAAV.RPE65-injected Rpe65$^{-/-}$ mouse retina at (A) 2 weeks post injection showing almost normal retinal morphology and (B) 18 months showing severe photoreceptor degeneration.

4. CONCLUSIONS

The work presented in this article describes the first stage of developing a functional small animal model for assessing the capacity of virus-mediated gene therapy in treating inherited retinal degeneration. The data presented demonstrates that the rAAV-mediated delivery of a normal, functional RPE65 gene to the retina of *Rpe65*$^{-/-}$ mice can have a positive therapeutic effect, as seen using both ERG and a behavioral analysis of the rAAV.RPE65-injected *Rpe65*$^{-/-}$ mice. Although the magnitude of the effects seen appears small, it should be noted that an area of only approximately 30% of one eye of each mouse was injected. Thus, in holistic terms, the therapeutic region was only 15% of the injected mice's total binocular visual field. In addition, it should also be remembered that, although promising, the field of ocular gene therapy is still in its infancy. Much more optimization is required, including increasing specificity of the rAAV localization, improving on transduction efficiency, quantifying, and perhaps augmenting the levels of RPE65 protein being expressed, and determining the optimal time in the animals' life to deliver the rAAV. Only when the technology has been fully optimized can the true potential of rAAV-mediated gene therapy be seen.

The therapeutic effect of rAAV.RPE65 injection was also seen at a morphological level, with a dramatic decrease in the extent of lipid inclusions within the RPE cells of injected *Rpe65*$^{-/-}$ mice. The presence of the lipid inclusions in the RPE cells of *Rpe65*$^{-/-}$ mice is due to retinyl ester accumulation caused by inhibition of the visual cycle (Redmond et al., 1998). This reduction in the amount of lipid inclusions suggests that rAAV.RPE65 is expressing biologically active RPE65 protein *in vivo*. Additionally, and more importantly, it implies that the presence of rAAV.RPE65 is leading to activation of the visual cycle. The suggestion that the visual cycle has been activated in treated *Rpe65*$^{-/-}$ mice is supported by the findings from the ERG and behavioral analysis. Both of these assays are a measure of the light sensitivity of the *Rpe65*$^{-/-}$ mice, a factor that is dependent upon phototransduction, and in turn correlates to a functioning visual cycle (Saari, 1994). However, although these data suggest that the visual cycle is functioning, the lack of a difference in the photoreceptor morphology indicates that the presence of rAAV.RPE65 cannot stop photoreceptor degeneration The data therefore implies that either the cascade of events that leads to photoreceptor degeneration is not reversible, or that the inadequate efficacy of our present gene therapy technologies may limit the therapeutic outcome.

In conclusion, the results obtained from the subretinal injection of rAAV.RPE65 into the retinas of *Rpe65*$^{-/-}$ mice are significant in their implications for the future treatments of inherited retinal degeneration. It is therefore with much anticipation that the future research into rAAV-mediated ocular gene therapy studies is awaited.

5. ACKNOWLEDGEMENTS

The authors wish to thank Dr Wei-Yong Shen for expert assistance with rAAV.GFP delivery and observation. The authors also acknowledge the support of the Foundation for Fighting Blindness, USA, Retina Australia and the National Health and Medical Research Council of Australia.

6. REFERENCES

Acland, G.M., Aguirre, G.D., Ray, J., Zhang, Q., Aleman, T.S., Cideciyan, A.V., Pearce-Kelling, S.E., Anand, V., Zeng, Y., Maguire, A.M., Jacobson, S.G., Hauswirth, W.W., and Bennett, J., 2001, Gene therapy restores vision in a canine model of childhood blindness, *Nat. Genet.* **28:**92-95.

Bennett, J. and Maguire, A.M., 2000, Gene therapy for ocular disease, *Mol. Ther.* **1:**501-505.

Campochiaro, P.A., 2002, Gene therapy for retinal and choroidal diseases, *Expert. Opin. Biol. Ther* **2:**537-544.

Grant, C.A., Ponnazhagan, S., Wang, X.S., Srivastava, A., and Li, T., 1997, Evaluation of recombinant adeno-associated virus as a gene transfer vector for the retina, *Curr. Eye Res.* **16:** 949-956.

Lorenz, B., Gyurus, P., Preising, M., Bremser, D., Gu, S., Andrassi, M., Gerth, C., and Gal, A., 2000, Early-onset severe rod-cone dystrophy in young children with RPE65 mutations, *Invest. Ophthalmol. Vis. Sci.* **41:**2735-2742.

Martin-Iverson, M.T., Stahl, S.M., and Iversen, S.D., 1988, Chronic administration of a selective dopamine D-2 agonist:factors determining behavioral tolerance and sensitization, *Psychopharmacology (Berl)* **95:**534-539.

Marlhens, F., Bareil, C., Griffoin, J.M., Zrenner, E., Amalric, P., Eliaou, C., Liu, S.Y., Harris, E., Redmond, T.M., Arnaud, B., Claustres, M., and Hamel, C.P., 1997, Mutations in RPE65 cause Leber's congenital amaurosis, *Nat. Genet.* **17:**139-141.

Narfstrom, K., Katz, M.L., Bragadottir, R., Seeliger, M., Boulanger, A., Redmond, T.M, Caro, L., Lai, C-M., Rakoczy, E., 2002, Functional and structural recovery of the retina after gene therapy in the RPE65 null mutation dog, *Invest. Opthalmol. Vis. Sci.* (In Press).

Redmond, T.M. and Hamel, C.P., 2000, Genetic analysis of RPE65: from human disease to mouse model, *Methods Enzymol.* **316:**705-724.

Redmond, T.M., Yu, S., Lee, E., Bok, D., Hamasaki, D., Chen, N., Goletz, P., Ma, J.X., Crouch, R.K., and Pfeifer, K., 1998, Rpe65 is necessary for production of 11-cis-vitamin A in the retinal visual cycle, *Nat. Genet.* **20:**344-351.

Rolling, F. and Samulski, R.J., 1995, AAV as a viral vector for human gene therapy. Generation of recombinant virus, *Mol. Biotechnol.* **3:**9-15.

Rolling, F., Shen, W.Y., Tabarias, H., Constable, I., Kanagasingam, Y., Barry, C.J., and Rakoczy, P.E., 1999, Evaluation of adeno-associated virus-mediated gene transfer into the rat retina by clinical fluorescence photography, *Hum. Gene Ther.* **10:** 641-648.

Saari, J.C., Retinoids in Photosensetive Systems in: *The Retinoids,* edited by Sporn, M.B., Roberts, A.B., and Goodman, D.S. (Raven Press, New York, 1994), pp. 351-384.

Seeliger, M.W., Grimm, C., Stahlberg, F., Friedburg, C., Jaissle, G., Zrenner, E., Guo, H., Reme, C.E., Humphries, P., Hofmann, F., Biel, M., Fariss, R.N., Redmond, T.M., and Wenzel, A., 2001, New views on RPE65 deficiency: the rod system is the source of vision in a mouse model of Leber congenital amaurosis, *Nat. Genet.* **29:**70-74.

Zolotukhin, S., Byrne, B.J., Mason, E., Zolotukhin, I., Potter, M., Chesnut, K., Summerford, C., Samulski, R.J., and Muzyczka, N., 1999, Recombinant adeno-associated virus purification using novel methods improves infectious titer and yield, *Gene Ther.* **6:**973-985.

MÜLLER CELL TRANSDUCTION BY AAV2 IN NORMAL AND DEGENERATIVE RETINAS

Fong-Qi Liang[1], Enrico Surace[2], Nadine S. Dejneka[2], Albert M. Maguire[2], and Jean Bennett[2]

1. INTRODUCTION

Gene therapy for retinal diseases has gained increasing attention in recent years (see review, Bennett and Maguire, 2000). Among various gene delivery systems, viral vectors such as adenovirus (Ad), adeno-associated virus (AAV) and lentivirus appear to be the most efficient vehicles for gene transfer. In comparison to the Ad and lentivirus, AAV offers several advantages, including nonpathogenicity, minimal immuogenicity, and stable transgene expression in a wide host range (Flotte and Carter, 1995; Bennett et al., 1999; Anand et al., 2000).

AAV preferentially transduces neural cells in the central nervous system (Bartlett et al., 1998). In the retina, AAV can efficiently target photoreceptor and retinal ganglion cells (RGC) (Bennett et al., 1997; Flannery et al., 1997). Our recent study showed that when AAV were administered intravitreally in degenerating rat retinas, they also transduced Muller cells in addition to the RGC (Liang et al., 2001b) .

The Muller cells are the most populous glia in the retina. They span nearly the full thickness of the retina, from the vitreal surface to the outer limiting membrane (OLM). The somata of these cells lie in the inner nuclei layer (INL). Their inwardly directed processes enlarge to become conical endfeet, which are incorporated into the inner limiting membrane (ILM). Muller cells play a major role in maintaining retinal structure and function (Newman and Reichenbach, 1996; Bringmann and Reichenbach, 2001). Recent studies suggest that Muller cells may mediate photoreceptor rescue by neurotrophic factors (Zack, 2000). To this end, Muller cells might become a useful target for gene transfer of neurotrophic factors. As a first step, we sought to determine whether AAV is able to efficiently transduce retinal Muller cells both *in vitro* and *in vivo*.

[1]Retina Foundation of the Southwest, 9900 N. Central Expressway, Suite 400, Dallas, TX 75231 and Dept. of Ophthalmology, University of Texas Southwestern Medical Center, Dallas, TX, USA; [2]Dept. of Ophthalmology, Scheie Eye Institute, University of Pennsylvania, Philadelphia, PA 19104, USA. Corresponding author: F. Qi-Liang, E-mail: fliang@retinafoundation.org

Retinal Degenerations: Mechanisms and Experimental Therapy
Edited by LaVail *et al.*, Kluwer Academic/Plenum Publishers, 2003

2. METHODS

2.1. rAAV Vector Construction and Production

The vectors were constructed and AAV were prepared as previously described (Liang et al., 2001a). The CNTF virus (rAAV.CNTF.GFP) carries two transgenes: a gene encoding a secretable form of murine CNTF under the transcriptional control of the immediate early promoter from cytomegalovirus (CMV) and a marker gene encoding an enhanced form of GFP linked by internal ribosomal entry site (IRES). Recombinant AAV was produced by transfecting 293cells with pAAV.CNTF.EGFP along with a helper plasmid (pAd.Help.Rep/Cap) that encodes adenovirus (E2a, E4, and VA RNA) and AAV (*Rep* and *Cap*) genes, and purified through a gradient of cesium chloride. The titer of virus was $2x10^{12}$. This virus was extensively characterized *in vitro* prior to injection into animals (Liang et al., 2001a).

rAAV.GFP, used as a control in this study, was produced by co-transfection with AAV plasmids and infection with Ad-LacZ or by triple transfection. The transgene cassette consists of *EGFP* under the control of the CMV promoter. The titer of this virus was $1.5x10^{12}$ particles/ml. There was no evidence of adenoviral contamination as measured by histochemical assay of b-galactesidase (which would reveal presence of LacZ in helper virus) (Bennett et al., 1997). Replication-competent AAV were present at a level of approximately 0.1% as assayed by Western blot analyses.

2.1.1. Animals

The wild-type adult CD-1 mice (n=5) and Sprague-dawley rats (n=7) were purchased from Charles River. *P23H* line 1 mutant rhodopsin transgenic rats (n=5) were kindly provided by Dr. M. LaVail. They were maintained as homozygotes. All procedures involving these animals were performed in accordance with the guidelines of the University of Pennsylvania for animal research.

2.1.2. Intravitreal and Subretinal Injections

At the age of approximately 1-month, animals were injected subretinally or intravitreally using a transcleral transchoroidal approach as described previously (Liang et al., 2000). One eye was injected with rAAV.CNTF.GFP or rAAV.GFP, while the contralateral eye received a sham injection. One microliter of viral solution was administered subretinally, thereby raising a dome-shaped retinal detachment; and 2 ul into the vitreous. After injection, Pred-G ointment (Allergan, Irvine, CA) was applied topically to prevent corneal desiccation and inflammation. All injections were monitored by direct visualization through an operating microscope.

2.2. Tissue Processing

Animals were sacrificed by CO2 asphyxiation 2 months post-injection. Prior to enucleation, eyes were marked by cautery at the limbus in the temporal quadrant. This mark served as a reference point for orientation of the eye during embedding. Eyes were fixed in 4% paraformaldehyde and 0.5% glutaraldehyde, and cryoprotected in PBS containing 20% and 30% sucrose. Eyes were then embedded in tissue freezing medium (Triangle Biomedical Sci., Durham, NC) and sectioned at -20°C in the horizontal meridian using a cryostat microtome (10 µm). Sections were coverslipped with Vectorshield containing DAPI or propidium iodine (Vector Lab) which stain nuclei only.

2.2.1. Muller Cell Cultures and AAV Infection

The immortalized rat Muller cells (rMC-1) were kindly provided by Dr. V. Sathy at Northwestern University, Chicago, IL. These cells have been well characterized (Sarthy et al., 1998) and was cultured in DMEM supplemented with 10% FBS and penicillin (100IU/ml) and streptomycin (10 ug/ml) at 37C in a 5% CO2 atmosphere. Muller cells were seeded in 2-chamber slides. At about 80% confluence, cells were infected by rAAV (2 ul) for 2-3 hrs under serum-free condition, then maintained in DMEM supplemented with 10% FBS. At various time-points, living cells were examined with an Olympus inverted light microscope with a FITC and rhodomine filters.

2.2.2. Immunohistochemistry

Frozen retinal sections or Muller cells were sequentially incubated in 5% blocking serum in 0.15M phosphate buffer (30 min), in primary antibody (rabbit anti-rat FGFR-1, 1:100, overnight at 4°C (Santa Cruz); rabbit anti-rat HSPG overnight at 4°C (Chemicon); or rabbit anti-human GFAP, 1:100, 2 hr at RT (Dako); in secondary antibody (biotinylated goat anti-rabbit IgG, Jackson ImmunoRes Labs) for 2 hr at RT, and in avidin-rhodamine (Vector Labs, Burlingame, CA) for 1 hr. Sections were examined using epifluorescent illumination with a dual filter allowing co-visualization of FITC and rhodamine. The specificity of immunostaining for FGFR-1, HSPG and GFAP was verified by omitting primary antibodies. No immunoreactivity was observed in retinal sections or Muller cells subjected to staining procedures in which primary antibodies were omitted.

3. RESULTS

3.1. Intravitreal Injection AAV.GFP Led To Transduction Of Muller Cells In Wild-Type Rats

To determine what types of retinal cells were transduced by rAAV after intravitreal administration, GFP fluorescence was examined histologically. Fluorescent microscopy showed that GFP was strongly expressed in retinal ganglion cells in both rAAV.CNTF.GFP and rAAV.GFP injected eyes (Fig. 1). Interestingly, some Muller cells also showed GFP expression, but the number of GFP-expressing Muller cells was less than that of GFP-expressing ganglion cells. GFP-expressing cells were located primarily in the temporal retina corresponding with the injection site. GFP fluorescence was also observed in the nerve fiber layer of temporal retina and the temporal side of optic nerve. Additionally, some RPE and photoreceptor cells showed GFP expression in the injection site. This was likely due to leakage of viral solution into the subretinal space during needle entry and withdraw, and subsequent infection of RPE and photoreceptors. Subretinal injection of AAV-GFP resulted in transduction of photoreceptor and RPE cells, but not Muller cells, which is consistent with other reports (Flannery et al., 1997; Ali et al., 1998). Lack of Muller cell transduction by AAV via the subretinal approach is likely due to the physical barrier of inner and outer segments of photoreceptor cells since the outward processes of Muller cells terminate in the outer limiting membrane.

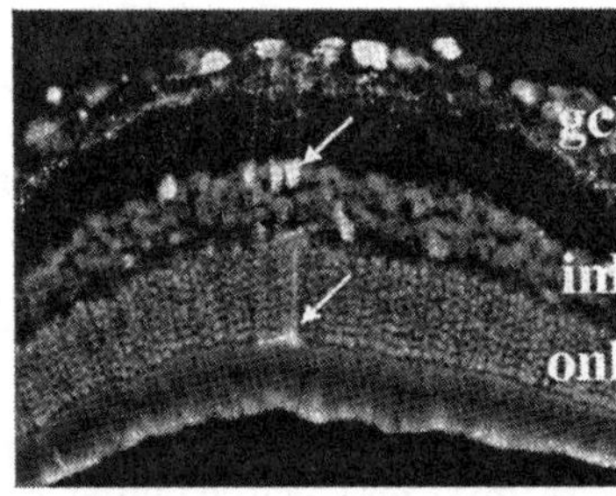

Figure 1. Presence of GFP in retinal cells 2 months after intravitreal injection of AAV-GFP in a Sprague-Dawley rat. Note GFP expression in ganglion cells and Muller cells (arrows).

3.2. Increased Muller Cell Transduction By AAV In Retinal Degenerative Rats

Muller cells are reactivated during retinal degeneration (Ekstrom et al., 1988). To see whether Muller cells are more susceptible to AAV transduction in the degenerative retina, 1-month old P23H transgenic rats were intravitreally injected with rAAV.CNTF.GFP or rAAV.GFP. Similar to the transduction in the wild-type Sprague-Dawley rats, GFP fluorescence was seen predominantly in the temporal retina in both rAAV.CNTF.GFP and rAAV.GFP injected eyes. In addition to ganglion cell transduction, strong GFP expression was observed in the Muller cells (Fig. 2). The number of GFP-expressing Muller cells appeared to be much greater than that in the wild-type rats. In addition, Muller cells become hypertrophic, with their apical processes extended into the subretinal space.

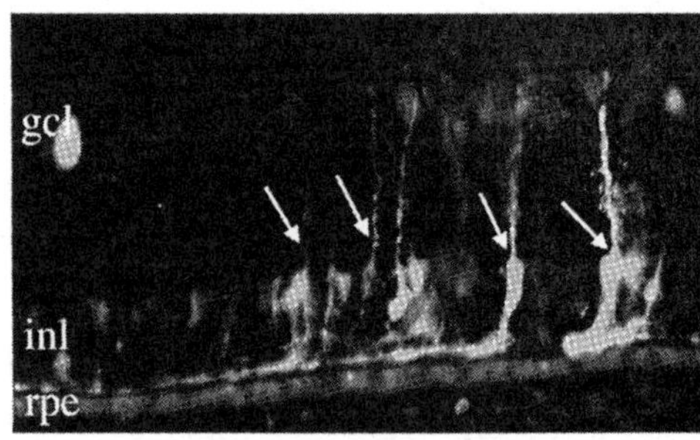

Figure 2. Presence of GFP in retinal cells 5 months after intravitreal injection of AAV.CNTF.GFP in a P23H rhodopsin transgenic rat. A greater number of Muller cells (arrows) showed GFP expression compared to retinal ganglion cells. Note that outer nuclear layer (onl) has completely degenerated.

3.3. Adenovirus Accelerates AAV-Mediated GFP Expression In The Retina

It is known that Ad can efficiently transduce Muller cells (Bennett et al., 1994), and Ad-mediated transgene expression is rapid. To determine if AAV can enter the Muller cells, a mixture of AAV-GFP and wild-type Ad (no transgene encoded) was co-injected into the vitreous of wild-type CD-1 mice. Animals were sacrificed 2 days post-injection. Fluorescent microscopy showed strong GFP expression exclusively in Muller cells (Fig. 3). Some RPE cells also expressed GFP at the injection site.

Figure 3. Muller cells transduction (arrowheads) 2 days after co-injection of AAV-GFP and wild-type Ad in a CD-1 mouse. GFP fluorescence was seen exclusively in the Muller cells.

3.4. AAV Is Able To Transduce Muller Cells *In Vitro*

At about 80% confluence, rMC-1 cells were infected with AAV.CNTF-GFP or AAV.GFP. The GFP fluorescence was observed at 16 hrs after infection. The number of GFP-expressing cells reached a maximal level (approximately 70% of total cells) at 3-5 days and remained stable for another 10 days (the end of experiment). The fluorescence was observed predominantly in cell body (Fig. 4).

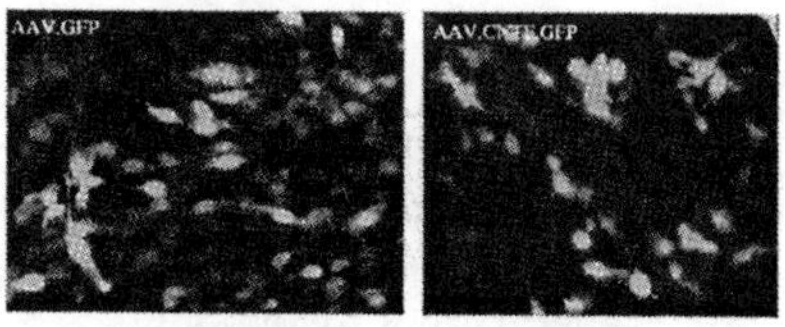

Figure 4. Presence of GFP in Muller cells in vitro 2 days after infection with AAV-GFP or AAV.CNTF.GFP .

3.5. Muller Cells Express AAV Receptors

In order for AAV to enter the cells, it has to bind two receptors, fibroblast growth factor receptor-1 (FGFR-1) and heparan sulfate proteoglycan (HSPG) (Summerford and Samulski, 1998; Qing et al., 1999). To determine if Muller cells express these AAV receptors, immunohistochemical staining was performed on cultured Muller cells. Prominent FGFR-1 and HSPG immunoreactivity was observed in the Muller cells (Fig. 5 A, C). The Muller cells incubated without primary antibodies showed no immunostaining (Fig. 5B, D).

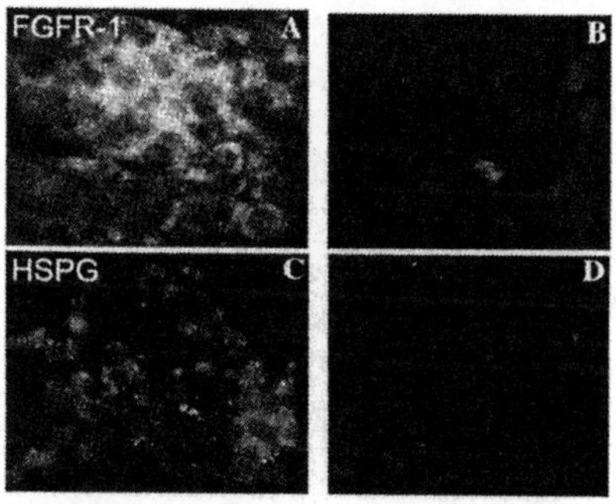

Figure 5. Immunohistochemical staining of FGFR-1 and HSPG in rMc-1 cells. Strong immunostaining of FGFR-1 (A) and HSPG (C) was present in these cells. No staining was observed when primary antibodies were omitted (B, D).

4. DISCUSSION

The present study demonstrates that intravitreal administration of rAAV enables transduction of retinal Muller cells in both normal and degenerative retinas. It appears that an increased number of Muller cells were transduced in the degenerative retina as compared to the normal retina. Muller cells became hypertrophic in the degenerative retina, with their apical processes filling the subretinal space. The ability of AAV to transduce the Muller cells in the retina was confirmed by its efficient transduction of rMc-1 cells *in vitro* and by the expression of FGFR-1 and HSPG receptors. Intravitreal injection of a mixture of AAV and Ad led to exclusive and early expression of GFP in the Muller cells, suggesting that Muller cells are able to uptake AAV. The ability of AAV to target Muller glial cells in the retina is consistent with other reports that rAAV are capable of targeting other glia such as oligodendrocytes and astrocytes (Chen et al., 1998; Mi et al., 1999).

AAV transduction of Muller cells requires intravitreal, but not subretinal injection. This is likely due to the anatomic location of Muller cells in the retina. Muller cells span nearly the full thickness of retina. Their endfeet form the internal limiting membrane (ILM) which borders the vitreous. The ILM is rich in heparin sulfate proteoglycan (Chai and Morris, 1999), a co-receptor for AAV-2. These features facilitate their physical contact with AAV after intravitreal injection. In contrast, AAV administered in the subretinal space first make contact with photoreceptor outer segments and RPE cells. The physical contact of AAV to Muller cells is kept apart by the inner and outer segments of photoreceptor cells since the apical processes of Muller cells are terminated in the outer limiting membrane.

The present study used AAV vectors carrying the CMV promoter. Although Muller cells can be transduced through intravitreal approach, retinal ganglion and horizontal/amacrine cells are also targeted. Transduction of these cells, particularly ganglion cells, may lead to unwanted effects, such that transgene would be transported from the retina to the brain, mirroring the normal neuroanatomical pathways taken by ganglion cell axons. We have demonstrated high levels of reporter gene expression in optic nerve and brain after intravitreal injection of AAV-GFP (Dudus et al., 1999). Thus, specific targeting of Muller cells with AAV by using Muller cell-specific promoters would minimize exposure of transgenes to other cell types.

Several lines of evidence suggest that Muller cells may be an important player in trophic factor mediated photoreceptor rescue. When the retina is damaged, including retinal degenerations, Muller cells become stimulated, and up-regulate trophic factor expression (Ekstrom et al., 1988; Wen et al., 1995). Following intravitreal injection of some trophic factors, Muller cells are consistently activated, but not photoreceptors (Wahlin et al., 2000), suggesting that these factors exert their protective effects through Muller cells. If photoreceptor rescue by trophic factors is through the Muller cells, then it may not be necessary and ideal to transduce photoreceptors with rAAV. Plus, photoreceptors are already compromised in retinal degenerative diseases, thus may not be the best source for gene transfer. In addition, transduction of photoreceptor cells requires subretinal administration of rAAV. This technique is technically difficult in rodents, and covers only small portions of retina. Muller cells cannot be targeted via this approach. In contrast, Muller cells can be easily targeted by the intravitreal approach, which is more efficient and less invasive, and can potentially target the entire retina. Thus the specific gene transfer to the Muller cells via the intravitreal approach would not only provide an effective therapy for photoreceptor degeneration, but also facilitate understanding the role of Muller cells in photoreceptor protection. Given the easy accessibility of these cells from the vitreous base and their role in mediating photoreceptor rescue, Muller cells may serve as a useful target the treatment of photoreceptor degeneration.

5. ACKNOWLEDGMENTS

This study was supported by 1F32 EY07073 (FQL), NIH R01 EY12156 (JB), the Foundation Fighting Blindness (JB), the Lois Pope LIFE Foundation (JB), the Ruth and Milton Steinbach Fund (JB), the Paul and Evanina Mackall Trust, and the F.M. Kirby Foundation. We thank Dr. K. Fisher at Tulane University for providing AAV.CNTF.GFP. We thank the National Eye Institute for the travel award provided to FQL to attend this meeting.

6. REFERENCES

Ali RR, Reichel MB, De Alwis M, Kanuga N, Kinnon C, Levinsky RJ, Hunt DM, Bhattacharya SS, Thrasher AJ, 1998, Adeno-associated virus gene transfer to mouse retina. *Hum Gene Ther* **9**:81-86.

Anand V, Chirmule N, Fersh M, Maguire AM, Bennett J, 2000, Additional transduction events after subretinal readministration of recombinant adeno-associated virus. *Hum Gene Ther* **11**:449-457.

Bartlett J, Samulski RJ, McCown T, 1998, selective and rapid uptake of AAV-2 in brain, *Hum Gene Ther* **9**:1181-1186.

Bennett J, Maguire AM, 2000, Gene therapy for ocular disease. *Mol Ther* **1**:501-505.

Bennett J, Duan D, Engelhardt JF, Maguire AM, 1997, Real-time, noninvasive in vivo assessment of adeno-associated virus- mediated retinal transduction. *Invest Ophthalmol Vis Sci* **38**:2857-2863.

Bennett J, Wilson J, Sun D, Forbes B, Maguire A, 1994, Adenovirus vector-mediated in vivo gene transfer into adult murine retina. *Invest Ophthalmol Vis Sci* **35**:2535-2542.

Bennett J, Maguire AM, Cideciyan AV, Schnell M, Glover E, Anand V, Aleman TS, Chirmule N, Gupta AR, Huang Y, Gao GP, Nyberg WC, Tazelaar J, Hughes J, Wilson JM, Jacobson SG, 1999, Stable transgene expression in rod photoreceptors after recombinant adeno-associated virus-mediated gene transfer to monkey retina. *Proc Natl Acad Sci USA* **96**:9920-9925.

Bringmann A, Reichenbach A, 2001, Role of Muller cells in retinal degenerations. Front Biosci **6**:E72-92.

Chai L, Morris J, 1999, heparan sulfate in the inner limiting membrane of embryonic chicken retina binds bFGF to promote axonal outgrowth. *Exp Neurol* **160**:175-185.

Chen H, McCarty D, Bruce A, Suzuki K, Suzuki K, 1998, Gene transfer and expression in oligodendrocytes under the control of mylin basic protein transcriptonal control region mediated by AAV. *Gene Ther* **5**:50-58.

Dudus L, Anand V, Chen S-J, Wilson J, Fisher KJ, Acland G, Maguire AM, Bennett J, 1999, Persistent transgene product in retina, optic nerve and brain after intraocular injection of rAAV. *Vision Res* **39**:2545-2554.

Ekstrom P, Sanyal S, Narfstrom K, Chader GJ, van Veen T, 1988, Accumulation of glial fibrillary acidic protein in Muller radial glia during retinal degeneration. *Invest Ophthalmol Vis Sci* **29**:1363-1371.

Flannery J, Zolotukin S, Vaquero M, LaVail M, Muzyczka N, Hauswirth W, 1997, Efficient photoreceptor-targeted gene expression in vivo by recombinant adeno-associated virus. *Proc Natl Acad Sci USA* **94**:6916-6921.

Flotte TR, Carter BJ, 1995, Adeno-associated virus vectors for gene therapy. Gene Ther **2**:357-362.

Liang F, Anand V, Maguire AM, Bennett J, eds, 2000) Intraocular delivery of recombinant virus: *Humana* press.

Liang F, Dejneka N, Cohen D, Krasnoperova N, Lem J, Maguire A, Dudus L, Fisher K, Bennett J, 2001a AAV-mediated delivery of CNTF prolongs photoreceptor survival in the rhodopsin knockout mouse. *Mol Ther* **3**:241-248.

Liang FQ, Aleman TS, Dejneka NS, Dudus L, Fisher KJ, Maguire AM, Jacobson SG, Bennett J, 2001b, Long-term protection of retinal structure but not function using RAAV.CNTF in animal models of retinitis pigmentosa. *Mol Ther* **4**:461-472.

Mi J, Chartterjee S, Wong K, Forbes C, Lawless G, Tobin A, 1999, rAAV drives constitutive production of glutamate decarboxylase in neural cell lines. *J Neurosci Res* **57**:137-148.

Newman E, Reichenbach A, 1996, The Muller cell: a functional element of the retina. *Trends Neurosci* **19**:307-312.

Qing K, Mah C, Hansen J, Zhou S, Dwarki V, Srivastava A, 1999, Human fibroblast growth factor receptor 1 is a co-receptor for infection by adeno-associated virus 2. *Nature Med* **5**:71-77.

Sarthy V, Brodjian S, Dutt K, Kennedy B, French R, Crabb J, 1998, Establishment and chracterization of a retinal Muller cell line. *Invest Ophthalmol Vis Sci* **39**:212-216.

Summerford C, Samulski RJ, 1998, Membrane-associated heparan sulfate proteoglycan is a receptor for AAV-2 virons. *J Virology* **72**:1438-1445.

Wahlin KJ, Campochiaro PA, Zack DJ, Adler R, 2000, Neurotrophic factors cause activation of intracellular signaling pathways in Muller cells and other cells of the inner retina, but not photoreceptors. *Invest Ophthalmol Vis Sci* **41**:927-936.

Wen R, Song Y, Cheng T, Matthes MT, Yasumura D, LaVail MM, Steinberg RH, 1995, Injury-induced upregulation of bFGF and CNTF mRNAS in the rat retina. *J Neurosci* **15**:7377-7385.

Zack DJ, 2000, Neurotrophic rescue of photoreceptors: are Muller cells the mediators of survival? *Neuron* **26**:285-286.

VIRUS-MEDIATED SECRETION GENE THERAPY – A POTENTIAL TREATMENT FOR OCULAR NEOVASCULARIZATION

Yvonne K.Y. Lai[1], Shiwani Sharma[2], Chooi-May Lai[3], Meliha Brankov[1], Ian J. Constable[3], and P. Elizabeth Rakoczy[3]

1. INTRODUCTION

Ocular neovascularization (NV) is a severe and major complication of many ocular conditions such as age-related macular degeneration, diabetic retinopathy, retinopathy of prematurity and retinal vein occlusion. The uncontrolled growth of blood vessels and its associated vascular hyperpermeability are the most common pathologic causes of vision loss in developed countries. Current conventional therapies can only provide symptomatic treatment for a limited subset of patients, often induce adverse effects and disease recurrence is common. Therefore, gene therapy aimed at blocking the underlying stimuli for new vessel growth is a potentially more effective treatment.

Vascular endothelial growth factor (VEGF) is believed to be the major stimulatory factor responsible for retinal and choroidal NV. Elevated levels of VEGF have been detected in ocular tissues of patients and animal models exhibiting ocular NV (Gilbert et al., 1998; Kvanta et al., 1996; Pe'er et al., 1996). However, a range of cells have been shown to contribute towards the upregulated VEGF in the retina, including retinal pigment epithelial (RPE) cells, fibroblasts, macrophages, neutrophils, Müller cells, ganglion cells, retinal capillary pericytes, and retinal or choroidal vascular endothelial cells (Shen et al., 1998; Blaauwgeers et al., 1999).

With such varied sources of VEGF, the ability to target and access these cells for treatment by cell-based strategies becomes a problem. To overcome these difficulties and achieve stable ocular anti-angiogenic therapy, we explored the potential of adenovirus (Ad)- and adeno-associated virus (AAV)-mediated secretion gene therapy in inhibiting ocular NV via the expression of sFlt-1, soluble receptor 1 for VEGF.

---

[1]Department of Molecular Ophthalmology, Lions Eye Institute, Nedlands, Western Australia; [2]Department of Ophthalmology, Flinders University, Bedford Park, South Australia; [3]Centre for Ophthalmology and Visual Science, University of Western Australia, Nedlands, Western Australia, Australia. Corresponding author: P. E. Rakoczy, E-mail: rakoczy@cyllene.uwa.edu.au

2. MATERIALS AND METHODS

2.1. Corneal NV Rat Model

The anterior chamber of the right eyes of non-pigmented RCS-rdy$^+$ rats were injected with either Ad-CMV.sflt (10^8 plaque forming units (pfu)/eye), AAV-CMV.sflt (10^8 particles/eye), or various Ad (10^8 pfu/eye), AAV (10^8 particles/eye) or vehicle controls. The central corneas were cauterized with silver nitrate at 1 day (Ad-CMV.sflt, n=11; Ad-RSV.βgal, n=6; Ad-Null, n=11; PBS vehicle, n=5) or 21 days post-injection (AAV-CMV.sflt, n=22; AAV-CMV.gfp, n=13; PBS vehicle, n=11; uninjected, n=11). Level of NV was graded 0 to 3 (increasing severity) by slit-lamp photography.

2.2. SFlt-1 Detection

In situ hybridization (ISH) was performed using digoxigenin-labeled antisense and sense sFlt-1 riboprobes. Frozen eye sections were fixed in 4% paraformaldehyde, dehydrated and then rehydrated in a graded methanol series (in PBS, 0.1% Triton X-100). The sections were treated with 100μg/ml pronase E, acetylated (0.25% acetic anhydride, 0.1M triethanolamine, pH 8) and dehydrated. ISH was performed overnight at 57°C in 50% formamide, 5x SSC, 100 μg/ml heparin, 1x Denhardts, 0.1% Tween 20, 0.1% CHAPS, 5mM EDTA, 1mg/ml tRNA and denatured labeled probes (0.5–3.5μg/ml). After the sections were washed several times, non-specific binding sites were blocked with TBT (50mM Tris-HCl, pH 7.5, 150mM NaCl, 0.1% Triton X-100), 0.2% BSA and 40% calf serum, and sections were incubated overnight with alkaline phosphatase-conjugated anti-digoxigenin antibody at 4°C. Signal was detected using NBT/BCIP as substrate. Using a similar protocol, ISH on AAV-CMV.sflt-infected and non-infected RPE cultures was performed directly in the culture dishes.

Western blot analysis was performed on conditioned media from RPE51 cultures as described previously (Lai et al., 2002). The cultures were infected with either 10^{12} particles (equivalent to 2 pmol) AAV-CMV.sflt, 10^8 pfu Ad-CMV.sflt, or transfected with 3 pmol plasmid pCMV.sflt. Briefly, sFlt-1 present in conditioned media was concentrated with heparin sepharose CL-6B beads (Pharmacia Biotech) for 4 h at 4°C. The beads were collected, washed with PBS and then boiled in 40 μl reducing sample buffer. An aliquot was electrophoresed on an 8% polyacrylamide gel and transferred onto a nitrocellulose membrane. Presence of sFlt-1 was detected by a monoclonal anti-Flt-1 antibody (1:500 Flt-11; Sigma) followed by incubation with horseradish peroxidase-conjugated sheep anti-mouse IgG antibody. Labelling was detected with the enhanced chemiluminescence system (Amersham).

3. RESULTS AND DISCUSSION

3.1. Comparison of sFlt-1 expression by Ad and AAV vectors

Four days following delivery of Ad-CMV.sflt into the anterior chamber of rat eyes, ISH was performed on the tissue sections. Using digoxigenin-labeled sFlt-1 antisense riboprobes at a concentration of 0.5 μg/ml, positive signal was observed predominantly in the outer region of the ciliary body (Fig. 1A). Occasional positive cells were detected in the corneal endothelium and iris (Figs. 1B and C). This ISH pattern is consistent with the expression of GFP following Ad-CMV.gfp injection into the anterior chamber as reported previously (Lai et al., 2001a).

Using the same ISH conditions, no positive cells were detected in uninjected eyes or in AAV-CMV.sflt-injected eyes at 21 days post-injection. Nevertheless, using RT-PCR, the presence of human sFlt-1 mRNA in the AAV-CMV.sflt-injected eyes was confirmed (data not shown). Further optimization of the ISH conditions indicated that the lack of signal in AAV-CMV.sflt-injected eyes was probably due to the low sensitivity of the method as incubation with a higher riboprobe concentration (3.5 μg/ml) and longer color development (10 hrs) successfully detected positive AAV-CMV.sflt-transduced cells *in vitro* (Fig. 1E).

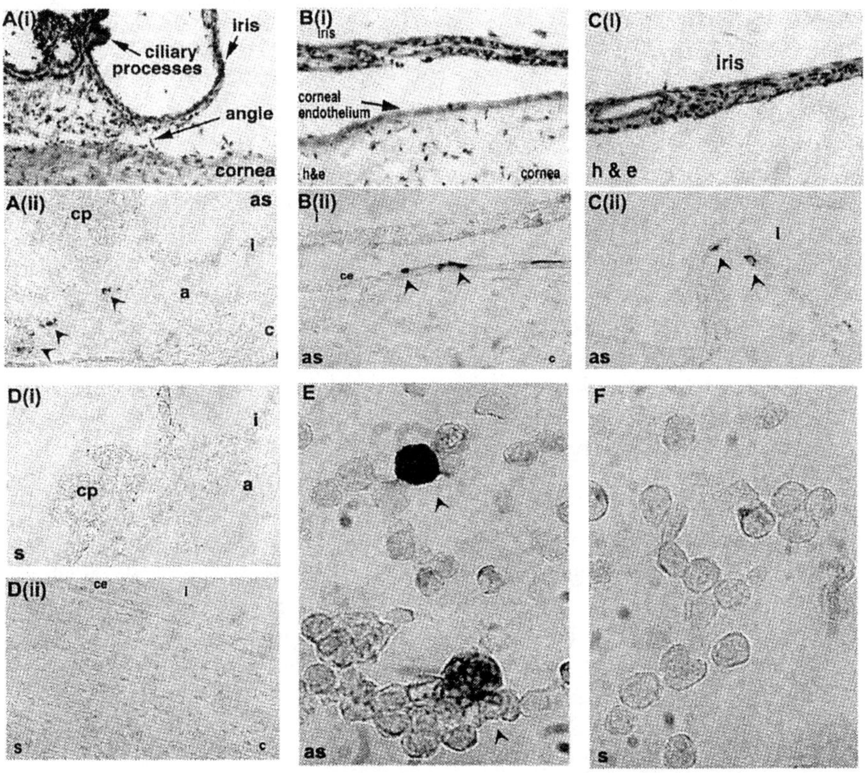

Figure 1. In situ hybridization of virus-mediated sFlt-1 in Ad-CMV.sflt-injected eyes (A-D) and AAV-CMV.sflt-infected RPE cultures (E and F). Positive cells (arrowheads) were observed in sections and cultures when hybridized with sFlt-1 antisense riboprobes (A-C(ii) and E). Sections stained with hematoxylin and eosin indicated that signal in anterior segment was present outside ciliary body, in iris and in corneal endothelium (A-C(i)). No staining was obtained following hybridization with sFlt-1 sense riboprobes of corresponding section of ciliary body (D(i)), iris and corneal endothelium (D(ii)) and AAV-CMV.sflt-infected cells (F). (as = antisense probe, s = sense probe, cp = ciliary processes, i = iris, a = angle, ce = corneal endothelium, c = cornea.)

This lower expression of sFlt-1 by the AAV vector compared to the Ad vector was also observed in the Western blot analysis (Fig. 2). The intensity of the sFlt-1 protein band from the conditioned medium of AAV-CMV.sflt-infected RPE cells (lane 3) was similar to that from plasmid-transfected cells (lane 2). However, both these sFlt-1 bands were much less intense than the signal from the Ad-CMV.sflt-infected culture (lane 1).

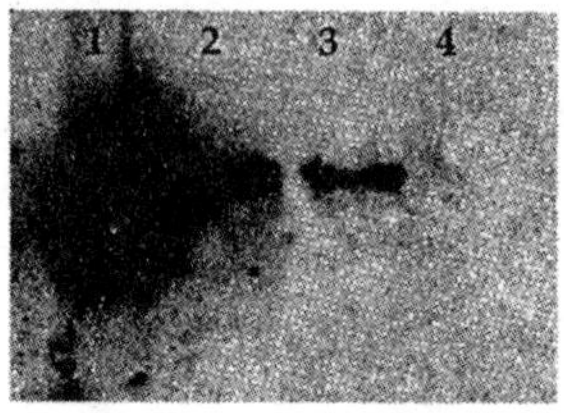

Figure 2. Western blot analysis showing different levels of secreted sFlt-1 protein in conditioned media from Ad- and AAV-infected RPE51 cells. The cultures were infected or transfected with 10^8 pfu Ad-CMV.sflt, 3 pmol pCMV.sflt, or 10^{12} particles AAV-CMV.sflt. Sample aliquots were loaded onto an 8% polyacrylamide gel: Ad-CMV.sflt (3μl, lane 1), pCMV.sflt (20μl, lane 2), AAV-CMV.sflt (20μl, lane 3) and uninfected RPE51 cells (20μl, lane 4).

The different levels of sFlt-1 expression between the Ad and AAV vectors could be due to several possibilities such as greater infectivity of cells by Ad, stronger transgene expression by Ad vectors, poor AAV genome conversion, or a higher Ad titer. Using flow cytometry, the transduction of RPE51 cells by Ad-CMV.gfp and AAV-CMV.gfp were compared. Ad and AAV have previously been shown to efficiently transduce RPE cells (da Cruz et al., 1997; Lai et al., 2001b). The mean fluorescence intensities of the two vectors were the same as measured by flow cytometry (Table 1) and also as observed by fluorescence microscopy indicating that the levels of transgene expression mediated by Ad and AAV vectors are similar, that is, Ad does not mediate stronger transgene expression. The transduction efficiencies were also similar but 10^6 transducing units (tu) of AAV-CMV.gfp were needed to achieve the same level as 10^5 pfu of Ad-CMV.gfp (Table 1) demonstrating the lower infectivity of AAV. There is at least a 100-fold difference between the physical particle and transducing titers of AAV (Zolotukhin et al., 1999). Therefore, a low transducing titer of AAV-CMV.sflt was used (~10^6 and 10^9 tu for injection and Western analysis, respectively) compared to 10^8 pfu of Ad-CMV.sflt. Together, these results suggest that the lower sFlt-1 expression mediated by AAV was due to a low titer and lower infectivity of the AAV vector.

Table 1. Transduction efficiency and mean fluorescence intensity of RPE51 cells infected with Ad-CMV.gfp and AAV-CMV.gfp

Vector	Infection titer per well	Transduction efficiency[a]	Mean fluorescence intensity[a,b]
Ad-CMV.gfp	10^5 pfu	15.25 ± 1.41 %	59.53 ± 1.54 %
AAV-CMV.gfp	10^6 tu	9.51 ± 2.17 %	60.91 ± 3.16 %

[a] 10^6 RPE51 cells were infected with Ad-CMV.gfp or AAV-CMV.gfp. The cells were harvested after 76 hrs post-infection for flow cytometry. The experiment was done in triplicates.

[b] Geometric mean was used as an approximate indicator of mean fluorescence intensity.

3.2. Suppression of ocular NV by Ad-CMV.sflt and AAV-CMV.sflt

Despite its lower sFlt-1 expression, injection with AAV-CMV.sflt reduced the relative mean NV score by at least 35% compared to cauterized control AAV-CMV.gfp-injected, PBS-injected and uninjected eyes (Fig. 3). The reduction in NV by AAV-CMV.sflt is comparable to the level achieved by Ad-CMV.sflt, which decreased the relative NV score by at least 37% and had fewer eyes showing severe NV compared to the cauterized control eyes (Figs. 3 and 4). However, there were differences in the cautery efficiency between the AAV and Ad groups, which were carried out at different times. Problems with controlling a consistent and uniform level of cautery appear to limit the efficiencies of this NV model. Nevertheless, the overall results clearly showed a lower proportion of AAV-CMV.sflt-injected eyes exhibiting severe NV compared to the control eyes (Fig. 4). Furthermore, we have shown that injection of AAV-CMV.sflt into the subretinal space of rat eyes, which were thereafter subjected to laser photocoagulation, reduced the number of leaky lesions by 19% (Lai et al., 2002).

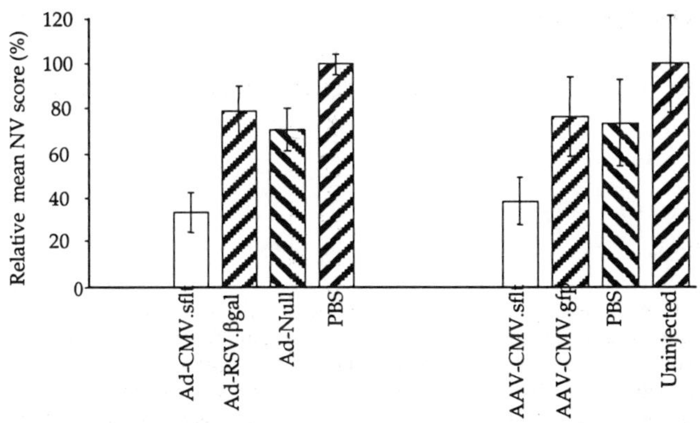

Figure 3. Partial inhibition of corneal NV by Ad- and AAV-CMV.sflt. The mean NV scores were standardized to give the relative mean NV score (the mean NV score divided by the mean NV score of the uninjected eyes). In group 1, the rat eyes were cauterized 1 day post-injection with Ad-CMV.sflt, Ad-RSV.βgal, Ad-Null or PBS. In group 2, cautery was done 21 days post-injection with AAV-CMV.sflt, AAV-CMV.gfp, PBS or uninjected.

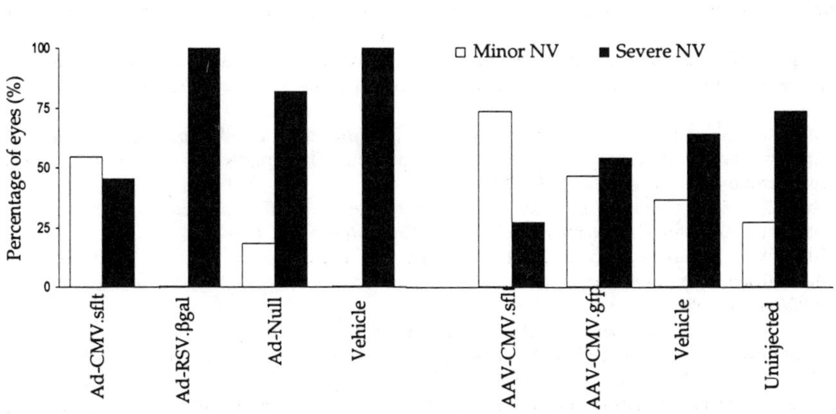

Figure 4. Percent cauterized eyes in each group showing minor NV (scored < 1) or severe NV (scored ≥ 1).

The reduction in NV by both AAV-CMV.sflt and Ad-CMV.sflt vectors indicate that there is sufficient diffusion or flux within the eye to distribute locally expressed sFlt-1. Furthermore, the similar levels of NV inhibition between the two vectors, despite the lower sFlt-1 expression shown by AAV, indicate that AAV vectors can produce and maintain sufficient sFlt-1 concentration to achieve an inhibitory effect. However, the lack of total inhibition by both vectors suggest that there might be an upper limit to neovascular inhibition achievable by sFlt-1 alone. Previous studies using other VEGF antagonists have also failed to completely block retinal or choroidal NV by achieving a maximum inhibition of only 50% (Aiello et al., 1995; Robinson et al., 1996). The results so far suggest that inhibition of VEGF alone is not sufficient to totally inhibit ocular NV and illustrates the complexity of the neovascular process. Nevertheless, our results show that in the eye, use of virus-mediated secretion gene therapy is a feasible approach for treatment of ocular diseases.

4. ACKNOWLEDGEMENTS

We thank the National Eye Institute for the travel award provided to YKYL to attend this meeting.

5. REFERENCES

Aiello, L.P., Pierce, E.A., Foley, E., Takagi, H., Chen, H., Riddle, L., Ferrara, N., King, G., and Smith, L., 1995, Suppression of retinal neovascularization *in vivo* by inhibition of vascular endothelial growth factor (VEGF) using soluble VEGF-receptor chimeric proteins, *Proc Natl Acad Sci USA* **92**:10457-10461.

Blaauwgeers, H.G.T., Holtkamp, G.M., Rutten, H., Witmer, A.N., Koolwijk, P., Partanen ,T.A., Alitalo, K., Kroon, M.E., Kijlstra, A., van Hinsbergh, V.W.M., and Schlingemann, R.O., 1999, Polarized vascular endothelial growth factor secretion by human retinal pigment epithelium and localization of vascular endothelial growth factor receptors on the inner choriocapillaris, *Am J Pathol.* **155**:421-428.

da Cruz, L., Rakoczy, P.E., Perricaudet, M., and Constable, I.J., 1996, Dynamics of gene transfer to retinal pigment epithelium, *Invest Ophthalmol Vis Sci.* **37**:2447-2454.

Gilbert, R.E., Vranes, D., Berka, J.L., Kelly, D.J., Cox, A., Wu, L.L., Stacker, S.A., and Cooper, M.E., 1998, Vascular endothelial growth factor and its receptors in control and diabetic rat eyes, *Lab Invest.* **78**:1017-1027.

Kvanta, A., Algvere, P., Berglin, L., and Seregard, S., 1996, Subfoveal fibrovascular membranes in age-related macular degeneration express vascular endothelial growth factor, *Invest Ophthalmol Vis Sci.* **37**:1929-1934.

Lai, C.M., Brankov, M., Zaknich, T., Lai, Y.K., Shen, W.Y., Constable, I.J., Kovesdi, I., and Rakoczy, P.E., 2001a, Inhibition of angiogenesis by adenovirus-mediated sFlt-1 expression in a rat model of corneal neovascularization, *Hum Gene Ther.* **12**:1299-1310.

Lai, Y.K., Rolling, F., Baker, E., and Rakoczy, P.E. 2001b, Kinetics of efficient recombinant adeno-associated virus transduction in retinal pigment epithelial cells, *Exp Cell Res.* **267**:184-192.

Lai, Y.K., Shen, W.Y., Brankov, M., Lai, C.M., Constable, I.J., and Rakoczy, P.E., 2002, Potential long-term inhibition of ocular neovascularization by recombinant adeno-associated virus-mediated secretion gene therapy, *Gene Ther.* **9**:804-813.

Pe'er, J., Gnessin, H., Folberg, R., Hemo, I., Itin, A., and Keshet, E., 1996, Upregulated expression of vascular endothelial growth factor in proliferative diabetic retinopathy, *Br J Ophthalmol.* **80**:241-245.

Robinson, G.S., Pierce, E.A., Rook, S.L., Foley, E., Webb, R., and Smith, L.E., 1996, Oligodeoxynucleotides inhibit retinal neovascularization in a murine model of proliferative retinopathy, *Proc Natl Acad Sci USA.* **93**: 4851-4856.

Shen, W.Y., Yu, M.J.T., Barry, C.J., Constable, I.J., and Rakoczy, P.E., 1998, Expression of cell adhesion molecules and vascular endothelial growth factor in experimental choroidal neovascularization in the rat, *Br J Ophthalmol.* **82:**1063-1071.

Zolotukhin, S., Byrne, B.J., Potter, M., Chesnut, K., Summerford, C., Samulski, R.J., and Muzyczka, N., 1999, Recombinant adeno-associated virus purification using novel methods improves infectious titer and yield, *Gene Ther.* **6:**973-985.

ABOUT THE EDITORS

Matthew M. LaVail, Ph.D., is Professor of Anatomy and Ophthalmology at the University of California, San Francisco School of Medicine. He received his Ph.D. degree in Anatomy (1969) from the University of Texas Medical Branch in Galveston and was subsequently a postdoctoral fellow at Harvard Medical School. Dr. LaVail was appointed Assistant Professor of Neurology-Neuropathology at Harvard Medical School in 1973. In 1976, he moved to UCSF, where he was appointed Associate Professor of Anatomy. He was appointed to his current position in 1982, and in 1988, he also became director of the Retinitis Pigmentosa Research Center at UCSF, later named the Kearn Family Center for the Study of Retinal Degeneration. Dr. LaVail has published extensively in the research areas of photoreceptor-retinal pigment epithelial cell interactions, retinal development, circadian events in the retina, genetics of pigmentation and ocular abnormalities, inherited retinal degenerations, light-induced retinal degeneration, and pharmaceutical and gene therapy for retinal degenerative diseases. He has identified several naturally occurring murine models of human retinal degenerations and has developed transgenic mouse and rat models of others. He is the author of more than 135 research publications and has edited ten books on inherited and environmentally induced retinal degenerations. Dr. LaVail has received the Fight for Sight Citation (1976); the Sundial Award from the Retina Foundation (1976); the Friedenwald Award from the Association for Research in Vision and Ophthalmology (ARVO, 1981); two Senior Scientific Investigators Awards from Research to Prevent Blindness (1988 and 1998); a MERIT Award from the National Eye Institute (1989); an Award for Outstanding Contributions to Vision Research from the Alcon Research Institute (1990); the Award of Merit from the Retina Research Foundation (1990); the first John A. Moran Prize for Vision Research from the University of Utah (1997); and the first Trustee Award from The Foundation Fighting Blindness (1998). He has served on the editorial board of *Investigative Ophthalmology and Visual Science* and is currently on the editorial board of *Experimental Eye Research*. Dr. LaVail has been an active participant in the program committee of ARVO and has served as a Trustee (Retinal Cell Biology Section) of ARVO. He has been a member of the program committee and a Vice President of the International Society for Eye research. He has also served on the Scientific Advisory Board of the Foundation Fighting Blindness since 1973.

Joe G. Hollyfield, Ph.D., is Director of Ophthalmic Research in the Cole Eye Institute at The Cleveland Clinic Foundation, Cleveland, Ohio. He received his Ph.D. from the University of Texas at Austin and did postdoctoral work at the Hubrecht Laboratory in Utrecht, The Netherlands. He has held faculty positions at Columbia University College of Physicians and Surgeons in New York City and at Baylor College of Medicine in Houston, TX. He was Director of the Retinitis Pigmentosa Research Center in The Cullen Eye Institute at Baylor from 1978 until his move to The Cleveland Clinic Foundation in 1995. He is currently Director of the Foundation Fighting Blindness Research Center at The Cleveland Clinic Foundation. Dr. Hollyfield has published over 170 papers in the area of cell and developmental biology of the retina and retinal pigment epithelium in both normal and retinal degenerative tissue. He has edited ten books, nine on retinal degenerations and one on the structure of the eye. Dr. Hollyfield has received the Marjorie W. Margolin Prize (1981, 1994), the Sam and Bertha Brochstein Award (1985) and the Award of Merit in Retina Research (1998) from the Retina Research Foundation; the Olga Keith Wiess Distinguished Scholars' Award (1981), two Senior Scientific Investigator Awards (1988, 1994) from Research to Prevent Blindness, Inc.; an award for Outstanding Contributions to Vision Research from the Alcon Research Institute (1987); the Distinguished Alumnus Award (1991) from Hendrix College, Conway, Arkansas; and the Endre A. Balazs Prize (1994) from the International Society for Eye Research (ISER). He is currently Editor-in-Chief of the journal, *Experimental Eye Research* published by Academic Press. Dr. Hollyfield has been active in the Association for Research in Vision and Ophthalmology (ARVO) serving on the Program Committee, as a Trustee and as President. He is also a past President and former Secretary of the International Society of Eye Research. He currently serves on the Scientific Advisory Boards of The Foundation Fighting Blindness, Research to Prevent Blindness, The Helen Keller Eye Research Foundation, The South Africa Retinitis Pigmentosa Foundation, and is Co-Chairman of the International Retinitis Pigmentosa Foundation Medical and Scientific Advisory Board.

Robert E. Anderson, M.D., Ph.D., is Professor and Chair of Cell Biology, Dean A. McGee Professor of Ophthalmology, and Adjunct Professor of Biochemistry & Molecular Biology and Geriatric Medicine at The University of Oklahoma Health Sciences Center in Oklahoma City, Oklahoma. He is also Director of Research at the Dean A. McGee Eye Institute. He received his Ph.D. in Biochemistry (1968) from Texas A&M University and his M.D. from Baylor College of Medicine in 1975. In 1968, he was a postdoctoral fellow at Oak Ridge Associated Universities. At Baylor, he was appointed Assistant Professor in 1969, Associate Professor in 1976, and Professor in 1981. He joined the faculty of the University of Oklahoma in January of 1995. Dr. Anderson has published extensively in the areas of lipid metabolism in the retina and biochemistry of retinal degenerations. He has edited eleven books, ten on retinal degenerations and one on the biochemistry of the eye. Dr. Anderson has received the Sam and Bertha Brochstein Award for Outstanding Achievement in Retina Research from the Retina Research Foundation (1980), and the Dolly Green Award (1982) and two Senior Scientific Investigator Awards (1990 and 1997) from Research to Prevent Blindness, Inc. He received an Award for Outstanding Contributions to Vision Research from the Alcon Research Institute (1985), and the Marjorie Margolin Prize (1994). He has served on the editorial boards of *Investigative Ophthalmology and Visual Science,*

Journal of Neuroscience Research, *Neurochemistry International*, *Current Eye Research*, and *Experimental Eye Research*. Dr. Anderson has received grants from the National Institutes of Health, The Retina Research Foundation, the Foundation Fighting Blindness, and Research to Prevent Blindness, Inc. He has been an active participant in the program committees of the Association for Research in Vision and Ophthalmology (ARVO) and was a trustee representing the Biochemistry and Molecular Biology section. He has served on the Vision Research Program Committee and Board of Scientific Counselors of the National Eye Institute and the Board of the Basic and Clinical Science Series of The American Academy of Ophthalmology. Dr. Anderson is a past Councilor and Treasurer, and is current President-Elect, of the International Society for Eye Research.

INDEX